D0986962

Plant Mitochondria

Plant Mitochondria

Edited by

DAVID C. LOGAN
School of Biology
University of St Andrews
St Andrews
Scotland
UK

Blackwell
Publishing

Editorial Offices:
Blackwell Publishing Ltd, 9600 Garsington Road, Oxford OX4 2DQ, UK
Tel: +44 (0)1865 776868
Blackwell Publishing Professional, 2121 State Avenue, Ames, Iowa 50014-8300, USA
Tel: +1 515 292 0140
Blackwell Publishing Asia Pty Ltd, 550 Swanston Street, Carlton, Victoria 3053, Australia
Tel: +61 (0)3 8359 1011

First published 2007 by Blackwell Publishing Ltd

ISBN: 978-1-4051-4939-6

Library of Congress Cataloging-in-Publication Data

Plant mitochondria / edited by David C. Logan.
p. cm.—(Annual plant reviews ; v. 31)
Includes bibliographical references.
ISBN: 978-1-4051-4939-6 (hardback : alk. paper)
1. Plant mitochondria. I. Logan. David C.

QK725.P476 2007
571.6′572—dc22

2006026442

A catalogue record for this title is available from the British Library

Set in 10/12 pt Times
by TechBooks, New Delhi, India
Printed and bound in Singapore
by Fabulous Printers Pte Ltd

The publisher's policy is to use permanent paper from mills that operate a sustainable forestry policy, and which has been manufactured from pulp processed using acid-free and elementary chlorine-free practices. Furthermore, the publisher ensures that the text paper and cover board used have met acceptable environmental accreditation standards.

For further information on Blackwell Publishing, visit our website:
www.blackwellpublishing.com

Contents

The colour plate section follows page 110

Contributors

Richard Berthomé Station de Génétique et d'Amélioration des Plantes, Institut Jean-Pierre Bourgin, INRA, route de Saint Cyr, 78026 Versailles cedex, France
Stefan Binder Molekulare Botanik, Universität Ulm, Albert-Einstein-Allee 11, 89069 Ulm, Germany
Egbert J. Boekema Department of Biophysical Chemistry, GBB, University of Groningen, Nijenborgh 4, 9747 AG Groningen, The Netherlands
Hans-Peter Braun Institute for Plant Genetics, Faculty of Natural Sciences, Universität Hannover, Herrenhäuser Str. 2, 30419 Hannover, Germany
Françoise Budar Station de Génétique et d'Amélioration des Plantes, Institut Jean-Pierre Bourgin, INRA, route de Saint Cyr, 78026 Versailles cedex, France
Mark Diamond School of Biology and Environmental Science, University College Dublin, Dublin 4, Ireland
Natalya V. Dudkina Department of Biophysical Chemistry, GBB, University of Groningen, Nijenborgh 4, 9747 AG Groningen, The Netherlands
Alisdair R. Fernie Max-Planck-Institut für Molekulare Pflanzenphysiologie, Am Mühlenberg 1, 14476 Potsdam-Golm, Germany
Dominique Gagliardi Institut de Biologie Moléculaire des Plantes – CNRS UPR2357, 12 rue du général Zimmer, 67084 Strasbourg cedex, France
Philippe Giegé Institut de Biologie Moléculaire des Plantes – CNRS UPR2357, 12 rue du général Zimmer, 67084 Strasbourg cedex, France
Elzbieta Glaser Department of Biochemistry and Biophysics, Stockholm University, Arrhenius Laboratories for Natural Sciences, 106 91 Stockholm, Sweden
Jesco Heinemeyer Institute for Plant Genetics, Faculty of Natural Sciences, Universität Hannover, Herrenhäuser Str. 2, 30419 Hannover, Germany
David C. Logan School of Biology, Sir Harold Mitchell Building, University of St Andrews, St Andrews, Fife KY16 9TH, Scotland, UK
Sally A. Mackenzie Department of Agronomy and Horticulture, University of Nebraska-Lincoln, P.O. Box 830665, Lincoln, NE 68583-0665, USA
Paul F. McCabe School of Biology and Environmental Science, University College Dublin, Dublin 4, Ireland
Ian M. Møller Department of Agricultural Sciences, Royal Veterinary and Agricultural University, Thorvaldsensvej 40, DK-187 Frederiksberg C, Denmark
Adriano Nunes-Nesi Max-Planck-Institut für Molekulare Pflanzenphysiologie, Am Mühlenberg 1, 14476 Potsdam-Golm, Germany
Iain Scott School of Biology, Sir Harold Mitchell Building, University of St Andrews, St Andrews, Fife KY16 9TH, Scotland, UK
James Whelan ARC Centre of Excellence in Plant Energy Biology, MCS Building M310, University of Western Australia, 35 Stirling Highway, Crawley 6009, Western Australia, Australia

Preface

Mitochondria have evolved from an α-proteobacterial endosymbiont into vital eukaryotic organelles. It is almost 60 years since mitochondria were discovered to be the sites of oxidative energy metabolism, and in the intervening years further discoveries have demonstrated that mitochondria perform a variety of fundamental, essential functions, in addition to energy production.

As with most areas of life science research, plant mitochondrial research has benefited greatly from the genomics revolution. The last decade has seen the publication of both the mitochondrial and nuclear genome sequences of several plant species, and the concomitant developments in functional genomics technology have underpinned some tremendous advances in our understanding of mitochondrial form and function. This book documents the complex biology of plant mitochondria from a cell biological, biochemical and molecular biological perspective.

Chapter 1 describes the emerging field of plant mitochondrial dynamics, encompassing mitochondrial division, fusion, cellular inheritance and distribution. The dynamic mitochondrial genome is discussed in the Chapter 2, including a discussion of the role of recombination in shaping mitochondrial genome structure. The mitochondrial genome remains the subject of Chapter 3, which provides a comprehensive discussion of plant mitochondrial genome expression and its control. Functional mitochondria require the controlled expression of mitochondrial and nuclear genes and, in the case of nuclear genes, import of the translated protein into mitochondria. The sequence and structural determinants for protein import into mitochondria are the subjects of Chapter 4, which also provides a full review of the protein import and processing machineries of plant mitochondria. Chapter 5 considers the interplay between nuclear and mitochondrial gene expression and how this is controlled to effect biogenesis of the large multisubunit respiratory complexes while the supramolecular structure of these complexes is discussed in Chapter 6. Mitochondrial electron transport is the subject of Chapter 7 with a focus on the unique NAD(P)H dehydrogenases of plant mitochondria. This chapter also addresses the unavoidable production of reactive oxygen species as a consequence of aerobic respiration, the damage caused by ROS and the ways by which mitochondria repair the damage. Despite the involvement of mitochondria in a number of fundamental processes, it is their central role in primary metabolism for which they are best known. Chapter 8 provides a comprehensive analysis of plant mitochondrial metabolism. Cytoplasmic male sterility (CMS) (Chapter 9) is a phenomenon of considerable agricultural importance and is also an intriguing model for studying mitochondrial–nuclear coordination. The book ends with a discussion of the emerging role of mitochondria in plant programmed cell death, bringing us full circle in the life of a plant mitochondrion, the organelle that provides not only the energy for life but also the trigger for death.

I take this opportunity to thank all the authors for their hard work in producing an excellent set of chapters, and also to thank the publishers for commissioning this volume and for their support during the editing process. Together I think we have produced an excellent volume that will provide researchers and plant science professionals with a valuable and timely monograph on plant mitochondria. I also record my thanks to Lydia, Peter and Florence for their unwavering support.

I would like to dedicate this book to Professor Christopher J. Leaver, FRS, FRSE, CBE, currently Sibthorpian Professor of Plant Science at the University of Oxford. The impact Chris has had in the field of plant science research in general, and plant mitochondrial research in particular, should not be underestimated. I was very lucky to start my research into plant mitochondria as a postdoctoral fellow with Chris just over 10 years ago, by which time Chris had been working on plant mitochondria for over 25 years. However, the reason this book is dedicated to Chris is neither because of his longevity in the field, nor does it simply reflect my gratitude to him for that first opportunity. Instead it is because, as an author, Chris is referenced in every chapter of this book (an honour he alone holds), with an average citation of four papers per chapter. While this is impressive in its own right, and indicative of the breadth of his research interests within the plant mitochondrial field, what is so much more impressive is that even as Chris approaches retirement (Autumn 2007), the majority of the references to his work are to papers he has published in the past 5 years. I am sure I write for all authors when I thank Chris for all he has done for the field of plant mitochondrial research. His input and presence will be sorely missed and we wish him a long and happy retirement.

David C. Logan

Annual Plant Reviews

A series for researchers and postgraduates in the plant sciences. Each volume in this series focuses on a theme of topical importance and emphasis is placed on rapid publication.

Titles in the series:

1. **Arabidopsis**
   Edited by M. Anderson and J.A. Roberts
2. **Biochemistry of Plant Secondary Metabolism**
   Edited by M. Wink
3. **Functions of Plant Secondary Metabolites and Their Exploitation in Biotechnology**
   Edited by M. Wink
4. **Molecular Plant Pathology**
   Edited by M. Dickinson and J. Beynon
5. **Vacuolar Compartments**
   Edited by D.G. Robinson and J.C. Rogers
6. **Plant Reproduction**
   Edited by S.D. O'Neill and J.A. Roberts
7. **Protein–Protein Interactions in Plant Biology**
   Edited by M.T. McManus, W.A. Laing and A.C. Allan
8. **The Plant Cell Wall**
   Edited by J.K.C. Rose
9. **The Golgi Apparatus and the Plant Secretory Pathway**
   Edited by D.G. Robinson
10. **The Plant Cytoskeleton in Cell Differentiation and Development**
    Edited by P.J. Hussey
11. **Plant–Pathogen Interactions**
    Edited by N.J. Talbot
12. **Polarity in Plants**
    Edited by K. Lindsey
13. **Plastids**
    Edited by S.G. Moller
14. **Plant Pigments and their Manipulation**
    Edited by K.M. Davies
15. **Membrane Transport in Plants**
    Edited by M.R. Blatt
16. **Intercellular Communication in Plants**
    Edited by A.J. Fleming
17. **Plant Architecture and Its Manipulation**
    Edited by C. Turnbull
18. **Plasmodeomata**
    Edited by K.J. Oparka
19. **Plant Epigenetics**
    Edited by P. Meyer

20. Flowering and Its Manipulation
Edited by C. Ainsworth

21. Endogenous Plant Rhythms
Edited by A.J.W. Hall and H.G. McWatters

22. Control of Primary Metabolism in Plants
Edited by W.C. Plaxton and M.T. McManus

23. Biology of the Plant Cuticle
Edited by M. Riederer and C. Müller

24. Plant Hormone Signaling
Edited by P. Hedden and S.G. Thomas

25. Plant Cell Separation and Adhesion
Edited by J.A. Roberts and Z. Gonzalez-Carranza

26. Senescence Processes in Plants
Edited by S. Gan

27. Seed Development, Dormancy and Germination
Edited by K. Bradford and H. Nonogaki

28. Plant Proteomics
Edited by C. Finnie

29. Regulation of Transcription in Plants
Edited by K.D. Grasser

30. Light and Plant Development
Edited by G.C. Whitelam and K.J. Halliday

31. Plant Mitochondria
Edited by D.C. Logan

32. Cell Cycle Control and Plant Development
Edited by D. Inzé

1 Mitochondrial dynamics: the control of mitochondrial shape, size, number, motility, and cellular inheritance

Iain Scott and David C. Logan

1.1 Introduction

Microscopy and sample preparation techniques were sufficiently advanced by the mid-nineteenth century to enable pioneering cell biologists to make the first descriptions of granular bodies within the eukaryotic cell (see Scott and Logan, 2004, for a brief history of cell biology). Although some of these granules were probably mitochondria, the various fixation and staining methods employed at the time made their unambiguous identification impossible (Hughes, 1959). As a result, it is impossible to ascribe the *in vivo* discovery of mitochondria to any one researcher; the first clear isolation and description of mitochondria is generally attributed to Kolliker, who, in 1888, removed the granular bodies from insect flight muscles. He also observed that the mitochondria became swollen in water and, importantly, were bound by a limiting membrane (Lehninger, 1964; Tribe and Whittaker, 1972).

The name *mitochondrion*, derived from the Greek *mitos* – a thread – and *chondros* – a grain – was coined in 1898 by Benda; the granules were previously called *sarcosomes*, *bioblasts* or *chondrioconts* by various researchers (Hughes, 1959; Tribe and Whittaker, 1972; Tzagoloff, 1982). While not being universally accepted at the time (Tribe and Whittaker, 1972), mitochondrion was an apposite name as it recognised the morphological heterogeneity that these organelles exhibit within and between cell types.

During the years 1880–1910, mitochondria were observed by a number of researchers in a variety of cell types, which naturally led to questions about the function of mitochondria. Some researchers, including Benda, postulated that these organelles were involved in fertility or inheritance due to their presence, in great numbers, in spermatozoa (Hughes, 1959). Others, such as Altmann in 1886, suggested that mitochondria were involved with cellular oxidation (Hughes, 1959). This idea gained momentum in the years following 1898, after Michaelis demonstrated that mitochondria were capable of producing an oxidation–reduction change in the vital stain Janus green B (Tribe and Whittaker, 1972). In 1913, Warburg demonstrated that the oxidation of metabolites was associated with insoluble, granular elements of the cell (Kennedy and Lehninger, 1949; Tribe and Whittaker, 1972), but did not link these observations to mitochondria. The primary role of mitochondria was finally established in 1949, when Kennedy and Lehninger showed that mitochondria are the site of oxidative energy metabolism (Kennedy and Lehninger, 1949).

Research into the morphology and behaviour of mitochondria *in vivo* began in the early twentieth century, when Lewis and Lewis (1914) described changes in mitochondrial size, shape, and position within the cytoplasm using a simple light microscope. Advances in microscopy, particularly the advent of phase contrast microscopy in 1934 (Ross, 1967), led to startling insights into mitochondrial movement within the cell. For example, in 1941 Michel was able to show for the first time that there was a rapid intracellular movement of mitochondria during meiosis (Ross, 1967). While many studies followed over the next two decades, the use of phase contrast for mitochondrial morphology studies fell out of favour, mainly as it does not provide a wholly unambiguous identification of cytoplasmic organelles (Bereiter-Hahn and Voth, 1994).

Fluorescence microscopy has become the predominant tool for visualising mitochondrial dynamics *in vivo* over the last 25 years (Bereiter-Hahn and Voth, 1994). Mitochondria-specific fluorochromes, mainly lipophilic cationic dyes such as rhodamine 123 (Johnson *et al.*, 1980), and Mito Tracker (Poot *et al.*, 1996), were developed to give organelles the ability to fluoresce *in situ*. Although much used, these dyes have the capacity to be cytotoxic and are susceptible to photobleaching under high-light conditions (Bereiter-Hahn and Voth, 1994). The application of green fluorescent protein (GFP)-based cell imaging (Chalfie *et al.*, 1994) enables the unambiguous analysis of mitochondria, in real-time, non-invasively, and in living tissue, through the expression of nuclear transgene fusions of GFP to N-terminal mitochondrial signal sequences. The first report on the use of GFP as a mitochondrial marker (mito-GFP) was published in 1995 by Rizzuto *et al.* (1995). Targeting of GFP to plant mitochondria was subsequently demonstrated independently by several research groups, each using different mitochondrial signal sequences: (i) the signal sequence of the yeast CoxIV subunit (Kohler *et al.*, 1997); (ii) the *Arabidopsis* F_1-ATPase gamma-subunit (Niwa *et al.*, 1999) or (iii) the *Arabidopsis* chaperonin-60 or *Nicotiana plumbaginifolia* F_1-ATPase beta-subunit (Logan and Leaver, 2000).

1.2 Endosymbiosis and mitochondrial evolution

It is now generally agreed that mitochondria evolved from free-living α-proteobacteria following a single endosymbiotic event over 1.5 billion years ago (Margulis, 1970; Martin and Muller, 1998; Gray *et al.*, 1999; Martin *et al.*, 2001). Since that time, eukaryotic cells have diversified into the many forms seen today.

1.2.1 Mitochondria and prokaryotic cell division

The endosymbiotic theory was first proposed in the 1890s by Altmann, who noted the similarity in the size and shape of mitochondria to free-living bacteria and suggested that mitochondria may be derived from prokaryotic ancestors (Munn, 1974; Tzagoloff, 1982). A similar observation was made by Mereschkowsky in 1905 to account for the presence of plastids in plants, stating that they were reduced forms of cyanobacteria enslaved within the cell (Mereschkowsky, 1905; Martin and Kowallik, 1999; Dyall *et al.*, 2004). The suggestion of an evolutionary link between plastids and free-living cyanobacteria led to the hypothesis that chloroplasts might share features of their division apparatus with modern free-living bacteria (Possingham and Lawrence, 1983; Osteryoung and McAndrew, 2001). Bacterial cytokinesis involves a suite of proteins, and the earliest acting and most phylogenetically widespread of these is the tubulin-like protein FtsZ (Kiefel *et al.*, 2004). FtsZ is found in nearly all prokaryotes and assembles into a large oligomeric structure forming a contractile ring around the interior surface of the cell membrane (Bi and Lutkenhaus, 1991). The first component of the chloroplast division apparatus to be identified was a homologue of bacterial FtsZ (Osteryoung and Vierling, 1995). Plant protein homologues of FtsZ have been shown to be targeted to the chloroplast (Osteryoung and Vierling, 1995; Osteryoung *et al.*, 1998; Osteryoung and McAndrew, 2001) to form a ring around the chloroplast midpoint (Vitha *et al.*, 2001), and in plants expressing antisense constructs of *FtsZ* genes, chloroplasts are larger than wild type and there are fewer per cell (Osteryoung *et al.*, 1998). Because of a shared prokaryotic origin, one can assume that at the times of their endosymbiosis the mechanisms of cell division in the α-proteobacterium mitochondrial ancestor and the cyanobacteria plastid ancestor were similar.

The evolution of mitochondria has resulted in the loss from yeast, and all higher eukaryotes of the prokaryotic genes originally involved in division of the symbiont. For example, while FtsZ is likely to be ubiquitous amongst prokaryotes, it is absent from the mitochondria of yeast and higher eukaryotic cells (Erickson, 1997; Lutkenhaus, 1998; Martin, 2000). Indeed, until 2000 no eukaryote had been identified that contained mitochondria using FtsZ as part of the division apparatus. The evolutionary missing link, a eukaryote with FtsZ-using mitochondria, was discovered by Beech *et al.* (2000). By screening a cDNA library of the unicellular chromophyte (stramenopile/heterokont) alga *Mallomonas splendens* with a fragment of the *ftsZ1* gene of the α-proteobacterium *Sinorhizobium meliloti* (Margolin *et al.*, 1991), Beech *et al.* (2000) identified a single-copy nuclear gene named *MsFtsZ-mt*. Phylogenetic analysis confirmed that MsFtsZ-mt is closely related to FtsZs of α-proteobacteria (Beech *et al.*, 2000). Later in 2000, a mitochondrial-type FtsZ homologue was reported in the genome of the red alga *Cyanidioschyzon merolae* (Takahara *et al.*, 2000). Phylogenetic analysis suggested that this gene, named *CmftsZ1*, is a mitochondrial-type *FtsZ* gene, and immunoblotting with anitsera raised against bacterial-expressed *CmftsZ1* demonstrated that the protein was located within mitochondria (Takahara *et al.*, 2000); subsequent research has identified a second *C. merolae* mitochondrial-type *FtsZ* gene (the genes are now named *CmftsZ1-1* and *CmftsZ1-2*) (Miyagishima *et al.*, 2004). Two mitochondrial-type *ftsZ* genes (named *fszA* and *fszB*) have also been discovered recently in the amoeboid protist *Dictyostelium discoideum* (Gilson *et al.*, 2003).

During the course of evolution, all mitochondrial-type *FtsZ* genes identified to date transferred from the mitochondrial genome to the nucleus and are now targeted back to mitochondria. The FtsZs in *C. merolae*, and *M. splendens*, and at least one of the two proteins in *D. discoideum* localise early on with dividing mitochondria as patches, or ring-like structures coincident with the division site (Beech *et al.*, 2000; Takahara *et al.*, 2000; Gilson *et al.*, 2003). For example, immunofluorescence localisation of MsFtsZ-mt in *M. splendens* demonstrated that MsFtsZ-mt was always associated with mitochondria in one of two locations: around the middle of the organelle or at the tips (Beech *et al.*, 2000). Genetic evidence for the involvement of mitochondrial-type FtsZ proteins in the control of mitochondrial division was provided by Gilson *et al.* (2003), who demonstrated that disruption of either *fszA* or *fszB* in *D. discoideum* caused the normally spherical, or rod-shaped mitochondria to be replaced with elongated tubular mitochondria, suggesting that fewer mitochondrial division events occurred in the mutants. FszA and FszB are differentially localized within *D. discoideum* mitochondria: FszA localises to presumptive constriction sites whereas FszB localises to the end of the organelle (Gilson *et al.*, 2003), together mimicking the localisation of MsFtsZ-mt in *M. splendens*.

The discovery of mitochondrial FtsZ genes in *M. splendens*, *C. merolae* and *D. discoideum* has plugged a gap in the evolution of the mitochondrion from its free-living bacterial past to its present-day central role in mitochondriate eukaryotes (Martin, 2000) but still leaves unanswered the question: why have higher eukaryotes lost mitochondrial FtsZ?

1.2.2 Plastids, the contractile ring, actomyosin and the loss of mitochondrial FtsZn

Mitochondrial *FtsZ* genes have been lost, independently, from both the higher plant lineage, after divergence of the green-plastid lineage from the red algae and Glaucophyta (Arimura and Tsutsumi, 2002; McFadden and Ralph, 2003), and from the yeast and animal lineages, before diversification of the opisthokonts (Arimura and Tsutsumi, 2002; McFadden and Ralph, 2003; Miyagishima *et al.*, 2003b; Kiefel *et al.*, 2004).

Like mitochondria, plastids have evolved from a prokaryotic endosymbiont, in this case a free-living cyanobacterium (Mereschkowsky, 1905; Margulis, 1970; Martin and Kowallik, 1999; McFadden, 2001; Palmer, 2003). As mentioned above, the evolutionary link to free-living cyanobacteria led to the suggestion that chloroplasts might share features of their division apparatus with modern free-living bacteria. This hypothesis was proven correct when the first component of the chloroplast division apparatus to be identified was a homologue of FtsZ (Osteryoung and Vierling, 1995). Subsequently, plastid-type *FtsZ* genes were identified in *M. splendens* (Beech *et al.*, 2000) and *C. merolae* (Takahara *et al.*, 2000) (see Section 1.2.1). In primitive plants, therefore, plastids and mitochondria divide using very similar mechanisms that include an FtsZ ring (Miyagishima *et al.*, 2003a,b; Nishida *et al.*, 2003, 2004).

It has been postulated that the dual role of FtsZ in mitochondrial and chloroplast division was a limiting factor in plant evolution and was unsuited to the more complex mitochondrial and plastidic dynamics in chlorophyte algae and higher plants (Logan, 2006a). The single cells of *C. merolae* contain only one chloroplast and one mitochondrion, and organelle division occurs sequentially prior to cell division (Miyagishima *et al.*, 1998); similarly, there is only one chloroplast in *M. splendens* (and 20–40 mitochondria; P.L. Beech, personal communication). In contrast, a typical *Arabidopsis* mesophyll cell contains approximately 120 chloroplasts (Pyke and Leech, 1994) and many hundreds of mitochondria. Furthermore, higher plant mitochondria are highly dynamic, undergoing frequent division and fusion events, even in non-dividing cells (Arimura and Tsutsumi, 2002; Logan, 2003; Logan *et al.*, 2003, 2004; Arimura *et al.*, 2004a). Logan (2006a) hypothesized that the loss of FtsZ as a component of the plant mitochondrial division apparatus in the green-plastid lineage removed an insurmountable barrier to the development of a complex organellar architecture, multicellularity, and therefore the evolution of the plant cell. Mitochondrial *FtsZ*-type genes have also been lost independently from the Apicomplexans – parasitic protozoans that contain a vestigial plastid, the apicoplast, which originated via secondary endosymbiosis of a photosynthetic alga (Gardner *et al.*, 2002; Kiefel *et al.*, 2004; Walter and McFadden, 2005). The use of FtsZ for division of both mitochondria and plastids would therefore appear to be an evolutionary dead end that has been avoided independently at least twice.

As explained above, mitochondrial FtsZ was also lost independently from a common ancestor of modern-day animals and fungi. As yeast and animals do not contain plastids, the above hypothesis for the loss of FtsZ from higher plant genomes is not relevant. Instead, an alternative hypothesis was proposed (Logan, 2006a) centred on the evolution of the actomyosin cytoskeleton: the evolutionary development of the actomyosin cytoskeleton underpins cell and organism motility and locomotion (Mitchison and Cramer, 1996). Mitochondrial movement in plants, budding yeast (*Saccharomyces cerevisiae*) and in at least some protists (e.g. *D. discoideum*) is based predominantly on actin microfilaments (Olyslaegers and Verbelen, 1998; Yaffe, 1999; Van Gestel *et al.*, 2002; Logan, 2003), whereas in animals, mitochondrial movement is predominantly microtubule based (Yaffe, 1999). It was suggested that the evolving role of actin and myosin as components of the contractile ring, formed during cytokinesis in yeast and animals, was incompatible with their roles as the predominant cytoskeletal proteins involved in mitochondrial movement. It therefore follows that these conflicting roles underpinned the evolution of the microtubule cytoskeleton as the predominant network for mitochondrial movement in filamentous yeast and animals. That mitochondria still move predominantly on actin in *S. cerevisiae*, although an actomyosin contractile ring functions during budding, is likely due to the importance of the actin cytoskeleton in the inheritance of mitochondria in this organism. Since FtsZ is structurally related to eukaryotic tubulin and believed to be their evolutionary progenitor (Margolin, 2005), Logan (2006a) suggested that the switch from actin to microtubules as the predominant mediator of mitochondrial motility may have been incompatible with the role of FtsZ in the division of mitochondria in close association with microtubules. In this hypothetical scenario, the switch to microtubules from actin thus underpins the evolution of the actomyosin cytoskeleton (muscle) and locomotion (Logan, 2006a).

1.3 Chondriome structure and organisation – the discontinuous whole

The higher plant chondriome (all the mitochondria of a cell collectively) is a highly dynamic structure composed predominantly of physically discrete organelles. This structure contrasts with that of most animal cell types and yeast cells where the chondriome is frequently organised into long tubules or reticula (Figure 1.1). Analysis of mitochondrial morphology *in vivo*, either by differential interference contrast microscopy, staining with low concentrations of DiOC6 (a fluorescent lipophilic dye) and fluorescence microscopy, or by fluorescence microscopy of transgenic tissue expressing mitochondrial-targeted GFP, has shown higher plant mitochondria to be highly pleomorphic, although most frequently they are spherical to sausage-shaped organelles. The typical organisation of the higher plant chondriome into a population of hundred to thousands of physically discrete organelles has a knock-on effect on the organisation of the mitochondrial genome.

The plant mitochondrial genome is a large, complex structure relative to the smaller, simpler mitochondrial genomes in other eukaryotes. For example, the mitochondrial genome of animals varies from around 15 to18 kbp (human = 16.6 kbp) and that of yeast from 18 to more than 100 kbp (*S. cerevisiae* = 75–85 kbp), whereas the mitochondrial genome of higher plants varies from 208 kbp in white mustard (*Brassica hirta*) (Palmer and Herbon, 1987) to an estimated 2500 kbp in muskmelon (*Cucumis melo*). However, the gene contents of the greatly expanded plant mitochondrial genomes approximate to those of ancestral protists (Gray *et al.*, 1999). A comparison of human and *Arabidopsis* mtDNA reveals that the relatively small human mitochondrial genome (16.6 kbp) encodes 13 polypeptides, while the relatively large *Arabidopsis* mitochondrial genome (366.9 kbp) encodes 33 polypeptides, meaning *Arabidopsis* mtDNA only encodes 2.5 times as many proteins as in humans even though the genome is 22 times as large. The relatively large sizes of plant mitochondrial genomes are attributed to a large increase in non-coding sequence, unidentified open reading frames (ORFs), introns and intron ORFs through horizontal gene transfer (Unseld *et al.*, 1997; Gray *et al.*, 1999; Palmer *et al.*, 2000).

The plant mitochondrial genome is composed of small circular, and large, circularly permutated DNA molecules (Lonsdale *et al.*, 1988; Bendich, 1993, 1996; Backert *et al.*, 1995; Oldenburg and Bendich, 2001) that are generated by active inter- and intramolecular homologous recombination (Lonsdale *et al.*, 1988). The organisation of the plant chondriome into a population of physically discrete organelles is likely to have a large influence on the organisation of the mitochondrial genome since recombination between mtDNA molecules in physically distinct mitochondria initially requires mitochondrial fusion, in addition to any other intramitochondrial processes. A high frequency of recombination results in the plant mitochondrial genome existing as a series of subgenomic, occasionally substoichiometric, DNA molecules that, once generated, may replicate autonomously (Lonsdale *et al.*, 1988; Small *et al.*, 1989; Janska and Mackenzie, 1993; Janska *et al.*, 1998; Abdelnoor *et al.*, 2003). Indeed, it has been demonstrated by means of a quantitative analysis

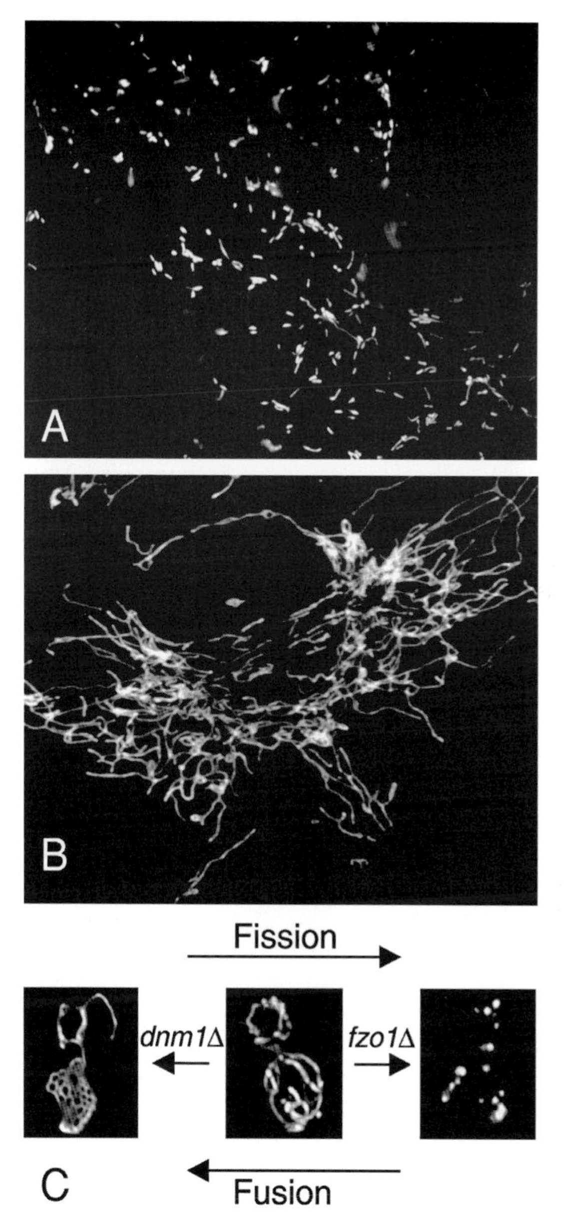

Figure 1.1 Chondriome structure in different eukaryotes. (A) Typical organisation of the higher plant chondriome as a discontinuous whole composed of generally physically discrete organelles. Image shows mitochondria expressing mito-GFP within the abaxial epidermal cell layer of a 14-day-old *Arabidopsis* seedling. (B) Reticular chondriome structure typical of mammalian cells, in this case a HeLa cell stained with anti-cytochrome *c* antibodies (image kindly provided by Dr. Mariusz Karbowski, NINDS, NIH, Bethesda, MD, USA). (C) Chondriome structure in budding yeast *S. cerevisiae* is maintained by a balance between fission and fusion. Loss of Dnm1p (*dnm1p*Δ) leads to net formation due to unopposed mitochondrial fusion (left); loss of Fzo1p (*fzo1p*Δ) leads to fragmentation due to unopposed mitochondrial fission (right) (Mozdy *et al.*, 2000). (Part C is reproduced from Mozdy *et al.*, 2000, by copyright permission of The Rockefeller University Press.)

of mtDNA content, mitochondrial genome size and mitochondrial number per cell that many physically discrete mitochondria within a plant cell contain less than a full genome (Lonsdale *et al.*, 1988). In order to explain the observed complexity of the mitochondrial genomes of higher plants, Lonsdale *et al.* (1988) proposed that the chondriome forms a *dynamic syncytium* and that the cellular mitochondrial population is panmictic, resulting in a state of recombinational equilibrium. One of us has since suggested (Logan, 2006b) that dynamic syncytium is better suited as a description of a predominantly reticular chondriome (such as in *S. cerevisiae* and many mammalian cell types) than as a description of the higher plant chondriome that normally comprises discrete organelles and rarely forms a syncytium. Instead, it was suggested that the higher plant chondriome should be termed a discontinuous whole (Logan, 2006b).

1.4 Control of mitochondrial number, size, and gross morphology

Mitochondria are pleomorphic organelles, and at any point in time the mitochondrial population of a given cell consists of a heterogeneous mix of mitochondrial morphologies. In higher plants, mitochondria are typically discrete, spherical to sausage-shaped organelles although more extreme morphologies are frequently seen. The number of mitochondria per cell is variable, depending on cell type and physiological state; however, a typical *Arabidopsis* mesophyll cell contains approximately 200–300 discrete mitochondria, while tobacco mesophyll protoplasts typically contain 500–600. As mentioned in the previous section, the structural organisation of the higher plant chondriome as a population of physically discrete organelles sets plants apart from many other eukaryotes. For example, in *S. cerevisiae* there are typically 5–10 tubular mitochondria forming an extended reticular network beneath the cell cortex (Stevens, 1977). In various mammalian cell types, the mitochondrial morphology is complex consisting of a mixture of small, discrete, spherical mitochondria and elongate tubular regions that often form a reticulum. Mitochondrial pleomorphism in mammalian cells is known to be affected by culture conditions since different research groups working on the same cell lines have described different morphologies (Karbowski and Youle, 2003).

1.4.1 Mitochondrial division

While observations of mitochondrial division have been made for over 30 years (Bereiter-Hahn and Voth, 1994), the genes, proteins and mechanisms underpinning this vital process have only been elucidated over the past decade. Much of this work has been carried out using budding yeast (*S. cerevisiae*).

The first component of the yeast mitochondrial division apparatus to be discovered, Dnm1p, is structurally similar to the dynamin-related GTPase proteins involved in membrane scission during endocytosis (Hermann *et al.*, 1997; Otsuga *et al.*, 1998). Dnm1p is targeted to mitochondrial division sites and believed to act as a mechano-enzyme, constricting and/or severing the mitochondrial membranes.

This dynamin-related protein requires at least three other interacting proteins to effect mitochondrial division. First, Fis1p is found evenly distributed across the mitochondrial outer surface (Mozdy *et al.*, 2000; Tieu and Nunnari, 2000) and may recruit Dnm1p to division sites, possibly using a tetratricopeptide repeat-binding (TPR) domain (Suzuki *et al.*, 2005). Second is Mdv1p (independently discovered by four groups, now commonly referred to as Mdv1p), which also localises to the mitochondrial outer membrane at division sites (Fekkes *et al.*, 2000; Mozdy *et al.*, 2000; Tieu and Nunnari, 2000). The third Dnm1p-interacting protein, Caf4p, was identified in an affinity screen aimed at identifying proteins interacting with Fis1p (Griffin *et al.*, 2005). Caf4p interacts with each of the other members of the mitochondrial division apparatus: the N-terminal half of Caf4p interacts with Fis1p and Mdv1p, and the C-terminal WD40 domain interacts with Dnm1p (Griffin *et al.*, 2005). These four proteins (Dnm1p, Fis1p, Mdv1p and Caf4p) form a complex on the mitochondrial outer membrane and act together during division of the organelle (Okamoto and Shaw, 2005). While several competing models exist to explain the formation and interaction of these proteins, it is clear that knocking out *DNM1*, *FIS1* or *MDV1* leads to a similar abnormal mitochondrial phenotype (a net-like sheet of mitochondrial tubules, caused by fusion in the absence of division), suggesting that functional copies of all three genes are required for effective mitochondrial division.

The role of dynamin-like proteins in mitochondrial division is highly conserved in eukaryotes (Smirnova *et al.*, 1998; Labrousse *et al.*, 1999; Arimura and Tsutsumi, 2002; Arimura *et al.*, 2004a; Logan *et al.*, 2004). Disruption of either of two *Arabidopsis* dynamin-like genes, *DRP3A* or *DRP3B*, results in an aberrant mitochondrial morphology characterised by an increase in the size of individual mitochondria and a concomitant decrease in the number of mitochondrial per cell (Figure 1.2) (Arimura and Tsutsumi, 2002; Arimura *et al.*, 2004a; Logan *et al.*, 2004; Scott, 2006). In agreement with the localisations of Dnm1p in yeast (Sesaki and Jensen, 1999; Legesse-Miller *et al.*, 2003) and Drp1 in mammals (Karbowski *et al.*, 2002), DRP3A and DRP3B localise to the tips and constriction sites of mitochondria (Plate 1.1) (Arimura and Tsutsumi, 2002; Arimura *et al.*, 2004a). The mutant mitochondria also have frequent constrictions along their length suggesting that they are arrested at the stage of membrane scission (Logan *et al.*, 2004; Scott, 2006). In addition to DRP3A and DRP3B, two other members of the *Arabidopsis* dynamin-like superfamily have been implicated in the control of mitochondrial morphology (Jin *et al.*, 2003). DRP1C and DRP1E, members of the DRP1 subfamily with closest homology to soybean phragmoplastin (Gu and Verma, 1996), were reported to partially locate to mitochondria, and disruption of either gene was reported to increase the proportion of mitochondria with an elongated morphology (Jin *et al.*, 2003). However, DRP1C and DRP1E have also been localised to the developing cell plate, and so these dynamin-like proteins may have a complex role in plant development.

Apart from dynamin-like proteins, only one other member of the plant mitochondrial division apparatus has been identified. We recently identified an *Arabidopsis* functional homologue of hFis1/Fis1p that we named *BIGYIN* (At3g57090) to reflect the mitochondria phenotype in homozygous T-DNA insertion mutants (Scott

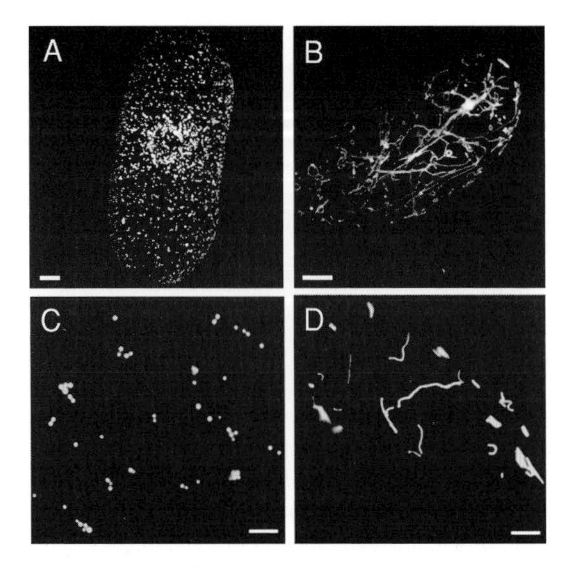

Figure 1.2 Mutation of the dynamin-like genes *DRP3A* and *DRP3B* affects mitochondrial morphology. (A) Wild-type tobacco BY-2 cell expressing mito-GFP. (B) Highly elongated mitochondria in a tobacco BY-2 cell expressing mito-GFP and a dominant negative mutant form of *Arabidopsis* DRP3A. (C) Mitochondria in a wild-type *Arabidopsis* epidermal cell expressing mito-GFP. (D) Elongate mitochondria in an epidermal cell of an *Arabidopsis* mutant homozygous for a T-DNA insertion in *DRP3B*. Scale bar = 10 μm in (A) and (B) and 5 μm in (C) and (D). (The images in (A) and (B) are reproduced with permission from Arimura *et al*., 2004a, copyright the Japanese Society of Plant Physiologists. (C) and (D) are reproduced from Scott *et al*., 2006 by permission of Oxford University Press.)

et al., 2006). Disruption of *BIGYIN* leads to a reduced number of mitochondria per cell, coupled to a large increase in the size of individual mitochondria relative to wild type, a phenotype very similar to that of *drp3a*, and *drp3b* mutants. The similarity between the *drp3a*, *drp3b* and *bigyin* mutant phenotypes reflects the

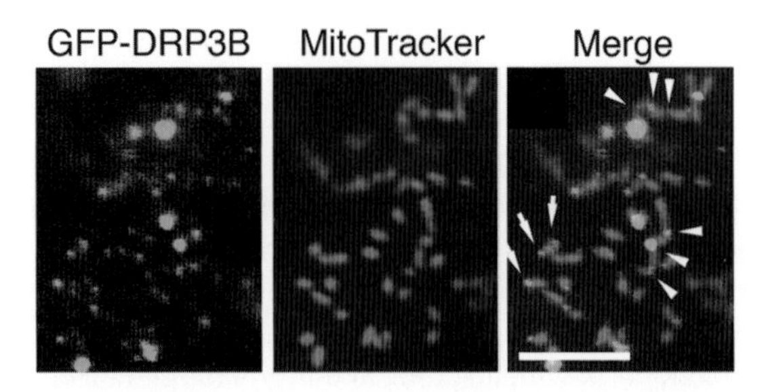

Plate 1.1 Localization of GFP-DRP3B with tips and constriction sites of mitochondria. BY-2 cells transiently expressing GFP-DRP3B, and stained with the mitochondrial marker MitoTracker, were examined by confocal laser scanning microscopy. High-magnification images of a part of a single BY-2 cell are shown. GFP-DRP3B (green) was localized in a punctate distribution and many foci colocalized with the tips (arrows) and constricted sites (arrow-heads) of mitochondria. Scale bar = 5 μm. (Reproduced from Arimura and Tsutsumi, 2002, copyright National Academy of Sciences, USA.) See colour plate section following p. 110 for colour version of this figure.

situation in yeast and mammalian cells where *Dnm1p/Drp1* and *Fis1/hFis1* mutants have indistinguishable mitochondrial morphologies. This suggests that, as in yeast and mammalian cells (Mozdy *et al.*, 2000; Tieu *et al.*, 2002; Stojanovski *et al.*, 2004), in plant cells dynamin-like (DRP3A, DRP3B) and Fis-type (BIGYIN) proteins act in the same pathway. The *Arabidopsis* genome contains two homologues of the yeast *FIS1* and human *hFIS1* genes, and analysis of their protein sequence shows that *BIGYIN* shares highest homology with *hFIS1* (26.7% identity, 48.3% similarity) and other *FIS1*-type genes from multicellular organisms, such as the *C. elegans FIS-2* gene (locus NM 001029389). Conversely, the second *Arabidopsis FIS1*-type gene, At5g12390, shares highest homology with yeast Fis1p (27.0% identity, 43.8% similarity). An analysis of the role of the second *Arabidopsis FIS1*-type gene, At5g12390, in mitochondrial fission is currently hampered by the lack of T-DNA mutants of this gene. However, future research using alternative reverse genetics approaches should help delineate the role of At5g12390 in mitochondrial dynamics and should reveal any redundancy between the two *Arabidopsis FIS1*-type genes. The *Arabidopsis* genome contains structural homologues of Caf4; however, to our knowledge, nothing is known concerning their function in mitochondrial fission. At the time of writing, no structural homologues of MDV1 have been identified in any multicellular eukaryote.

1.4.2 Mitochondrial fusion

Phase-contrast microscopy observations of mitochondrial behaviour indicated that fusion took place when the tip of one organelle came into close contact with another, leading to a physical bond forming which was maintained during intracellular movement (Bereiter-Hahn and Voth, 1994). The first genetic component of the mitochondrial fusion machinery was identified in the fruit fly, *Drosophila melanogaster*, during defective spermatogenesis. In wild-type *D. melanogaster*, the mitochondria in spermatids fuse to form a spherical structure called a Nebenkern. In flies defective for the *fuzzy onions* (*fzo*) gene, the formation of the Nebenkern is incomplete due to a lack of mitochondrial fusion, leading to a mass of unfused mitochondria. The blurred, spherical appearance of the Nebenkern in the mutants was likened to a fuzzy onion (Hales and Fuller, 1997).

Analyses of yeast *FZO1* deletion mutants confirmed the role of this protein in mitochondrial fusion. Fzo1p regulates mitochondrial fusion, and loss of Fzo1p function causes fragmentation of the mitochondrial network and loss of mtDNA (Hermann *et al.*, 1998; Rapaport *et al.*, 1998). Fzo1p is an outer-membrane protein containing a cytosolic-facing GTPase domain, essential for its function, and two transmembrane domains that anchor it to the mitochondrion (Hermann *et al.*, 1998). In a defining study, Sesaki and Jensen (1999) demonstrated that Dnm1p and Fzo1p acted in opposing pathways to maintain yeast chondriome structure. In *fzo1p dnm1p* double mutants, the wild-type yeast mitochondrial phenotype is restored. This analysis elegantly demonstrated that a delicate balance of fission and fusion is required to maintain yeast mitochondrial shape, size and number (Sesaki and Jensen, 1999).

The yeast mitochondrial fusion apparatus contains at least two more vital components, Mgm1p and Ugo1p. Mgm1p (mitochondrial genome maintenance 1) is a dynamin-like GTPase that resides in the intermembrane space (Wong *et al.*, 2000). The phenotype of *mgm1p* mitochondria is similar to that in *fzo1p* mutants, where the mitochondrial reticulum becomes fragmented, indicating that Mgm1p plays a role in mitochondrial fusion (Wong *et al.*, 2000, 2003). Again, like *fzo1p* mutants *mgm1p* mutants lose mtDNA, and this phenotype, along with alterations in mitochondrial morphology, can be suppressed by disrupting the division protein Dnm1p (Wong *et al.*, 2003). Mgm1p was identified independently by three groups (Herlan *et al.*, 2003; McQuibban *et al.*, 2003; Sesaki *et al.*, 2003a) as a substrate of a yeast rhomboid-type protease named Rbd1p (rhomboid), Pcp1p (processing of cytochrome *c* peroxidase; Esser *et al.*, 2002) or Ugo2p (ugo is Japanese for fusion; Sesaki and Jensen, 2001), and it was shown that cleavage of Rbd1p regulates inner-membrane remodelling (Herlan *et al.*, 2003; McQuibban *et al.*, 2003). Rbd1p contains six transmembrane domains and is embedded in the inner mitochondrial membrane (McQuibban *et al.*, 2003). Upon import of Mgm1p precursor, the N-terminal hydrophobic region becomes tethered in the inner membrane at the site of the first transmembrane domain, by what is assumed to be a translocation–arrest mechanism, leaving the N-terminal mitochondrial targeting presequence exposed to the matrix (Herlan *et al.*, 2003). Cleavage by the matrix processing peptidase generates what is called the large isoform of Mgm1p (l-Mgm1p) (Herlan *et al.*, 2003; Sesaki *et al.*, 2003a). Next, l-Mgm1p is further translocated into the matrix, and the second transmembrane domain becomes inserted into the inner membrane, whereupon it undergoes further proteolytic cleavage by Rbd1p producing a smaller isoform, s-Mgm1p, which is released into the intermembrane space where it becomes associated with either the outer- or inner mitochondrial membrane (Herlan *et al.*, 2003). Both isoforms function in the maintenance of mitochondrial morphology and respiratory competence, but the mechanism controlling the ratio of l-Mgm1p to s-Mgm1p is unknown (Herlan *et al.*, 2003). The exact role of the two isoforms in mitochondrial fusion is unclear: fusion can occur after mating in cells lacking Rbd1p (and hence without s-Mgm1p) and so it appears that the presence of the long isoform alone is sufficient to allow fusion (Sesaki *et al.*, 2003b).

The only other crucial component of the yeast fusion apparatus identified to date is Ugo1p. This protein was isolated in a screen for mutants exhibiting a mitochondrial morphology similar to *fzo1p* mutants. Mitochondria in *ugo1p* mutants lose mtDNA and have a fragmented morphology that, like *fzo1p* and *mgm1p* mutants, can be rescued by disrupting *DNM1* (Sesaki and Jensen, 2001). *UGO1* encodes a short 58-kDa outer-membrane protein, with a cytosolic N-terminus and a C-terminus in the intermembrane space (Sesaki and Jensen, 2001).

The roles played by Fzo1p, Mgm1p and Ugo1p in the fusion of mitochondria remain unclear; however, study of their structure and localisation provides clues to their function. It is believed that Fzo1p, Mgm1p and Ugo1p form a complex to effect mitochondrial fusion, with Ugo1p playing a particularly important role in mediating interactions between Fzo1p and Mgm1p (Sesaki and Jensen, 2004; Okamoto and Shaw, 2005). Ugo1p acts as a bridge, binding both Fzo1p and Mgm1p, using its cytosolic and intermembrane-space domains, respectively. The

Fzo1p-Mgm1p-Ugo1p fusion complex helps to connect the inner- and outer membranes by drawing the Fzo1p and Mgm1p GTPases together (Wong *et al.*, 2003; Sesaki and Jensen, 2004). While the GTPase domains of Fzo1p and Mgm1p are not required for the construction of the complex (Wong *et al.*, 2003; Sesaki and Jensen, 2004), GTPase function in both proteins is required for correct membrane fusion (Hermann *et al.*, 1998; Wong *et al.*, 2003). As Mgm1p is associated with the inner membrane or intermembrane space and Fzo1p is an integral outer membrane protein, it may be tempting to speculate that these two proteins control fusion in their respective membranes, powered by GTP hydrolysis, after the complex with Ugo1p has formed. However, it has been shown that the inner membrane-localised Mgm1p is required for fusion of the outer membrane, indicating that the proteins must act in concert, rather than separately, in controlling membrane fusion (Sesaki *et al.*, 2003a).

At the time of writing, no plant genes mediating mitochondrial fusion have been identified. There are *Arabidopsis* homologues of Mgm1p within the large family of dynamin-like proteins, but the protein with the highest homology to Mgm1p is DRP3A, which is believed to be involved in mitochondrial fission. However, if the parsimonious explanation of the role of DRP3A as provided by Arimura *et al.* (2004) and Logan *et al.* (2004) is incorrect, it is possible that DRP3A functions as a negative regulator of mitochondrial fusion; although this role would contrast with the role of Mgm1p and its mammalian homologue OPA1 (Misaka *et al.*, 2002; Okamoto and Shaw, 2005).

Although no genetic components of the plant mitochondrial fusion apparatus have been identified, there is no doubt that plant mitochondria fuse. Movies showing mitochondria fusing *in vivo* in a variety of plants have been available for a number of years (see movies at www.plantcellbiologyoncd.com), but it is only in the past couple of years that attempts have been made to investigate the process in detail. In an elegant study, Arimura *et al.* (2004) used a photoconvertible fluorescent protein called Kaede that can be induced to change irreversibly from green to red upon exposure to light of 350–400 nm. Using transient expression of mitochondria-targeted Kaede in onion cells, Arimura *et al.* (2004) were able to convert half of the mitochondria to fluoresce red and then visualise fusion between red and green mitochondria through the appearance of yellow mitochondria due to mixing of the matrix-localised fluorescent proteins (Plate 1.2A). Although in onion cells the fusion was often transient lasting only a few seconds, there was sufficient fusion between the mitochondria to convert them all to yellow within 1–2 h (Arimura *et al.*, 2004a).

Extensive fusion of mitochondria, called massive mitochondrial fusion (MMF), also occurs prior to cell division and after protoplast fusion (Plate 1.2B) (Sheahan *et al.*, 2005). MMF is followed by numerous fission events, which re-fragment the chondriome prior to redistribution of the newly mixed mitochondrial population throughout the cytoplasm and subsequent cytokinesis (Sheahan *et al.*, 2005). A similar situation is known to occur in yeast during mating, when, upon cell fusion, the mitochondria fuse mixing their DNA and matrix proteins (Nunnari *et al.*, 1997; Okamoto *et al.*, 1998). The MMF, which occurs within 4–8 h of the initiation of protoplast culture, requires an inner-membrane electrochemical gradient, cytoplasmic protein synthesis and an intact microtubule cytoskeleton; MMF did not require ATP or an intact actin cytoskeleton (Sheahan *et al.*, 2005).

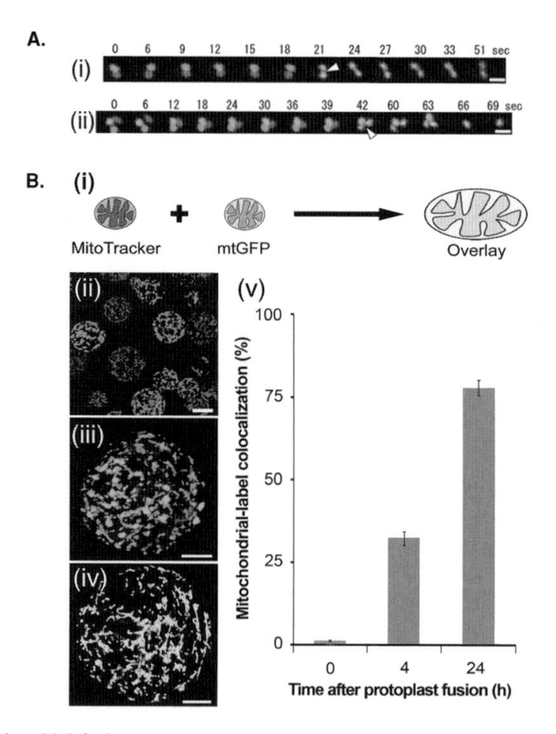

Plate 1.2 Mitochondrial fusion. (A) Mitochondrial fusion in onion bulb epidermal cells. (i) Mitochondrial fusion. After 21 s of observation, the previously physically discrete red or green mitochondria fuse, demonstrated by the appearance of yellow mitochondria. (ii) Mitochondrial fusion and fission. Three mitochondria are shown and fusion between the lower two organelles occurs between 39 and 42 s after the start of observation. This fusion event is only transient and by the 60-s time point the newly fused mitochondria have divided. Arrowheads indicate constrictions at the point of fusion. Scale bars = 2 μm. (Reproduced from Arimura *et al.*, 2004b, copyright National Academy of Sciences, USA.) (B) Mitochondria fuse during early protoplast culture. (i) Schematic representation of mitochondrial fusion as demonstrated by colocalization of green (mtGFP) and red (MitoTracker) mitochondrial labels to produce a yellow signal. (ii) Mixture of MitoTracker-stained (red) and mtGFP-expressing (green) protoplasts. (iii) Limited colocalization of label after 4-h protoplast culture. (iv) Almost complete colocalization after 24 h indicated that fusion occurred throughout the population of mitochondria. (v) Quantification of colocalization revealed that mitochondrial fusion was three-quarters complete in cultured protoplasts 24 h after protoplast fusion. Scale bar = 20 μm in (ii) and 10 μm in (iii) and (iv). (Reproduced from Sheahan *et al.*, 2005, copyright Blackwell Publishing Ltd.) See colour plate section following p. 110 for colour version of this figure.

1.5 Mitochondrial movement and cellular inheritance

1.5.1 Mitochondrial movement and the cytoskeleton

The cytoskeleton plays a hugely important role in the maintenance of normal mitochondrial morphology, both in terms of regulating the distribution of a mitochondrial population within a cell and in determining the shape and size of individual organelles.

The cytoskeleton forms a network throughout the cytoplasm along which organelles move, either by direct attachment to a growing microtubule or actin filament or by using the network as 'tracks' along which the organelle is transported using motor proteins (Catlett and Weisman, 2000; Wada and Suetsugu, 2004). In contrast to mitochondria in mammals, and most yeast, but similar to mitochondria in *S. cerevisiae*, plant mitochondria move predominantly on the actin cytoskeleton (Olyslaegers and Verbelen, 1998; Van Gestel *et al.*, 2002), although movement on and tethering to microtubules probably also occurs, depending on developmental state and physiology (Van Gestel *et al.*, 2002).

Mitochondrial movement on actin filaments is predominantly associated with actin binding and polymerisation (Simon *et al.*, 1995; Catlett and Weisman, 2000). Although mitochondrial membranes have been shown to exhibit motor activity when bound to actin, this activity does not appear to be reliant on the traditional actin motor protein, myosin (Simon *et al.*, 1995). In yeast, myosin gene deletion does not affect the velocity of mitochondrial movement (Simon *et al.*, 1995), and the disruption of type V myosin, important for mitochondrial inheritance (Itoh *et al.*, 2001), does not affect either mitochondrial movement or their association with actin filaments (Boldogh *et al.*, 2004). Rather than involving myosin, mitochondrial movement along actin filaments appears to be powered by another force generator, the ARP2/3 complex (Boldogh *et al.*, 2001). Disruption of either *ARC15* or *ARP2*, two subunits of the complex, leads to a decrease in mitochondrial motility and alterations in mitochondrial morphology (Boldogh *et al.*, 2001). The ARP2/3 complex binds newly polymerised actin and promotes actin nucleation (Pollard and Beltzner, 2002), which generates forces to facilitate mitochondrial movement (Boldogh *et al.*, 2005). While there are homologues of the yeast ARP2/3-complex genes in higher plants (McKinney *et al.*, 2002), it is not known whether they function in actin-based movement of mitochondria.

In contrast to actin-based movement, mitochondrial movement along microtubules is closely linked to motor proteins, especially kinesin, which binds cargo for transport and is powered by ATP hydrolysis (Yaffe, 1999). Two kinesin proteins have been identified in the movement of mitochondria along microtubules in mammalian cells, KIF1B (Nangaku *et al.*, 1994) and KIF5B (Tanaka *et al.*, 1998). KIF1B is an N-terminal motor protein which co-localises with mitochondria *in vivo* and has the capacity to bind and transport mitochondria along microtubules *in vitro* (Nangaku *et al.*, 1994). KIF5B, a ubiquitous kinesin heavy chain protein (the kinesin complex has two light and two heavy chains), associates with mitochondria during subcellular fractionation (Tanaka *et al.*, 1998). Disruption of the *kif5B* gene leads to the clustering of mitochondria around the nucleus and causes mortality in mouse embryos (Tanaka *et al.*, 1998). One isoform of the kinesin light chain has also been shown to associate with mitochondria and is implicated as the cargo-binding domain of the complex (Khodjakov *et al.*, 1998).

1.5.2 Chondriome structure and the cell cycle

It has been known for some time that there is a correlation between the phase of the cell cycle, and the number, and mass of the mitochondrial population (James and Bowman, 1981). During the human cell cycle, mitochondria switch between two predominant morphological states (Barni *et al.*, 1996; Karbowski *et al.*, 2001;

Margineantu *et al.*, 2002). During the G1 phase of the cell cycle, mitochondria fuse to form reticula bringing the number of individual organelles to half the number prior to M phase (Karbowski *et al.*, 2001). As cells proceed from G1 to S phase, mitochondrial numbers increase due to fragmentation (fission) of the mitochondrial reticula (Barni *et al.*, 1996; Karbowski *et al.*, 2001; Margineantu *et al.*, 2002).

These cell cycle-dependent changes in mitochondrial morphology are also evident in plants and algae. Using tobacco plants, Sheahan *et al.* (2004) showed that the size and morphology of mitochondria in cultured protoplasts varied with the cell cycle. Mitochondria are typically small and numerous when observed after protoplast isolation, but within 4 h of protoplast culture, the majority of mitochondria undergo MMF (see Section 1.4.2; Sheahan *et al.*, 2005) to form a reticulum. While the volume of mitochondria (measured by GFP fluorescence) remained static, the number of individual organelles fell slightly for the first 24 h, indicating a net rise in fusion (Sheahan *et al.*, 2004). From 48 h onwards, the reticulum began to fragment leading to a rise in the number of physically discrete mitochondria (Sheahan *et al.*, 2004). This process continued so that by 72 h of culture (when protoplasts are ready to divide), there was a net doubling of the number of physically discrete mitochondria in the population (Sheahan *et al.*, 2004). Mitochondrial fusion, enabling the exchange, and complementation of mtDNA molecules both during the MMF (Sheahan *et al.*, 2005) and more frequent, but less extensive, fusion events that occur in non-dividing cells (Arimura *et al.*, 2004b) is inherent in Lonsdale's dynamic syncytium hypothesis (Lonsdale *et al.*, 1988). Mitochondrial fusion overcomes the physical barrier to the exchange and complementation of mtDNA molecules that accompanied the organisation of the higher plant chondriome into physically discrete organelles. The opposing processes of fission and fusion are therefore responsible for maintenance of the plant chondriome as a discontinuous whole. Following mixing of the mitochondrial population as a result of the MMF event, and subsequent division, the physically discrete mitochondria distribute in an ordered, actin filament -dependent manner to ensure unbiased portioning into daughter cells.

These data are in agreement with a previous report, which demonstrated that mitochondrial morphology in tobacco varies with the cell cycle and physiological status (Stickens and Verbelen, 1996). Dividing cells were characterised by numerous small, round mitochondria, while expanding cells mainly contained longer, vermiform organelles (Stickens and Verbelen, 1996). In the unicellular algae *C. merolae*, division of the single mitochondrion appears to be integrated in the process of cell division (Nishida *et al.*, 2005). Nishida and colleagues (2005) demonstrated, using immunofluorescence and electron microscopy, that there is a close association between dividing mitochondria and microtubules attached to spindle poles, indicating a relationship between mitochondrial segregation and mitosis.

1.5.3 Mitochondrial inheritance in yeast

The cycle of mitochondrial fusion, followed by organelle division immediately prior to cell division, appears to be a general characteristic of mitochondrial biology in higher eukaryotes. In yeast, the process of mitochondrial inheritance during budding has been extensively studied and is dependent on aspects such as the cytoskeleton, mitochondrial division and mitochondrial fusion (Hermann and Shaw, 1998; Catlett and Weisman, 2000). Briefly, during the budding process in *S. cerevisiae*, the mitochondrial reticulum moves towards the intended bud site and a single tubule moves into the newly formed bud. This movement continues until mother and daughter have an equal mitochondrial complement, at which point cytokinesis can occur (Catlett and Weisman, 2000; Boldogh *et al.*, 2005). Study of the genetic control of this phenomenon over the past 15 years has revealed many of the genes involved, mostly discovered through research into mitochondrial morphology mutants.

The first mutants defective in mitochondrial inheritance were isolated during a microscopic screen of temperature-sensitive yeast cell lines, searching for individuals that did not pass mitochondria to the daughter bud prior to cell division (McConnell *et al.*, 1990). Many of these mutant lines exhibited an altered mitochondrial morphology and were named *mdm* (mitochondrial distribution and morphology) (Table 1.1). Initial work on these cell lines indicated that the mutant genes fell into two broad categories: genes that encoded integral outer mitochondrial membrane proteins and genes encoding cytosolic proteins (Yaffe, 1999).

The cytosolic proteins are believed to be mainly associated with the cytoskeleton and are typified by three proteins: Mdm1p, Mdm14p and Mdm20p (Yaffe, 1999). Mdm1p is an intermediate filament-like protein, and disruptions in *MDM1* lead to a breakdown of the yeast mitochondrial reticulum, resulting in numerous small organelles (McConnell *et al.*, 1990; Fisk and Yaffe, 1997). Mdm14p, which contains a coiled-coil interaction domain, is involved in both nuclear and mitochondrial inheritance and mutations of this gene lead to aggregated mitochondria (Shepard and Yaffe, 1997). Finally, Mdm20p, which also has a coiled-coil domain, is involved in the organisation and assembly of the actin cytoskeleton (Hermann *et al.*, 1997). While the gross mitochondrial morphology of this mutant was unchanged relative to wild type, Mdm20p-defective cells had no transmission of mitochondria to daughter buds (Hermann *et al.*, 1997).

The second group of proteins was found to be integral constituents of the outer mitochondrial membrane, including Mdm10p, Mmm1p (maintenance of mitochondrial morphology) and Mdm12p (Yaffe, 1999). Mutations in *MDM10, MMM1* or *MDM12* led to a similar phenotype – greatly enlarged, spherical organelles instead of the mitochondrial tubules in wild type (Burgess *et al.*, 1994; Sogo and Yaffe, 1994; Berger *et al.*, 1997). Both Mdm10p and Mmm1p have large cytosolic domains, which may indicate that they mediate the interaction of mitochondria with other cellular components (Yaffe, 1999). The function of these proteins is currently unresolved. However, there are three main theories regarding the role of these proteins in inheritance and morphology: (i) mediating mitochondrial attachment to the actin cytoskeleton, (ii) the tubulation of mitochondrial membranes, and (iii) the anchoring of mtDNA nucleoids (Boldogh *et al.*, 2003, 2005; Okamoto and Shaw, 2005).

Table 1.1 Genes involved in mitochondrial dynamics

Gene	Organism	Location	Mutant mitochondrial phenotype	Protein properties/Role	Reference	*Arabidopsis* homologue[a]
BIGYIN (BGY1)	*A. thaliana*	Mitochondrial outer membrane (I. Sparkes, unpublished data)	Reduced number of mitochondria per cell, individual spherical mitochondria approximately twice as large as wild type	Orthologue of *S. cerevisiae FIS1*. Interacts with dynamin in mitochondrial division (?)	Scott *et al.* (2006)	The role of a second *Arabidopsis FIS1*-type homologue, At5g12390 is unknown, although the protein does localise to mitochondria (I. Sparkes, unpublished data).
BMT1	*A. thaliana*	Not known	Reduced number of mitochondria per cell, individual spherical mitochondria approximately twice as large as wild type	?, not *BIGYIN* or *DRP3A* or *DRP3B*	Logan *et al.* (2003)	–
BMT2	*A. thaliana*	Not known	Fewer but larger mitochondria per cell	Not known	D. C. Logan and A.K̃. Tobin (unpublished data)	–
DNM1/Drp1/DRP-1	*S. cerevisiae/ Homo sapiens*	Cytoplasm	Netlike sheet of interconnected tubules	Dynamin-related GTPase/ membrane fission	Otsuga *et al.* (1998), Smirnova *et al.* (2001)	*DRP3A/B* orthologues, see below.
DRP3A (formerly *ADL2a*)	*A. thaliana*	Cytoplasm, recruited to mitochondrial constriction/division sites	Mitochondria form long tubules with many constrictions and protuberances (matrixules)	Dynamin-related GTPase/ membrane fission (?)	Logan *et al.* (2004), Arimura *et al.* (2004a)	DRP3A is one of a large family of dynamin-like proteins in *Arabidopsis*

DRP3B (formerly *ADL2b*)	*A. thaliana*	Cytoplasm, recruited to mitochondrial constriction/division sites	Mitochondria form long interconnected tubules	Dynamin-related GTPase/ membrane fission (?)	Arimura and Tsutsumi (2002)	See above.
CAF4	*S. cerevisiae*	Cytosolic, associates with mitochondrial outer membrane	No change, partially rescues *mdv1p* mutants	WD-40 repeat protein, mediates Fis1p/Dnm1p interactions	Griffin *et al.* (2005)	Homology to WD-40 repeat proteins
cluA/CLU1	*D. discoideum/S. cerevisiae*	Cytoplasm	Clusters of mitochondria. In yeast they collapse to side of cell	Kinesin-like domain/outer membrane fission	Zhu *et al.* (1997), Fields *et al.* (1998), Fields *et al.* (2002)	*FMT*, see below.
FMT	*A. thaliana*	Not known	Clusters of mitochondria	TPR-domain; homologue of *cluA* and *CLU1*	Logan *et al.* (2003)	There is a small gene family of similar-sized TPR-containing proteins in *Arabidopsis*
FIS1 (synonym: *MDV2*)/*hFIS1*	*S. cerevisiae/H. sapiens*	Mitochondrial outer membrane	Net-like sheet of interconnected tubules	Integral membrane protein. Interacts with dynamin and Mdv1p (*S. cerevisiae*only) in fission process	Tieu and Nunnari 2000), Mozdy *et al.* (2000), Tieu *et al.* (2002), Yoon *et al.* (2003), Suzuki *et al.* (2003, 2005)	*BIGYIN* (At3g57090 and At5g12390)
fzo/FZO1/Mfn1/Mfn2	*D. melanogaster/S. cerevisiae/H. sapiens*	Mitochondrial outer membrane	Aberrant mitochondrial fusion; fragmentation of tubules	GTPase/mitochondrial fusion	Hales and Fuller (1997), Rapaport *et al.* (1998), Hermann *et al.*(1998), Santel and Fuller (2001), Rojo *et al.* (2002)	*FLZ* (At1g03160). Determinant of thylakoid and chloroplast morphology. FZL reported to have no affect on mitochondria (Gao *et al.*, 2006)

(*Continued*)

Table 1.1 Genes involved in mitochondrial dynamics (*Continued*)

Gene	Organism	Location	Mutant mitochondrial phenotype	Protein properties/Role	Reference	*Arabidopsis* homologue[a]
KIF5B	*Mus musculus*	Cytoplasm, mitochondrial associated	Mitochondria collapse around nucleus	Conventional kinesin heavy chain protein, attachment to cytoskeleton	Tanaka *et al.* (1998)	There are many kinesin-like proteins in *Arabidopsis*. Closest to KIF5B is At3g63480, 3e-71
MGM1/Opa1	*S. cerevisiae/H. sapiens*	Inter-membrane space	Mitochondria fragmentation. Loss of mtDNA. Defective transmission to daughter buds	Dynamin-related protein/inner-membrane modelling, may interact with Ugo1p and Fzo1p	Alexander *et al.* (2000), Misaka *et al.* (2002), Jones and Fangman (1992), Guan *et al.* (1993), Shepard and Yaffe (1997, 1999), Delettre *et al.*(2000), Misaka *et al.* (2002), Wong *et al.* (2000, 2003)	Dynamin-likegenes in Arabidopsis, closest is DRP3A, At4g33650
MDM1	*S. cerevisiae*	Cytoplasm	Fragmentation of tubules. Defective transmission to daughter buds	Intermediate filament-like	McConnell and Yaffe (1992)	At2g15900, 0.029
MDM10	*S. cerevisiae/ Podospora anserina*	Mitochondrial outer membrane	Large spherical mitochondria. Defective transmission to daughter buds	Integral membrane protein	Sogo and Yaffe (1994), Jamet-Vierny *et al.* (1997)	None
MDM12	*S. cerevisiae*	Mitochondrial outer membrane	Large spherical mitochondria. Defective transmission to daughter buds	Integral membrane protein	Berger *et al.* (1997)	None
MDM14	*S. cerevisiae*	Cytoplasm	Mitochondria aggregate. Defective transmission to daughter buds	Coiled-coil domain	Shepard and Yaffe (1997)	None

MDM20	*S. cerevisiae*	Cytoplasm	Defective transmission to daughter buds	Coiled-coil domain/disrupts actin cables	Hermann *et al.* (1997)	None
MDM30	*S. cerevisiae*	Cytoplasm ?	Fragmented or aggregated; few short tubules	?	Dimmer *et al.* (2002)	None
MDM31	*S. cerevisiae*	Predicted mitochondrial inner-membrane protein	Compact mitochondrial aggregates	?	Dimmer *et al.* (2002)	None
MDM32	*S. cerevisiae*	Predicted mitochondrial inner-membrane protein	Compact mitochondrial aggregates	?	Dimmer *et al.* (2002)	None
MDM33	*S. cerevisiae*	Mitochondrial inner-membrane protein	Giant ring-like mitochondria	Coiled-coil domains, part of high molecular weight complex/ putatively involved ininner-membrane fission	Dimmer *et al.* (2002); Messerschmitt *et al.*(2003)	At1g54560, 5e-05
MDM34	*S. cerevisiae*	Cytoplasm ?	Spherical mitochondria	?	Dimmer *et al.* (2002)	None
MDM35	*S. cerevisiae*	Cytoplasm ?	Spherical mitochondria	?	Dimmer *et al.* (2002)	At4g33100, 7e-07
MDM36	*S. cerevisiae*	Cytoplasm ?	Mitochondrial tubules aggregate/collapse to one side of cell	?	Dimmer *et al.* (2002)	None
MDM38	*S. cerevisiae*	Predicted mitochondrial inner-membrane protein	Lasso-like mitochondria	Calcium binding protein	Dimmer *et al.* (2002)	At3g59820, 3e-54; At1g65540, 1e-52
MDM39	*S. cerevisiae*	Predicted integral membrane protein – no mitochondrial targeting presequence	Fragmented mitochondrial tubules	?	Dimmer *et al.* (2002)	None
MDV1 (synonyms: *FIS2, GAG3, NET2*)	*S. cerevisiae*	Cytoplasm/associates with mitochondrial outer membrane	Net-like sheet of interconnected tubules	Predicted coiled-coil and seven WD-40 repeats/interacts with Dnm1p and Fis1p in fission process	Fekkes *et al.* (2000), Mozdy *et al.* (2000), Tieu and Nunnari (2000), Cerveny *et al.* (2001), Tieu *et al.* (2002)	Homology to WD-40-repeat containing proteins

(*Continued*)

Table 1.1 Genes involved in mitochondrial dynamics (*Continued*)

Gene	Organism	Location	Mutant mitochondrial phenotype	Protein properties/Role	Reference	*Arabidopsis* homologue[a]
MMM1	*S. cerevisiae*	Mitochondrial outer membrane	Large spherical mitochondria. Defective transmission to daughter buds	Integral membrane protein	Burgess *et al.* (1994)	None
MMT1	*A. thaliana*	?	Much larger and smaller mitochondria in same cell. Giant chloroplasts	?	Logan *et al.* (2003)	—
MMT2	*A. thaliana*	?	Much larger and smaller mitochondria in same cell. Altered thylakoid morphology	?	Logan *et al.* (2003)	—
MsftsZ-mt/CmftsZ1-1/CmftsZ1-2/fszA/fszB	*M. splendens/C. merolae/D. discoideum*	Mitochondrial division sites and tips	Elongated tubular mitochondria (*D. discoideum*)	Tubulin-like, ring-forming protein, organelle constriction	Beech *et al.* (2000), Takahara *et al.* (2000), Gilson *et al.* (2003)	Chloroplast *FtsZ1*(At5g55280) and *FtsZ2* (At2g36250) division genes (McAndrew *et al.*, 2001)
NMT	*A. thaliana*	?	Mitochondria form extensive interconnected network	?	Logan *et al.* (2003)	—
RBD1 (synonyms: *MDM37, PCP1, UGO2)/ hPARL*	*S. cerevisiae/H. sapiens*	Mitochondrial inner membrane	Fragmented mitochondria, aggregated	Rhomboid-like protein, processing of Mgm1p	Dimmer *et al.* (2002), Herlan *et al.* (2003), McQuibban *et al.* (2003), Sesaki *et al.* (2003a,b)	Rhomboid domain proteins, closest is At1g18600
UGO1	*S. cerevisiae*	Mitochondrial outer membrane	Fragmentation of mitochondrial tubules	Involved in mitochondrial fusion	Sesaki and Jensen (2001)	None

[a] Homologues were identified from the literature or by BLASTP search at NCBI (www.ncbi.nlm.nih.gov/blast).

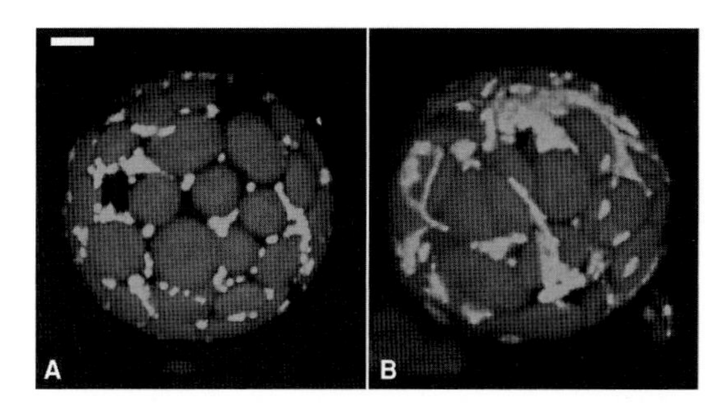

Plate 1.3 Mitochondrial distribution in the wild type and in the *friendly* (*fmt*) mutant of *Arabidopsis*. (A) Protoplast of wild-type *Arabidopsis* expressing mito-GFP, displaying the characteristic even distribution of mitochondria (green) amongst the cortical chloroplasts (red autofluorescence). (B) Large clusters of mitochondria in the *friendly* mutant. Scale bar = 5 μm. See colour plate section following p. 110 for colour version of this figure.

1.5.4 Arabidopsis *friendly mitochondria mutant and the cellular distribution of mitochondria*

Mutations of the *Arabidopsis Friendly Mitochondria* (*FMT*) gene lead to a grossly altered cellular distribution of mitochondria (Logan *et al.*, 2003). Disruption of *FMT* causes the mitochondria to form large clusters of tens or hundreds of organelles, although some mitochondria remain apparently normally distributed as singletons throughout the cytoplasm (Plate 1.3; Logan *et al.*, 2003). *FMT* is a conserved eukaryotic gene but, apart from a short TPR domain that is thought to function in protein–protein interactions, the FMT protein has no homology to proteins of known function. Disruption of *FMT* homologues in *D. discoideum* (*cluA*) or *S. cerevisiae* (*CLU1*) also causes aberrant mitochondrial phenotypes (Zhu *et al.*, 1997; Fields *et al.*, 1998). In the *cluA*$^-$ mutant of *D. discoideum* the mitochondria cluster near the cell centre (Zhu *et al.*, 1997), while in the *S. cerevisiae clu1* mutant the mitochondrial tubules collapse to one side of the cell (Fields *et al.*, 1998). The only clues to the function of FMT come from the mitochondrial phenotype and the presence of the TPR domain. Mitochondrial association with microtubules has been shown to involve the microtubule-specific motor protein, kinesin (Khodjakov *et al.*, 1998) that binds cargo at the TPR domains in the kinesin light chains (Stenoien and Brady, 1997; Verhey *et al.*, 2001). On the basis of the above information, it was suggested that FMT is involved in the interaction of mitochondria with the microtubule cytoskeleton (Logan *et al.*, 2003). This hypothesis for FMT function is supported by the observation that the pattern of mitochondrial distribution in *fmt* mutants mimics the effect of latrunculin-B, which promotes the rapid depolymerisation of the actin cytoskeleton (D. C. Logan, unpublished data). The *Arabidopsis* genome contains 61 identified kinesin-like genes, the highest number in any sequenced eukaryotic genome (Reddy and Day, 2001). However, no heavy chains have been identified in any plant, although some light chains have been predicted in the *Arabidopsis* genome (Reddy and Day, 2001). In contrast, there are many microtubule-associated processes unique to plants that are likely to require additional plant-specific microtubule-associated proteins including motors (Reddy and Day, 2001; Hussey, 2004). Yeast-two-hybrid screens are currently being performed to identify proteins interacting with FMT *in vivo*.

1.6 Other plant mitochondrial dynamics mutants

There are few *Arabidopsis* homologues of animal or yeast proteins playing a direct role in mitochondrial dynamics (Logan, 2003). In some cases, this can be explained by the fundamental differences between yeast and plant mitochondrial morphology and reproductive biology. For example, the yeast protein Mdm1p is involved in the maintenance of the tubular mitochondrial morphology (McConnell and Yaffe, 1992) so it is not surprising that there is no homologue in *Arabidopsis* where the numerous discrete mitochondria alternate between spherical and sausage-shaped structures. Similarly, in contrast to animal and plant cells, the budding yeast *S. cerevisiae* proliferates by budding whereby the mother cell produces a daughter bud that grows and eventually becomes an independent cell. An essential part of this process is the transport of mitochondria and other organelles into the daughter bud. Mmm1p, Mdm10p, Mdm12p and Mdm20p are all involved in the transmission of mitochondria to the daughter buds (Yaffe, 1999), and since cell proliferation in plants occurs by cell division, these genes are not required. Equally, the differences between yeast and plant cells mean that there are likely to be many proteins involved in mitochondrial dynamics that are specific to either multicellular organisms or are plant specific. To identify the genes, proteins and mechanisms controlling mitochondrial dynamics in higher plants, an ethyl methane sulphonate (EMS)-mutagenised *Arabidopsis* population expressing mitochondria-targeted GFP (Logan and Leaver, 2000) was screened for altered mitochondrial shape, size, number and distribution mutants (Logan *et al.*, 2003). Seven viable mutants with distinct mitochondrial phenotypes were identified (including *fmt*, see above) from a population of approximately 9500 individuals. In the *mmt1* (first *motley mitochondria* mutant), the mitochondrial population is highly heterogeneous varying in size from one quarter to four times the average plan area of wild-type mitochondria. The size distribution of chloroplasts is also affected in the mutant; chloroplast plan areas in the mutant range from 4 to 240 times the plan area in the wild type (Logan *et al.*, 2003). The *mmt2* (second *motley mitochondria* mutant) contains a highly heterogeneous mitochondrial population similar to *mmt1*. Although gross chloroplast morphology remains normal in *mmt2*, transmission electron microscopy (TEM) demonstrated that the internal structure of the chloroplasts is severely altered. Chloroplasts in the *mmt2* mutant contain a large number of electron dense particles and a mass of densely packed membranes instead of the normal morphology of granal stacks connected by stromal lamellae. Mitochondria in *bmt* (*big mitochondria* mutant) have plan areas approximately two- to four times wild type, and there are approximately half as many per microscope field-of-view. The presence of long, interconnected mitochondrial tubules extending to many tens of micrometers in length characterises the *nmt* (*network mitochondria* mutant). Examination of leaf tissue of *nmt* plants under the TEM showed that the reticular morphology was not maintained in the fixed tissue; instead the mitochondrial tubules fragmented to form organelles as small as one-sixteenth the plan area of those in wild-type cells. Gene mapping of the *mmt1, mmt2, bmt*, and *nmt* mutants indicates that these four loci are novel genes involved in mitochondrial dynamics since no mitochondrial development genes/mutants have been mapped to the regions containing the mutant loci, nor are any obvious candidate genes in these regions (Logan *et al.*, 2003).

In an attempt to identify additional *Arabidopsis* mitochondrial morphology mutants, Feng *et al.* (2004) generated and screened an independent EMS-mutagenised population. From 19 000 M2 individuals, 17 mutant lines were identified; in all cases the mitochondria were either longer or larger than wild type (Feng *et al.*, 2004). Gene mapping of the mutant loci in 7 of the 17 lines demonstrated that at least 4 different loci were involved. The mutant locus in three lines mapped close to *DRP3A*, two close to *bmt* and one close to *nmt*. The one remaining mutant locus was mapped to the long arm of chromosome 4, a locus not previously implicated in the control of mitochondrial morphology, nor containing homologues of any genes known to affect mitochondrial morphology (Feng *et al.*, 2004). None of the mutant loci identified in this screen have been cloned.

1.7 Metabolic control of mitochondrial morphology and motility

The metabolic status of the cell is known to affect mitochondrial dynamics (Lloyd, 1974; Bereiter-Hahn and Voth, 1983). These effects, which have been suggested to be indicative of the rapid adaptation of mitochondrial function to a changing environment (Bereiter-Hahn and Voth, 1983), include alterations to the gross morphology of individual mitochondria, the structural organisation of the chondriome as a whole, and the mitochondrial membrane ultrastructure.

The conformation of the inner membrane is believed to be continuously variable between two extremes, orthodox and condensed, depending on the energy state of the mitochondrion, and has been shown to affect the external morphology and motility of mitochondria (Bereiter-Hahn and Voth, 1983; Logan, 2006b). Changes in the external morphology of mitochondria, the bending, branching, formation and retraction of localised protrusions (matrixules, Logan *et al.*, 2004) that are typical of mitochondria in living cells, have all been ascribed to the rearrangement of cristae (Bereiter-Hahn and Voth, 1994). However, the extent to which these shape changes are truly intrinsic or involve the activity of molecular motors on the cytoskeleton is not known. Bereiter-Hahn and Voth (1983) analysed shape changes and motility of mitochondria in endothelial cells from *Xenopus laevis* tadpole hearts. In the condensed state mitochondria are immobile, while in the orthodox state they are motile (Bereiter-Hahn and Voth, 1983). Inhibition of electron transport or oxidative phosphorylation causes a decrease in mitochondrial motility and a concomitant transition to the condensed conformation (Bereiter-Hahn and Voth, 1983). Injection of ADP, which induces extreme condensation, also immobilises mitochondria. In addition to their effect on mitochondrial motility, inhibitors of electron transport induce the formation of large disc-shaped mitochondria; an identical morphology is seen in tissues under anoxic conditions.

Changes to the gaseous environment in and around the cell can have a profound effect on mitochondrial morphology and structure. Growth under elevated CO_2 concentrations leads to a 1.3- to 3-fold increase in the number of mitochondria per cell in some plant species (Robertson *et al.*, 1995; Griffin *et al.*, 2001). However, changes in CO_2 concentration had no apparent effect on mitochondrial ultrastructure (Robertson *et al.*, 1995). Culturing yeast cells anaerobically results in a vast increase in mitochondrial size, with individual organelles being around 10 times the size of those found in aerobic cultures (Lloyd, 1974). Concomitant with this change is a large decrease in mitochondrial number per cell, with anaerobically cultured yeast containing 20 times fewer organelles, the net result of which is a 50% reduction of mitochondrial volume in anaerobic cells (Lloyd, 1974). In addition to changes in size and number, mitochondria in anaerobically grown yeast contain no cristae (Lloyd, 1974).

Maintaining tobacco cell cultures under anoxic conditions produces a similar effect on mitochondrial morphology. Using pure nitrogen to purge oxygen from the culture medium, Van Gestel and Verbelen (2002) showed that anoxia results in the formation of giant mitochondria, some 80-nm long, which eventually form a reticulum similar to the chondriome structure in *S. cerevisiae*. This process is wholly reversible since increasing the oxygen concentration of the medium returns the organelles to their wild-type state (Van Gestel and Verbelen, 2002). Low oxygen pressure, achieved by mounting cells at high density under a coverslip on a microscope slide, also induces the formation of disc-like mitochondria in tobacco suspension cultured cells (Van Gestel and Verbelen, 2002). Over a time period of 4 h (shorter at higher cell densities), the normally discrete mitochondria (0.5–5 μm in length) have fused to form a reticulum composed of linear and ring-shaped tubular sections interspersed with large plate-like structures (Van Gestel and Verbelen, 2002). Mitochondria in *Arabidopsis* leaf epidermal cells have been observed undergoing similar morphological transitions during prolonged (40 min) incubation of sections of leaf between slide and coverslip (D. C. Logan, unpublished observations). Unlike the *Xenopus* mitochondria (see above), tobacco suspension cell mitochondria did not change morphology in response to respiration inhibitors or uncouplers (KCN, dinitrophenol or carbonyl cyanide m-chlorophenylhydrazone), nor did oxidative stress induced by paraquat, menadion, hydrogen peroxide or $CuSO_4$ induce changes in the normal mitochondrial morphology (Van Gestel and Verbelen, 2002). Van Gestel and Verblen (2002) suggest that this may be due to up-regulation of the alternative respiratory pathway which has been suggested to mitigate against reactive oxygen species (ROS) damage in plant cells (see Chapter 8). However, paraquat and hydrogen peroxide do induce a change in the mitochondrial morphology in *Arabidopsis* leaf epidermal cells and mesophyll protoplasts. Within 4 h of exposing *Arabidopsis* mesophyll protoplasts to ROS (paraquat, H_2O_2), mitochondria in 60–70% of protoplasts undergo a morphology transition characterised by swelling to at least double their volume (a similar effect of paraquat and H_2O_2 was reported by Yoshinaga *et al.*, 2005). No significant cell death occurs over the following 20 h, but after 48 h of ROS treatment 70–90% of protoplasts have died, all containing mitochondria that underwent a morphology transition; 10–30% of protoplasts that remain alive after treatment contain mitochondria of normal morphology (Scott, 2006). Excess oxygen has a remarkable effect on Drosophila flight muscle mitochondrial ultrastructure. Placing Drosophila flight muscles under hyperoxic (100% [v/v] oxygen) conditions results in a re-arrangement of the cristae, which form into a 'swirl'-like pattern (Walker and Benzer, 2004). These changes are believed to be caused by an increase in ROS leading to degeneration of the mitochondria and cell death (Walker and Benzer, 2004).

The effect of the metabolic status of the mitochondrion on mitochondrial morphology and motility has been suggested to help ensure that the mitochondria are located where they are needed. Association of mitochondria with energy-requiring structures or organelles has been well described in a variety of systems (see Munn, 1974; Tyler, 1992; Bereiter-Hahn and Voth, 1994). One classic example is the formation of the Nebenkern, a collar around the sperm axoneme formed during spermatogenesis, and comprising two giant mitochondria formed by repeated fusion events (Hales and Fuller, 1997). In plant tissues containing chloroplasts, visualisation of mitochondria stained with $DiOC_6$ or expressing GFP has shown the frequent close proximity of these two organelles (Stickens and Verbelen, 1996; Logan and Leaver, 2000). It is assumed that this facilitates exchange of respiratory gases, and possibly metabolites, although direct evidence for this is lacking. In characean internode cells, it has been suggested that the spatiotemporal distribution of mitochondria within the cell promotes their association with chloroplasts (Foissner, 2004). A further example of mitochondrial association with energy-consuming structures is the association of mitochondria with the endoplasmic reticulum. One explanation for this association has recently been gaining acceptance. It has been demonstrated in HeLa cells that there are micro-domains of the mitochondrial reticulum where it is in very close contact (<60 nm) with the endoplasmic reticulum (Rizzuto *et al.*, 1998). The functional significance of these micro-domains has been explained on the basis of Ca^{2+} dynamics (Rutter and Rizzuto, 2000). For example, localised agonist-induced release of Ca^{2+} from the endoplasmic reticulum may stimulate uptake into the closely associated mitochondria where the transient increase in Ca^{2+} may modulate mitochondrial function.

1.8 Concluding remarks

Mitochondria evolved from free-living α-proteobacteria, and subsequent evolution, has resulted in the development of significant differences in chondriome structure and in the genes required to maintain chondriome structure between plant, yeast and animal lineages. Despite all the differences between plant mitochondria and those of other eukaryotes, some aspects of mitochondrial dynamics have been conserved, including the maintenance of the chondriome by the dual processes of fission and fusion and the use of dynamin-like and Fis1-type proteins for organelle division. In contrast, no genes involved in plant mitochondrial fusion have been identified.

One of the main conclusions that can be reached from recent research on mitochondrial dynamics is the clear interplay between mitochondrial form and function. For example, studies into programmed cell death (see Chapter 10) have uncovered the vital role played by mitochondrial dynamics in these processes. As we learn more about the genes, proteins and mechanisms controlling mitochondrial dynamics, we can expect to discover many more instances where mitochondrial form and function are integrated.

Acknowledgements

The authors were supported by research grants (DCL), a studentship (IS) and Wain Fellowship (IS) from the BBSRC. The authors record their gratitude to Dr Alyson K. Tobin, School of Biology, University of St Andrews for her support.

References

Abdelnoor RV, Yule R, Elo A, Christensen AC, Meyer-Gauen G and Mackenzie SA (2003) Substoichiometric shifting in the plant mitochondrial genome is influenced by a gene homologous to MutS. *Proc Natl Acad Sci USA* **100**, 5968–5973.

Alexander C, Votruba M, Pesch UE, *et al.* (2000) OPA1, encoding a dynamin-related GTPase, is mutated in autosomal dominant optic atrophy linked to chromosome 3q28. *Nat Genet* **26**, 211–215.

Arimura S, Aida GP, Fujimoto M, Nakazono M and Tsutsumi N (2004a) *Arabidopsis* dynamin-like protein 2a (ADL2a), like ADL2b, is involved in plant mitochondrial division. *Plant Cell Physiol* **45**, 236–242.

Arimura S and Tsutsumi N (2002) A dynamin-like protein (ADL2b), rather than FtsZ, is involved in *Arabidopsis* mitochondrial division. *Proc Natl Acad Sci USA* **99**, 5727–5731.

Arimura S, Yamamoto J, Aida GP, Nakazono M and Tsutsumi N (2004b) Frequent fusion and fission of plant mitochondria with unequal nucleoid distribution. *Proc Natl Acad Sci USA* **101**, 7805–7808.

Backert S, Dorfel P and Borner T (1995) Investigation of plant organellar DNAs by pulsed-field gel-electrophoresis. *Curr Genet* **28**, 390–399.

Barni S, Sciola L, Spano A and Pippia P (1996) Static cytofluorometry and fluorescence morphology of mitochondria and DNA in proliferating fibroblasts. *Biotech Histochem* **71**, 66–70.

Beech PL, Nheu T, Schultz T, *et al.* (2000) Mitochondrial FtsZ in a chromophyte alga. *Science* **287**, 1276–1279.

Bendich AJ (1993) Reaching for the Ring – the Study of Mitochondrial Genome Structure. *Curr Genet* **24**, 279–290.

Bendich AJ (1996) Structural analysis of mitochondrial DNA molecules from fungi and plants using moving pictures and pulsed-field gel electrophoresis. *J Mol Biol* **255**, 564–588.

Bereiter-Hahn J and Voth M (1983) Metabolic control of shape and structure of mitochondria in situ. *Biol Cell* **47**, 309–322.

Bereiter-Hahn J and Voth M (1994) Dynamics of mitochondria in living cells: shape changes, dislocations, fusion, and fission of mitochondria. *Microsc Res Tech* **27**, 198–219.

Berger KH, Sogo LF and Yaffe MP (1997) Mdm12p, a component required for mitochondrial inheritance that is conserved between budding and fission yeast. *J Cell Biol* **136**, 545–553.

Bi E and Lutkenhaus J (1991) Ftsz ring structure associated with division in *Escherichia coli*. *Nature* **354**, 161–164.

Boldogh IR, Fehrenbacher KL, Yang HC and Pon LA (2005) Mitochondrial movement and inheritance in budding yeast. *Gene* **354**, 28–36.

Boldogh IR, Nowakowski DW, Yang HC, *et al.* (2003) A protein complex containing Mdm10p,

Mdm12p, and Mmm1p links mitochondrial membranes and DNA to cytoskeleton-based segregation machinery. *Mol Biol Cell* **14**, 4618–4627.
Boldogh IR, Ramcharan SL, Yang HC and Pon LA (2004) A type V myosin (Myo2p) and a Rab-like G-protein (Ypt11p) are required for retention of newly inherited mitochondria in yeast cells during cell division. *Mol Biol Cell* **15**, 3994–4002.
Boldogh IR, Yang HC, Nowakowski WD, *et al.* (2001) Arp2/3 complex and actin dynamics are required for actin-based mitochondrial motility in yeast. *Proc Natl Acad Sci USA* **98**, 3162–3167.
Burgess SM, Delannoy M and Jensen RE (1994) MMM1 encodes a mitochondrial outer membrane protein essential for establishing and maintaining the structure of yeast mitochondria. *J Cell Biol* **126**, 1375–1391.
Catlett NL and Weisman LS (2000) Divide and multiply: organelle partitioning in yeast. *Curr Opin Cell Biol* **12**, 509–516.
Cerveny KL, McCaffery JM and Jensen RE (2001) Division of mitochondria requires a novel DNM1-interacting protein, Net2p. *Mol Biol Cell* **12**, 309–321.
Chalfie M, Tu Y, Euskirchen G, Ward WW and Prasher DC (1994) Green fluorescent protein as a marker for gene-expression. *Science* **263**, 802–805.
Delettre C, Lenaers G, Griffoin JM, *et al.* (2000) Nuclear gene OPA1, encoding a mitochondrial dynamin-related protein, is mutated in dominant optic atrophy. *Nat Genet* **26**, 207–210.
Dimmer KS, Fritz S, Fuchs F, *et al.* (2002) Genetic basis of mitochondrial function and morphology in *Saccharomyces cerevisiae*. *Mol Biol Cell* **13**, 847–853.
Dyall SD, Brown MT and Johnson PJ (2004) Ancient invasions: from endosymbionts to organelles. *Science* **304**, 253–257.
Erickson HP (1997) FtsZ, a tubulin homologue in prokaryote cell division. *Trends Cell Biol* **7**, 362–367.
Esser K, Tursun B, Ingenhoven M, Michaelis G and Pratje E (2002) A novel two-step mechanism for removal of a mitochondrial signal sequence involves the mAAA complex and the putative rhomboid protease Pcp1. *J Mol Biol* **323**, 835–843.
Fekkes P, Shepard KA and Yaffe MP (2000) Gag3p, an outer membrane protein required for fission of mitochondrial tubules. *J Cell Biol* **151**, 333–340.
Feng XG, Arimura S, Hirano HY, Sakamoto W and Tsutsumi N (2004) Isolation of mutants with aberrant mitochondrial morphology from *Arabidopsis* thaliana. *Genes Genet Syst* **79**, 301–305.
Fields SD, Arana Q, Heuser J and Clarke M (2002) Mitochondrial membrane dynamics are altered in cluA – mutants of Dictyostelium. *J Musc Res Cell Motil* **23**, 829–838.
Fields SD, Conrad MN and Clarke M (1998) The *S. cerevisiae* CLU1 and *D. discoideum* cluA genes are functional homologues that influence mitochondrial morphology and distribution. *J Cell Sci* **111**, 1717–1727.
Foissner I (2004) Microfilaments and microtubules control the shape, motility, and subcellular distribution of cortical mitochondria in characean internodal cells. *Protoplasma* **224**, 145–157.
Gao H, Sage TL and Osteryoung KW (2006) FZL, an FZO-like protein in plants, is a determinant of thylakoid and chloroplast morphology. *Proc Natl Acad Sci USA* **103**, 6759–6764.
Gardner MJ, Hall N, Fung E, *et al.* (2002) Genome sequence of the human malaria parasite Plasmodium falciparum. *Nature* **419**, 498–511.
Gilson PR, Yu XC, Hereld D, *et al.* (2003) Two Dictyostelium orthologs of the prokaryotic cell division protein FtsZ localize to mitochondria and are required for the maintenance of normal mitochondrial morphology. *Eukaryotic Cell* **2**, 1315–1326.
Gray MW, Burger G and Lang BF (1999) Mitochondrial evolution. *Science* **283**, 1476–1481.
Griffin EE, Graumann J and Chan DC (2005) The WD40 protein Caf4p is a component of the mitochondrial fission machinery and recruits Dnm1p to mitochondria. *J Cell Biol* **170**, 237–248.
Griffin KL, Anderson OR, Gastrich MD, *et al.* (2001) Plant growth in elevated CO_2 alters mitochondrial number and chloroplast fine structure. *Proc Natl Acad Sci USA* **98**, 2473–2478.
Guan K, Farh L, Marshall TK and Deschenes RJ (1993) Normal mitochondrial structure and genome maintenance in yeast requires the dynamin-like product of the MGM1 gene. *Curr Genet* **24**, 141–148.

Gu XJ and Verma DPS (1996) Phragmoplastin, a dynamin-like protein associated with cell plate formation in plants. *EMBO J* **15**, 695–704.

Hales KG and Fuller MT (1997) Developmentally regulated mitochondrial fusion mediated by a conserved, novel, predicted GTPase. *Cell* **90**, 121–129.

Herlan M, Vogel F, Bornhovd C, Neupert W and Reichert AS (2003) Processing of Mgm1 by the rhomboid-type protease Pcp1 is required for maintenance of mitochondrial morphology and of mitochondrial DNA. *J Biol Chem* **278**, 27781–27788.

Hermann GJ, King EJ and Shaw JM (1997) The yeast gene, MDM20, is necessary for mitochondrial inheritance and organization of the actin cytoskeleton. *J Cell Biol* **137**, 141–153.

Hermann GJ and Shaw JM (1998) Mitochondrial dynamics in yeast. *Ann Rev Cell Dev Biol* **14**, 265–303.

Hermann GJ, Thatcher JW, Mills JP, *et al.* (1998) Mitochondrial fusion in yeast requires the transmembrane GTPase Fzo1p. *J Cell Biol* **143**, 359–373.

Hughes A (1959) *A History of Cytology.* Abelard-Schuman, London.

Hussey PJ (2004) *The Plant Cytoskeleton in Cell Differentiation and Development.* Blackwell, Oxford.

Itoh R, Fujiwara M, Nagata N and Yoshida S (2001) A chloroplast protein homologous to the eubacterial topological specificity factor minE plays a role in chloroplast division. *Plant Physiol* **127**, 1644–1655.

James TW and Bohman R (1981) Proliferation of mitochondria during the cell cycle of the human cell line (HL-60). *J Cell Biol* **89**, 256–260.

Jamet-Vierny C, Contamine V, Boulay J, Zickler D and Picard M (1997) Mutations in genes encoding the mitochondrial outer membrane proteins Tom70 and Mdm10 of Podospora anserina modify the spectrum of mitochondrial DNA rearrangements associated with cellular death. *Mol Cell Biol* **17**, 6359–6366.

Janska H and Mackenzie SA (1993) Unusual mitochondrial genome organization in cytoplasmic male-sterile common bean and the nature of cytoplasmic reversion to fertility. *Genetics* **135**, 869–879.

Janska H, Sarria R, Woloszynska M, Arrieta-Montiel M and Mackenzie SA (1998) Stoichiometric shifts in the common bean mitochondrial genome leading to male sterility and spontaneous reversion to fertility. *Plant Cell* **10**, 1163–1180.

Jin JB, Bae HJ, Kim SJ, *et al.* (2003) The *Arabidopsis* dynamin-like proteins ADL1C and ADL1E play a critical role in mitochondrial morphogenesis. *Plant Cell* **15**, 2357–2369.

Jones BA and Fangman WL (1992) Mitochondrial DNA maintenance in yeast requires a protein containing a region related to the GTP-binding domain of dynamin. *Genes Dev* **6**, 380–389.

Johnson LV, Walsh ML and Chen LB (1980) Localization of mitochondria in living cells with rhodamine 123. *Proc Natl Acad Sci USA* **77**, 990–994.

Karbowski M, Lee YJ, Gaume B, *et al.* (2002) Spatial and temporal association of Bax with mitochondrial fission sites, Drp1, and Mfn2 during apoptosis. *J Cell Biol* **159**, 931–938.

Karbowski M, Spodnik JH, Teranishi M, *et al.* (2001) Opposite effects of microtubule-stabilizing and microtubule-destabilizing drugs on biogenesis of mitochondria in mammalian cells. *J Cell Sci* **114**, 281–291.

Karbowski M and Youle RJ (2003) Dynamics of mitochondrial morphology in healthy cells and during apoptosis. *Cell Death Differentiation* **10**, 870–880.

Kennedy EP and Lehninger AL (1949) Oxidation of fatty acids and tricarboxylic acid cycle intermediates by isolated rat liver mitochondria. *J Biol Chem* **179**, 957–972.

Khodjakov A, Lizunova EM, Minin AA, Koonce MP and Gyoeva FK (1998) A specific light chain of kinesin associates with mitochondria in cultured cells. *Mol Biol Cell* **9**, 333–343.

Kiefel BR, Gilson PR and Beech PL (2004) Diverse eukaryotes have retained mitochondrial homologues of the bacterial division protein FtsZ. *Protist* **155**, 105–115.

Kohler RH, Zipfel WR, Webb WW and Hanson MR (1997) The green fluorescent protein as a marker to visualize plant mitochondria *in vivo*. *Plant J* **11**, 613–621.

Labrousse AM, Zappaterra MD, Rube DA and Van Der Bliek AM (1999) *C. elegans* dynamin-related protein DRP-1 controls severing of the mitochondrial outer membrane. *Mol Cell* **4**, 815–826.

Legesse-Miller A, Massol RH and Kirchhausen T (2003) Constriction and Dnm1p recruitment are distinct processes in mitochondrial fission. *Mol Biol Cell* **14**, 1953–1963.
Lehninger AL (1964) *The Mitochondrion.* W A Benjamin, Inc., New York.
Lewis MR and Lewis WH (1914) Mitochondria (and other cytoplasmic structures) in tissue cultures. *Am J Anat* **17**, 339–401.
Lloyd D (1974) *The Mitochondria of Microorganisms.* Academic Press, London.
Logan DC (2003) Mitochondrial dynamics. *New Phytol* **160**, 463–478.
Logan DC (2006a) Plant mitochondrial dynamics. *Biochim Biophys Acta* **1763**, 430–441.
Logan DC (2006b) The mitochondrial compartment. *J Exp Bot* **57**, 1225–1243.
Logan DC and Leaver CJ (2000) Mitochondria-targeted GFP highlights the heterogeneity of mitochondrial shape, size and movement within living plant cells. *J Exp Bot* **51**, 865–871.
Logan DC, Scott I and Tobin AK (2003) The genetic control of plant mitochondrial morphology and dynamics. *Plant J* **36**, 500–509.
Logan DC, Scott I and Tobin AK (2004) ADL2a, like ADL2b, is involved in the control of higher plant mitochondrial morphology. *J Exp Bot* **55**, 783–785.
Lonsdale DM, Brears T, Hodge TP, Melville SE and Rottmann WH (1988) The plant mitochondrial genome – homologous recombination as a mechanism for generating heterogeneity. *Philos T Roy Soc B* **319**, 149–163.
Lutkenhaus J (1998) The regulation of bacterial cell division: a time and place for it. *Curr Opin Microbiol* **1**, 210–215.
Margineantu DH, Cox WG, Sundell L, Sherwood SW, Beechem JA and Capaldi RA (2002) Cell cycle dependent morphology changes and associated mitochondrial DNA redistribution in mitochondria of human cell lines. *Mitochondrion* **1**, 425–435.
Margolin W (2005) FtsZ and the division of prokaryotic cells and organelles. *Nat Rev Mol Cell Biol* **6**, 862–871.
Margolin W, Corbo JC and Long SR (1991) Cloning and characterization of a rhizobium-meliloti homolog of the *Escherichia-coli* cell-division gene ftsz. *J Bacteriol* **173**, 5822–5830.
Margulis L (1970) *Origin of Eukaryotic Cells: Evidence and Research Implications for a Theory of the Origin and Evolution of Microbial, Plant, and Animal Cells on the Precambrian Earth.* Yale University Press, New Haven.
Martin W (2000) Evolutionary biology – A powerhouse divided. *Science* **287**, 1219–1219.
Martin W, Hoffmeister M, Rotte C and Henze K (2001) An overview of endosymbiotic models for the origins of eukaryotes, their ATP-producing organelles (mitochondria and hydrogenosomes), and their heterotrophic lifestyle. *Biol Chem* **382**, 1521–1539.
Martin W and Kowallik KV (1999) Annotated English translation of Mereschkowsky's 1905 paper 'Uber Natur und Ursprung der Chromatophoren im Pflanzenreiche. *Eur J Phycol* **34**, 287–295.
Martin W and Muller M (1998) The hydrogen hypothesis for the first eukaryote. *Nature* **392**, 37–41.
McAndrew RS, Froehlich JE, Vitha S, Stokes KD and Osteryoung KW (2001) Colocalization of plastid division proteins in the chloroplast stromal compartment establishes a new functional relationship between FtsZ1 and FtsZ2 in higher plants. *Plant Physiol* **127**, 1656–1666.
McConnell SJ, Stewart LC, Talin A and Yaffe MP (1990) Temperature-sensitive yeast mutants defective in mitochondrial inheritance. *J Cell Biol* **111**, 967–976.
McConnell SJ and Yaffe MP (1992) Nuclear and mitochondrial inheritance in yeast depends on novel cytoplasmic structures defined by the MDM1 protein. *J Cell Biol* **118**, 385–395.
McFadden GI (2001) Primary and secondary endosymbiosis and the origin of plastids. *J Phycol* **37**, 951–959.
McFadden GI and Ralph SA (2003) Dynamin: the endosymbiosis ring of power? *Proc Natl Acad Sci USA* **100**, 3557–3559.
McKinney EC, Kandasamy MK and Meagher RB (2002) *Arabidopsis* contains ancient classes of differentially expressed actin-related protein genes. *Plant Physiol* **128**, 997–1007.
McQuibban GA, Saurya S and Freeman M (2003) Mitochondrial membrane remodelling regulated by a conserved rhomboid protease. *Nature* **423**, 537–541.

Mereschkowsky C (1905) Uber natur und ursprung der chromatophoren in pfanzenreich. *Biol Centralbl* **25**.
Messerschmitt M, Jakobs S, Vogel F, *et al.* (2003) The inner membrane protein Mdm33 controls mitochondrial morphology in yeast. *J Cell Biol* **160**, 553–564.
Misaka T, Miyashita T and Kubo Y (2002) Primary structure of a dynamin-related mouse mitochondrial GTPase and its distribution in brain, subcellular localization, and effect on mitochondrial morphology. *J Biol Chem* **277**, 15834–15842.
Mitchison TJ and Cramer LP (1996) Actin-based cell motility and cell locomotion. *Cell* **84**, 371–379.
Miyagishima S, Nishida K, Mori T, *et al.* (2003a) A plant-specific dynamin-related protein forms a ring at the chloroplast division site. *Plant Cell* **15**, 655–665.
Miyagishima SY, Itoh R, Toda K, Takahashi H, Kuroiwa H and Kuroiwa T (1998) Orderly formation of the double ring structures for plastid and mitochondrial division in the unicellular red alga *Cyanidioschyzon merolae*. *Planta* **206**, 551–560.
Miyagishima SY, Nishida K and Kuroiwa T (2003b) An evolutionary puzzle: chloroplast and mitochondrial division rings. *Trends Plant Sci* **8**, 432–438.
Miyagishima SY, Nozaki H, Nishida K, Nishida K, Matsuzaki M and Kuroiwa T (2004) Two types of FtsZ proteins in mitochondria and red-lineage chloroplasts: The duplication of FtsZ is implicated in endosymbiosis. *J Mol Evol* **58**, 291–303.
Mozdy AD, McCaffery JM and Shaw JM (2000) Dnm1p GTPase-mediated mitochondrial fission is a multi-step process requiring the novel integral membrane component Fis1p. *J Cell Biol* **151**, 367–380.
Munn EA (1974) *The Structure of Mitochondria*. Academic Press, London.
Nangaku M, Sato-Yoshitake R, Okada Y, *et al.* (1994) KIF1B, a novel microtubule plus end-directed monomeric motor protein for transport of mitochondria. *Cell* **79**, 1209–1220.
Nishida K, Misumi O, Yagisawa F, Kuroiwa H, Nagata T and Kuroiwa T (2004) Triple immunofluorescent labeling of FtsZ, dynamin, and EF-Tu reveals a loose association between the inner and outer membrane mitochondrial division machinery in the red alga Cyanidioschyzon merolae. *J Histochem Cytochem* **52**, 843–849.
Nishida K, Takahara M, Miyagishima S, Kuroiwa H, Matsuzaki M and Kuroiwa T (2003) Dynamic recruitment of dynamin for final mitochondrial severance in a primitive red alga. *Proc Natl Acad Sci USA* **100**, 2146–2151.
Nishida K, Yagisawa F, Kuroiwa H, Nagata T and Kuroiwa T (2005) Cell cycle-regulated, microtubule-independent organelle division in *Cyanidioschyzon merolae*. *Mol Biol Cell* **16**, 2493–2502.
Niwa Y, Hirano T, Yoshimoto K, Shimizu M and Kobayashi H (1999) Non-invasive quantitative detection and applications of non-toxic, S65T-type green fluorescent protein in living plants. *Plant J* **18**, 455–463.
Nunnari J, Marshall WF, Straight A, Murray A, Sedat JW and Walter P (1997) Mitochondrial transmission during mating in *Saccharomyces cerevisiae* is determined by mitochondrial fusion and fission and the intramitochondrial segregation of mitochondrial DNA. *Mol Biol Cell* **8**, 1233–1242.
Okamoto K, Perlman PS and Butow RA (1998) The sorting of mitochondrial DNA and mitochondrial proteins in zygotes: Preferential transmission of mitochondrial DNA to the medial bud. *J Cell Biol* **142**, 613–623.
Okamoto K and Shaw JM (2005) Mitochondrial morphology and dynamics in yeast and multicellular eukaryotes. *Ann Rev Genet* **39**, 503–536.
Oldenburg DJ and Bendich AJ (2001) Mitochondrial DNA from the liverwort Marchantia polymorpha: Circularly permuted linear molecules, head-to-tail concatemers, and a 5′ protein. *J Mol Biol* **310**, 549–562.
Olyslaegers G and Verbelen JP (1998) Improved staining of F-actin and co-localization of mitochondria in plant cells. *J Microsc* **192**, 73–77.
Osteryoung KW and McAndrew RS (2001) The plastid division machine. *Annu Rev Plant Physiol Plant Mol Biol* **52**, 315–333.

Osteryoung KW, Stokes KD, Rutherford SM, Percival AL and Lee WY (1998) Chloroplast division in higher plants requires members of two functionally divergent gene families with homology to bacterial ftsZ. *Plant Cell* **10**, 1991–2004.

Osteryoung KW and Vierling E (1995) Conserved cell and organelle division. *Nature* **376**, 473–474.

Otsuga D, Keegan BR, Brisch E, *et al.* (1998) The dynamin-related GTPase, Dnm1p, controls mitochondrial morphology in yeast. *J Cell Biol* **143**, 333–349.

Palmer JD (2003) The symbiotic birth and spread of plastids: How many times and whodunit? *J Phycol* **39**, 4–11.

Palmer JD, Adams KL, Cho YR, Parkinson CL, Qiu YL and Song KM (2000) Dynamic evolution of plant mitochondrial genomes: Mobile genes and introns and highly variable mutation rates. *Proc Natl Acad Sci USA* **97**, 6960–6966.

Palmer JD and Herbon LA (1987) Unicircular structure of the *Brassica hirta* mitochondrial genome. *Curr Genet* **11**, 565–570.

Pollard TD and Beltzner CC (2002) Structure and function of the Arp2/3 complex. *Curr Opin Structural Biol* **12**, 768–774.

Poot M, Zhang YZ, Kramer JA, *et al.* (1996) Analysis of mitochondrial morphology and function with novel fixable fluorescent stains. *J Histochem Cytochem* **44**, 1363–1372.

Possingham JV and Lawrence ME (1983) Controls to plastid division. *Int Rev Cytol* **84**, 1–56.

Pyke KA and Leech RM (1994) A genetic analysis of chloroplast division and expansion in *Arabidopsis* thaliana. *Plant Physiol* **104**, 201–207.

Rapaport D, Brunner M, Neupert W and Westermann B (1998) Fzo1p is a mitochondrial outer membrane protein essential for the biogenesis of functional mitochondria in *Saccharomyces cerevisiae*. *J Biol Chem* **273**, 20150–20155.

Reddy AS and Day IS (2001) Kinesins in the *Arabidopsis* genome: a comparative analysis among eukaryotes. *BMC Genomics* **2**, 2.

Rizzuto R, Brini M, Pizzo P, Murgia M and Pozzan T (1995) Chimeric green fluorescent protein as a tool for visualizing subcellular organelles in living cells. *Curr Biol* **5**, 635–642.

Rizzuto R, Pinton P, Carrington W, *et al.* (1998) Close contacts with the endoplasmic reticulum as determinants of mitochondrial Ca^{2+} responses. *Science* **280**, 1763–1766.

Robertson EJ, Williams M, Harwood JL, Lindsay JG, Leaver CJ and Leech RM (1995) Mitochondria increase three-fold and mitochondrial proteins and lipid change dramatically in postmeristematic cells in young wheat leaves grown in elevated CO_2. *Plant Physiol* **108**, 469–474.

Rojo M, Legros F, Chateau D and Lombes A (2002) Membrane topology and mitochondrial targeting of mitofusins, ubiquitous mammalian homologs of the transmembrane GTPase Fzo. *J Cell Sci* **115**, 1663–1674.

Ross KFA (1967) *Phase Contrast and Interference Microscopy for Cell Biologists.* Edward Arnold, London.

Rutter GA and Rizzuto R (2000) Regulation of mitochondrial metabolism by ER Ca^{2+} release: an intimate connection. *Trends Biochem Sci* **25**, 215–221.

Santel A and Fuller MT (2001) Control of mitochondrial morphology by a human mitofusin. *J Cell Sci* **114**, 867–874.

Scott, I (2006) *The Control of Mitochondrial Morphology and Dynamics in Arabidopsis Thaliana.* PhD Thesis, University of St Andrews, St Andrews, Scotland, UK.

Scott I and Logan DC (2004) The birth of cell biology. *New Phytol* **163**, 7–9.

Scott I, Tobin AK and Logan DC (2006) BIGYIN, an orthologue of human and yeast FIS1 genes functions in the control of mitochondrial size and number in *Arabidopsis* thaliana. *J Exp Bot* **57**, 1275–1280.

Sesaki H and Jensen RE (1999) Division versus fusion: Dnm1p and Fzo1p antagonistically regulate mitochondrial shape. *J Cell Biol* **147**, 699–706.

Sesaki H and Jensen RE (2001) UGO1 encodes an outer membrane protein required for mitochondrial fusion. *J Cell Biol* **153**, 635–635.

Sesaki H and Jensen RE (2004) Ugo1p links the Fzo1p and Mgm1p GTPases for mitochondrial fusion. *J Biol Chem* **279**, 28298–28303.

Sesaki H, Southard SM, Hobbs AEA and Jensen RE (2003a) Cells lacking Pcp1p/Ugo2p, a rhomboid-like protease required for Mgm1p processing, lose mtDNA and mitochondrial structure in a Dnm1p-dependent manner, but remain competent for mitochondrial fusion. *Biochem Biophys Res Commun* **308**, 276–283.

Sesaki H, Southard SM, Yaffe MP and Jensen RE (2003b) Mgm1p, a dynamin-related GTPase, is essential for fusion of the mitochondrial outer membrane. *Mol Biol Cell* **14**, 2342–2356.

Sheahan MB, McCurdy DW and Rose RJ (2005) Mitochondria as a connected population: ensuring continuity of the mitochondrial genome during plant cell dedifferentiation through massive mitochondrial fusion. *The Plant J* **44**, 744–755.

Sheahan MB, Rose RJ and McCurdy DW (2004) Organelle inheritance in plant cell division: the actin cytoskeleton is required for unbiased inheritance of chloroplasts, mitochondria and endoplasmic reticulum in dividing protoplasts. *Plant J* **37**, 379–390.

Shepard KA and Yaffe MP (1997) Genetic and molecular analysis of Mdm14p and Mdm17p, proteins involved in organelle inheritence. *Mol Biol Cell* **8** (Suppl. S), 2585.

Shepard KA and Yaffe MP (1999) The yeast dynamin-like protein, Mgm1p, functions on the mitochondrial outer membrane to mediate mitochondrial inheritance. *J Cell Biol* **144**, 711–719.

Simon VR, Swayne TC and Pon LA (1995) Actin-dependent mitochondrial motility in mitotic yeast and cell-free systems – identification of a motor-activity on the mitochondrial surface. *J Cell Biol* **130**, 345–354.

Small I, Suffolk R and Leaver CJ (1989) Evolution of plant mitochondrial genomes via substoichiometric intermediates. *Cell* **58**, 69–76.

Smirnova E, Griparic L, Shurland D-L and Van Der Bliek (2002) Dynamin-related protein Drp1 is required for mitochondrial division in mammalian cells. *Mol Biol Cell* **12**, 2245–2256.

Smirnova E, Shurland DL, Ryazantsev SN and Van Der Bliek AM (1998) A human dynamin-related protein controls the distribution of mitochondria. *J Cell Biol* **143**, 351–358.

Sogo LF and Yaffe MP (1994) Regulation of mitochondrial morphology and inheritance by Mdm10p, a protein of the mitochondrial outer membrane. *J Cell Biol* **126**, 1361–1373.

Stenoien DL and Brady ST (1997) Immunochemical analysis of kinesin light chain function. *Mol Biol Cell* **8**, 675–689.

Stevens BJ (1977) Variation in number and volume of the mitochondria in yeast according to growth conditions. *Biol Cell* **28**, 37–56.

Stickens D and Verbelen JP (1996) Spatial structure of mitochondria and ER denotes changes in cell physiology of cultured tobacco protoplasts. *Plant J* **9**, 85–92.

Stojanovski D, Koutsopoulos OS, Okamoto K and Ryan MT (2004) Levels of human Fis1 at the mitochondrial outer membrane regulate mitochondrial morphology. *J Cell Sci* **117**, 1201–1210.

Suzuki M, Jeong SY, Karbowski M, Youle RJ and Tjandra N (2003) The solution structure of human mitochondria fission protein Fis1 reveals a novel TPR-like helix bundle. *J Mol Biol* **334**, 445–458.

Suzuki M, Neutzner A, Tjandra N and Youle RJ (2005) Novel structure of the N terminus in yeast Fis1 correlates with a specialized function in mitochondrial fission. *J Biol Chem* **280**, 21444–21452.

Takahara M, Takahashi H, Matsunaga S, *et al.* (2000) A putative mitochondrial ftsZ gene is present in the unicellular primitive red alga *Cyanidioschyzon merolae*. *Mol Gen Genet* **264**, 452–460.

Tanaka Y, Kanai Y, Okada Y, *et al.* (1998) Targeted disruption of mouse conventional kinesin heavy chain, kif5B, results in abnormal perinuclear clustering of mitochondria. *Cell* **93**, 1147–1158.

Tieu Q and Nunnari J (2000) Mdv1p is a WD repeat protein that interacts with the dynamin-related GTPase, Dnm1p, to trigger mitochondrial division. *J Cell Biol* **151**, 353–365.

Tieu Q, Okreglak V, Naylor K and Nunnari J (2002) The WD repeat protein, Mdv1p, functions as a molecular adaptor by interacting with Dnm1p and Fis1p during mitochondrial fission. *J Cell Biol* **158**, 445–452.

Tribe M and Whittaker P (1972) *Chloroplasts and Mitochondria*. Edward Arnold, London.

Tyler DD (1992) *The Mitochondrion in Health and Disease*. VCH, New York, Cambridge.

Tzagoloff A (1982) *Mitochondria*. Plenum Press, New York.

Unseld M, Marienfeld JR, Brandt P and Brennicke A (1997) The mitochondrial genome of *Arabidopsis* thaliana contains 57 genes in 366,924 nucleotides. *Nat Genet* **15**, 57–61.

Van Gestel K, Kohler RH and Verbelen JP (2002) Plant mitochondria move on F-actin, but their positioning in the cortical cytoplasm depends on both F-actin and microtubules. *J Exp Bot* **53**, 659–667.

Van Gestel K and Verbelen JP (2002) Giant mitochondria are a response to low oxygen pressure in cells of tobacco (Nicotiana tabacum L.). *J Exp Bot* **53**, 1215–1218.

Verhey KJ, Meyer D, Deehan R, *et al.* (2001) Cargo of kinesin identified as JIP scaffolding proteins and associated signaling molecules. *J Cell Biol* **152**, 959–970.

Vitha S, McAndrew RS and Osteryoung KW (2001) FtsZ ring formation at the chloroplast division site in plants. *J Cell Biol* **153**, 111–120.

Wada M and Suetsugu N (2004) Plant organelle positioning. *Curr Opin Plant Biol* **7**, 626–631.

Walker DW and Benzer S (2004) Mitochondrial "swirls" induced by oxygen stress and in the Drosophila mutant hyperswirl. *Proc Natl Acad Sci USA* **101**, 10290–10295.

Walter RF and McFadden GI (2005) The apicoplast: a review of the derived plastid of apicomplexan parasites. *Curr Issues Mol Biol* **7**, 57–79.

Wong ED, Wagner JA, Gorsich SW, McCaffery JM, Shaw JM and Nunnari J (2000) The dynamin-related GTPase, Mgm1p, is an intermembrane space protein required for maintenance of fusion competent mitochondria. *J Cell Biol* **151**, 341–352.

Wong ED, Wagner JA, Scott SV, *et al.* (2003) The intramitochondrial dynamin-related GTPase, Mgm1p, is a component of a protein complex that mediates mitochondrial fusion. *J Cell Biol* **160**, 303–311.

Yaffe MP (1999) The machinery of mitochondrial inheritance and behavior. *Science* **283**, 1493–1497.

Yoon Y, Krueger EW, Oswald BJ and McNiven MA (2003) The mitochondrial protein hFis1 regulates mitochondrial fission in mammalian cells through an interaction with the dynamin-like protein DLP1. *Mol Cell Biol* **23**, 5409–5420.

Yoshinaga K, Arimura SI, Niwa Y, Tsutsumi N, Uchimiya H and Kawai-Yamada M (2005) Mitochondrial behaviour in the early stages of ROS stress leading to cell death in *Arabidopsis* thaliana. *Ann Bot* **96**, 337–342.

Zhu Q, Hulen D, Liu T and Clarke M (1997) The cluA- mutant of Dictyostelium identifies a novel class of proteins required for dispersion of mitochondria. *Proc Natl Acad Sci USA* **94**, 7308–7313.

2 The unique biology of mitochondrial genome instability in plants

Sally A. Mackenzie

2.1 Introduction

Mitochondria participate in a broad repertoire of biochemical functions in plant and animal cells that are essential for life and increasingly well characterized. However, despite the many conserved biochemical functions, one of the most enigmatic features of mitochondria, in both plants and animals, is their remarkably dissimilar genomes. While the general features of mitochondrial genomes have been known for several years, including DNA sequence and gene coding capacities, debate continues with regard to their *in vivo* structure, recombination activity and control by the nucleus. Here the focus is primarily on the plant mitochondrial genome and its influence on plant development and evolution, with reference to potential parallels with animal mitochondrial systems in which an understanding of one can provide insight to the other.

2.2 Unusual and dynamic nature of the plant mitochondrial genome

The evolutionary process that has given rise to present-day mitochondria and their genomes is thought to have initiated from a single endosymbiotic event involving an alpha-proteobacterial progenitor (Gray, 1999; Gray *et al.*, 1999; Lang *et al.*, 1999). This process included massive gene transfers to the nucleus and culminated, for most animal lineages, in the retention of approximately 13 protein-coding mitochondrial genes, together with various components of translation (see Chapter 3). Within plant lineages, the gene transfer process continues to present day, with most mitochondrial genomes containing 54–57 known genes, and a variable number of putative open reading frames (Adams and Palmer, 2003). Although mitochondrial gene number is relatively stable among plant families, genome size is not. Plant mitochondria have, for the most part, tolerated the accumulation of intergenic DNA sequences, derived from inter-organellar DNA transfer, duplication activity and some from unknown origin, as a surprisingly large (>50%) proportion of their genomes (Marienfeld *et al.*, 1999; Kubo *et al.*, 2000; Handa 2003; Clifton *et al.*, 2004). This expansion of plant mitochondrial genome size appears to have been accompanied by two other, presumably related features. The introduction of extraneous sequences to the mitochondrial genome has given rise to split gene structures, some requiring *trans*-splicing of their transcripts for expression (see Chapter 4). Sequences present within the plant mitochondrial genome also display recombination activity that creates considerable variation in genome organization (for review, see Knoop, 2004). All of these features distinguish plant mitochondrial genomes and appear to be the consequence of reduced constraints on DNA recombination activity.

In most plant species to date, mitochondrial DNA physical mapping and sequencing efforts define a comprehensive, circular genome structure, which, in earlier studies, was termed the *master chromosome*, thought to encompass the entire mitochondrial genomic complement (Lonsdale *et al.*, 1988). To date, the identification of a master molecule remains elusive, although there has been some evidence to suggest that such a form would likely reside within meristem cells where most active mitochondrial DNA replication takes place (Sakai *et al.*, 2004). Within vegetative tissues, the plant mitochondrial genome's physical structure appears to exist largely as a heterogeneous population of linear, often branched molecules, with smaller than genome-size circular molecules present in low abundance (Bendich, 1996). Recombination permits the subdivision of the genome into a multipartite, highly redundant organization. This more complex genome structure, relative to that first predicted, has underscored the importance of recombination for genome maintenance. As circularly permuted, large, linear molecules, the genome likely replicates via rolling circle mechanisms based on electron microscopic analyses (Backert and Borner, 2000). In addition, recombination-mediated replication initiation by strand invasion occurs (Manchekar *et al.*, 2006), although few details of DNA replication initiation have been confirmed biochemically in higher plant mitochondria.

Physical mapping studies predict a genome that actively subdivides by homologous recombination to create highly redundant subgenomic forms (Fauron *et al.*, 1995). This recombination occurs at high frequency between relatively large, repeated sequences in the genome. This type of recombination generally allows detection of both recombination products in approximately equal stoichiometry to the parental forms when examined by physical mapping (Figure 2.1). In contrast, a second type of recombination characterizes plant mitochondria. Sporadic, low-frequency illegitimate recombination occurs, generally at smaller repeats (Andre *et al.*, 1992, Marienfeld *et al.*, 1997) and nearly always results in the formation of only one of the predicted recombination products (see Figure 2.1). In early studies, it was thought that the second predicted product might simply have lacked an active replication origin and was subsequently lost. However, a single recombination product appears to be the rule for these infrequent events, suggesting that this activity may occur by an asymmetric recombination mechanism.

The control of recombination activity observed in plant mitochondria is not well understood. For instance, it is not clear whether recombinant products subsequently replicate autonomously or whether recombination is occurring in all plant cell types. While it is assumed that recombination activity is controlled by nuclear-encoded factors, the role of the nucleus in directly regulating recombination activity and the nuclear components required for this process have not been defined.

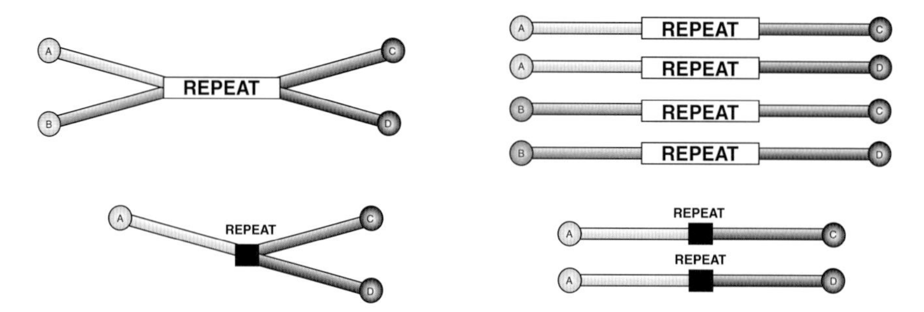

Figure 2.1 Two types of recombinationally active repeat sequences within the plant mitochondrial genome. Large repeats mediating homologous recombination result in the predicted two parental and two recombinant forms in approximately equal stoichiometries. The small repeat sequences that mediate lower frequency nonhomologous recombination generally result in only two products, one parental and one recombinant.

A circular mitochondrial genome structure of approximately 16 kbp is predicted to exist in mammalian mitochondria based on both physical mapping and electron microscopy evidence. It appears that recombination activity may also exist at low frequency within mammalian mitochondrial genomes under particular conditions (Kajander *et al.*, 2001; Kraytsberg *et al.*, 2004). Nevertheless, one key distinguishing feature that appears to separate mammalian and plant mitochondria, likely accounting for the observed genome size expansion, rapid changes in genome structure and emergence of novel split and chimeric gene structures, seems to be the apparent vulnerability of plant mitochondrial DNA to illegitimate recombination activity.

2.3 Mitochondrial DNA recombination is controlled by the nucleus

Evidence of nuclear influence on mitochondrial genome stability first emerged indirectly in plant cell cultures, where mitochondrial rearrangements often arise in apparent response to relaxation of nuclear control *in vitro*. It is common to observe the emergence of novel DNA polymorphisms in lines following *in vitro* culture. In some cases the DNA polymorphisms that emerge are highly reproducible or represent conformations that are detectable at much lower levels in the original material (Ozias-Akins *et al.*, 1988; Shirzadegan *et al.*, 1991; Vitart *et al.*, 1992). This observation provided the first indication that the *in vitro* growth conditions could permit amplification of rearrangements already existing in low abundance within the plant, and the observed DNA rearrangements are less random than was initially surmised.

Cytoplasmic male sterility (CMS) mutations that block a plant's ability to produce or shed viable pollen are generally the result of mitochondrial illegitimate recombination activity, or non-homologous end-joining to create novel, dominant, chimeric gene forms. These DNA rearrangements can arise in response to alloplasmy – a process of nuclear substitution into a foreign cytoplasm by recurrent backcrossing – or by plant cell hybridization (Hanson and Bentolila, 2004). CMS mutations have also been detected at low frequency within fertile plant lines in natural populations, which is likely the outcome of particular crosses.

In a report in 1987, the plant mitochondrial genome was shown to be composed of a collection of DNA configurations existing in vastly different relative concentrations (Small *et al.*, 1987). It was later suggested that genomic shifts could occur in the relative composition of the genome such that previously substoichiometric forms could be rapidly amplified. This phenomenon, termed substoichiometric shifting (SSS), has been observed in a number of plant species, including tobacco (Kanazawa *et al.*, 1994), common bean (Janska *et al.*, 1998), maize (Escote-Carlson, 1990; Fauron *et al.*, 1995) and *Brassica* (Belaoui *et al.*, 1998), and is generally associated with a phenotypic transition from male sterile to fertile in CMS mutants. In tobacco, distinct mitochondrial DNA configurations were observed in undifferentiated green callus tissue and regenerated plants, with the transition between these two mitochondrial forms shown to be reversible. In the above examples, these SSS events appear to involve the rapid amplification of existing substoichiometric DNA forms.

The SSS process is controlled by nuclear genotype. In the common bean, where CMS is associated with expression of the novel mitochondrial sequence *orf239* (Johns *et al.*, 1992), a permanent condition of fertility restoration occurs with introduction, by crossing, of the nuclear gene *Fr*. The F_2 segregation for *Fr* and fertility restoration is accompanied by the cosegregation of a mitochondrial SSS event, resulting in copy number suppression of the mitochondrial DNA molecule that contains *orf239* (Mackenzie and Chase, 1990). Once this mitochondrial rearrangement occurs, it is permanent and independent of *Fr* genotype.

In *Arabidopsis*, mutation within the nuclear gene originally designated *chloroplast mutator* (*CHM*) also induces a mitochondrial SSS event (Sakamoto *et al.*, 1996). In this case, the phenotypic consequence is a green-white variegation of the plant, with no clear evidence of male sterility. The SSS process is apparently a reversible phenomenon within the mitochondrial genome; while the SSS event observed in *Arabidopsis* involves the selective *amplification* of a substoichiometric mitochondrial DNA molecule in response to *CHM* mutation, the common bean SSS event involves the directed copy-number *suppression* of *orf239*.

Two observations suggest that the functions of *Fr* and *CHM* are distinct. Although the *Fr* gene product has not yet been determined, marker-based analysis indicates that *Fr* and *CHM* map to distinct locations within the nuclear genome (He *et al.*, 1995; S. Mackenzie and C.E. Vallejos, unpublished data). Furthermore, analysis of the mitochondrial SSS event in each case suggests that *Fr* may influence replication of the DNA molecule containing *orf239* (Mackenzie and Chase, 1990; Janska *et al.*, 1998), while the *CHM* product likely participates in suppressing *de novo* recombination events or the maintenance of their products. *CHM* has been shown to encode a protein homologous to the product of *MutS* in *Escherichia coli* (Abdelnoor *et al.*, 2003). MutS is a mismatch repair protein, and several *MutS* homologous (*MSH*) loci have been identified to function not only in nuclear mismatch repair but also in suppression of nonhomologous recombination in eukaryotes (Kunkel and Erie, 2005). With identification of the *CHM* product, the gene has now been redesignated *Msh1*.

The mitochondrial MSH1 protein appears to be absent from the mammalian proteome, but is present in fungi (Reenan and Kolodner, 1992), and is well conserved in higher plants (Abdelnoor *et al.*, 2006). In *Arabidopsis*, *Msh1* is a single-copy gene. Phenotypic analysis of *msh1* null mutants in *Arabidopsis*, along with studies of *msh1* mutants in yeast (Sia and Kirkpatrick, 2005), suggests that the gene may no longer function in DNA mismatch repair. High amino acid conservation in plant MSH1 proteins has allowed identification of six protein domains likely to be important for function. Out of these, two are characteristic of *MutS* homologous proteins: a DNA-binding and mismatch recognition domain (I) and an ATP-binding and hydrolysis domain (V), both of which are shown by mutational analysis to be essential for MSH1 protein function (Abdelnoor *et al.*, 2006). The C-terminal domain VI was identified as a GIY-YIG type homing endonuclease. Homing endonucleases are generally thought to function in the excision and transposition of introns (Belfort and Roberts, 1997). Presumably, a gene fusion event occurred during the evolution of this gene, permitting a protein previously functioning in mismatch repair to acquire a novel role, or a function previously carried out by two components to now be carried out by one.

A mitochondrial *MutS*-related protein has not been found in animal lineages, although one group stands as the exception. The soft corals (phylum Cnidaria, class Anthozoa, subclass Octocorallia) encode, within their mitochondrial genomes, a protein with *MutS*-like features, including well-conserved DNA-binding and ATPase domains (Pont-Kingdon *et al.*, 1998). Remarkably, this particular *MutS* homolog also encodes an additional C-terminal domain derived from a homing endonuclease. However, the coral *Msh1* C-terminal endonuclease is of the HNH-type, not the GIY-YIG type characteristic of plants. This structural correlation between the coral protein and *Msh1* of higher plants is stunning, and the two *Msh1* C-terminal extensions (HNH- or GIY-YIG-type) have been interpreted as evidence of convergent evolution within the two distinct lineages (Abdelnoor *et al.*, 2006). It has been suggested that the fusion of a *MutS* homolog to an endonuclease might represent an adaptation to facilitate the cleavage reaction required for nucleotide substitution in mismatch repair (Malik and Henikoff, 2000) or to abort illegitimate recombination. In *E. coli*, *MutS* provides mismatch recognition, but a separate protein component, *MutH*, is required to provide the endonuclease activity (Schofield and Hseih, 2003).

The surprising similarity between the corals and higher plant proteins represents just one of several intriguing mitochondrial similarities between the two lineages. The corals are distinguished among animals by possessing an unusually low mitochondrial mutation frequency (Shearer *et al.*, 2002), as is the case in plants (Wolfe *et al.*, 1987). However, no direct association has yet been shown in either case between the presence of MSH1 and this feature. Moreover, the Cnidarians may be undergoing similar mitochondrial genome structural transitions as in plants, retaining a partially or fully circular genome structure in the Anthozoan class, encompassing the corals, while assuming a linear mitochondrial genome in related motile Hydrozoa, Scyphozoa and Cubozoa classes (Bridge *et al.*, 1992). The corals represent the only known animal lineage to have acquired mitochondrial gene introns (Pont-Kingdon *et al.*, 1998; van Oppen *et al.*, 2002), another feature shared with plant mitochondria. The degree to which the Cnidarian mitochondrial genome undergoes homologous or illegitimate recombination can now be investigated with recently emerging mitochondrial genome sequences from this group.

In a recent review, Lynch *et al.* (2006) postulate that the many striking differences in size, structure, and behavior that distinguish plant mitochondrial genomes from animal mitochondrial genomes are the likely consequence of the dramatic differences in mitochondrial mutation rates in plants relative to animals. Lynch *et al.* (2006) hypothesize that features of mitochondrial genome evolution are largely defined by the distinct differences in mutation pressure experienced in animal versus plant populations. This inference, together with direct experimental analyses, would suggest that *Msh1*, with likely functions in illegitimate recombination suppression and possibly mismatch repair, may have played a key role in the divergence of plant and animal mitochondrial genome evolution.

2.4 Recombination versus replication in plant mitochondrial genome dynamics

The mitochondrial phenomenon of SSS appears unique to higher plants. The process is rapid, directing the amplification or loss of a segment of the mitochondrial genome within a single plant generation. What has made this process so interesting is that it is nuclear-controlled and directly influences plant phenotype, and so may be subject to selection at the population level. While physical mapping studies have allowed identification of the mitochondrial subgenomic molecules modulated by the process and the number of reports of SSS have made clear the prevalence of the process in both laboratory (Kanazawa *et al.*, 1994; Gutierres *et al.*, 1997; Belaoui *et al.*, 1998) and natural populations (Andersson, 1999; Arrieta-Montiel *et al.*, 2001), two questions remain unanswered. The first is whether SSS occurs by differential replication or *de novo* recombination. The second is why plants would adopt such a mechanism.

Recent studies suggest that the answer to the first question is ‘both’. In common bean, where nuclear gene *Fr* effects the shifting of mitochondrial sequence *orf239*, an analysis shows that *orf239* contains unique sequences present nowhere else in the genome (Johns *et al.*, 1992). Therefore, it is likely that the sequence resides on an autonomously replicating molecule, and *Fr* is thought to interfere with its replication or transmission (Mackenzie and Chase, 1990; Janska *et al.*, 1998). Similar arguments for influencing replication have been made for several other mitochondrial mutations as well (Fauron *et al.*, 1995; Mackenzie and McIntosh, 1999).

However, evidence suggests that the SSS process effected by *Msh1* involves *de novo* recombination. The *Arabidopsis* repeat involved in *msh1*-associated recombination has been identified (Sakamoto *et al.*, 1996; Shedge *et al.*, submitted). RNAi-suppression of *Msh1* expression in various plant species results in reproducible mitochondrial rearrangements, with more than one related rearrangement being derived from a single original RNAi transformant in some cases (Sandhu *et al.*, submitted). This observation, and the absence of any corresponding detectable band in the nontransformed nonshifted control, implies that *de novo* illegitimate recombination, or the stable maintenance of its products, occurs with reduction of the MSH1 protein.

An association exists between mitochondrial SSS and control of CMS expression in higher plants. Gynodioecy as a reproductive system in natural populations results in a mixture of hermaphrodite and female (male sterile) plants to facilitate both self-pollination and outcrossing in dynamic equilibrium (Ehlers *et al.*, 2005). The spontaneous emergence of CMS individuals in such populations is facilitated by the SSS process. *Phaseolus vulgaris*, the common bean, is thought to be a predominantly self-pollinating species. However, a broad survey of *Phaseolus* germplasm reveals the presence of the CMS determinant, *orf239*, in all lines tested to date. The *orf239* sequence, present at high levels in 10% of the lines tested, was found as a well-conserved sequence at substoichiometric levels in the remainder (Arrieta-Montiel *et al.*, 2001).

Based on these observations, and the recent identification of *Msh1* in *Arabidopsis*, it was possible to test for a direct relationship between the natural induction of CMS and allelic variation or modulation of *Msh1*. Transgene-mediated RNA interference was used to downregulate *Msh1* in normal, fertile tobacco and tomato lines. Two independent experiments, each involving several independent transformants, consistently resulted in the emergence of a maternally inherited male-sterile phenotype with evidence of mitochondrial DNA rearrangements and subtle to intense green-white leaf variegation of regenerated plants, similar to that observed in *Arabidopsis msh1* mutants (Sandhu *et al.*, submitted). The observation of heritable male sterility in these experiments confirms the association between *Msh1* and mitochondrial genome stability and suggests that male sterility mutations are readily induced in response to *Msh1* mutation.

From a practical perspective, the induction of CMS using RNAi-mediated suppression of *Msh1* provides a means of testing for CMS mutants in several agriculturally valuable crops where a stable source of CMS is currently not available but the potential value of hybrids is great. Observations in tomato and tobacco transgenic lines suggest that the derived mitochondrial rearrangement remains stable independent of *Msh1* genotype. Consequently, it may be feasible to release the novel CMS sources as nontransgenic lines, an advantage in particular crops with high export value.

Based on previous analyses of CMS-associated mutations in plants, it appears unlikely that transgenic CMS induction occurs by *de novo* nonhomologous, intergenic recombination to create new CMS-inducing gene chimeras. However, the sterility-associated sequences have not yet been characterized. Most CMS-inducing sequences described to date have involved numerous recombination events, often including introduction of nonmitochondrial DNA sequences (Schnable and Wise, 1998). More likely, RNAi suppression of *Msh1* results in the amplification of a subgenomic form derived as a single event. With more detailed analysis of the sterility-causing sequence in these transgenic lines, it should be feasible to determine whether this gene configuration is already present at substoichiometric levels in the original wild-type line.

2.5 Nuclear regulation of mitochondrial DNA maintenance

A search for additional components of plant mitochondrial DNA maintenance has resulted in some interesting observations. The first is that genes encoding components of mitochondrial DNA and RNA metabolism do not appear to be randomly distributed throughout the genome; rather, many are linked primarily on Chromosome 3 in the *Arabidopsis* genome, with a second group on Chromosome 5 (Elo *et al.*, 2003). This general clustering has facilitated the discovery of additional mitochondrially targeted proteins involved in DNA maintenance but poorly or incompletely annotated. A large proportion of the genes identified to be involved in mitochondrial DNA maintenance is dual targeted to the plastid or duplicated to plastid-targeted forms (Elo *et al.*, 2003; Mackenzie, 2005). Linkage of these loci may have facilitated their nuclear integration and adaptation for nuclear expression and protein targeting. Physical linkage may also facilitate their coordinated regulation in response to cellular cues during organelle biogenesis.

Investigation of some of the nuclear-encoded components of mitochondrial DNA maintenance uncovered a second gene that maintains genome stability. Three loci in the *Arabidopsis* genome show significant homology to *RecA*, a gene in *E. coli* involved in homologous recombination. The first, located on Chromosome 1 and designated *RecA1*, is plastid targeted and essential; mutations at this locus are lethal (Shedge *et al.*, submitted). A second homolog, encoded on Chromosome 2 and designated *RecA2*, is dual targeted to mitochondria and plastids, and disruption of this locus also results in lethality. However, *RecA3*, located on Chromosome 3 and encoding a mitochondrially targeted protein, displays distinct protein sequence modifications in key functional domains as well as a C-terminal truncation of the protein. Disruption of this locus results in reproducible patterns of mitochondrial DNA rearrangement (Shedge *et al.*, submitted).

Detailed analysis of the *recA3* mutant suggests that the gene product participates in suppression of illegitimate recombination in a manner that can be distinguished from the action of *Msh1*. Mitochondrial rearrangements effected by mutation at *RecA3* appear to be reversible, in contrast to those observed in *msh1* mutants, so that the DNA rearrangement phenotype of a *recA3* mutant demonstrates Mendelian rather than cytoplasmic inheritance. Yet, similarity in *recA3*- and *msh1*-induced mitochondrial DNA rearrangements suggested that the two gene products might interact; so *msh1/msh1-recA3/recA3* double mutants were generated and examined for phenotype. The double mutant phenotype is striking: dramatically slowed plant growth, evidenced by reduced mitotic index, small plant stature and 80–90% increase in generation time (Shedge *et al.*, submitted). Double mutants display markedly increased sensitivity to environmental stress and greatly reduced seed set and germination. The mitochondrial genome of the double mutant displays the pattern of SSS polymorphisms characteristic of *msh1* mutants, with several novel DNA bands. Yet, reduction in mitochondrial DNA concentration is not evident, and oxygen uptake in the mutants appears normal. Global gene expression changes in the double mutant, assessed by microarray analysis of mature, nonflowering plants, suggest that these plants expend considerable effort to enhance mitochondrial functions while modulating gene expression to reduce plant growth rate. Nuclear-encoded mitochondrial proteins constitute nearly 10% of the upregulated genes in the double mutant, while transcription factors, signal transduction mediators and cell cycle-related loci appear to comprise the most significant class of downregulated genes (V. Shedge and S.A. Mackenzie, unpublished results).

Thus, while genetic evidence indicates that *Msh1* and *RecA3* do not act within the same pathway, they do appear to play complementary roles in the regulation of illegitimate recombination and maintenance of the recombinant products within the genome.

2.6 Cellular and developmental responses to mitochondrial genome rearrangement in plants

The profound changes in growth phenotype occurring in *msh1/recA3* double mutants are indicative of the importance of controlling illegitimate recombination activity in plant mitochondria. The SSS phenomenon is most apparent during reproductive development, based on polymerase chain reaction assays for the detected genomic shift (Arrieta-Montiel *et al.*, 2001; A. Elo and S. Mackenzie, unpublished data). The *RecA3* and *Msh1* gene products are also highest in their expression in reproductive tissues (Shedge *et al.*, submitted).

Mitochondrial genome integrity clearly influences overall cellular function and growth rate. While mitochondrial DNA levels remain approximately normal in these mutants, their rearranged configuration triggers a striking reduction in cell cycle and overall plant-growth rate. One possible signal linking growth rate to mitochondrial genome status is redox regulation, and a number of redox-responsive proteins, including alternative oxidase (*AOX1*), demonstrate marked upregulation in the double mutant. Redox status has been shown previously to influence cell cycle rate in plants (Jiang *et al.*, 2003). A second possible signal may be ATP levels, also likely to influence growth rate.

In maize, examples exist of nuclear-mitochondrial interactions leading to impaired regulation of nonhomologous recombination activity. Nonchromosomal stripe and P2-induced abnormal phenotypes have been shown to arise in association with aberrant mitochondrial recombination activity in particular nuclear-cytoplasmic combinations (Newton and Coe, 1986; Kuzmin *et al.*, 2005). These aberrant growth phenotypes again reflect the importance of suppressing mitochondrial ectopic recombination activity to maintain cellular homeostasis.

Nonhomologous recombination activity in plant mitochondria gives rise to novel gene chimeras, producing numerous examples of CMS mutations. These mutations have served to underline the intrinsic relationship between mitochondrial function and pollen development. A broad range of phenotypes characterizes CMS mutants, including premature tapetal breakdown, incomplete cytokinesis during microsporogenesis and postmeiotic microspore abortion (Conley and Hanson, 1995; Hanson and Bentolila, 2004). CMS mutations also influence homeotic programs within the developing flower to alter male reproductive development (Zubko, 2004). Yet, only one CMS mutation (maize cms-T) has caused clear pleiotropic effects on the plant at other stages of development (Rhoads *et al.*, 1995). The multifaceted nature of mitochondrial influence on male reproductive development implies that the recombination processes leading to CMS mutations have been subject to intensive selection.

2.7 Plant mitochondrial genome and its adaptations

If expansion in the size of the plant mitochondrial genome, by massive sequence duplications and foreign DNA insertions, has allowed for aberrant, potentially deleterious recombination activity to occur, why has this genome expansion been so prevalent within the plant kingdom? Interesting speculation on this issue has emerged recently. Given the unusually low rate of mitochondrial gene mutation in higher plants, relative to that of animals, genome expansion may have been tolerated as a nonadaptive feature (Lynch *et al.*, 2006). With the introduction of intergenic sequences the potential for ectopic recombination activity increases, leading to gene expression patterns that influence the reproductive capacity of the population. Very similar illegitimate recombination activity at small repeated sequences appears to be responsible for both the creation of CMS mutations, as well as their regulation within the population by SSS activity. Thus, the adaptive value of illegitimate recombination activity at the population level appears significant.

While it is not yet clear whether *Msh1* contributes to the unusually low gene mutation rate of plant mitochondria, the MSH1 protein possesses features that are unique and well conserved within the plant kingdom (Abdelnoor *et al.*, 2006). The same appears to be the case for RECA3 (Shedge *et al.*, submitted). At least one additional nuclear gene in *Arabidopsis*, displaying mitochondrial DNA-binding and influencing ectopic recombination activity, has also been reported (V. Zaegel, M. Vermel, B. Guermann, J.M. Gualberto and P. Imbault, unpublished results). During their evolution, higher plants appear to have implemented a specialized, multicomponent system for the regulation of mitochondrial nonhomologous recombination. Evidence to date suggests that this genetic system is assembled as nonredundant components so that allelic variation in any single component permits a tolerable level of illegitimate recombination, evident as sporadic CMS induction, or SSS activity, but depletion of more than one component results in significant genome instability, as evidenced in the *msh1/recA3* double mutant.

While speculative at present, this model provides important opportunities for further study. For example, an important feature of the SSS process is that it does not involve random recombination at multiple sites; recombination activity controlled by *Msh1* or *RecA3* appears to involve only particular repeated sequences in a highly reproducible pattern. This was also the case in RNAi-mediated transgenic suppression of *Msh1* in other plant species (Sandhu *et al.*, submitted). It should now be possible to compare the structure of these *Msh1*-responsive mitochondrial regions in various plant species to better understand the nature of double-strand breakage underlying SSS. It should also be feasible to assess frequencies of allelic variation at *Msh1* and *RecA3* within natural plant (and perhaps coral) populations as a means of determining the extent to which such nuclear-mitochondrial genetic interactions may influence natural reproductive behavior in populations.

Acknowledgements

I wish to thank Alan Christensen for helpful comments on this chapter and for the immensely valuable collaboration that we share. Work from my lab presented here is supported by grants from the National Science Foundation, Department of Energy and Biotechnology Research Development Corporation.

References

Abdelnoor RV, Christensen AC, Mohammed S, Munoz-Castillo B, Moriyama H and Mackenzie SA (2006) Mitochondrial genome dynamics in plants and animals: convergent gene fusions of a *MutS* homologue. *J Molec Evol* **63**, 165–173.

Abdelnoor RV, Yule R, Elo A, Christensen A, Meyer-Gauen G and Mackenzie S (2003) Substoichiometric shifting in the plant mitochondrial genome is influenced by a gene homologous to *MutS*. *Proc Natl Acad Sci USA* **100**, 5968–5973.

Adams KL and Palmer JD (2003) Evolution of mitochondrial gene content: gene loss and transfer to the nucleus. *Mol Phylogent Evol* **29**, 380–395.

Andersson H (1999) Female and hermaphrodite flowers on a chimeric gynomonoecious *Silene vulgaris* plant produce offspring with different genders: a case of heteroplasmic sex determination. *J Hered* **90**, 563–563.

Andre C, Levy A and Walbot V (1992) Small repeated sequences and the structure of plant mitochondrial genomes. *Trends Genet* **8**, 128–32.

Arrieta-Montiel M, Lyznik A, Woloszynska M, Janska H, Tohme J and Mackenzie S (2001) Tracing evolutionary and developmental implications of mitochondrial genomic shifting in the common bean. *Genetics* **158**, 851–864.

Backert S and Borner T (2000) Phage T4-like intermediates of DNA replication and recombination in the mitochondria of the higher plant *Chenopodium album* (L.). *Curr Genet* **37**, 304–314.

Belfort M and Roberts RJ (1997) Homing endonucleases: keeping the house in order. *Nucleic Acids Res* **25**, 3379–3388.

Bellaoui M, Martin-Canadell A, Pelletier G and Budar F (1998) Low-copy-number molecules are produced by recombination, actively maintained and can be amplified in the mitochondrial genome of Brassicaceae: relationship to reversion of the male sterile phenotype in some cybrids. *Mol Gen Genet* **257**, 177–185.

Bendich AJ (1996) Structural analysis of mitochondrial DNA molecules from fungi and plants using moving pictures and pulsed field gel electrophoresis. *J Mol Biol* **255**, 564–588.

Bridge D, Cunningham CW, Schierwater B, DeSalle R and Buss LW (1992) Class-level relationships in the phylum Cnidaria: evidence from mitochondrial genome structure. *Proc Natl Acad Sci USA* **89**, 8750–8753.

Clifton SW, Minx P, Fauron CM-R, *et al.* (2004) Sequence and comparative analysis of the maize NB mitochondrial genome. *Plant Physiol* **136**, 3486–3503.

Conley CA and Hanson MR (1995) How do alterations in plant mitochondrial genomes disrupt pollen development? *J Bioenerg Biomembr* **27**, 447–457.

Ehlers BK, Maurice S and Bataillon T (2005) Sex inheritance in gynodioecious species: a polygenic view. *Proc Biol Sci* **272**, 1795–1802.

Elo A, Lyznik A, Gonzalez DO, Kachman SD and Mackenzie S (2003) Nuclear genes encoding mitochondrial proteins for DNA and RNA metabolism are clustered in the *Arabidopsis* genome. *Plant Cell* **15**, 1619–1631.

Escote-Carlson LJ, Gabay-Laughnan S and Laughnan JR (1990) Nuclear genotype affects mitochondrial genome organization of CMS-S maize. *Mol Gen Genet* **223**, 457–464.

Fauron C, Casper M, Gao Y and Moore B (1995) The maize mitochondrial genome: dynamic, yet functional. *Trends Genet* **11**, 228–235.

Gray MW (1999) Evolution of organellar genomes. *Curr Opin Genet Dev* **9**, 678–687.

Gray MW, Burger G and Lang BF (1999) Mitochondrial evolution. *Science* **283**, 1476–1481.

Gutierres S, Lelandais C, Paepe RD, Vedel F and Chetrit P (1997) A mitochondrial sub-stoichiometric orf87-nad3-nad1 exonA co-transcription unit present in Solanaceae was amplified in the genus *Nicotiana*. *Curr Genet* **31**, 55–62.

Handa H (2003) The complete nucleotide sequence and RNA editing content of the mitochondrial genome of rapeseed (Brassica napus L): comparative analysis of the mitochondrial genomes of rapeseed and *Arabidopsis* thaliana. *Nucleic Acids Res* **31**, 5907–5916.

Hanson MR and Bentolila S (2004) Interactions of mitochondrial and nuclear genes that affect male gametophyte development. *Plant Cell* **16**, S154–S169.

He S, Lu Z-H, Vallejos CE and Mackenzie S (1995) Pollen fertility restoration by nuclear gene *Fr* in cms common bean: an *Fr* linkage map and the mode of *Fr* action. *Theor Appl Genet* **90**, 1056–1062.

Janska H, Sarria R, Woloszynska M, Arrieta-Montiel M and Mackenzie S (1998) Stoichiometric shifts in the common bean mitochondrial genome leading to male sterility and spontaneous reversion to fertility. *Plant Cell* **10**, 1163–1180.

Jiang K, Meng YL and Feldman LJ (2003) Quiescent center formation in maize roots is associated with an auxin-regulated oxidizing environment. *Development* **130**, 1429–1438.

Johns C, Lu M, Lyznik A and Mackenzie SA (1992) A mitochondrial DNA sequence is associated with abnormal pollen development in cytoplasmic male sterile bean plants. *Plant Cell* **4**, 435–449.

Kajander OA, Karhunen PJ, Holt IJ and Jacobs HT (2001) Prominent mitochondrial DNA recombination intermediates in human heart muscle. *EMBO Rep* **2**, 1007–1012.

Kanazawa A, Tsutsumi N and Hirai A (1994) Reversible changes in the composition of the population of mtDNAs during dedifferentiation and regeneration in tobacco. *Genetics* **138**, 865–870.

Knoop V (2004) The mitochondrial DNA of land plants: peculiarities in phylogenetic perspective. *Curr Genet* **46**, 123–139.

Kraytsberg Y, Schwartz M, Brown TA, *et al.* (2004) Recombination of human mitochondrial DNA. *Science* **304**, 981.

Kubo T, Nishizawa S, Sugawara A, Itchoda N, Estiati A and Mikami T (2000) The complete nucleotide sequence of the mitochondrial genome of sugar beet (Beta vulgaris L) reveals a novel gene for rRNA(Cys)GCA. *Nucleic Acids Res* **28**, 2571–2576.

Kunkel TA and Erie DA (2005). DNA mismatch repair. *Annu Rev Biochem* **74**, 681–710.

Kuzmin EV, Duvick DN and Newton KJ (2005) A mitochondrial mutator system in maize. *Plant Physiol* **137**, 779–789.

Lang BF, Gray MW and Burger G (1999) Mitochondrial genome evolution and the origin of eukaryotes. *Annu Rev Genet* **33**, 351–397.
Lonsdale DM, Brears T, Hodge TP, Melville SE and Rottmann WH (1988) The plant mitochondrial genome: homologous recombination as a mechanism for generating heterogeneity. *Philos Trans Royal Soc Lond B* **319**, 149–163.
Lynch M, Koskella B and Schaack S (2006) Mutation pressure and the evolution of organelle genomic architecture. *Science* **311,** 1727–1730.
Mackenzie SA (2005) Plant organellar protein targeting: a traffic plan still under construction. *Trends Cell Biol* **10**, 548–554.
Mackenzie SA and Chase CD (1990) Fertility restoration is associated with loss of a portion of the mitochondrial genome in cytoplasmic male-sterile common bean. *Plant Cell* **2**, 905–912.
Mackenzie S and McIntosh L (1999) Higher plant mitochondria. *Plant Cell* **11**, 571–586.
Malik HS and Henikoff S (2000) Dual recognition–incision enzymes might be involved in mismatch repair and meiosis. *Trends Biochem Sci* **25**, 414–418.
Manchekar M, Scissum-Gunn K, Song D, Khazi F, McLean SL and Nielsen BL (2006) DNA recombination activity in soybean mitochondria. *J Mol Biol* **356**, 288–299.
Marienfeld, JR, Unseld M, Brandt P and Brennicke A (1997) Mosaic open reading frames in the *Arabidopsis thaliana* mitochondrial genome. *Biol Chem* **378**, 859–862.
Marienfeld JR, Unseld M and Brennicke A (1999) The mitochondrial genome of *Arabidopsis* is composed of both native and immigrant information. *Trends Plant Sci* **4**, 495–502.
Newton KJ and Coe EH (1986) Mitochondrial DNA changes in abnormal growth (nonchromosomal stripe) mutants of maize. *Proc Natl Acad Sci USA* **83**, 7363–7366.
Ozias-Akins P, Tabaeizadeh Z, Pring DR and Vasil IK (1988) Preferential amplification of mitochondrial DNA fragments in somatic hybrids of the Graminae. *Curr Genet* **13**, 241–245.
Pont-Kingdon G, Okada NA, Macfarlane JL, *et al.* (1998) Mitochondrial DNA of the coral *Sarcophyton glaucum* contains a gene for a homologue of bacterial MutS: a possible case of gene transfer from the nucleus to the mitochondrion. *J Mol Evol* **46**, 419–431.
Reenan RAG and Kolodner RD (1992) Characterization of insertion mutations in the *Saccharomyces cerevisiae MSH1* and *MSH2* genes: evidence for separate mitochondrial and nuclear functions. *Genetics* **132**, 975–985.
Rhoads DM, Levings CS, III and Siedow JN (1995) URF13, a ligand-gated, pore-forming receptor for T-toxin in the inner membrane of cms-T mitochondria. *J Bioenerg Biomembr* **27**, 437–445.
Sakai A, Takano H and Kuroiwa T (2004) Organelle nuclei in higher plants: structure, composition, function and evolution. *Int Rev Cytol* **238**, 59–118.
Sakamoto W, Kondo H, Murata M and Motoyoshi F (1996) Altered mitochondrial gene expression in a maternal distorted leaf mutant of *Arabidopsis* induced by chloroplast mutator. *Plant Cell* **8**, 1377–1390.
Sandhu APS, Abdelnoor RV and Mackenzie SA (submitted) Transgenic induction of mitochondrial rearrangements for cytoplasmic male sterility in crop plants.
Schnable PS and Wise RP (1998) The molecular basis of cytoplasmic male sterility and fertility restoration. *Trends Plant Sci* **3**, 175–180.
Schofield MJ and Hseih P (2003) DNA mismatch repair: molecular mechanisms and biological function. *Annu Rev Microbiol* **57**, 579–608.
Shearer TL, Van Oppen MJ, Romano SL and Worheide G (2002) Slow mitochondrial DNA sequence evolution in the Anthozoa (Cnidaria). *Mol Ecol* **11**, 2475–2487.
Shedge V, Arrieta-Montiel M, Christensen AC and Mackenzie SA (submitted) Plant mitochondrial recombination surveillance requires novel RecA and MutS homologues.
Shirzadegan M, Palmer JD, Christey M and Earle ED (1991) Patterns of mitochondrial DNA instability in *Brassica campestris* cultured cells. *Plant Mol Biol* **16**, 21–37.
Sia EA and Kirkpatrick DT (2005) The yeast MSH1 gene is not involved in DNA repair or recombination during meiosis. *DNA Repair* **4**, 253–261.
Small ID, Isaac PG and Leaver CJ (1987) Stoichiometric differences in DNA molecules containing the atpA gene suggest mechanisms for the generation of mitochondrial genome diversity in maize. *EMBO J* **6**, 865–869.

van Oppen MJ, Catmull J, McDonald BJ, Hislop NR, Hagerman PJ and Miller DJ (2002) The mitochondrial genome of *Acropora tenuis* (Cnidaria; Scleractinia) contains a large group I intron and a candidate control region. *J Mol Evol* **55**, 1–13.

Vitart V, De Paepe R, Mathieu C, Chetrit P and Vedel F (1992) Amplification of substoichiometric recombinant mitochondrial DNA sequences in a nuclear, male sterile mutant regenerated from protoplast culture in Nicotiana sylvestris. *Mol Gen Genet* **233**, 193–200.

Wolfe KH, Li WH and Sharp PM (1987) Rates of nucleotide substitution vary greatly among plant mitochondrial, chloroplast and nuclear DNAs. *Proc Natl Acad Sci USA* **84**, 9054–9058.

Zubko MK (2004) Mitochondrial tuning fork in nuclear homeotic functions. *Trends Plant Sci* **9**, 61–64.

3 Expression of the plant mitochondrial genome

Dominique Gagliardi and Stefan Binder

3.1 Introduction

Like almost all eukaryotic organisms, higher plants contain mitochondria and these mitochondria are indispensable for life. Mitochondria are essential as the reaction compartments for many metabolic pathways and also for energy conservation by oxidative phosphorylation. But what is special about plant mitochondria? In contrast to their counterparts in other eukaryotes, plant mitochondria host a number of unique pathways, including, for example, parts of photorespiration and distinct steps of biotin, as well as of folate biosynthesis (Mouillon *et al.*, 2002; Raghavendra and Padmasree, 2003; Pinon *et al.*, 2005; see Chapter 8). Plant mitochondria also contain alternative oxidases, and NAD(P)H dehydrogenases for uncoupling NAD(P)H oxidation from proton pumping (Considine *et al.*, 2002; Juszczuk and Rychter, 2003; Rasmusson *et al.*, 2004; see Chapter 7). But, apart from these special metabolic features, plant mitochondria are characterized by the size and structure of their genomes (see Chapter 2). With up to 2.4 Mb these are extremely large, and their organization is complex due to high recombination activity involving direct and inverted repeats (Levings and Brown, 1989; Bonen and Brown, 1993; Mackenzie and McIntosh, 1999). The exact structure of a mitochondrial genome, if there is one at all, is still unclear as is its influence on gene expression.

In plant mitochondrial genomes, up to 60 genes encode a set of about 30 proteins, 15–20 tRNAs and 3 rRNAs. These sets differ slightly between different species, but the products encoded are directly or indirectly exclusively involved in ATP production by oxidative phosphorylation (Unseld *et al.*, 1997; Kubo *et al.*, 2000; Notsu *et al.*, 2002; Handa, 2003; Clifton *et al.*, 2004; Ogihara *et al.*, 2005; Sugiyama *et al.*, 2005). The genes are usually spread over the complete genome, giving rise to both mono- and polycistronic primary transcripts. These transcripts undergo a series of processing steps such as RNA editing, the generation of secondary 5′ and 3′ ends and the removal of group II intron, until they obtain their final mature form (Figure 3.1). Good progress has been made in recent years in understanding some mechanisms underpinning these processes and in identifying the participating proteins. However, it is still unclear how or whether these processes are networked, or interdependent, and whether they directly influence translation, and thus gene expression. Finally, RNA stability also seems to influence gene expression, bringing RNA degradation processes into the business of gene expression regulation (Giegé *et al.*, 2000; Leino *et al.*, 2005). Taken together, these processes mean that in plant mitochondria a tremendous effort is required for the expression of a comparatively small number of genes. In this chapter we summarize recent advances

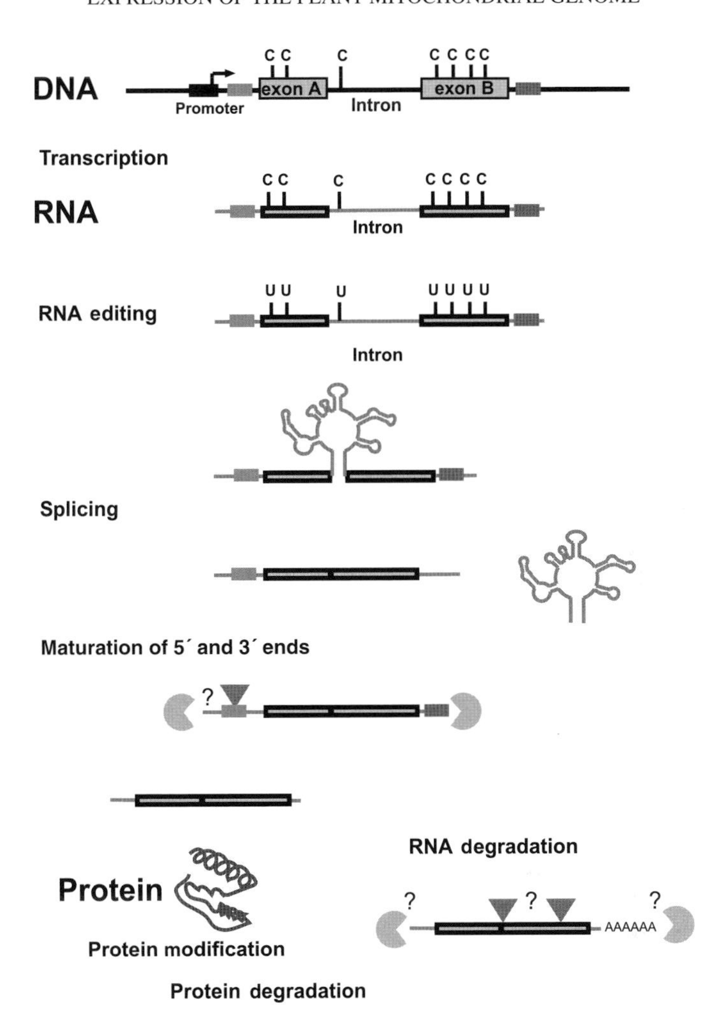

Figure 3.1 Gene expression in plant mitochondria. A complex series of processing steps is required for maturation and generation of translatable transcripts. After transcription, the primary transcripts undergo RNA editing, intron splicing and generation of secondary 5′ and 3′ ends. RNA degradation is initiated by polyadenylation, and RNA is most likely degraded by endo- and/or exoribonucleases.

made in the understanding of the processes involved in plant mitochondrial gene expression.

3.2 Identification of the pentatricopeptide repeats protein family: a key discovery toward the understanding of plant mitochondrial gene expression

Sequencing of the *Arabidopsis* nuclear genome enabled the identification of several large gene families that are either plant specific, or at least overrepresented in plants as compared with other organisms. One of the largest of these gene families is defined by the presence of repeats of a degenerate motif of 35 amino acids (Small and Peeters, 2000). These pentatricopeptide repeats (PPR) are present in about 450 and 650 proteins in *Arabidopsis* and *Oryza sativa*, respectively (Lurin *et al.*, 2004). By contrast, only a few PPR genes exist in nonplant eukaryotes, and none in prokaryotes, apart from a single exception most likely due to a horizontal transfer event (Lurin *et al.*, 2004). About two thirds of PPR proteins are predicted to be targeted to organelles and, to date, the few available functional analyses of individual PPR protein point to their involvement in posttranscriptional processes in organelles. These processes include editing, processing, translation and stability of mitochondrial or chloroplast transcripts. PPR proteins are thus likely to intervene in many steps of plant mitochondrial genome expression.

3.2.1 Structure of PPR proteins

The characteristic motif of 35 amino acids is present in 2–26 copies in *Arabidopsis* PPR proteins (Small and Peeters, 2000). However, PPR motifs are only loosely conserved in sequence, and their identification is further complicated by the existence of short and long variants of the PPR motif. These short and long PPR-like motifs (PPR-like S and L, respectively) contain 31 and 35–36 amino acids, respectively (Lurin *et al.*, 2004). However, prediction of PPR motifs can be validated by the fact that true PPR motifs are usually found as tandem arrays. This feature, as well as the consensus sequence itself, reveals a link between PPR and tetratricopeptide repeat (TPR) motifs (Small and Peeters, 2000), the latter being involved in protein–protein interactions (Das *et al.*, 1998). Based on the TPR motif structure, each PPR motif is predicted to fold into two α-helices and the tandem PPR motifs likely form a superhelix. This structure delimits a hydrophilic groove, positively charged at its base, which has been hypothesized as possibly interacting with RNA (Small and Peeters, 2000). This hypothesis is gaining support considering the increasing number of PPR proteins involved in RNA metabolism. However, the possibility that PPR motifs could also mediate protein–protein interactions cannot be ruled out.

In *Arabidopsis*, PPR proteins can be subdivided into two subfamilies (Lurin *et al.*, 2004): about half of the PPR proteins display series of only ‘true’ PPR motifs (P subfamily), whereas for the other PPR proteins, series of PPR motifs (P), long (L) and short (S) PPR-like motifs are observed (PLS subfamily). Members of the PLS subfamily can be further classified into subgroups based on the absence (PLS subgroup) or presence of conserved C-terminal domains (E, E+ and DYW subgroups). A strict ordering of these domains is observed as the only possible combinations are no extra C-terminal domain (PLS subgroup), E domain (E subgroup), E:E+ domains (E+ subgroup) and E:E+:DYW domains (DYW subgroup). These domains are not found in nonplant organisms and are specific to PPR genes in *Arabidopsis*, with the notable exception of a gene consisting of a DYW motif (At1g47580). The molecular role of these domains is unknown at present.

The current model regarding PPR function proposes that the series of PPR motifs recognizes a specific RNA sequence. When present, the different C-terminal domains either recruit or directly perform a molecular posttranscriptional process.

3.2.2 Functions of PPR proteins

PPR proteins have been genetically linked to a variety of posttranscriptional processes in plant and nonplant organelles. Human LRPPRC, *Saccharomyces cerevisiae* PET309 and *Neurospora crassa* CYA-5 proteins are PPR proteins necessary for the translation of *cox1* mRNA, and possibly for the maturation, and/or stability of this mRNA (Coffin *et al.*, 1997; Manthey *et al.*, 1998; Xu *et al.*, 2004). Chloroplast RNA metabolism is also affected by mutations in *PPR* genes. The *CRP1* maize gene is required for translation of *petA* and *petD* mRNA as well as for the processing of *petD* mRNA from its precursor (Fisk *et al.*, 1999). Also in maize, the *PPR2* gene is essential for the synthesis or assembly of some component(s) of the plastid translation machinery (Williams and Barkan, 2003), and the *CRP1* gene is required for activating translation of *petA* and *psaC* mRNA (Schmitz-Linneweber *et al.*, 2005). In *Arabidopsis*, the HCF152 PPR protein is necessary for the accumulation of *petB* mRNA excised from its large primary transcript (Meierhoff *et al.*, 2003). Also in *Arabidopsis*, the *PGR3* PPR gene is necessary for the stability of the tricistronic transcript from the *petL* operon (Yamazaki *et al.*, 2004). Transcript processing, stability and translation are also under the specific control of PPR proteins in plant mitochondria (detailed below). Most importantly, the *CCR4* gene has been linked recently with the editing of the translation initiation codon of *ndhD* transcript in chloroplasts (Kotera *et al.*, 2005).

PPR proteins could be part of large protein complexes, as in the case of CRP1 (Fisk *et al.*, 1999), or bind RNA on their own, as shown for HCF152 that exists as a homodimer in chloroplasts (Meierhoff *et al.*, 2003). For both HCF152 and CRP1, direct interaction between the PPR protein and their target RNA has been proven experimentally (Meierhoff *et al.*, 2003; Nakamura *et al.*, 2003; Schmitz-Linneweber *et al.*, 2005).

Recent studies have also demonstrated that PPR proteins are implicated in plant mitochondrial gene expression. Indeed, restorer to fertility genes (*Rf*) have been cloned in several cytoplasmic male sterility (CMS) systems and found to encode PPR proteins. The CMS phenotype, widely used in crop plant breeding, is due to mitochondrial mutations that prevent the production of functional pollen (see Chapter 9). So far, the CMS phenotype has always been associated with the expression of novel mitochondrial genes. The nuclear genes able to restore pollen production are named *Rf* genes. This restoration occurs either by compensating for the mitochondrial defect due to the mutation or by altering the expression of the novel gene associated with the CMS phenotype. Identifying *Rf* genes is therefore a very useful strategy to understand different aspects of plant genome expression. In recent years, several *Rf* genes encoding PPR proteins have been identified in Petunia, radish (and its counterpart in rapeseed), rice and sorghum (Bentolila *et al.*, 2002; Brown *et al.*, 2003; Desloire *et al.*, 2003; Kazama and Toriyama, 2003; Koizuka *et al.*, 2003; Akagi *et al.*, 2004; Komori *et al.*, 2004; Klein *et al.*, 2005; Wang *et al.*, 2006). Similar to the situation in chloroplasts, mitochondrial PPR proteins influence transcript profiles and/or amounts of the deleterious protein associated with CMS (Hanson and Bentolila, 2004).

In conclusion, several PPR proteins have now been linked with different steps of plant organellar gene expression such as the maturation, editing and stability of RNA and translation. PPR proteins are encoded by one of the largest gene families in plants and it seems possible that these proteins are involved in most, if not all, of the complex posttranscriptional processes in plant organelles. The discovery of PPR proteins undoubtedly represents a milestone in our understanding of plant organellar gene expression.

3.3 Transcription in higher plant mitochondria

3.3.1 Conserved and variable transcription units in plant mitochondria

The plant mitochondrial genome is a large, complex structure relative to the smaller, simpler mitochondrial genomes in other eukaryotes (see Chapter 2), although the gene content of the plant mitochondrial genome is not proportionately larger. Sequence analyses of the complete mitochondrial genomes of seven seed plant species allow some conclusions to be made about the distribution and clustering of the mitochondrial-encoded genes (Unseld *et al.*, 1997; Kubo *et al.*, 2000; Notsu *et al.*, 2002; Handa, 2003; Clifton *et al.*, 2004; Ogihara *et al.*, 2005; Sugiyama *et al.*, 2005). There are very few gene clusters conserved in all species investigated. These are 18S-5S rRNAs, *nad*3-*rps*12 and *rps*3-*rpl*16 with the latter being disrupted in sugar beet, where *rpl*16 is absent from mtDNA (Kubo *et al.*, 2000; Satoh *et al.*, 2004). However, when these genes are present in mitochondria, there seems to be a clear preference for cotranscription. There is a reasonable explanation for a more or less stoichiometric transcription of those genes whose products are part of the same complex, i.e. the ribosome in the case of 18S-5S rRNAs and *rps*3-*rpl*16. The clear preference for cotranscription of *nad*3-*rps*12, encoding subunits of different complexes, is puzzling but seems not to be obligatory. While this dicistronic cluster is conserved in many different plant species, there is an additional *rps*12 independent gene copy present in *Nicotiana sylvestris* (Lelandais *et al.*, 1996). Moreover, in *Oenothera* mtDNA *nad*3 is cotranscribed with the upstream-located *rpl*5 and the downstream *rps*12 is only present as a pseudo gene, while a complete locus can be found in the nucleus (Grohmann *et al.*, 1992; Schuster, 1993). This indicates that for *nad*3 no obligatory cotranscription with *rps*12 exists and differences in terms of upstream-cotranscribed sequences are accepted. Taken together, these data show that 18S-5S rRNA genes represent the sole true obligatory polycistronic transcription unit in seed plants; this arrangement is even conserved in *Marchantia polymorpha* (Oda *et al.*, 1992). The comparison of the complete mtDNA sequences from seed plant species also shows that the gene orders are highly variable, even between closely related species, e.g. *Arabidopsis* and rapeseed (Unseld *et al.*, 1997; Handa, 2003). The genes also vary in terms of their transcription within mono- or polycistronic units, in terms of spacer sequences and in the sequence contexts in the flanking regions. Thus the mitochondrial genetic system exhibits enough flexibility to maintain many new gene orders, and transcription units, and to compensate for the consequences arising from those. While there may be insufficient data for definitive statements, three factors are likely to underpin this flexibility. First, there is probably enough redundancy in transcriptional control, i.e. a large number of equally distributed promoters so that after recombination a given gene is still transcribed. Second, the extensive posttranscriptional machineries might compensate for the aberrant transcriptional levels and guarantee an adequate amount of translatable steady-state transcripts. Third, there might be another level of expression control that could even-out imbalances in transcript levels. An example of this flexibility is the expression of the *cox*3 gene in different *Arabidopsis* ecotypes. Different genomic environments are present upstream of the *cox*3 gene in ecotypes C24, Columbia (Col) and Landsberg *erecta* (L*er*). Most likely different primary transcripts are generated, which are processed to identical mature forms in all three ecotypes, and an additional slightly larger RNA in C24 (see Section 3.4.1). Despite these differences, very similar amounts of *cox*3 mRNAs are present in the steady-state pool (Forner *et al.*, 2005).

But what makes cotranscription of 18S-5S rRNAs indispensable? Generation of stoichiometric amounts might be an argument, but many different genes whose products are part of the same multisubunit complex are transcribed in many different contexts, and at different levels. In addition, 26S rRNA is, in all species investigated, exclusively encoded and transcribed from another locus. This might indicate that cotranscription of 18S-5S rRNAs might be important for further maturation. It is possible that one or several processing steps required to generate the mature end products of these RNAs has to occur cotranscriptionally (Perrin *et al.*, 2004a).

3.3.2 How many different conserved and nonconserved promoter structures are there in plant mitochondria?

Since the discovery of mitochondrial DNA, researchers have addressed the question of how this genetic information is expressed and how this expression is regulated. In many genetic systems, such as in bacteria and in the nuclei of eukaryotes, regulation of transcription is the main means to regulate gene expression. Thus, with mitochondria, research has been undertaken to study transcription initiation and to identify mitochondrial promoters.

The complete mitochondrial genome sequences clearly show that multiple transcription units exist (Unseld *et al.*, 1997; Notsu *et al.*, 2002; Handa, 2003; Clifton *et al.*, 2004; Ogihara *et al.*, 2005; Sugiyama *et al.*, 2005). These are a consequence of genome size and the number of genes encoded, with average gene densities of 4.1–9.8 kb per gene. The uneven distribution further increases distances between individual transcription units with, for example, up to 57 kb of noncoding spacers in *Arabidopsis* (Unseld *et al.*, 1997). Thus multiple promoters are required to drive transcription of mitochondrial genes.

To identify and characterize mitochondrial transcription initiation sites, *in vitro* capping approaches have been used (Mulligan *et al.*, 1988a,b; Brown *et al.*, 1991; Covello and Gray, 1991; Binder and Brennicke, 1993). This technique takes advantage of the selective *in vitro* labeling of 5′ ends derived from *de novo* transcription initiation events. These termini contain triphosphates instead of monophosphates present at 5′ ends generated by posttranscriptional processing. This discrimination is necessary since complex transcription patterns are frequently found for mitochondrial genes, even if these are transcribed as monocistronic units. This complexity is usually explained by different 5′ ends, derived from either transcription initiation or processing.

The *in vitro* capping studies identified many different transcription initiation sites upstream of different genes in various plant species. Conserved promoter motifs were deduced by comparison of the surrounding sequences and considered to be at least part of the promoter elements required for function. The 5′-CRTA-3′ tetranucleotide has been identified as a core element of mitochondrial promoters in both mono- and dicotyledon species (Mulligan *et al.*, 1988a,b; Brown *et al.*, 1991; Covello and Gray, 1991; Mulligan *et al.*, 1991). In the dicotyledons, this motif is part of the so-called conserved nonanucleotide motif (CNM) 5′-CRTAAGAGA-3′ (Binder and Brennicke, 1993; Binder *et al.*, 1994b). Functional studies using *in vitro* transcription systems not only confirmed the importance of the conserved motifs but also clearly identified further essential nucleotide identities at positions upstream and downstream of these motifs (Hanic-Joyce and Gray, 1991; Rapp and Stern, 1992; Rapp *et al.*, 1993; Binder *et al.*, 1995; Caoile and Stern, 1997; Dombrowski *et al.*, 1999). The sequences required for full-promoter activity are 17- (maize *atp*1, −12 to +5), 18- (pea *atp*9, −14 to +4) and 26-bp (maize *cox*3) long single entities that generally extend beyond the transcription start site (+1). *In vitro* transcription analyses of site-directed mutagenized maize *atp*1 and pea *atp*9 promoter sequences identified several nucleotide identities important for promoter function (Rapp *et al.*, 1993; Binder *et al.*, 1995; Dombrowski *et al.*, 1999; Hoffmann and Binder, 2002). In the pea *atp*9 promoter, the exchange of nucleotide identities that are highly conserved in promoters from other dicotyledons can lead to a severe loss of promoter function *in vitro*. These nucleotides cluster near the transcription start site (positions −3, −2, +1), a situation that is very similarly found in the maize *atp*1 promoter although the identities found in these positions differ between maize and pea. Also common to both promoters is the importance of a conserved adenosine at position −12. In the maize *atp*1 promoter, this position is part of an A-rich upstream domain, while in pea *atp*9 this position is located in the A/T-box. In pea, the exchange of these A/Ts from one strand to the other almost completely abolished promoter function, strongly suggesting that it is not the A/T content *per se* that is important but rather certain nucleotides identities. The functionally characterized structure of the pea *atp*9 promoter, which was found to be conserved in other genes and species, was used to screen the *Arabidopsis* complete mitochondrial genome sequence. About 30 potential promoters were found, both at reasonable distances upstream of coding regions as well as in the middle of large noncoding stretches (Giese *et al.*, 1996; Dombrowski *et al.*, 1998). Although these would be sufficient for transcription of all mitochondrial genes, the uneven distribution left a considerable number of genes without a promoter, raising the possibility that other promoter structures exist. This has been supported by the description of several other promoters that do not fit with characterized CNM (Binder *et al.*, 1994b; Remacle and Marechal-Drouard, 1996; Fey and Marechal-Drouard, 1999).

Common to both the maize *atp*1 and the pea *atp*9 promoters is the exchangeability of several nucleotide identities within a functionally defined sequence without any substantial negative influence on promoter activity. In pea this is also true for highly conserved nucleotides (−7, −1, +2) so that there seems to be no strict correlation between evolutionary conservation and functional importance. Two reasons for this apparent discrepancy have been considered. Either these nucleotide identities fulfill functions *in vivo* that cannot be followed *in vitro* or the number of promoter sequences used to define the consensus sequence was too small. While the former explanation cannot be addressed with present technologies, the latter might be true, at least partially, as detailed below.

A recent analysis of the transcription initiation sites of 12 genes in *Arabidopsis* confirmed experimentally that some of the predicted CNMs are active (Kühn *et al.*, 2005). By applying 5′-RACE and *in vitro* capping a second CNM, 5′-CGTATATAA-3′ was found upstream of several genes. This motif (referred to as CNM-type 2) also contains a 5′-CRTA-3′ tetranucleotide as in the previously characterized nonanucleotide 5′-CRTAAGAGA-3′ (now designated CNM-type 1). The transcription start site is found at the same distance relative to the 5′-CRTA-3′ motif but is exclusively an A instead of a G in the CNM-type 1. Besides the conserved promoter, many other transcription initiation sites have been determined but sequences surrounding these sites show only a relaxed conservation, containing the tetranucleotide motifs 5′-ATTA-3′, or 5′-RGTA-3′, or no conserved motif at all.

Taken together, the analyses performed to date reveal the existence of at least two conserved promoter motifs in *Arabidopsis* and probably other dicotyledonous species. Considering that the study performed by Kühn *et al.* (2005) comprises *only* 12 genes, or transcription units, one might assume that another motif will emerge when the number of promoters analysed increases. Functional studies are now required to tell us more about the similarities or differences between the distinct conserved motifs and about the other nonconserved initiation sites. In monocotyledons, promoters seem to be more heterogeneous; however, an analysis of a large number of genes and promoters in one species would contribute to a clearer picture of the conservation of possibly different, relative to dicotyledons, promoter structures of this subclass of plants (Mulligan *et al.*, 1991).

Common to monocotyledons and dicotyledons are multiple promoters upstream of almost every gene, or transcription unit investigated. The reason for this is unclear, but tissue-specific preferences for certain motifs or individual promoters seem, according to the study in *Arabidopsis,* rather unlikely (Kühn *et al.*, 2005).

3.3.3 What proteins are necessary for transcription of mitochondrial DNA?

It is now clear that in mitochondria in dicotyledons, two or even more classes of promoters exist raising the question as to whether these are activated by a different set of proteins. However, this question cannot be answered at present since our knowledge about the proteins involved in transcription, especially initiation, is still incomplete, although good progress has been made in the recent years.

In different plant species, several genes encoding phage-type RNA polymerases have been identified (Weihe, 2004). For instance, in *Arabidopsis* three *RPO* genes have been found whose products are either exclusively transported into mitochondria (AtRPOT;1 or AtRPOTm), into chloroplasts (AtRPOT;3 or AtRPOTp) or into both types of organelle (AtRPOT;2 or AtRPOTmp) (Hedtke *et al.*, 1997, 1999, 2000). In the moss *Physcomitrella patens*, two RNA polymerase genes have been found whose gene products are dually targeted to mitochondria and chloroplasts (Kabeya *et al.*, 2002; Richter *et al.*, 2002). Common to all dually targeted RNA polymerases are two in frame AUG codons in the N-terminal regions that could both serve as translation initiation codons. In the case of AtRPOT;2, translation can start at the first AUG-generating proteins targeted to mitochondria and plastids (Hedtke *et al.*, 2000). In *N. sylvestris* dual targeting might arise from alternative translation initiation on either AUG on a single transcript generating two different proteins with differing N-termini (Kobayashi *et al.*, 2002). In *P. patens*, however, somewhat contradictory results have been obtained. While Richter *et al.* (2002) found both proteins to be transported into mitochondria and chloroplasts, Kabeya *et al.* (2002) found exclusive targeting to mitochondria. This exclusive targeting was recently confirmed by immunodetection of these proteins in the respective organelle as determined previously by green fluorescent protein tagging (Kabeya and Sato, 2005). A reinvestigation of AtRPOT;2 also revealed contradicting results. As with the *P. patens* results, these studies claimed that in this case only the second AUG is used *in vivo*, generating a polypeptide that is exclusively imported into mitochondria (Kabeya and Sato, 2005). This conclusion is substantiated by expression studies of the three *RPO* genes in *Arabidopsis* (Emanuel *et al.*, 2005). AtRPOT;1 and AtRPOT;2 show very similar spatiotemporal transcription patterns that are clearly different from the pattern of the plastid-located AtRPOT;3. However, there is speculation that AtRPOT;2 is active in non-green parts of the plant while AtRPOT;3 is active in green tissues. In addition, analysis of a T-DNA insertion line, which most likely abolishes expression of AtRPOT;2, clearly supports a function of this protein in plastids. In these plants, which display slow growth, short roots and hypocotyls and a delayed greening phenotype, aberrant accumulation of several plastid RNAs is observed while no major changes in mitochondrial RNAs are detected (Baba *et al.*, 2004). However, independent from the question of whether AtRPOT;2 is also transported to chloroplasts, it is clear that there are two different RNA polymerases in mitochondria in dicotyledons. This seems to be different in monocotyledons. In this subclass, only two *RPO* genes are found coding for proteins that are either sorted to mitochondria or chloroplasts (Young *et al.*, 1998; Chang *et al.*, 1999; Ikeda and Gray, 1999b; Emanuel *et al.*, 2004).

The relative importance of the two mitochondrial RNA polymerases in dicotyledons is presently unclear. A preference of each polymerase for distinct sets of promoter seems to be rather unlikely since there is no major effect on mitochondrial transcription in an AtrpoT;2 mutant (Baba *et al.*, 2004). Also, transcription is not substantially different between AtRPOT;1 and AtRPOT;2 such that preferential activity of one of the two mitochondrial enzymes in certain tissues or at a specific developmental stage is also unlikely. Clearly more research is required to address this question.

The question of whether there are cofactors for the mitochondrial RNA polymerases remains more or less open. Considering that the homologous RNA polymerases in *S. cerevisiae* and humans also need two different types of transcription factor, one would assume that this is also the case in plants (Masters *et al.*, 1987; Diffley and Stillman, 1991; Jang and Jaehning, 1991; Parisi and Clayton, 1991; Tiranti *et al.*, 1997; Falkenberg *et al.*, 2002; McCulloch *et al.*, 2002). Approaches to purify such proteins from plants have so far been unsatisfactory. In pea, 43- and 32-kDa proteins binding to the CNM-type 1 pea *atp*9 promoter have been purified (Hatzack *et al.*, 1998). While the larger of these proteins was later found to be an isovaleryl-CoA dehydrogenase functioning in leucine catabolism (Daschner *et al.*, 1999, 2001), the identity of the smaller protein remained ambiguous. Similarly, a 63-kDa protein stimulating transcription initiation *in vitro* has been purified from wheat (Ikeda and Gray, 1999a). This protein shows highest similarity to PPR proteins (especially to *Arabidopsis* PPR protein At1g15480), which were suggested to have a function in RNA metabolism (Small and Peeters, 2000; Lurin *et al.*, 2004) so that the unambiguous function of this protein as a transcription factor *in vivo* is still somewhat unclear. An alternative approach for the identification of mitochondrial transcription factors in plant mitochondria is to search for *Arabidopsis* or rice gene homologues of the above-mentioned yeast or mammalian transcription factors in the *Arabidopsis* and rice genomes (*Arabidopsis* Genome Initiative, 2000; Goff *et al.*, 2002; Yu *et al.*, 2002). In yeast and mammals, two classes of mitochondrial transcription factors have been described: (i) the mtTFA proteins (also known as ABF2 and TFAM) harboring two HMG boxes and (ii) the mtTFB factors (mtTF1 in yeast, or TFB1M and TFB2M in human), which are related to rRNA methyltransferases (Schubot *et al.*, 2001; McCulloch *et al.*, 2002). While no mtTFA homologue has been detected, there are at least two good candidates for rRNA methyltransferase open reading frames (ORFs), i.e. At2g47420 and At5g66360 (Weihe, 2004). The latter has been found to encode a protein that is sorted to mitochondria, and both mtTFB-like factors display unspecific DNA-binding activity *in vitro*. Clearly more studies are necessary to elucidate the function of these proteins, or to identify, and characterize other mitochondrial transcription factors. Such factors could potentially mediate promoter-specific binding of the RNA polymerase and maybe certain factors bind to distinct promoter classes. However, it is possible that RNA polymerase initiates transcription alone, recognizing the real promoter, or at least preferred initiation sites such as the 'nonconserved promoters' (Kühn *et al.*, 2005).

Besides the proteins engaged in RNA synthesis and transcription initiation, a large number of genes encoding proteins with clear similarity to transcription termination factors is present in plants (Linder *et al.*, 2005). These proteins have initially been identified in human and other animals where they cause termination of transcription after the gene for the large ribosomal RNA (Kruse *et al.*, 1989; Loguercio Polosa *et al.*, 1999; Roberti *et al.*, 2003). In humans, this leucine zipper-containing factor is also known to stimulate transcription initiation at the so-called H1 site, contributing to an overrepresentation of rRNAs in the mtRNA steady-state pool. It simultaneously binds to the transcription initiation and termination sites in the same DNA molecule forming a loop, and this concurrent binding is a prerequisite for stimulation (Martin *et al.*, 2005). In *Arabidopsis* and rice, 33 and 27 genes respectively have been found to encode potential transcription termination factors which are, *in silico,* predominantly targeted to mitochondria and chloroplasts (Linder *et al.*, 2005). The function of these proteins in plants is so far unknown and could be termination of RNA synthesis at many different transcription units. However, it is also tempting to speculate that the plant proteins may also have an important role in transcription initiation, similar to the function of the human protein. A full analysis of this group of proteins will be necessary to elucidate their function.

3.3.4 Is gene expression in plant mitochondria regulated at the transcriptional level?

Most of the proteins encoded by the plant mitochondrial DNA are part of larger complexes that also contain nuclear-encoded subunits. It is thus assumed that the expression of the genes encoded in the different compartments is somehow coordinated to guarantee the generation of correct stoichiometric amounts of the different subunits. While many nuclear encoded genes are regulated on the transcriptional level, the regulation of mitochondrially encoded genes is less clear.

Different studies investigating the rate of transcription *in organello* have been performed with maize and *Arabidopsis* mitochondria. In one of the initial reports, it was found that in maize rDNA is transcribed at rates five- to tenfold higher than the proteins coding genes *atp*1, *atp*6 and *cox*2 (Finnegan and Brown, 1990). Similar results were obtained in a different report using the same system (Mulligan *et al.*, 1991). Again strongest transcription is found for rRNA genes, being 2- to 14-fold higher than protein-coding genes. In another report, the *rps*12 gene was found to be one of the genes that are most strongly transcribed. The relative transcriptional rates for several genes were found to be substantially different from the previous reports, most likely caused by the general difficulties with the *in organello* experiments (Muise and Hauswirth, 1992).

A comprehensive run-on transcription study, covering all assigned mitochondrially encoded genes in *Arabidopsis,* also detected distinct transcription rates among the individual genes, even if they encode components of the same multisubunit complexes (Giegé *et al.*, 2000). These differences are, at least partially, counterbalanced in the steady-state RNA pool, most likely by posttranscriptional processes and different RNA stabilities. In contrast to all previous reports, in which rRNA genes are generally among the most strongly transcribed, no enhanced transcription of these genes was found, although the 26S rRNA gene is among the strongest transcribed. However, it is clear that in the case of these structural RNAs the clear overrepresentation in the steady-state pool is a result of posttranscriptional processes, most likely increased stability (Giegé *et al.*, 2000). This is confirmed in another report investigating run-on transcription in *Arabidopsis*, *Brassica napus*, a cytoplasmic male sterile line obtained from protoplast fusion of both species as well as the fertility restored line of it (Leino *et al.*, 2005). Although rDNA transcription was found two- to fivefold higher than protein-coding genes, the enhanced overrepresentation of these RNAs in the steady-state pool is rather due to posttranscriptional events (Leino *et al.*,

2005). This report revealed that several parameters contribute to the final amounts of distinct RNAs in the steady-state pool. The comparison of mitochondrial transcriptional rates between *Arabidopsis* and *B. napus* identified species-specific rates for genes like *cox*1, *nad*4L, *nad*9, *ccm*B, *rps*7 and *rrn*5, which are most likely determined by different promoter strength in the mitochondrial DNA. However, some genes have different transcription rates in the alloplasmic line with the same promoter, revealing the influence of the nuclear background on the transcription rate. This also holds true for RNA turnover, which was found to be species-specific and dependent on the nuclear background. The counterbalancing mechanism seems to be less effective in the CMS line and is most likely responsible for this phenotype. Thus, it seems clear that both transcriptional rates and counterbalancing posttranscriptional mechanisms can be different between species, and both processes not only are due to *cis* elements in the mtDNA but also depend on nuclear factors.

The parameters determining promoter strength *in vivo* are unclear. *In vitro*, the exchange of certain nucleotide identities in the core promoter region strongly influences the strength of the promoter, and it seems reasonable that differing sequences within these regions determine the promoter strength *in vivo* (Rapp *et al.*, 1993; Hoffmann and Binder, 2002). However, there also appear to be long-range influences due to the genomic context. This has been found for two identical maize *cox*2 promoters, which are present in different genomic environments (Lupold *et al.*, 1999a). The promoters are located upstream of a direct repeat which is flanked downstream either by the *cox*2 gene or by a noncoding region. The promoter upstream of *cox*2 was disproportionally used at higher rates suggesting that promoter activity is influenced by its genomic context, indicating that genomic recombination may influence gene expression in plant mitochondria. How nuclear factors differentially influence transcription of different genes is completely unknown, but this observation indicates that gene-specific factors might exist.

Almost nothing is known about tissue-specific expression of mitochondrial genes and whether this might be achieved via transcriptional levels. A few reports indicated that there are tissue- or cell-specific differences and these might be due to transcriptional levels (Topping and Leaver, 1990; Li *et al.*, 1996); however, the recent comprehensive analysis of mitochondrial transcription initiation sites in *Arabidopsis* revealed no differences in the use of various promoters between leaves and flowers (Kühn *et al.*, 2005). Most likely, posttranscriptional processes play a dominant role for tissue-specific differences in the steady-state levels of mitochondrial transcripts (Monéger *et al.*, 1994; Smart *et al.*, 1994; Gagliardi and Leaver, 1999).

Coordinated regulation of gene expression in the nucleus and mitochondria at the transcriptional (or transcript) level has not yet been identified. A recent analysis, in which regulation of gene expression in response to sugar starvation was investigated, revealed that mitochondrial gene expression remains more or less unaffected at the transcriptional, as well as posttranscriptional, and translational levels (Giegé *et al.*, 2005). In contrast, the nuclear-encoded components of the ATPase are downregulated, becoming the rate-liming factor in the assembly of new complexes. Correct stoichiometric proportions seem to be achieved at the assembly processes itself, and the mitochondrial proteins probably present in excess are degraded. However, it is still unknown whether mitochondrial gene expression might respond or be regulated when other stimuli are applied (see Chapter 5).

Taken together, it is still unclear whether there is real regulation at the level of transcription initiation, i.e. change of transcription rates in response to certain stimuli, and whether these are gene-specific or rather affect up- and downregulation of transcription of the complete mitochondrial genome.

3.4 Splicing

3.4.1 Intron types

Organellar genomes from land plants are characterized by the presence of a relatively large number of group II introns compared to fungal-mitochondrial and bacterial genomes (Bonen and Vogel, 2001; Lambowitz and Zimmerly, 2004). The 20–23 group II introns found in higher plant mitochondria are all located in protein-coding genes. One noticeable exception is the presence of a group II intron in the *trnA* gene in several species including rough lemon (*Citrus jambhiri* Lush.) and the recently sequenced wheat mitochondrial genome (Ohtani *et al.*, 2002; Ogihara *et al.*, 2005). However, this *trnA* gene corresponds to an insertion of plastid DNA into the mitochondrial genome, and about one third of group II introns are located in tRNA genes in plastids.

Although the presence of numerous group II introns in higher plant mitochondrial genomes is the rule, several genera of land plants also contain a group I intron located in the *cox1* gene (Cho *et al.*, 1998). However, there is no direct correlation between phylogeny and the occurrence of this intron, indicating that it has been acquired by horizontal transfer, probably from a fungal donor. Remarkably, based on the frequency of the invasion of this group I intron, it has been estimated that this event occurred more than one thousand times during angiosperm evolution (Cho *et al.*, 1998).

3.4.2 Cis- *and* trans-*splicing*

Most plant mitochondrial introns are arranged in a *cis* configuration, i.e. the intron and flanking exons are encoded as a single RNA molecule. However, in some cases *trans*-splicing is also required to produce mature, functional mRNAs. Indeed, each of the *nad1*, *nad2* and *nad5* genes is fragmented in the mitochondrial genomes of flowering plants. The resulting RNA molecules, transcribed from the different loci, are joined by *cis*- and *trans*-splicing (Chapdelaine and Bonen, 1991; Knoop *et al.*, 1991; Pereira de Souza *et al.*, 1991; Sutton *et al.*, 1991; Wissinger *et al.*, 1991). For instance, the *nad5* gene is encoded by five exons (a–e): exons a and b as well as d and e are separated by an intron in *cis*, whereas two *trans*-splicing events allow the incorporation of the small exon c of 22 nt into *nad5* mRNA

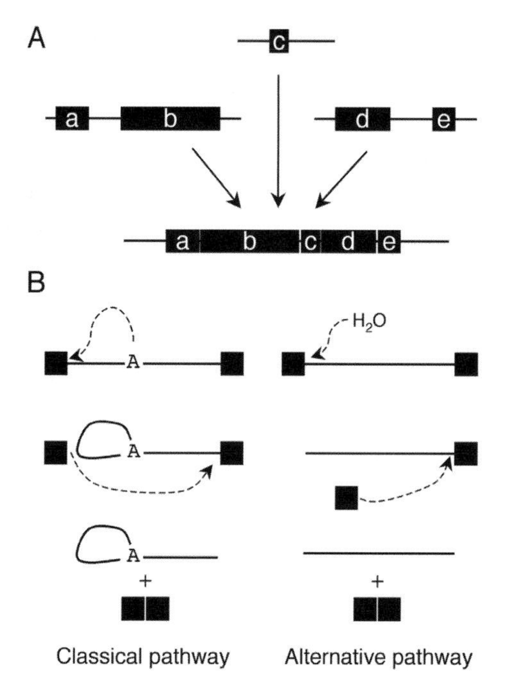

Figure 3.2 Intron splicing mechanism in plant mitochondria. Exons and introns are shown as thick and thin lines, respectively. (A) *Cis*- and *trans*-splicing of *nad5* transcripts. Mature *nad5* mRNAs are generated from three precursor molecules via two *cis*- and two *trans*-splicing events. (B) Possible splicing mechanism. In the classical pathway (left), the intron is spliced by two transesterification reactions and is released as a lariat. In the alternative pathway (right), the attack on the first exon–intro boundary is hydrolytic; the second step of the splicing reaction remaining unchanged. The intron is released as a linear molecule.

(Figure 3.2A). Mitochondrial *trans*-spliced genes evolved from genes containing usual *cis*-spliced introns (Malek and Knoop, 1998). Mitochondrial DNA recombination events likely led to the fragmentation of these ancestoral genes without affecting splicing capability.

3.4.3 Splicing mechanism

Group II introns are large catalytic RNAs organized in six domains (d1–d6) extending from a central hub (Michel and Ferat, 1995; Bonen and Vogel, 2001; Lambowitz and Zimmerly, 2004). However, the aforementioned *trans*-splicing phenomenon demonstrates that some introns disrupted in d4 are still functional, provided that the introns' secondary and tertiary structures can be maintained. The size of group II introns can vary from a few hundred nucleotides to several kilobases. Group II introns exhibit few conserved nucleotides but a number of features are frequently observed. These include specific interdomain interactions (e.g. d1-d5, d2-d3) and a bulged adenosine in d6 (Michel and Ferat, 1995; Bonen and Vogel, 2001; Lambowitz and Zimmerly, 2004). This bulged adenosine is involved in the conventional splicing mechanism involving two transesterification steps. First, the 2′ hydroxyl of the bulged adenosine attacks the 5′ splice site. Second, a nucleophile attack from the liberated 3′ hydroxyl group of the first exon is performed on the 3′ splice site. The intron is thus released as a lariat (Figure 3.2B). However, the bulged adenosine is missing in some plant mitochondrial introns, and therefore an alternative pathway must also operate. In this pathway, the first step is hydrolytic (i.e. water is the attacking nucleophile) rather than a transesterification, the second step remaining unchanged. In this alternative pathway, the intron is thus released as a linear molecule (Figure 3.2B). This alternative hydrolytic pathway was demonstrated as naturally occurring *in vivo* in the case of the chloroplast *trnV* intron (Vogel and Borner, 2002). As several plant mitochondrial introns lack the bulged adenosine in d6, the alternative hydrolytic pathway might also operate in plant mitochondria (Bonen and Vogel, 2001). In both pathways, specificity of the splicing junction is determined by short-sequence elements of the intron (named EBS for exon-binding site) that base pair with exon sequences (named IBS for Intron Binding Site) to position the active site of the intron.

Experimental approaches to study splicing mechanism have mainly been performed using certain yeast group II introns that can self-splice *in vitro*, albeit under nonphysiological conditions, and bacterial group II introns (Michel and Ferat, 1995). As none of the plant mitochondrial group II introns were shown to self-splice *in vitro*, the splicing mechanism was mostly inferred by sequence comparison between plant organelles and other genetic systems and by cloning splicing intermediates (Vogel and Borner, 2002). However, the splicing mechanism can now be directly experimentally investigated since methods for exogenous gene expression in isolated plant mitochondria have been developed (Farré and Araya, 2001; Koulintchenko *et al.*, 2003). To date, these *in organello* experiments using the *Triticum tymopheevi cox2* intron have demonstrated that the interaction between the EBS and IBS is essential for splicing and EBS and IBS can be functionally replaced by complementary sequences. Also, the d6 domain and the context of the bulged adenosine in d6 of this intron are required for efficient splicing (Farré and Araya, 2001, 2002). In addition, this technique enables study of the splicing of noncognate RNA substrates thereby allowing the future characterization of species-specific determinants of splicing (Staudinger and Kempken, 2003; Choury *et al.*, 2005). The influence of editing on splicing can also be tested in this system. Indeed, several editing sites have been characterized within introns and hypothesized to affect splicing efficiency by improving base pairing and helicity in domains d5 and d6 of group II intron. However, cloning of spliced introns that were not fully edited suggested that editing at some sites might not be an absolute prerequisite to splicing (Knoop *et al.*, 1991; Wissinger *et al.*, 1991; Binder *et al.*, 1992; Lippok *et al.*, 1994; Carrillo and Bonen, 1997; Carrillo *et al.*, 2001). This was demonstrated for the editing site present in IBS1 of the *Triticum timopheevi cox2* gene, which has no influence on splicing efficiency when expressed *in organello* (Farré and Araya, 2002). Only a limited number of such experiments have been performed to date, but already the results demonstrate the potential of this experimental approach to provide a better understanding of the splicing of mitochondrial group II introns.

3.4.4 Splicing machinery

Despite the catalytic nature of group II introns, splicing requires protein factors *in vivo*. These proteins can be classified into one of two groups: intron-encoded or host-encoded. Intron ORFs are common in bacterial and fungal group II introns and correspond to maturases involved in the splicing (usually intron-specific), and often the mobility of the intron (Lambowitz and Zimmerly, 2004). These maturases are rare in plant mitochondria and chloroplasts, which contain only one representative each, matR and matK, respectively. It has yet to be demonstrated whether matR or matK are involved in organelle-intron splicing. However, a functional role for both proteins is likely. Both matK and matR are conserved in plant species. matK is even present as a free-standing ORF in the genome of *Epifagus virginiana*, a non-photosynthetic plant with reduced plastid genome that lacks the *trnK* gene. matK could be necessary for the splicing of the remaining group II introns still present in this reduced genome (Ems *et al.*, 1995). In addition, disruption of chloroplast protein synthesis in barley mutants prevents the splicing of some group II introns (actually the ones related to the *trnK* intron) (Hess *et al.*, 1994; Hubschmann *et al.*, 1996; Vogel *et al.*, 1997, 1999). Unless another, so far unidentified, maturase is encoded in the chloroplast genome, it is likely that preventing matK synthesis results in the observed loss of splicing. Evidence of a functional role for matR is more circumstantial. In addition to its conservation among plant species, the *matr* gene has been shown to be produced from its own promoter in wheat (Farré and Araya, 1999) and its mRNA is edited at several positions that increase homology with fungal and bacterial maturases (Thomson *et al.*, 1994; Bégu *et al.*, 1998). These data support the hypothesis that matR is a functional protein in plant mitochondria.

Whether or not matR is involved in the splicing of one or several introns in plant mitochondria, it is now certain that other factors are encoded in the nucleus of higher plants. For instance, the *ms1* mutant in *N. sylvestris* is defective in the splicing of the first intron of the *nad4* transcript (Brangeon *et al.*, 2000). Although not identified yet, the existence of the *MS1* gene was the first proof that a nuclear-encoded factor can control a splicing event in plants. Several candidate genes have also been identified in plant nuclear genomes. First, there are four nuclear genes conserved between *Arabidopsis* and rice, which encode maturase-related proteins predicted to be imported into mitochondria (Mohr and Lambowitz, 2003). Second, among the four nucleus-encoded proteins (CRS1, CRS2, CAF1 and CAF2) involved in intron splicing in maize chloroplasts, three of them (CRS2, CAF1 and CAF2) have paralogs in the *Arabidopsis* nuclear genome predicted to be imported into mitochondria (Ostheimer *et al.*, 2003).

The investigation of all these candidate genes will likely lead to the characterization of nuclear-encoded factors involved in mitochondrial group II intron splicing.

3.5 Exchange of nucleotide identity by RNA editing

RNA editing is one of the most unusual posttranscriptional processes, which changes the genetic information content at the level of RNA. In general, this process is classified into two types: (i) processes that insert or delete nucleotides at distinct sites within an RNA precursor (deletion/insertion editing) and (ii) mechanisms which modify nucleotides identities within an RNA molecule (modification editing) (Grosjean and Benne, 1998; Brennicke *et al.*, 1999; Mulligan, 2004). RNA editing in plant mitochondria is a typical editing process of the modification type, in which cytidines (C) are most likely deaminated to uridines (U). Since its discovery in 1989 (Covello and Gray, 1989; Gualberto *et al.*, 1989; Hiesel *et al.*, 1989), RNA editing has been intensively studied in terms of its distribution within the plant kingdom, function and consequences, biochemical reaction mechanisms and the requirements for the specification of editing sites.

3.5.1 Distribution of RNA editing in the plant kingdom

The analysis of RNA editing in different lineages of the plant kingdom revealed several interesting features about its distribution. Analysis of the mitochondrial genome of *Chara vulgaris* and comparison of the deduced amino acids sequences with those of land plant mitochondria failed to provide any compelling evidence for the presence of potential editng sites in *C. vulgaris* mtDNA (Turmel *et al.*, 2003). Likewise no evidence for RNA editing has been found in *nad*5 sequences of other green algae (Steinhauser *et al.*, 1999) suggesting that mitochondrial transcripts are not edited in this algal lineage. This together with the observation that several species of the Marchantiales, which are considered to branch at the very early root of the land plant tree, displays RNA editing at least in mitochondria allowed the conclusion that this process is established in the early evolution of land plants (Malek *et al.*, 1996; Steinhauser *et al.*, 1999; Knoop, 2004). There are also substantial differences observed between different lineages of the land plants. While in mitochondria of seed plants C to U editing is found almost exclusively, the reverse direction of U to C exchanges is more frequently observed in nonseed plants; to an extreme extent in both mitochondria and chloroplasts of hornwort (Kugita *et al.*, 2003; Knoop, 2004; Joel Duff, 2006). In principle, both C to U and U to C could be performed by the same set of enzymes in reactions with antipodal directions. However, the preferential occurrence of U to C editing events in more primitive plants, especially in hornwort, suggests that these plants have a special apparatus for amination of uridines within an RNA (Mulligan, 2004).

The C to U editing observed in plant mitochondria is also found in plastids, although with much lower frequency (Mulligan, 2004). While the latter contain only about 30 different sites, between 400 and 500 editing sites are detected in mitochondria of a single plant (Giegé and Brennicke, 1999; Notsu *et al.*, 2002; Handa, 2003). Despite the difference in the number of sites, many common features suggest that some of the proteins required for RNA might be shared between the two organelles.

3.5.2 *Function and consequences of RNA editing*

The question of whether RNA editing has a particular function is still difficult to answer. Along with its discovery, it was realized that RNA editing alters the proteins sequence so that it is evolutionarily better conserved (Covello and Gray, 1989; Gualberto *et al.*, 1989; Hiesel *et al.*, 1989). The function behind such an increase in conservation has not yet been determined for plant mitochondria but has been elegantly demonstrated for chloroplasts. Bock *et al.* (1994b) reported that transplastomic tobacco was unable to edit a spinach-derived psbF editing site, which is pre-edited in the tobacco chloroplast genome. The requirement for editing is apparent from the photosystem II-deficient phenotype, with slow-growing pale-green plants. However, this obvious correction of 'mutations' would be dispensable if these mutations were to be eliminated at the genomic level, like other unfavorable mutations. Even if such mutations were present in a fraction of the mtDNA in one organelle, it would have no consequences. This is particularly substantiated by the fact that partially edited mRNAs are translated (Lu and Hanson, 1996; Phreaner *et al.*, 1996), but these translation products were never found at their final destination in the multisubunit respiratory complexes. Only proteins translated from fully edited mRNA are present in the complexes, indicating that mechanisms exist that prevent the incorporation of 'mutated' proteins (Grohmann *et al.*, 1994; Lu and Hanson, 1994). In addition, the functional requirement for RNA editing seems questionable considering the fact that several RNAs are found fully edited at a rate of a little more than 50%. Thus RNA editing, which requires an expensive enzymatic apparatus, might have other, as yet unidentified, functions.

Apart from its influence on codon capacity, RNA editing has other consequences for translation. It is clearly indispensable for translation by the creation of the AUG start codon from ACG. This has been found in several instances in both mitochondria and plastids and provides an additional control level for gene expression (Chapdelaine and Bonen, 1991; Hoch *et al.*, 1991). However, editing also creates stop codons and is thus required for translation of proteins with correct C-termini. Other essential RNA-editing events are observed in ferns in which internal, premature stop codons are eliminated by editing, thereby restoring conserved reading frames (Malek *et al.*, 1996).

C to U editing is also found in various tRNAs in different plant species (Maréchal-Drouard *et al.*, 1993; Binder *et al.*, 1994a; Maréchal-Drouard *et al.*, 1996a,b). The nucleotide identify is already changed in the precursor molecule and was later found to be a prerequisite for processing (detailed in Section 3.6.2) (Marchfelder *et al.*, 1996; Maréchal-Drouard *et al.*, 1996a,b; Kunzmann *et al.*, 1998). The function of another editing event, creating a noncanonical U–U pair in $tRNA^{Cys}$, is unclear. It is not necessary for further processing of the 5′ or 3′ extremities or aminoacylation of the tRNA. The uridine is later modified to a pseudo uridine, but the function of this modification is also unclear (Fey *et al.*, 2000, 2002). Studies of the tRNAs further demonstrated that other mismatches in the paired region of a tRNA are not corrected by RNA editing, indicating that mismatches might be not the sole parameter determining an editing site (Schock *et al.*, 1998).

RNA editing has also been analysed with respect to potential interaction with intron splicing (Sutton *et al.*, 1991; Yang and Mulligan, 1991). The analysis of unspliced and spliced *cox*2 transcripts in maize showed that there is no requirement for splicing for RNA editing and vice versa. As mentioned above (Section 3.4.4), editing of intron sequences is unlikely to be necessary for intron splicing. Thus there is no evidence for a distinct interaction or interdependence between these processes.

3.5.3 Mechanism and biochemistry of the RNA-editing process in higher plant mitochondria

The simplest way to form a uridine from a cytidine is a de- or transamination reaction. In this case the sugar backbone of the RNA molecule remains intact, and this is indeed what occurs in plant mitochondria (Rajasekhar and Mulligan, 1993; Yu and Schuster, 1995). Since no requirements for typical cosubstrates of transamination reactions are observed, it is most likely that a deaminase similar to the RNA-specific cytidine deaminase APOBEC-1 operating in the *apo*B RNA editing in mammals could be involved in modificational editing in plants. Several cytidine deaminases are encoded by a gene family in *Arabidopsis*, one of which has been analysed in detail. This protein, called AtCDA1, has cytidine deamination activity but does not seem to be localized in mitochondria and was therefore considered not to be involved in RNA editing in this organelle (Faivre-Nitschke *et al.*, 1999). Thus another cytidine deaminase, which strongly deviates from the classical type of this enzyme involved in nucleotide metabolism, must be considered for RNA editing. This cytidine deaminase cannot be identified in the *Arabidopsis* genome sequence by similarity searches, at least not at a first glance. In addition, biochemical data obtained with an *in vitro* editing system revealed that zinc, usually required by the classical type of cytidine deaminase and RNA-editing deaminases like APOBEC-1, seems not to be required for editing in plant mitochondria, indicating that an alternative type of protein may be involved (Takenaka and Brennicke, 2003).

No other proteins involved in plant mitochondrial RNA editing have been identified. The need for NTP in the *in vitro* editing reaction indicates the participation of an RNA helicase, an observation that was similarly made in studies of plastid editing *in vitro* (Takenaka and Brennicke, 2003; Hegeman *et al.*, 2005). However, the respective proteins have not been identified.

Hints toward the identity of such proteins might come from plastid editing. Besides proteins that have been cross-linked to editing sites containing RNAs (Hirose and Sugiura, 2001; Miyamoto *et al.*, 2004), a PPR protein was recently identified in a genetic approach searching for mutants impaired in plastid NDH activity (Kotera *et al.*, 2005). In the *ccr4* mutant, editing of the *ndh*D translation initiation codon (ACG to AUG) is specifically impaired, while other sites investigated remain unaffected. Although the exact function of CCR4 is unclear, this study provided clear evidence that this PPR protein is a site-specific factor required for the editing of a single site. Considering the large number of PPR proteins encoded in plants, it is possible that these might serve as specificity, or at least auxiliary proteins (Small and Peeters, 2000; Lurin *et al.*, 2004).

3.5.4 Upstream cis *elements specify editing sites*

If the PPR proteins are indeed some kind of specificity factors, it is still unclear how they specifically interact with distinct editing sites. The question of how the specificity is realized has been addressed using different experimental approaches. *In organello* systems, first established for wheat mitochondria (Farré and Araya, 2001), and later also for maize mitochondria (Staudinger and Kempken, 2003), have been shown to be useful to identify and characterize *cis* elements required for editing (Farré *et al.*, 2001; Staudinger *et al.*, 2005). Similarly, *in vitro* systems have been used to narrow down sequences essential for editing (Takenaka and Brennicke, 2003; Takenaka *et al.*, 2004). Essentially all these analyses revealed that sequences upstream of the editing site are generally important while distinct regions downstream are needed in some instances (Figure 3.3). In the wheat *cox*2 mRNA, two different regions were found to be critical for editing of site C259 – a 16-bp upstream element, which can be replaced by the same sequence element from another editing site, and a site-specific six-nucleotide 3′ element (Farré *et al.*, 2001). A similar result was found for site C77 in the same transcript, which is even edited when inserted into another sequence context indicating that further distant sequences seem to have no influence (Choury *et al.*, 2004). However, when a *Sorghum atp*6 transcript is expressed in maize mitochondria the mRNA is not edited at all, although the editing sites and their surrounding sequences are identical to the endogenous maize *atp*6 RNA. It was concluded that distant regions might influence the editing process (Staudinger and Kempken, 2003; Staudinger *et al.*, 2005). In pea, only *cis* elements upstream of an editing site in the *atp*9 mRNA seem to be important. Here about 20 nucleotides are essential, while a stretch of a further 20 nucleotides, up to position –40, are necessary for full editing efficiency (Takenaka and Brennicke, 2003; Neuwirt *et al.*, 2005). Identical upstream requirements are found for editing

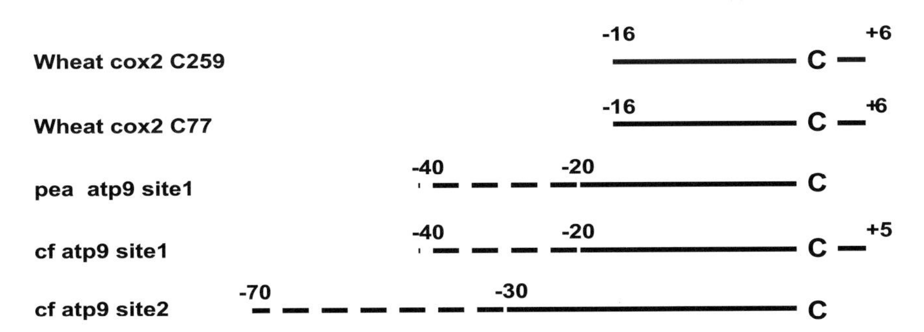

Figure 3.3 *Cis* element requirements of different RNA editing sites. Sequence elements essential for editing or stimulating regions that have been functionally characterized by *in organello* and *in vitro* editing systems (Farré *et al.*, 2001; Takenaka and Brennicke, 2003; Choury *et al.*, 2004; Takenaka *et al.*, 2004; Neuwirt *et al.*, 2005; van der Merwe *et al.*, 2006). Two sites in wheat *cox*2 (C259 and C77), two sites in cauliflower *atp*9 (cf, sites 1 and 2) and a single site (site 1) in pea *atp*9 have been investigated. Essential *cis* elements are shown as continuous lines, regions enhancing RNA-editing activity *in vitro* are shown as dashed lines. Sequences downstream of cauliflower *atp*9 site 2 have not been analysed. Numbering is given with respect to the editing site (C).

site 1 in the cauliflower *atp*9 mRNA, but in this case five nucleotides downstream of the editing site are also important. The exchange of individual nucleotide identities within these short elements leads to a reduction of about 80% of editing efficiency (Neuwirt *et al.*, 2005).

When editing sites are located very close to each other (within 30 nt), both sites carry independent 20–30 nucleotide-long specificity elements; however, the removal of sequences upstream of the 5′ located site also reduces editing efficiency of the downstream site (van der Merwe *et al.*, 2006). Similar site-specific requirements have also been found in plastids both by *in vitro* and *in vivo* analysis in transplastomic tobacco (Chaudhuri *et al.*, 1995; Bock *et al.*, 1996; Chaudhuri and Maliga, 1996; Reed *et al.*, 2001). This might further indicate that the editing mechanism and its specificity determinants are very similar in both plant organelles.

Upstream *cis* elements are likely to be recognized by site-specific factors that could be *trans*-acting RNAs, or *trans*-acting proteins, both directing the 'editosome' to the editing site. The nature of the specificity factors is still a matter of debate, but PPR proteins are currently considered to be the best candidates (Small and Peeters, 2000; Lurin *et al.*, 2004; Kotera *et al.*, 2005).

3.6 5′ and 3′ processing of plant mitochondrial transcripts

Although a few 5′ extremities of mitochondrial transcripts correspond to transcription initiation sites, and therefore do not require further 5′ processing, there is no example to date that a mature 3′ extremity is generated by transcription termination (despite the probable presence of mTERF homologues in plant mitochondria, see Section 3.3.3). Therefore, probably all mitochondrial RNAs, whether tRNAs, rRNAs or mRNAs, have to be processed from longer precursor molecules.

3.6.1 5′ and 3′ processing machinery

Theoretically, three types of activities could be responsible for generating mature 5′ and/or 3′ extremities from larger precursor RNAs: 5′ to 3′ exoribonucleases, endoribonucleases and 3′ to 5′ exoribonucleases. Numerous examples of processing of tRNAs, rRNAs and even dicistronic mRNAs demonstrate the involvement of endonucleases in maturation of 5′ and 3′ ends in plant mitochondria. However, to date, no gene encoding a plant mitochondrial endonuclease has been identified.

The existence of mitochondrial 3′ to 5′ exoribonucleases has also been inferred from *in vitro* processing or degradation experiments (Bellaoui *et al.*, 1997; Dombrowski *et al.*, 1997; Gagliardi and Leaver, 1999; Gagliardi *et al.*, 2001). Recently, two genes encoding exonucleases with homologies to bacterial polynucleotide phosphorylase (PNPase, At5g14580) and RNaseII (At5g02250) have been characterized in *Arabidopsis* (Perrin *et al.*, 2004b). While mtPNPase is localized exclusively in mitochondria, RNaseII is targeted to both mitochondria and chloroplasts (Perrin *et al.*, 2004b; Bollenbach *et al.*, 2005). The mitochondrial PNPase is essential for viability in contrast to another PNPase-like protein present in

Arabidopsis encoded by the *At3g03710* gene and targeted to chloroplast (Walter *et al.*, 2002). Conversely, RNaseII seems dispensable for mitochondrial function whereas it is required for proper chloroplast biogenesis (Bollenbach *et al.*, 2005). RNaseII is able to degrade unstructured RNA and plays a role in the final trimming of 3′ extremities (Perrin *et al.*, 2004b; Bollenbach *et al.*, 2005). PNPase appears to be a key player in plant mitochondrial RNA maturation and degradation (Perrin *et al.*, 2004a,b; Holec *et al.*, 2006). Both processes seem to be triggered by polyadenylation in plant mitochondria (detailed below).

To date, there is no experimental evidence that a 5′ to 3′ exoribonuclease exists in plant mitochondria. In fact, all classes of RNA that have been characterized as quickly turned over in wild-type plants are stabilized in the absence of PNPase (Perrin *et al.*, 2004a,b; Holec *et al.*, 2006). A 5′ to 3′ exoribonuclease cannot, therefore, compensate for the lack of PNPase, and thus its existence is doubtful. Nevertheless, this hypothesis is moderated by two possibilities: (i) the hypothetical 5′ to 3′ exoribonuclease could be in association with PNPase, and downregulation of the latter would result in the disappearance of both proteins or (ii) the putative 5′ to 3′ exoribonuclease could have specific, not yet identified substrates.

Clearly, a complete characterization of the processing machinery is required for a proper understanding of 5′ and 3′ processing events in plant mitochondria. However, as discussed below, progress is being made toward the identification of specific *cis*- or *trans* factors that specify endonucleolytic cuts or stop 3′ to 5′ exonucleolytic digestion.

3.6.2 tRNA processing

Production of a mature functional tRNA requires several posttranscriptional processes that include maturation of the 5′ and 3′ extremities, nucleotide modifications (including C to U editing in some cases) and addition of the 3′ terminal CCA nucleotides that are not genomically encoded in plant mitochondria. Several of these processes have been investigated in detail using efficient *in vitro* maturation systems.

Plant mitochondrial *trn* genes are transcribed either as a single unit or as clusters of *trn* genes, or even cotranscribed with protein-coding genes. In any case, both 5′ and 3′ extremities are generated by endonucleolytic cuts as shown using wheat, *Oenothera*, potato and pea mitochondrial protein extracts (Hanic-Joyce and Gray, 1990; Marchfelder *et al.*, 1990; Marchfelder and Brennicke, 1994; Marchfelder *et al.*, 1996; Maréchal-Drouard *et al.*, 1996a; Kunzmann *et al.*, 1998). An endonucleolytic cut occurring precisely at the 5′ end of the mature tRNA is performed by an RNase P-like activity, provided that the tRNA is folded in its typical clover-leaf structure in the precursor. Maturation of the 3′ end of the tRNA is performed by two distinct activities. First, an RNase Z-type endonuclease cuts the tRNA precursor exactly 3′ to the discriminator nucleotide, or one nucleotide downstream (Kunzmann *et al.*, 1998). Second, the 3′ terminal CCA nucleotides are added by a nucleotidyl transferase activity that has been observed in several plant mitochondrial extracts (e.g. Hanic-Joyce and Gray, 1990; Kunzmann *et al.*, 1998). It seems that there is no general rule about the order of 5′ and 3′ processing since, depending on the substrate, and/or on the species, both events are either independent, or the 5′ processing has to precede 3′ maturation (Hanic-Joyce and Gray, 1990; Kunzmann *et al.*, 1998). Due to the presence of tRNA-like structures in some plant mitochondrial transcripts, the tRNA processing machinery might also be involved in the maturation of other types of RNAs (Hanic-Joyce *et al.*, 1990; Spencer *et al.*, 1992).

C to U conversions were detected in a restricted number of plant mitochondrial tRNAs and were also referred to as editing by analogy to the numerous C to U conversions observed for plant mitochondrial mRNAs. Editing corrects mismatches in the acceptor stem of $tRNA^{Phe}$, and the different paired region of $tRNA^{His}$ in larch, while in $tRNA^{Cys}$ of *Oenothera* and potato C to U editing creates a noncanonical U–U mismatch from a C–U mismatch (Maréchal-Drouard *et al.*, 1993, 1996b; Binder *et al.*, 1994a). Editing of both $tRNA^{Phe}$ and $tRNA^{His}$ is an absolute prerequisite for 5′ and 3′ processing, probably because in absence of editing the tRNA precursor cannot adopt the proper structure recognized by the tRNA processing machinery (Marchfelder *et al.*, 1996; Maréchal-Drouard *et al.*, 1996a,b; Kunzmann *et al.*, 1998). These results were obtained *in vitro* but were recently confirmed by expressing either an edited or unedited versions of the larch $tRNA^{His}$ in transformed potato mitochondria. The unedited version of the larch $tRNA^{His}$ was expressed but not edited (likely because of the lack of the corresponding larch *trans* factor in potato mitochondria), and only the edited version could undergo maturation *in organello* (Placido *et al.*, 2005). There is one case in which editing is not necessary for processing as the C to U conversion in $tRNA^{Cys}$ could represent a first step in the modification of this nucleotide which is further changed to a pseudouridine (Fey *et al.*, 2000). None of the classical (i.e. non C-to-U editing) nucleotide modifications present *in vivo* seem to be essential for processing as the *in vitro* processing experiments were performed with *in vitro* transcribed transcripts.

Surprisingly, not a single gene involved in any of the different steps of plant mitochondrial tRNA processing has been cloned. However, several candidate genes, for instance encoding putative mitochondrial nucleotidyl transferases, have now been identified (Martin and Keller, 2004).

3.6.3 rRNA processing

Plant mitochondrial genomes contain three *rrn* genes coding for 26S rRNA, 18S rRNA and 5S rRNA. The gene copy number varies between plant species, but in all cases 18S rRNA and 5S rRNA are cotranscribed whereas the 26S rRNA is produced from a distinct locus (see Section 3.3.1).

The 5′ mature extremity of rRNA can be established by transcription initiation as in the case of the potato 26S rRNA (Binder *et al.*, 1994b). However, transcription is frequently initiated upstream of the mature rRNA. For instance, transcription starts at ca 900 and 150 nt upstream of the mature 5′ ends of the 26S and 18S rRNAs, respectively, in *Arabidopsis* (there are additional, but less active, promoters for the *rrn18* gene) (Kühn *et al.*, 2005). The maturation of the 5′ ends is then likely performed by a yet unidentified endonuclease. Such endonucleolytic cuts have recently been mapped for the 18S and 5S rRNAs in *Arabidopsis* (Figure 3.4).

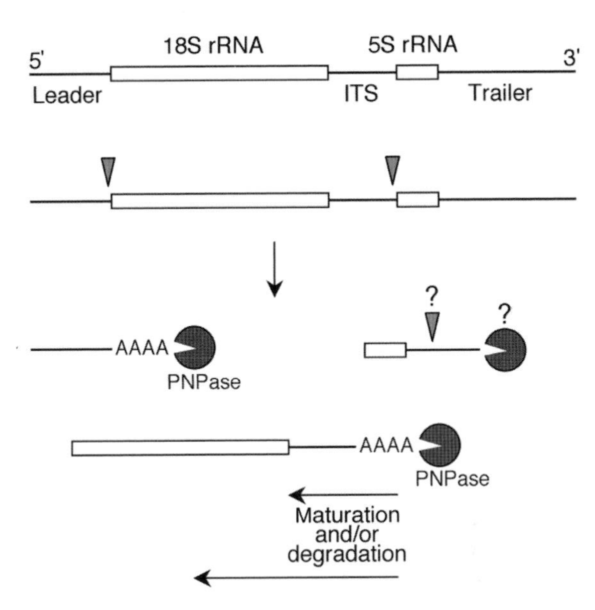

Figure 3.4 Maturation of 18S and 5S rRNAs in *Arabidopsis*. The primary 18S-5S rRNA transcript is composed of a leader sequence, the 18S rRNA, an ITS, the 5S rRNA and a trailer sequence of undefined length. Endonucleolytic cleavages are indicated by triangles. The RNA fragments generated by endonucleolytic cuts are polyadenylated and subsequently degraded or matured by PNPase. The 5S rRNA maturation pathway is unknown at present and could involve endonuclease(s) and/or exonuclease(s).

For both rRNAs, the endonuclease(s) cuts either exactly at the mature 5′ end of the rRNA or between one or two nucleotides upstream of this mature end (Perrin *et al.*, 2004a).

Very little information is available regarding 3′ end maturation of rRNA. As transcription proceeds beyond the 26S and 5S rRNAs, their mature 3′ ends could be produced either by sole exonucleolytic trimming or by a combination of endo- and exonuclease activities. To date, no evidence for a precise endonucleolytic cut to generate the 3′ end of any plant mitochondrial rRNA has been reported. However, much more is known about the maturation of the 18S rRNA 3′ extremity. The primary transcript contains the leader sequence, the 18S rRNA, the intergenic transcribed sequence (ITS) between the 18S and 5S rRNAs, the 5S rRNA and a trailer sequence of unknown length (Figure 3.4). Endonucleolytic cuts are made at the 5′ ends of both the 18S and 5S rRNAs to generate a secondary precursor corresponding to the 18S rRNA and the ITS. The length of the ITS can vary from plant to plant but is typically between 100 and 150 nt. When PNPase, the major mitochondrial exoribonuclease, is downregulated, the ITS is no longer removed from the 18S rRNA precursor (Perrin *et al.*, 2004a). This strongly suggests that PNPase is essential for initiating the exonucleolytic maturation of the 18S rRNA 3′ extremity. It is unknown at present whether PNPase is sufficient for, or only initiates, this maturation step. However, it is unlikely to be the sole player in this maturation step, as it is expected that a *trans* factor is required to stop the progression of the PNPase, and thereby define the 3′ mature end of the 18S rRNA. Although not supported by any experimental data, such *trans* factors could be the ribosomal proteins themselves during ribosome assembly. This mechanism would provide an elegant way to adjust the amounts of ribosomal proteins and rRNA necessary for ribosome biogenesis.

3.6.4 mRNA processing

In plant mitochondria coding sequences are highly conserved between plant species, mitochondrial genes being some of the most slowly evolving genes. By contrast, 5′ and 3′ untranslated region (UTR) are extremely diverse. This situation is in striking contrast with yeast mitochondrial mRNAs that all end with a conserved dodecamer sequence involved in 3′ processing (Hofmann *et al.*, 1993). In addition, plant mitochondrial mRNA profiles can be complex depending on the plant species. This complexity is explained by the existence of several mature 5′ and 3′ ends, the presence of unspliced introns and polycistronic transcripts that exhibit diverse levels of processing into distinct, monocistronic mRNA.

3.6.4.1 Processing of 5′ extremities

The 5′ end of mRNA can be generated either by transcription initiation or by processing. Distinction between these two possibilities is therefore necessary to map a true 5′ processing site. Primary and processed transcripts can be discriminated because of the presence of tri- or monophosphates, respectively, at their 5′ ends. Experimentally, the distinction is achieved by *in vitro* capping experiments using guanylyltransferase, or by a modified 5′ RACE polymerase chain reaction (PCR) technique (Kühn *et al.*, 2005).

As mentioned above, no experimental data support the existence of a 5′ to 3′ exoribonuclease in plant mitochondria. Therefore, processing of 5′ extremities from primary transcripts is most likely performed by endoribonuclease(s). In any case, mRNAs processed from multicistronic transcripts must have their 5′ extremities generated by endoribonucleases. Maturation by endonucleolytic cuts implies that these cleavages are precisely controlled, and therefore *cis* elements must be recognized by the endonucleases. The *cis* elements are either sequence and/or structural elements. For instance, the tRNA-processing machinery could be involved in the maturation of cotranscribed mRNAs and tRNAs, or tRNA-like genes (e.g. Makaroff *et al.*, 1989; Singh and Brown, 1991). In this case, the structure of the tRNAs or tRNA-like genes would be sufficient to promote maturation of the cotranscribed mRNA. However, primary sequences are also likely to be determinants for 5′ processing for most transcripts. Given the diversity of 5′ UTR sequences, a common motif that could be involved in general 5′ processing is unlikely. Nevertheless, short conserved sequences were occasionally observed 5′ to several transcripts such as 26S rRNA and mRNAs (Maloney *et al.*, 1989; Schuster and Brennicke, 1989). Processing at these putative sites could now be tested using the recent techniques developed for expressing foreign DNA sequences *in organello* (Farré and Araya, 2001; Koulintchenko *et al.*, 2003).

There are numerous examples of 5′ processing of a particular mRNA or a small set of transcripts that are dependent on a nuclear gene or locus (e.g. Singh *et al.*, 1996; Ohtani *et al.*, 2002; Hanson and Bentolila, 2004; Wang *et al.*, 2006). Therefore, the *cis* elements involved in 5′ processing are likely recognized by specific factors encoded in the nucleus. The identification of these specific factors was initiated while studying the effect of restorer genes on the expression of CMS-associated genes. Although restorer genes can affect mitochondrial translation and protein stability, novel transcript patterns are often observed in restorer background (Hanson and Bentolila, 2004). For instance, the processing of CMS-linked transcripts is affected upon restoration in maize CMS-S and CMS-T, *Brassica nap* and *pol* CMS, petunia, sorghum and rice CMS (Pruitt and Hanson, 1991; Singh and Brown, 1991; Iwabuchi *et al.*, 1993; Tang *et al.*, 1996; Wise *et al.*, 1996; Dill *et al.*, 1997; Li *et al.*, 1998; Pring *et al.*, 1998; Menassa *et al.*, 1999; Wen and Chase, 1999; Wise *et al.*, 1999). It should be noted that the observed altered transcript profile may or may not be directly responsible for fertility restoration itself (Hanson and Bentolila, 2004). Also, when the processing of several transcripts is affected upon restoration, it is unclear whether the maturation of all the transcripts is dependent on a single gene or if several genes affecting processing are closely linked at the same locus (e.g. Singh *et al.*, 1996; Li *et al.*, 1998; Tang *et al.*, 1998; Bentolila *et al.*, 2002; Akagi *et al.*, 2004). Members of the PPR protein family are good candidates for promoting the alternative transcript pattern observed upon restoration of fertility in some CMS systems (Hanson and Bentolila, 2004). Indeed, a PPR protein is responsible for the processing of CMS-associated RNAs in CMS BT (Boro II) rice (Kazama and Toriyama, 2003; Akagi *et al.*, 2004; Komori *et al.*, 2004; Wang *et al.*, 2006). Although a genetic link between PPR proteins and altered transcript profile is now established, it remains to be demonstrated at the molecular level how the PPR proteins affect the processing event.

Processing can also be influenced by sequences upstream of 5′ extremities. Indeed, although the mature *cox3* main 5′ end is conserved between L*er*, Col and C24 *Arabidopsis* ecotypes, an additional 5′ extremity is observed in C24 (Forner *et al.*, 2005). Because the immediate sequence region of this novel extremity is identical in the three ecotypes, it was expected that the nuclear background was responsible for this discrepancy. Surprisingly, the appearance of this additional 5′ extremity is maternally inherited, and therefore, likely due to the mitochondrial genome. Because the sequence of the mitochondrial genomes differs by only about 140 bp upstream of the additional 5′ end, these results indicate that upstream sequences can influence 5′ processing by a yet unknown mechanism.

In conclusion, the intricate mechanism of 5′ processing of plant mitochondrial mRNAs begins to be unravelled. The 5′ ends are likely matured by a yet unidentified endoribonuclease. Specificity factors are likely to be members of the PPR family. Such PPR proteins could specify processing of a single or limited number of transcripts. Several of these types of PPR proteins could be encoded by genes closely linked at a same locus. In addition, mitochondrial sequences upstream of the 5′ ends could in some cases influence the processing site.

3.6.4.2 Processing of 3′ extremities

Processing of 3′ extremities is still a puzzling phenomenon in plant mitochondria. In theory, 3′ ends could be generated by transcription termination, by endoribonuclease(s) and/or exoribonuclease(s). Recently, mTERF homologues predicted to be imported in both chloroplasts and mitochondria have been identified in *Arabidopsis* (see Section 3.3) (Linder *et al.*, 2005). Since mTERF is a transcription terminator in animal mitochondria, it is possible that the *Arabidopsis* mTERF homologues maintain this function and terminate transcription in plant mitochondria. Therefore, creation of a mature 3′ end by transcription termination can be envisaged but has not yet been demonstrated. On the contrary, experimental data suggest so far that transcription proceeds well beyond what will become a mature 3′ end. This was first hinted at by the fact that large regions beyond *rrn* genes can be transcribed *in organello* and *in vivo* (Finnegan and Brown, 1990). Naturally, reinitiation of transcription beyond the *rrn* genes could be an alternative explanation of these results. However, *in organello* rates of transcription immediately 5′ and 3′ to the mature end region are identical for the *Arabidopsis rrn26* gene, suggesting no termination of transcription at the mature 3′ end of the 26S rRNA (S. Holec and D. Gagliardi, unpublished results). Moreover, *in vitro* transcription experiments have demonstrated that a double-stem loop structure present at the 3′ end of pea *atp9* is not an efficient transcription terminator (Dombrowski *et al.*, 1997). In addition, large pretranscripts for *atp9* and *orfB* could be detected in wild-type *Arabidopsis* mitochondria (Perrin *et al.*, 2004b). Taken together, these data indicate that either transcription likely proceeds beyond the mature 3′ ends of mitochondrial RNAs or if transcription termination occurs, this is a rather inefficient process at, or near, the mature ends of RNAs.

Consequently, 3′ ends are, in all probability, matured from larger precursors. The large *atp9* and *orfB* mRNA precursors detected at low abundance in wild-type plants accumulate in the absence of PNPase (Perrin *et al.*, 2004b). Therefore, PNPase is clearly involved in the degradation of these transcripts. However, what remains to be demonstrated is whether PNPase is acting only in removing these ‘aberrant’ transcripts (see Section 3.7.3) or if this exonucleolytic degradation is an integral part of the maturation pathway (Perrin *et al.*, 2004b). Maturation of transcripts from large precursors by PNPase has recently been demonstrated to occur in chloroplasts (Walter *et al.*, 2002). Also supporting the hypothesis of exonucleolytic maturation is the fact that a putative maturation by-product created by an endoribonucleolytic cut close to the mature end was not detected in PNP-plants. However, it cannot be excluded that such maturation by-products could be quickly turned over independently of PNPase. Therefore, although the involvement of PNPase in 3′ maturation processes is possible, participation of endoribonuclease cannot be ruled out at this stage.

Whether the first step of maturation is exonucleolytic or endonucleolytic, some mRNAs may require a second step in 3′ processing consisting of exonucleolytic trimming as was initially suggested by *in vitro* processing experiments using pea *atp9* (Dombrowski *et al.*, 1997). Indeed, this was shown *in vivo* for *atp9* mRNA in *Arabidopsis* (Perrin *et al.*, 2004b). In the absence of RNaseII, about half of the *atp9* mRNA molecules harbor small extensions of nearly 10–12 nucleotides. These extensions are located downstream of a double-stem loop structure predicted to fold at the end of the mature *atp9* mRNA. *In vitro* experiments demonstrated that purified RNaseII is able to remove the extensions efficiently thereby generating the mature 3′ end (Perrin *et al.*, 2004b). Therefore, RNaseII is probably involved in trimming small nucleotide extensions left after the first maturation step. RNaseII is also targeted to chloroplasts where it performs the same molecular role of trimming precursors (Bollenbach *et al.*, 2005).

Extremities at the 3′ end must be defined by sequence or structural elements recognized by the processing machinery. Single- or double-stem loop structures can be found at the 3′ ends of certain mRNAs. When present, these structures clearly influence processing and stability of the mRNAs by stopping 3′ to 5′ exonucleases (Kaleikau *et al.*, 1992; Bellaoui *et al.*, 1997; Dombrowski *et al.*, 1997; Kuhn *et al.*, 2001; Perrin *et al.*, 2004b). However, it is unlikely that these secondary structures are the sole determinant for 3′ processing as numerous equally stable secondary structures exist in the large precursors and these are not recognized as processing signals. In addition, numerous mRNAs in plant mitochondria do not harbor any significant predictable secondary structure at their mature 3′ ends. It is therefore likely that specific *trans* factors are needed to specify 3′ mature ends, but to our knowledge none have been characterized yet.

Following processing, the 3′ extremities of transcripts can still be modified by the addition of nongenomically encoded nucleotides. This phenomenon was observed while characterizing the 3′ extremities of transcripts following RNA circularization or ligation of an anchor oligonucleotide prior to reverse transcription and PCR amplification (e.g. Williams *et al.*, 2000; Kuhn *et al.*, 2001; Kuhn and Binder, 2002). Although only a few mitochondrial RNAs have been analysed in this way, it is clear that some mRNAs can contain a few nongenomically encoded nucleotides at their 3′ extremities. For instance, about 95% of clones corresponding to truncated maize *rps12* mRNA contained one to four nongenomically encoded C or A residues (Williams *et al.*, 2000). About 60% of maize truncated *cox2* mRNA were also modified by the addition of one to five nucleotides, mostly C and A but also G and U nucleotides (Williams *et al.*, 2000). Addition of nongenomically encoded nucleotides was also observed at mature 3′ extremities of maize and pea *atp9* transcripts (Williams *et al.*, 2000; Kuhn *et al.*, 2001). It is unclear at present whether these added nucleotides have any biological roles. These modifications could reflect the accessibility of the 3′ ends to the modifying enzyme(s). The prevalence of CCA triplets suggests that some of these nucleotide extensions could be generated by the nucleotidyl transferase involved in the final maturation step of tRNA 3′ extremities.

3.7 Stabilization and degradation of RNA

In any genetic system, steady-state levels of functional RNA are defined by an equilibrium between the rates of transcription and degradation. Although transcription rates can vary severalfolds between plant mitochondrial genes, depending on their respective promoter strength, RNA stability is crucial in determining respective transcript steady-state levels. Indeed, there is no strict correlation between *in organello* run-on transcription rates of mitochondrial genes and steady-state levels of the corresponding transcripts in *Arabidopsis* and *B. napus* (Giegé *et al.*, 2000; Leino *et al.*, 2005) (see Section 3.4). Differential stability is expected when comparing rRNA and mRNA but is also observed between different mRNAs. As expected, *cis*- and *trans* factors are required to specify stability and turnover of transcripts.

3.7.1 Cis *and* trans *elements stabilizing mRNA*

Secondary structures forming single or double-stem loops are present at the 3′ ends of certain mitochondrial mRNAs. These structures play a role as processing signals but can also influence stability. Indeed, in rice and an alloplasmic wheat line, stability of *cob* mRNA is influenced by the presence or absence of such secondary structures at the 3′ end of alternative transcripts (Saalaoui *et al.*, 1990; Kaleikau *et al.*, 1992). This phenomenon was also observed for *orf138* mRNA in rapeseed (Bellaoui *et al.*, 1997), and stabilization of pea *atp9* transcripts by a terminal secondary structure has been shown *in vitro* (Dombrowski *et al.*, 1997; Kuhn *et al.*, 2001). It is likely that secondary structures present at the 3′ extremities impede the progression of 3′ to 5′ exoribonucleases such as RNaseII and PNPase. Different secondary structures at transcripts 3′ ends could therefore explain some observed differences in stability between mRNAs. However, secondary structure cannot be the sole feature responsible for stabilizing a transcript. Indeed, some transcripts are stable without exhibiting obvious, predictable stable secondary structure at their 3′ end. Also, the stability of a given transcript can depend on the nuclear background, and/or on the tissue considered, indicating the involvement of specific *trans* factor(s). For instance, the accumulation of CMS-associated transcripts in sunflower and *B. napus* is preferentially reduced in anthers of fertility-restored plants (Monéger *et al.*, 1994; Geddy *et al.*, 2005). Also, in CMS plants obtained by *Arabidopsis* and *B. napus* protoplast fusion, the stability of certain mitochondrial *Arabidopsis* transcripts is affected by the *B. napus* nuclear genome (Leino *et al.*, 2005). Recently, a PPR protein affecting the stability of the CMS-associated transcript in rice Boro II CMS (*B-atp6* mRNA) was characterized (Wang *et al.*, 2006). Whether this PPR protein actively destabilizes the *B-atp6* transcript, or corresponds to a defective allele of a stabilizing factor, is still unresolved. Although its precise mechanism of action is unknown, this PPR protein represents the first example of a *trans* factor affecting the stability of a mitochondrial transcript.

3.7.2 Mechanism of RNA degradation

The general machinery for RNA degradation must act in concert with factors specifying stability or instability. As mentioned above, there are several examples of transcript-specific destabilization in CMS systems. These preferential degradations could be due to specific *trans* factors that actively initiate degradation, for instance by specifying endonucleolytic cuts in a particular transcript. Alternatively, the preferential degradation could be a consequence of a defective function such as failure to initiate translation, or to stabilize a particular transcript. The precise mechanism of destabilization is likely to be solved in the near future once the *trans* factor has been identified such as the rice PPR protein resulting in *B-atp6* transcript destabilization (Wang *et al.*, 2006).

In recent years, progress has been made in understanding a general mechanism of RNA degradation in plant mitochondria. This mechanism involves polyadenylation and PNPase. By contrast to nuclear-encoded mRNAs, mature, functional mRNAs are not constitutively polyadenylated in plant mitochondria. Nevertheless, addition of poly(A) tails has been reported for a number of mRNAs in different plant species such as sunflower, maize, pea, *Oenothera* and *Arabidopsis* (Gagliardi and Leaver, 1999; Lupold *et al.*, 1999b; Gagliardi *et al.*, 2001; Kuhn *et al.*, 2001; Perrin *et al.*, 2004a,b; Holec *et al.*, 2006). Both *in vivo* and *in vitro* experiments revealed that poly(A) tails can trigger RNA degradation (Gagliardi and Leaver, 1999; Gagliardi *et al.*, 2001; Kuhn *et al.*, 2001; Perrin *et al.*, 2004a,b) as previously reported in *Escherichia coli* and chloroplasts (Dreyfus and Regnier, 2002; Bollenbach *et al.*, 2004). Although some internal sites of polyadenylation have been characterized in plant mitochondrial transcripts, the majority of poly(A) sites are observed at, or near, 3′ ends of mature RNAs in wild-type plants (Gagliardi and Leaver, 1999; Lupold *et al.*, 1999b; Gagliardi *et al.*, 2001; Kuhn *et al.*, 2001; Perrin *et al.*, 2004a,b). When stable stem loops are present at the 3′ ends, repetitive cycles of polyadenylation-degradation are needed to overcome these secondary structures (Kuhn *et al.*, 2001). However, stable secondary structures do not seem to impede the progression of 3′ to 5′ exonuclease(s) when present within the RNA substrate (Gagliardi *et al.*, 2001). This information coupled with the fact that poly(A) sites are found mainly at 3′ ends indicate that mRNA degradation could be performed by an efficient 3′ to 5′ exonucleolytic pathway. Of course, endonuclease(s) are also likely to be involved in mitochondrial RNA degradation but have not yet been characterized.

As polyadenylated mRNAs have been characterized in several plant species including mono- and dicotyledons, polyadenylation appears conserved between plant species. The exact ratio of polyadenylated versus nonpolyadenylated mRNA has not been determined so far in wild-type plants. However, it is thought to be rather low as polyadenylated RNAs were not, or rarely, observed when cloning 3′ extremities of mRNAs. The size and composition of mitochondrial poly(A) tails have not been determined precisely due to this low abundance of polyadenylated RNAs in wild-type plant mitochondria. Upon downregulation of PNPase, about 80% of unprocessed *atp9* and *orfB* mRNAs harbored poly(A) tails from 5 to 17 nucleotides in length, mostly composed of adenosines (Perrin *et al.*, 2004b). However, these data were obtained by RT-PCR following circularization of the RNA with T4 RNA ligase, and it cannot be excluded that this experimental approach displays a bias for shorter poly(A) tails. In addition, it is possible that the sizes of poly(A) tails obtained could be influenced by the absence of PNPase.

It is likely that poly(A) tails ‘attract’ PNPase to initiate degradation. This has not been demonstrated empirically with the plant mitochondrial PNPase but can be reasonably hypothesized from studies performed with the bacterial and chloroplastic PNPases (Lisitsky *et al.*, 1997; Yehudai-Resheff *et al.*, 2003). In addition, polyadenylated RNAs accumulate upon downregulation of PNPase, suggesting that PNPase can degrade polyadenylated RNAs (Perrin *et al.*, 2004a,b; Holec *et al.*, 2006). It is unknown at present whether poly(A) tails perform a role(s) other than triggering RNA degradation.

The enzyme(s) responsible for polyadenylating mitochondrial transcripts has not yet been identified. However, recent bioinformatics studies identified several genes encoding candidate poly(A) polymerases predicted to be targeted to plant organelles (Martin and Keller, 2004). Functional studies should reveal whether these candidates are truly involved in plant mitochondrial RNA polyadenylation. Inactivation of the poly(A) polymerase gene(s) will also reveal whether or not a poly(A) independent pathway for RNA degradation exists in plant mitochondria. This is likely to be the case for tRNAs as although polyadenylation of mature tRNAs is known in chloroplast (Komine *et al.*, 2000), it has not been detected so far in plant mitochondria.

There is no evidence that RNaseII participates in RNA degradation in *Arabidopsis*. However, RNase II degrades poly(A) tails efficiently, and therefore could possibly have an antagonist role to PNPase, i.e. RNaseII could stabilize PNPase RNA substrates (Perrin *et al.*, 2004a). Such a phenomenon has been described in *E. coli* (Marujo *et al.*, 2000; Mohanty and Kushner, 2003). Also, there is no reported evidence for a 5′ to 3′ degradation pathway in plant mitochondria either by a 5′ to 3′ exonuclease or by a 5′ to 3′ wave of endonucleolytic cuts. However, to our knowledge this possibility has not been tested experimentally.

3.7.3 Roles of RNA degradation

RNA degradation is not only important to specify steady-state levels of mRNAs and rRNAs but also to eliminate maturation by-products. Indeed, processing of mitochondrial transcripts from longer precursors generates RNA species such as transcribed intergenic regions, leader and trailer sequences, and introns. PNPase appears indispensable in removing 18S and 26S rRNA, and tRNA maturation by-products as these RNAs accumulate strongly upon downregulation of PNPase (Perrin *et al.*, 2004a; Holec *et al.*, 2006). Leader sequences generated by an endonuclease, or intergenic transcribed sequences excised by two endonucleolytic cuts, accumulate as full-length RNA species in PNP-plants (where PNPase is silenced by cosuppression) suggesting that their degradation is mainly exonucleolytic and performed by PNPase (Perrin *et al.*, 2004a; Holec *et al.*, 2006).

RNA degradation is also essential for removing cryptic transcripts. Indeed, regions of the mitochondrial genome are actively transcribed but no corresponding transcripts accumulate in wild-type plant mitochondria. For instance, large regions around rRNAs in the maize mitochondrial genome are transcribed but the corresponding RNAs are quickly degraded (Finnegan and Brown, 1990). In *Arabidopsis*, transcripts corresponding to regions without any known genes are actively produced and degraded by PNPase (Holec *et al.*, 2006). Some of these transcripts are produced following duplication of active promoters, such as the one from the *rrn26* gene, or rearrangements of the mitochondrial genome leading to the creation of expressed chimeric genes (Holec *et al.*, 2006). A somehow relaxed control of transcription in plant mitochondria may also explain the generation of transcripts in noncoding regions or in the antisense orientation of known genes, both types of transcripts being degraded by PNPase (Holec *et al.*, 2006).

In conclusion, RNA degradation is involved not only in the turn-over of functional transcripts in plant mitochondria but also an essential process in shaping the final transcriptome of plant mitochondria by removing cryptic, nonfunctional transcripts.

3.8 Translation and posttranslational control

Expression of the few plant mitochondrial mRNAs necessitates a full translation machinery, i.e. ribosomes, translation factors (such as initiation and elongation but probably also termination factors), tRNAs and aminoacyl-tRNA synthetases. In addition, *trans* factors can probably regulate translation in a gene-specific manner. Once translated, the proper function of mitochondrial proteins can also require posttranslational modifications. In addition, specific protein degradation is an integral part of the regulation of plant mitochondrial genome expression. As for transcriptional or posttranscriptional processes, plant mitochondria rely heavily on nuclear-encoded components for translation and posttranslational control.

3.8.1 Translational machinery

During evolution, genes involved in translation in the original endosymbiont have been, for the most part, transferred to the nucleus, or lost if dispensable (Palmer *et al.*, 2000; Adams and Palmer, 2003). The still ongoing process of gene transfer is particularly obvious for genes encoding translation components (Adams and Palmer, 2003), which vary according to species. However, several main features are conserved in higher plants:

1. All three rRNAs are encoded in plant mitochondria.
2. All translation factors such as elongation factor Tu (Kuhlman and Palmer, 1995) as well as aminoacyl-tRNA synthetases are encoded in the nucleus. Interestingly, in *Arabidopsis* all aminoacyl-tRNA synthetases are dual targeted, i.e. shared between the chloroplast and mitochondria or between the cytoplasm and mitochondria (Duchene *et al.*, 2005). An extreme case is represented by an alanyl-tRNA synthetase present in all three cellular compartments where translation occurs (Duchene *et al.*, 2005).
3. Higher plant mitochondrial genomes encode a variable set of both *native* and chloroplast-like *trn* genes, corresponding to either the original *trn* genes or to the chloroplast genes that have been inserted into the mitochondrial genome, respectively. However, *trn* genes in the mitochondrial genome represent only an incomplete set of the tRNAs necessary for translation. As a consequence,

the *missing* tRNAs are imported from the cytoplasm. Therefore, tRNAs operating in plant mitochondria are from three different genetic origins. Their respective proportion varies from plant to plant. Studies aimed at pinpointing the tRNA determinants necessary of import revealed that the cognate aminoacyl-tRNA synthetase might be involved (Dietrich *et al.*, 1996) but is certainly not sufficient for the import of a particular tRNA as sequence elements in the tRNA are also required (Brubacher-Kauffmann *et al.*, 1999; Mireau *et al.*, 2000; Delage *et al.*, 2003; Laforest *et al.*, 2005; Salinas *et al.*, 2005).

4. Some subunits of mitoribosomes are encoded in mitochondria but the numbers vary between plant species. Most original genes encoding ribosomal proteins have been transferred to the nucleus where they have acquired sequences coding for not only mitochondrial targeting signals but also sometimes additional functional domains (Sanchez *et al.*, 1996). Original genes have also been lost as they have been functionally replaced by a novel sequence acquired by another ribosomal protein gene. Gene loss is also explained by substitution by genes originally coding for the cytoplasmic, or chloroplast counterpart (Adams *et al.*, 2002; Mollier *et al.*, 2002). The precise composition of plant mitoribosomes has not been elucidated but is likely to be determined soon given the recent advances in plant mitochondrial proteomics.

3.8.2 Translational control

AUG is the typical initiation codon in plant mitochondria. However, several alternative initiation codons are possibly also used (Bock *et al.*, 1994a; Thomson *et al.*, 1994; Siculella *et al.*, 1996; Dong *et al.*, 1998). It is unclear at present whether there is an mRNA surveillance system that would prevent translation of aberrant transcripts. However, incompletely edited transcripts have been frequently found associated with ribosomes, suggesting that such a mechanism might not exist in plant mitochondria (Lu and Hanson, 1994, 1996; Lu *et al.*, 1996; Phreaner *et al.*, 1996).

No Shine-Dalgarno-like sequences seem to be present upstream of initiation codons. Conserved sequence blocks have been identified in a limited number of mRNAs (Pring *et al.*, 1992). Their location is consistent with a possible role in translation initiation, which cannot be experimentally confirmed due to technique limitations. Indeed, no efficient *in vitro* translation system exists for plant mitochondrial transcripts, and the expression levels of *in organello*-expressed transgenes are too low at present to study translation.

Trans factors activating translation of a particular mRNA are expected to exist in plant mitochondria and may be members of the PPR family, as recently described in chloroplasts (Schmitz-Linneweber *et al.*, 2005). Although the precise action of some restorer genes has not been determined, those encoding PPR proteins may be involved in the translational control of CMS-associated transcripts (Hanson and Bentolila, 2004) (see Chapter 9).

3.8.3 Posttranslational control

Posttranslational control of gene expression in mitochondria can be exerted either by protein modifications or by protein degradation. Protein modifications include various posttranslational covalent, or noncovalent, modifications and, in plant mitochondria, their analysis has only recently been started (Millar *et al.*, 2005). Mitochondrial-encoded proteins can also be processed from larger precursors (Figueroa *et al.*, 2000; Perrotta *et al.*, 2002). Protein degradation appears to be a crucial control point in plant mitochondrial genome expression as it is in chloroplasts (Adam, 2000). In addition to their expected turnover, proteins are also targeted for degradation when unassembled or damaged, for instance by reactive oxygen species (Moller and Kristensen, 2004; Giegé *et al.*, 2005). In addition to the proteases involved in processing of imported precursors, and the degradation of the excised targeting sequences (see Chapter 4), mitochondria contain a whole assortment of proteases that include members of the Clp, Lon and FtsH families, the latter also being called the AAA protease family (Sarria *et al.*, 1998; Halperin *et al.*, 2001; Kolodziejczak *et al.*, 2002; Peltier *et al.*, 2004; Urantowka *et al.*, 2005). Mitochondrial gene expression can be regulated by specific protein degradation as highlighted in two CMS systems. In common bean, the CMS-associated protein ORF239 only accumulates in reproductive tissues. However, ORF239 is produced in all tissues but degraded selectively in vegetative organs, likely by a Lon-type protease (Sarria *et al.*, 1998). In rapeseed, restoration of fertility is accompanied by a reduction in ORF138 protein abundance, which is associated with the Ogura CMS. As neither the level of *orf138* transcript nor its association with ribosomes is affected by the restorer gene *Rfo*, this restorer probably triggers ORF138 degradation (Bellaoui *et al.*, 1999).

3.9 Concluding remarks

Only a few years ago, not a single gene involved in plant mitochondrial gene expression had been identified. The situation has since improved considerably. A bioinformatic analysis of the nuclear genome sequences of *Arabidopsis* and rice has generated a wealth of information that has enabled identification of homologues of proteins known to participate in gene expression in other organisms, or cellular compartments. Novel, plant-specific proteins, such as the PPR protein family, have been identified by such *in silico* approaches. In addition, classical biochemical strategies have enabled the characterization of novel types of plant-specific mitochondrial proteins (e.g. Vermel *et al.*, 2002). Last, but not least, recent successes in identifying restorer genes in different CMS systems demonstrate that CMS remains a wonderful tool to help understand different aspects of mitochondrial gene expression during development (see Chapter 9). Comprehensive functional analysis of these newly identified genes will result in a considerable advancement in our knowledge of mitochondrial genome expression.

Another recent revolution has been the setting up of an *in vitro* editing system and techniques for foreign gene expression *in organello*. Although still in their infancy, these methods have already proven their potential in determining mechanistic aspects of gene expression. However, the long-awaited proper *in vivo* mitochondrion transformation protocol is still elusive.

Considerable progress has been made in identifying components of the mitochondrial genome-expression machinery. This knowledge will be useful in understanding whether or not expression of the mitochondrial genome is integrated into cellular processes such as the response to stress, interorganelle communication and development and if it is, then how this integration is manifested.

Acknowledgements

Work in DG's laboratory is supported by the Centre National de la Recherche Scientifique (CNRS, France) and an ACI JC grant from the French Ministry of Research. Research in SB's laboratory is funded by the Deutsche Forschungsgemeinschaft and the Rudolf und Clothilde Eberhardt-Stiftung.

References

Adam Z (2000) Chloroplast proteases: possible regulators of gene expression? *Biochimie* **82**, 647–654.

Adams KL, Daley DO, Whelan J and Palmer JD (2002) Genes for two mitochondrial ribosomal proteins in flowering plants are derived from their chloroplast or cytosolic counterparts. *Plant Cell* **14**, 931–943.

Adams KL and Palmer JD (2003) Evolution of mitochondrial gene content: gene loss and transfer to the nucleus. *Mol Phylogenet Evol* **29**, 380–395.

Akagi H, Nakamura A, Yokozeki-Misono Y, *et al.* (2004) Positional cloning of the rice Rf-1 gene, a restorer of BT-type cytoplasmic male sterility that encodes a mitochondria-targeting PPR protein. *Theor Appl Genet* **108**, 1449–1457.

Arabidopsis Genome Initiative (2000) Analysis of the genome sequence of the flowering plant *Arabidopsis* thaliana. *Nature* **408**, 796–815.

Baba K, Schmidt J, Espinosa-Ruiz A, *et al.* (2004) Organellar gene transcription and early seedling development are affected in the rpoT;2 mutant of *Arabidopsis*. *Plant J* **38**, 38–48.

Bégu D, Mercado A, Farré JC, *et al.* (1998) Editing status of mat-r transcripts in mitochondria from two plant species: C-to-U changes occur in putative functional RT and maturase domains. *Curr Genet* **33**, 420–428.

Bellaoui M, Grelon M, Pelletier G and Budar F (1999) The restorer Rfo gene acts post-translationally on the stability of the ORF138 Ogura CMS-associated protein in reproductive tissues of rapeseed cybrids. *Plant Mol Biol* **40**, 893–902.

Bellaoui M, Pelletier G and Budar F (1997) The steady-state level of mRNA from the Ogura cytoplasmic male sterility locus in Brassica cybrids is determined post-transcriptionally by its 3′ region. *EMBO J* **16**, 5057–5068.

Bentolila S, Alfonso AA and Hanson MR (2002) A pentatricopeptide repeat-containing gene restores fertility to cytoplasmic male-sterile plants. *Proc Natl Acad Sci USA* **99**, 10887–10892.

Binder S and Brennicke A (1993) Transcription initiation sites in mitochondria of Oenothera berteriana. *J Biol Chem* **268**, 7849–7855.

Binder S, Hatzack F and Brennicke A (1995) A novel pea mitochondrial *in vitro* transcription system recognizes homologous and heterologous mRNA and tRNA promoters. *J Biol Chem* **270**, 22182–22189.

Binder S, Marchfelder A and Brennicke A (1994a) RNA editing of tRNA(Phe) and tRNA(Cys) in mitochondria of Oenothera berteriana is initiated in precursor molecules. *Mol Gen Genet* **244**, 67–74.

Binder S, Marchfelder A, Brennicke A and Wissinger B (1992) RNA editing in *trans*-splicing intron sequences of nad2 mRNAs in Oenothera mitochondria. *J Biol Chem* **267**, 7615–7623.

Binder S, Thalheim C and Brennicke A (1994b) Transcription of potato mitochondrial 26S rRNA is initiated at its mature 5′ end. *Curr Genet* **26**, 519–523.

Bock H, Brennicke A and Schuster W (1994a) Rps3 and rpl16 genes do not overlap in Oenothera mitochondria: GTG as a potential translation initiation codon in plant mitochondria? *Plant Mol Biol* **24**, 811–818.

Bock R, Hermann M and Kossel H (1996) *In vivo* dissection of *cis*-acting determinants for plastid RNA editing. *EMBO J* **15**, 5052–5059.

Bock R, Kossel H and Maliga P (1994b) Introduction of a heterologous editing site into the tobacco plastid genome: the lack of RNA editing leads to a mutant phenotype. *EMBO J* **13**, 4623–4628.

Bollenbach TJ, Lange H, Gutierrez R, Erhardt M, Stern DB and Gagliardi D (2005) RNR1, a 3′-5′ exoribonuclease belonging to the RNR superfamily, catalyzes 3′ maturation of chloroplast ribosomal RNAs in *Arabidopsis* thaliana. *Nucleic Acids Res* **33**, 2751–2763.

Bollenbach TJ, Schuster G and Stern DB (2004) Cooperation of endo- and exoribonucleases in chloroplast mRNA turnover. *Prog Nucleic Acid Res Mol Biol* **78**, 305–337.

Bonen L and Brown GG (1993) Genetic pasticity and its consequences: perspectives on gene organization and expression in plant mitochondria. *Can J Bot* **71**, 645–660.

Bonen L and Vogel J (2001) The ins and outs of group II introns. *Trends Genet* **17**, 322–331.

Brangeon J, Sabar M, Gutierres S, *et al.* (2000) Defective splicing of the first nad4 intron is associated with lack of several complex I subunits in the *Nicotiana sylvestris* NMS1 nuclear mutant. *Plant J* **21**, 269–280.

Brennicke A, Marchfelder A and Binder S (1999) RNA editing. *FEMS Microbiol Rev* **23**, 297–316.

Brown GG, Auchincloss AH, Covello PS, Gray MW, Menassa R and Singh M (1991) Characterization of transcription initiation sites on the soybean mitochondrial genome allows identification of a transcription-associated sequence motif. *Mol Gen Genet* **228**, 345–355.

Brown GG, Formanova N, Jin H, *et al.* (2003) The radish Rfo restorer gene of Ogura cytoplasmic male sterility encodes a protein with multiple pentatricopeptide repeats. *Plant J* **35**, 262–272.

Brubacher-Kauffmann S, Marechal-Drouard L, Cosset A, Dietrich A and Duchene AM (1999) Differential import of nuclear-encoded tRNAGly isoacceptors into solanum Tuberosum mitochondria. *Nucleic Acids Res* **27**, 2037–2042.

Caoile AG and Stern DB (1997) A conserved core element is functionally important for maize mitochondrial promoter activity *in vitro*. *Nucleic Acids Res* **25**, 4055–4060.

Carrillo C and Bonen L (1997) RNA editing status of nad7 intron domains in wheat mitochondria. *Nucleic Acids Res* **25**, 403–409.

Carrillo C, Chapdelaine Y and Bonen L (2001) Variation in sequence and RNA editing within core domains of mitochondrial group II introns among plants. *Mol Gen Genet* **264**, 595–603.

Chang CC, Sheen J, Bligny M, Niwa Y, Lerbs-Mache S and Stern DB (1999) Functional analysis of two maize cDNAs encoding T7-like RNA polymerases. *Plant Cell* **11**, 911–926.

Chapdelaine Y and Bonen L (1991) The wheat mitochondrial gene for subunit I of the NADH dehydrogenase complex: a *trans*-splicing model for this gene-in-pieces. *Cell* **65**, 465–472.

Chaudhuri S, Carrer H and Maliga P (1995) Site-specific factor involved in the editing of the psbL mRNA in tobacco plastids. *EMBO J* **14**, 2951–2957.

Chaudhuri S and Maliga P (1996) Sequences directing C to U editing of the plastid psbL mRNA are located within a 22 nucleotide segment spanning the editing site. *EMBO J* **15**, 5958–5964.

Cho Y, Qiu YL, Kuhlman P and Palmer JD (1998) Explosive invasion of plant mitochondria by a group I intron. *Proc Natl Acad Sci USA* **95**, 14244–14249.

Choury D, Farré JC, Jordana X and Araya A (2004) Different patterns in the recognition of editing sites in plant mitochondria. *Nucleic Acids Res* **32**, 6397–6406.

Choury D, Farré JC, Jordana X and Araya A (2005) Gene expression studies in isolated mitochondria: Solanum tuberosum rps10 is recognized by cognate potato but not by the transcription, splicing and editing machinery of wheat mitochondria. *Nucleic Acids Res* **33**, 7058–7065.

Clifton SW, Minx P, Fauron CM, *et al.* (2004) Sequence and comparative analysis of the maize NB mitochondrial genome. *Plant Physiol* **136**, 3486–3503.

Coffin JW, Dhillon R, Ritzel RG and Nargang FE (1997) The *Neurospora crassa* cya-5 nuclear gene encodes a protein with a region of homology to the *Saccharomyces cerevisiae* PET309 protein and is required in a post-transcriptional step for the expression of the mitochondrially encoded COXI protein. *Curr Genet* **32**, 273–280.

Considine MJ, Holtzapffel RC, Day DA, Whelan J and Millar AH (2002) Molecular distinction between alternative oxidase from monocots and dicots. *Plant Physiol* **129**, 949–953.

Covello PS and Gray MW (1989) RNA editing in plant mitochondria. *Nature* **341**, 662–666.

Covello PS and Gray MW (1991) Sequence analysis of wheat mitochondrial transcripts capped *in vitro*: definitive identification of transcription initiation sites. *Curr Genet* **20**, 245–251.

Das AK, Cohen PW and Barford D (1998) The structure of the tetratricopeptide repeats of protein phosphatase 5: implications for TPR-mediated protein–protein interactions. *EMBO J* **17**, 1192–1199.

Daschner K, Couee I and Binder S (2001) The mitochondrial isovaleryl-coenzyme a dehydrogenase of *Arabidopsis* oxidizes intermediates of leucine and valine catabolism. *Plant Physiol* **126**, 601–612.

Daschner K, Thalheim C, Guha C, Brennicke A and Binder S (1999) In plants a putative isovaleryl-CoA-dehydrogenase is located in mitochondria. *Plant Mol Biol* **39**, 1275–1282.

Delage L, Duchene AM, Zaepfel M and Marechal-Drouard L (2003) The anticodon and the D-domain sequences are essential determinants for plant cytosolic tRNA(Val) import into mitochondria. *Plant J* **34**, 623–633.

Desloire S, Gherbi H, Laloui W, *et al.* (2003) Identification of the fertility restoration locus, Rfo, in radish, as a member of the pentatricopeptide-repeat protein family. *EMBO Rep* **4**, 588–594.

Dietrich A, Marechal-Drouard L, Carneiro V, Cosset A and Small I (1996) A single base change prevents import of cytosolic tRNA(Ala) into mitochondria in transgenic plants. *Plant J* **10**, 913–918.

Diffley JF and Stillman B (1991) A close relative of the nuclear, chromosomal high-mobility group protein HMG1 in yeast mitochondria. *Proc Natl Acad Sci USA* **88**, 7864–7868.

Dill CL, Wise RP and Schnable PS (1997) Rf8 and Rf* mediate unique T-urf13-transcript accumulation, revealing a conserved motif associated with RNA processing and restoration of pollen fertility in T-cytoplasm maize. *Genetics* **147**, 1367–1379.

Dombrowski S, Brennicke A and Binder S (1997) 3′-Inverted repeats in plant mitochondrial mRNAs are processing signals rather than transcription terminators. *EMBO J* **16**, 5069–5076.

Dombrowski S, Hoffmann M, Guha C and Binder S (1999) Continuous primary sequence requirements in the 18-nucleotide promoter of dicot plant mitochondria. *J Biol Chem* **274**, 10094–10099.

Dombrowski S, Hoffmann M, Kuhn J, Brennicke A and Binder S (1998) On mitochondrial promoters in *Arabidopsis* thaliana and other flowering plants. In: *Plant Mitochondria: From Gene to Function.* (eds Moller IM, Gardeström P, Glimelius K and Glaser E), pp. 165–170. Backhuys Publications, Leiden, The Netherlands.

Dong FG, Wilson KG and Makaroff CA (1998) The radish (Raphanus sativus L.) mitochondrial cox2 gene contains an ACG at the predicted translation initiation site. *Curr Genet* **34**, 79–87.

Dreyfus M and Regnier P (2002) The poly(A) tail of mRNAs: bodyguard in eukaryotes, scavenger in bacteria. *Cell* **111**, 611–613.

Duchene AM, Giritch A, Hoffmann B, *et al.* (2005) Dual targeting is the rule for organellar aminoacyl-tRNA synthetases in *Arabidopsis* thaliana. *Proc Natl Acad Sci USA* **102**, 16484–16489.

Emanuel C, von Groll U, Muller M, Borner T and Weihe A (2005) Development- and tissue-specific expression of the RpoT gene family of *Arabidopsis* encoding mitochondrial and plastid RNA polymerases. *Planta* **223**, 998–1009.

Emanuel C, Weihe A, Graner A, Hess WR and Borner T (2004) Chloroplast development affects expression of phage-type RNA polymerases in barley leaves. *Plant J* **38**, 460–472.

Ems SC, Morden CW, Dixon CK, Wolfe KH, dePamphilis CW and Palmer JD (1995) Transcription, splicing and editing of plastid RNAs in the nonphotosynthetic plant *Epifagus virginiana. Plant Mol Biol* **29**, 721–733.

Faivre-Nitschke SE, Grienenberger JM and Gualberto JM (1999) A prokaryotic-type cytidine deaminase from *Arabidopsis* thaliana gene expression and functional characterization. *Eur J Biochem* **263**, 896–903.

Falkenberg M, Gaspari M, Rantanen A, Trifunovic A, Larsson NG and Gustafsson CM (2002) Mitochondrial transcription factors B1 and B2 activate transcription of human mtDNA. *Nat Genet* **31**, 289–294.

Farré JC and Araya A (1999) The mat-r open reading frame is transcribed from a non-canonical promoter and contains an internal promoter to co-transcribe exons nad1e and nad5III in wheat mitochondria. *Plant Mol Biol* **40**, 959–967.

Farré JC and Araya A (2001) Gene expression in isolated plant mitochondria: high fidelity of transcription, splicing and editing of a transgene product in electroporated organelles. *Nucleic Acids Res* **29**, 2484–2491.

Farré JC and Araya A (2002) RNA splicing in higher plant mitochondria: determination of functional elements in group II intron from a chimeric cox II gene in electroporated wheat mitochondria. *Plant J* **29**, 203–213.

Farré JC, Leon G, Jordana X and Araya A (2001) *cis* Recognition elements in plant mitochondrion RNA editing. *Mol Cell Biol* **21**, 6731–6737.

Fey J and Marechal-Drouard L (1999) Expression of the two chloroplast-like tRNA(Asn) genes in potato mitochondria: mapping of transcription initiation sites present in the trnN1-trnY-nad2 cluster and upstream of trnN2. *Curr Genet* **36**, 49–54.

Fey J, Tomita K, Bergdoll M and Marechal-Drouard L (2000) Evolutionary and functional aspects of C-to-U editing at position 28 of tRNA(Cys)(GCA) in plant mitochondria. *RNA* **6**, 470–474.

Fey J, Weil JH, Tomita K, *et al.* (2002) Role of editing in plant mitochondrial transfer RNAs. *Gene* **286**, 21–24.

Figueroa P, Holuigue L, Araya A and Jordana X (2000) The nuclear-encoded SDH2-RPS14 precursor is proteolytically processed between SDH2 and RPS14 to generate maize mitochondrial RPS14. *Biochem Biophys Res Commun* **271**, 380–385.

Finnegan PM and Brown GG (1990) Transcriptional and post-transcriptional regulation of rna levels in maize mitochondria. *Plant Cell* **2**, 71–83.

Fisk DG, Walker MB and Barkan A (1999) Molecular cloning of the maize gene crp1 reveals similarity between regulators of mitochondrial and chloroplast gene expression. *EMBO J* **18**, 2621–2630.

Forner J, Weber B, Wietholter C, Meyer RC and Binder S (2005) Distant sequences determine 5′ end formation of cox3 transcripts in *Arabidopsis* thaliana ecotype C24. *Nucleic Acids Res* **33**, 4673–4682.

Gagliardi D and Leaver CJ (1999) Polyadenylation accelerates the degradation of the mitochondrial mRNA associated with cytoplasmic male sterility in sunflower. *EMBO J* **18**, 3757–3766.

Gagliardi D, Perrin R, Marechal-Drouard L, Grienenberger JM and Leaver CJ (2001) Plant mitochondrial polyadenylated mRNAs are degraded by a 3′- to 5′-exoribonuclease activity, which proceeds unimpeded by stable secondary structures. *J Biol Chem* **276**, 43541–43547.

Geddy R, Mahe L and Brown GG (2005) Cell-specific regulation of a *Brassica napus* CMS-associated gene by a nuclear restorer with related effects on a floral homeotic gene promoter. *Plant J* **41**, 333–345.

Giegé P and Brennicke A (1999) RNA editing in *Arabidopsis* mitochondria effects 441 C to U changes in ORFs. *Proc Natl Acad Sci USA* **96**, 15324–15329.

Giegé P, Hoffmann M, Binder S and Brennicke A (2000) RNA degradation buffers asymmetries of transcription in *Arabidopsis* mitochondria. *EMBO Rep* **1**, 164–170.

Giegé P, Sweetlove LJ, Cognat V and Leaver CJ (2005) Coordination of nuclear and mitochondrial genome expression during mitochondrial biogenesis in *Arabidopsis*. *Plant Cell* **17**, 1497–1512.

Giese A, Thalheim C, Brennicke A and Binder S (1996) Correlation of nonanucleotide motifs with transcript initiation of 18S rRNA genes in mitochondria of pea, potato and *Arabidopsis*. *Mol Gen Genet* **252**, 429–436.

Goff SA, Ricke D, Lan TH, *et al.* (2002) A draft sequence of the rice genome (*Oryza sativa* L. ssp. japonica). *Science* **296**, 92–100.

Grohmann L, Brennicke A and Schuster W (1992) The mitochondrial gene encoding ribosomal protein S12 has been translocated to the nuclear genome in Oenothera. *Nucleic Acids Res* **20**, 5641–5646.

Grohmann L, Thieck O, Herz U, Schroder W and Brennicke A (1994) Translation of nad9 mRNAs in mitochondria from Solanum tuberosum is restricted to completely edited transcripts. *Nucleic Acids Res* **22**, 3304–3311.

Grosjean H and Benne R (1998) *Modification and Editing of RNA*. ASM Press, Washington, DC.

Gualberto JM, Lamattina L, Bonnard G, Weil JH and Grienenberger JM (1989) RNA editing in wheat mitochondria results in the conservation of protein sequences. *Nature* **341**, 660–662.

Halperin T, Zheng B, Itzhaki H, Clarke AK and Adam Z (2001) Plant mitochondria contain proteolytic and regulatory subunits of the ATP-dependent Clp protease. *Plant Mol Biol* **45**, 461–468.

Handa H (2003) The complete nucleotide sequence and RNA editing content of the mitochondrial genome of rapeseed (*Brassica napus* L.): comparative analysis of the mitochondrial genomes of rapeseed and *Arabidopsis* thaliana. *Nucleic Acids Res* **31**, 5907–5916.

Hanic-Joyce PJ and Gray MW (1990) Processing of transfer RNA precursors in a wheat mitochondrial extract. *J Biol Chem* **265**, 13782–13791.

Hanic-Joyce PJ and Gray MW (1991) Accurate transcription of a plant mitochondrial gene *in vitro*. *Mol Cell Biol* **11**, 2035–2039.

Hanic-Joyce PJ, Spencer DF and Gray MW (1990) *In vitro* processing of transcripts containing novel tRNA-like sequences ('t-elements') encoded by wheat mitochondrial DNA. *Plant Mol Biol* **15**, 551–559.

Hanson MR and Bentolila S (2004) Interactions of mitochondrial and nuclear genes that affect male gametophyte development. *Plant Cell* **16** (Suppl.), S154–S169.

Hatzack F, Dombrowski S, Brennicke A and Binder S (1998) Characterization of DNA-binding proteins from pea mitochondria. *Plant Physiol* **116**, 519–528.

Hedtke B, Borner T and Weihe A (1997) Mitochondrial and chloroplast phage-type RNA polymerases in *Arabidopsis*. *Science* **277**, 809–811.

Hedtke B, Borner T and Weihe A (2000) One RNA polymerase serving two genomes. *EMBO Rep* **1**, 435–440.

Hedtke B, Meixner M, Gillandt S, Richter E, Borner T and Weihe A (1999) Green fluorescent protein as a marker to investigate targeting of organellar RNA polymerases of higher plants *in vivo*. *Plant J* **17**, 557–561.

Hegeman CE, Hayes ML and Hanson MR (2005) Substrate and cofactor requirements for RNA editing of chloroplast transcripts in *Arabidopsis in vitro*. *Plant J* **42**, 124–132.

Hess WR, Hoch B, Zeltz P, Hubschmann T, Kossel H and Borner T (1994) Inefficient rpl2 splicing in barley mutants with ribosome-deficient plastids. *Plant Cell* **6**, 1455–1465.

Hiesel R, Wissinger B, Schuster W and Brennicke A (1989) RNA editing in plant mitochondria. *Science* **246**, 1632–1634.

Hirose T and Sugiura M (2001) Involvement of a site-specific *trans*-acting factor and a common RNA-binding protein in the editing of chloroplast mRNAs: development of a chloroplast *in vitro* RNA editing system. *EMBO J* **20**, 1144–1152.

Hoch B, Maier RM, Appel K, Igloi GL and Kossel H (1991) Editing of a chloroplast mRNA by creation of an initiation codon. *Nature* **353**, 178–180.

Hoffmann M and Binder S (2002) Functional importance of nucleotide identities within the pea atp9 mitochondrial promoter sequence. *J Mol Biol* **320**, 943–950.

Hofmann TJ, Min J and Zassenhaus HP (1993) Formation of the 3′ end of yeast mitochondrial mRNAs occurs by site-specific cleavage two bases downstream of a conserved dodecamer sequence. *Yeast* **9**, 1319–1330.

Holec S, Lange H, Kühn K, Alioua M, Börner T and Gagliardi D (2006) Relaxed transcription in *Arabidopsis* mitochondria is counterbalanced by RNA stability control mediated by polyadenylation and PNPase. *Mol Cell Biol* **26**, 2869–2876.

Hubschmann T, Hess WR and Borner T (1996) Impaired splicing of the rps12 transcript in ribosome-deficient plastids. *Plant Mol Biol* **30**, 109–123.

Ikeda TM and Gray MW (1999a) Characterization of a DNA-binding protein implicated in transcription in wheat mitochondria. *Mol Cell Biol* **19**, 8113–8122.

Ikeda TM and Gray MW (1999b) Identification and characterization of T3/T7 bacteriophage-like RNA polymerase sequences in wheat. *Plant Mol Biol* **40**, 567–578.

Iwabuchi M, Kyozuka J and Shimamoto K (1993) Processing followed by complete editing of an altered mitochondrial atp6 RNA restores fertility of cytoplasmic male sterile rice. *EMBO J* **12**, 1437–1446.

Jang SH and Jaehning JA (1991) The yeast mitochondrial RNA polymerase specificity factor, MTF1, is similar to bacterial sigma factors. *J Biol Chem* **266**, 22671–22677.

Joel Duff R (2006) Divergent RNA editing frequencies in hornwort mitochondrial nad5 sequences. *Gene* **366**, 285–291.

Juszczuk IM and Rychter AM (2003) Alternative oxidase in higher plants. *Acta Biochim Pol* **50**, 1257–1271.

Kabeya Y, Hashimoto K and Sato N (2002) Identification and characterization of two phage-type RNA polymerase cDNAs in the moss *Physcomitrella patens*: implication of recent evolution of nuclear-encoded RNA polymerase of plastids in plants. *Plant Cell Physiol* **43**, 245–255.

Kabeya Y and Sato N (2005) Unique translation initiation at the second AUG codon determines mitochondrial localization of the phage-type RNA polymerases in the moss *Physcomitrella patens*. *Plant Physiol* **138**, 369–382.

Kaleikau EK, Andre CP and Walbot V (1992) Structure and expression of the rice mitochondrial apocytochrome b gene (cob-1) and pseudogene (cob-2). *Curr Genet* **22**, 463–470.

Kazama T and Toriyama K (2003) A pentatricopeptide repeat-containing gene that promotes the processing of aberrant atp6 RNA of cytoplasmic male-sterile rice. *FEBS Lett* **544**, 99–102.

Klein RR, Klein PE, Mullet JE, Minx P, Rooney WL and Schertz KF (2005) Fertility restorer locus Rf1 of sorghum (Sorghum bicolor L.) encodes a pentatricopeptide repeat protein not present in the colinear region of rice chromosome 12. *Theor Appl Genet* **111**, 994–1012.

Knoop V (2004) The mitochondrial DNA of land plants: peculiarities in phylogenetic perspective. *Curr Genet* **46**, 123–139.

Knoop V, Schuster W, Wissinger B and Brennicke A (1991) *Trans* splicing integrates an exon of 22 nucleotides into the nad5 mRNA in higher plant mitochondria. *EMBO J* **10**, 3483–3493.

Kobayashi Y, Dokiya Y, Kumazawa Y and Sugita M (2002) Non-AUG translation initiation of mRNA encoding plastid-targeted phage-type RNA polymerase in *Nicotiana sylvestris*. *Biochem Biophys Res Commun* **299**, 57–61.

Koizuka N, Imai R, Fujimoto H, *et al.* (2003) Genetic characterization of a pentatricopeptide repeat protein gene, orf687, that restores fertility in the cytoplasmic male-sterile Kosena radish. *Plant J* **34**, 407–415.

Kolodziejczak M, Kolaczkowska A, Szczesny B, *et al.* (2002) A higher plant mitochondrial homologue of the yeast m-AAA protease. Molecular cloning, localization, and putative function. *J Biol Chem* **277**, 43792–43798.

Komine Y, Kwong L, Anguera MC, Schuster G and Stern DB (2000) Polyadenylation of three classes of chloroplast RNA in Chlamydomonas reinhadtii. *RNA* **6**, 598–607.

Komori T, Ohta S, Murai N, *et al.* (2004) Map-based cloning of a fertility restorer gene, Rf-1, in rice (Oryza sativa L.). *Plant J* **37**, 315–325.

Kotera E, Tasaka M and Shikanai T (2005) A pentatricopeptide repeat protein is essential for RNA editing in chloroplasts. *Nature* **433**, 326–330.

Koulintchenko M, Konstantinov Y and Dietrich A (2003) Plant mitochondria actively import DNA via the permeability transition pore complex. *EMBO J* **22**, 1245–1254.

Kruse B, Narasimhan N and Attardi G (1989) Termination of transcription in human mitochondria: identification and purification of a DNA binding protein factor that promotes termination. *Cell* **58**, 391–397.

Kubo T, Nishizawa S, Sugawara A, Itchoda N, Estiati A and Mikami T (2000) The complete nucleotide sequence of the mitochondrial genome of sugar beet (Beta vulgaris L.) reveals a novel gene for tRNA(Cys)(GCA). *Nucleic Acids Res* **28**, 2571–2576.

Kugita M, Kaneko A, Yamamoto Y, Takeya Y, Matsumoto T and Yoshinaga K (2003) The complete nucleotide sequence of the hornwort (Anthoceros formosae) chloroplast genome: insight into the earliest land plants. *Nucleic Acids Res* **31**, 716–721.

Kuhlman P and Palmer JD (1995) Isolation, expression, and evolution of the gene encoding mitochondrial elongation factor Tu in *Arabidopsis* thaliana. *Plant Mol Biol* **29**, 1057–1070.

Kuhn J and Binder S (2002) RT-PCR analysis of 5′ to 3′-end-ligated mRNAs identifies the extremities of cox2 transcripts in pea mitochondria. *Nucleic Acids Res* **30**, 439–446.

Kuhn J, Tengler U and Binder S (2001) Transcript lifetime is balanced between stabilizing stem-loop structures and degradation-promoting polyadenylation in plant mitochondria. *Mol Cell Biol* **21**, 731–742.

Kühn K, Weihe A and Borner T (2005) Multiple promoters are a common feature of mitochondrial genes in *Arabidopsis*. *Nucleic Acids Res* **33**, 337–346.

Kunzmann A, Brennicke A and Marchfelder A (1998) 5′ end maturation and RNA editing have to precede tRNA 3′ processing in plant mitochondria. *Proc Natl Acad Sci USA* **95**, 108–113.

Laforest MJ, Delage L and Marechal-Drouard L (2005) The T-domain of cytosolic tRNAVal, an essential determinant for mitochondrial import. *FEBS Lett* **579**, 1072–1078.

Lambowitz AM and Zimmerly S (2004) Mobile group II introns. *Annu Rev Genet* **38**, 1–35.

Leino M, Landgren M and Glimelius K (2005) Alloplasmic effects on mitochondrial transcriptional activity and RNA turnover result in accumulated transcripts of *Arabidopsis* orfs in cytoplasmic male-sterile *Brassica napus*. *Plant J* **42**, 469–480.

Lelandais C, Gutierres S, Mathieu C, *et al.* (1996) A promoter element active in run-off transcription controls the expression of two cistrons of nad and rps genes in *Nicotiana sylvestris* mitochondria. *Nucleic Acids Res* **24**, 4798–4804.

Levings CS, III and Brown GG (1989) Molecular biology of plant mitochondria. *Cell* **56**, 171–179.

Li XQ, Jean M, Landry BS and Brown GG (1998) Restorer genes for different forms of brassica cytoplasmic male sterility map to a single nuclear locus that modifies transcripts of several mitochondrial genes. *Proc Natl Acad Sci USA* **95**, 10032–10037.

Li XQ, Zhang M and Brown GG (1996) Cell-specific expression of mitochondrial transcripts in maize seedlings. *Plant Cell* **8**, 1961–1975.

Linder T, Park CB, Asin-Cayuela J, *et al.* (2005) A family of putative transcription termination factors shared amongst metazoans and plants. *Curr Genet* **48**, 265–269.

Lippok B, Brennicke A and Wissinger B (1994) Differential RNA editing in closely related introns in Oenothera mitochondria. *Mol Gen Genet* **243**, 39–46.

Lisitsky I, Kotler A and Schuster G (1997) The mechanism of preferential degradation of polyadenylated RNA in the chloroplast. The exoribonuclease 100RNP/polynucleotide phosphorylase displays high binding affinity for poly(A) sequence. *J Biol Chem* **272**, 17648–17653.

Loguercio Polosa P, Roberti M, Musicco C, Gadaleta MN, Quagliariello E and Cantatore P (1999) Cloning and characterisation of mtDBP, a DNA-binding protein which binds two distinct regions of sea urchin mitochondrial DNA. *Nucleic Acids Res* **27**, 1890–1899.

Lu B and Hanson MR (1994) A single homogeneous form of ATP6 protein accumulates in petunia mitochondria despite the presence of differentially edited atp6 transcripts. *Plant Cell* **6**, 1955–1968.

Lu B and Hanson MR (1996) Fully edited and partially edited nad9 transcripts differ in size and both are associated with polysomes in potato mitochondria. *Nucleic Acids Res* **24**, 1369–1374.

Lu B, Wilson RK, Phreaner CG, Mulligan RM and Hanson MR (1996) Protein polymorphism generated by differential RNA editing of a plant mitochondrial rps12 gene. *Mol Cell Biol* **16**, 1543–1549.

Lupold DS, Caoile AG and Stern DB (1999a) Genomic context influences the activity of maize mitochondrial cox2 promoters. *Proc Natl Acad Sci USA* **96**, 11670–11675.

Lupold DS, Caoile AG and Stern DB (1999b) Polyadenylation occurs at multiple sites in maize mitochondrial cox2 mRNA and is independent of editing status. *Plant Cell* **11**, 1565–1578.

Lurin C, Andres C, Aubourg S, *et al.* (2004) Genome-wide analysis of *Arabidopsis* pentatricopeptide repeat proteins reveals their essential role in organelle biogenesis. *Plant Cell* **16**, 2089–2103.

Mackenzie S and McIntosh L (1999) Higher plant mitochondria. *Plant Cell* **11**, 571–586.

Makaroff CA, Apel IJ and Palmer JD (1989) The atp6 coding region has been disrupted and a novel reading frame generated in the mitochondrial genome of cytoplasmic male-sterile radish. *J Biol Chem* **264**, 11706–11713.

Malek O and Knoop V (1998) *Trans*-splicing group II introns in plant mitochondria: the complete set of *cis*-arranged homologs in ferns, fern allies, and a hornwort. *RNA* **4**, 1599–1609.
Malek O, Lattig K, Hiesel R, Brennicke A and Knoop V (1996) RNA editing in bryophytes and a molecular phylogeny of land plants. *EMBO J* **15**, 1403–1411.
Maloney AP, Traynor PL, Levings CS, III and Walbot V (1989) Identification in maize mitochondrial 26S rRNA of a short 5′-end sequence possibly involved in transcription initiation and processing. *Curr Genet* **15**, 207–212.
Manthey GM, Przybyla-Zawislak BD and McEwen JE (1998) The *Saccharomyces cerevisiae* Pet309 protein is embedded in the mitochondrial inner membrane. *Eur J Biochem* **255**, 156–161.
Marchfelder A and Brennicke A (1994) Characterization and partial purification of tRNA processing activities from potato mitochondria. *Plant Physiol* **105**, 1247–1254.
Marchfelder A, Brennicke A and Binder S (1996) RNA editing is required for efficient excision of tRNA(Phe) from precursors in plant mitochondria. *J Biol Chem* **271**, 1898–1903.
Marchfelder A, Schuster W and Brennicke A (1990) *In vitro* processing of mitochondrial and plastid derived tRNA precursors in a plant mitochondrial extract. *Nucleic Acids Res* **18**, 1401–1406.
Maréchal-Drouard L, Cosset A, Remacle C, Ramamonjisoa D and Dietrich A (1996a) A single editing event is a prerequisite for efficient processing of potato mitochondrial phenylalanine tRNA. *Mol Cell Biol* **16**, 3504–3510.
Maréchal-Drouard L, Kumar R, Remacle C and Small I (1996b) RNA editing of larch mitochondrial tRNA(His) precursors is a prerequisite for processing. *Nucleic Acids Res* **24**, 3229–3234.
Maréchal-Drouard L, Ramamonjisoa D, Cosset A, Weil JH and Dietrich A (1993) Editing corrects mispairing in the acceptor stem of bean and potato mitochondrial phenylalanine transfer RNAs. *Nucleic Acids Res* **21**, 4909–4914.
Martin G and Keller W (2004) Sequence motifs that distinguish ATP(CTP):tRNA nucleotidyl transferases from eubacterial poly(A) polymerases. *RNA* **10**, 899–906.
Martin W (2005) Molecular evolution: lateral gene transfer and other possibilities. *Heredity* **94**, 565–566.
Martin M, Cho J, Cesare AJ, Griffith JD and Attardi G (2005) Termination factor-mediated DNA loop between termination and initiation sites drives mitochondrial rRNA synthesis. *Cell* **123**, 1227–1240.
Marujo PE, Hajnsdorf E, Le Derout J, Andrade R, Arraiano CM and Regnier P (2000) RNase II removes the oligo(A) tails that destabilize the rpsO mRNA of *Escherichia coli*. *RNA* **6**, 1185–1193.
Masters BS, Stohl LL and Clayton DA (1987) Yeast mitochondrial RNA polymerase is homologous to those encoded by bacteriophages T3 and T7. *Cell* **51**, 89–99.
McCulloch V, Seidel-Rogol BL and Shadel GS (2002) A human mitochondrial transcription factor is related to RNA adenine methyltransferases and binds S-adenosylmethionine. *Mol Cell Biol* **22**, 1116–1125.
Meierhoff K, Felder S, Nakamura T, Bechtold N and Schuster G (2003) HCF152, an *Arabidopsis* RNA binding pentatricopeptide repeat protein involved in the processing of chloroplast psbB-psbT-psbH-petB-petD RNAs. *Plant Cell* **15**, 1480–1495.
Menassa R, L'Homme Y and Brown GG (1999) Post-transcriptional and developmental regulation of a CMS-associated mitochondrial gene region by a nuclear restorer gene. *Plant J* **17**, 491–499.
Michel F and Ferat JL (1995) Structure and activities of group II introns. *Annu Rev Biochem* **64**, 435–461.
Millar AH, Heazlewood JL, Kristensen BK, Braun HP and Moller IM (2005) The plant mitochondrial proteome. *Trends Plant Sci* **10**, 36–43.
Mireau H, Cosset A, Marechal-Drouard L, Fox TD, Small ID and Dietrich A (2000) Expression of *Arabidopsis* thaliana mitochondrial alanyl-tRNA synthetase is not sufficient to trigger mitochondrial import of tRNAAla in yeast. *J Biol Chem* **275**, 13291–13296.
Miyamoto T, Obokata J and Sugiura M (2004) A site-specific factor interacts directly with its cognate RNA editing site in chloroplast transcripts. *Proc Natl Acad Sci USA* **101**, 48–52.
Mohanty BK and Kushner SR (2003) Genomic analysis in *Escherichia coli* demonstrates differential roles for polynucleotide phosphorylase and RNase II in mRNA abundance and decay. *Mol Microbiol* **50**, 645–658.

Mohr G and Lambowitz AM (2003) Putative proteins related to group II intron reverse transcriptase/maturases are encoded by nuclear genes in higher plants. *Nucleic Acids Res* **31**, 647–652.

Moller IM and Kristensen BK (2004) Protein oxidation in plant mitochondria as a stress indicator. *Photochem Photobiol Sci* **3**, 730–735.

Mollier P, Beate H, Cédrig D and Small I (2002) The gene encoding *Arabidopsis* thaliana mitochondrial ribosomal protein S13 is a recent duplication of the gene encoding plastid S13. *Curr Genet* **40**, 405–409.

Monéger F, Smart CJ and Leaver CJ (1994) Nuclear restoration of cytoplasmic male sterility in sunflower is associated with the tissue-specific regulation of a novel mitochondrial gene. *EMBO J* **13**, 8–17.

Mouillon JM, Ravanel S, Douce R and Rebeille F (2002) Folate synthesis in higher-plant mitochondria: coupling between the dihydropterin pyrophosphokinase and the dihydropteroate synthase activities. *Biochem J* **363**, 313–319.

Muise RC and Hauswirth WW (1992) Transcription in maize mitochondria: effects of tissue and mitochondrial genotype. *Curr Genet* **22**, 235–242.

Mulligan RM (2004) RNA editing in plant organelles. In: *Molecular Biology and Biotechnology of Plant Organelles* (eds Daniell H and Chase C), pp. 239–260. Springer, Dordrecht, the Netherlands.

Mulligan RM, Lau GT and Walbot V (1988a) Numerous transcription initiation sites exist for the maize mitochondrial genes for subunit 9 of the ATP synthase and subunit 3 of cytochrome oxidase. *Proc Natl Acad Sci USA* **85**, 7998–8002.

Mulligan RM, Leon P and Walbot V (1991) Transcriptional and posttranscriptional regulation of maize mitochondrial gene expression. *Mol Cell Biol* **11**, 533–543.

Mulligan RM, Maloney AP and Walbot V (1988b) RNA processing and multiple transcription initiation sites result in transcript size heterogeneity in maize mitochondria. *Mol Gen Genet* **211**, 373–380.

Nakamura T, Meierhoff K, Westhoff P and Schuster G (2003) RNA-binding properties of HCF152, an *Arabidopsis* PPR protein involved in the processing of chloroplast RNA. *Eur J Biochem* **270**, 4070–4081.

Neuwirt J, Takenaka M, Van Der Merwe JA and Brennicke A (2005) An *in vitro* RNA editing system from cauliflower mitochondria: editing site recognition parameters can vary in different plant species. *RNA* **11**, 1563–1570.

Notsu Y, Masood S, Nishikawa T, *et al.* (2002) The complete sequence of the rice (Oryza sativa L.) mitochondrial genome: frequent DNA sequence acquisition and loss during the evolution of flowering plants. *Mol Genet Genomics* **268**, 434–445.

Oda K, Yamato K, Ohta E, *et al.* (1992) Gene organization deduced from the complete sequence of liverwort *Marchantia polymorpha* mitochondrial DNA. A primitive form of plant mitochondrial genome. *J Mol Biol* **223**, 1–7.

Ogihara Y, Yamazaki Y, Murai K, *et al.* (2005) Structural dynamics of cereal mitochondrial genomes as revealed by complete nucleotide sequencing of the wheat mitochondrial genome. *Nucleic Acids Res* **33**, 6235–6250.

Ohtani K, Yamamoto H and Akimitsu K (2002) Sensitivity to Alternaria alternata toxin in citrus because of altered mitochondrial RNA processing. *Proc Natl Acad Sci USA* **99**, 2439–2444.

Ostheimer GJ, Williams-Carrier R, Belcher S, Osborne E, Gierke J and Barkan A (2003) Group II intron splicing factors derived by diversification of an ancient RNA-binding domain. *EMBO J* **22**, 3919–3929.

Palmer JD, Adams KL, Cho Y, Parkinson CL, Qiu YL and Song K (2000) Dynamic evolution of plant mitochondrial genomes: mobile genes and introns and highly variable mutation rates. *Proc Natl Acad Sci USA* **97**, 6960–6966.

Parisi MA and Clayton DA (1991) Similarity of human mitochondrial transcription factor 1 to high mobility group proteins. *Science* **252**, 965–969.

Peltier JB, Ripoll DR, Friso G, *et al.* (2004) Clp protease complexes from photosynthetic and non-photosynthetic plastids and mitochondria of plants, their predicted three-dimensional structures, and functional implications. *J Biol Chem* **279**, 4768–4781.

Pereira de Souza A, Jubier MF, Delcher E, Lancelin D and Lejeune B (1991) A *trans*-splicing model for the expression of the tripartite nad5 gene in wheat and maize mitochondria. *Plant Cell* **3**, 1363–1378.
Perrin R, Lange H, Grienenberger JM and Gagliardi D (2004a) AtmtPNPase is required for multiple aspects of the 18S rRNA metabolism in *Arabidopsis* thaliana mitochondria. *Nucleic Acids Res* **32**, 5174–5182.
Perrin R, Meyer EH, Zaepfel M, *et al.* (2004b) Two exoribonucleases act sequentially to process mature 3′-ends of atp9 mRNAs in *Arabidopsis* mitochondria. *J Biol Chem* **279**, 25440–25446.
Perrotta G, Grienenberger JM and Gualberto JM (2002) Plant mitochondrial rps2 genes code for proteins with a C-terminal extension that is processed. *Plant Mol Biol* **50**, 523–533.
Phreaner CG, Williams MA and Mulligan RM (1996) Incomplete editing of rps12 transcripts results in the synthesis of polymorphic polypeptides in plant mitochondria. *Plant Cell* **8**, 107–117.
Pinon V, Ravanel S, Douce R and Alban C (2005) Biotin synthesis in plants. The first committed step of the pathway is catalyzed by a cytosolic 7-keto-8-aminopelargonic acid synthase. *Plant Physiol* **139**, 1666–1676.
Placido A, Gagliardi D, Gallerani R, Grienenberger JM and Marechal-Drouard L (2005) Fate of a larch unedited tRNA precursor expressed in potato mitochondria. *J Biol Chem* **280**, 33573–33579.
Pring DR, Chen W, Tang HV, Howad W and Kempken F (1998) Interaction of mitochondrial RNA editing and nucleolytic processing in the restoration of male fertility in sorghum. *Curr Genet* **33**, 429–436.
Pring DR, Mullen JA and Kempken F (1992) Conserved sequence blocks 5′ to start codons of plant mitochondrial genes. *Plant Mol Biol* **19**, 313–317.
Pruitt KD and Hanson MR (1991) Transcription of the Petunia mitochondrial CMS-associated Pcf locus in male sterile and fertility-restored lines. *Mol Gen Genet* **227**, 348–355.
Raghavendra AS and Padmasree K (2003) Beneficial interactions of mitochondrial metabolism with photosynthetic carbon assimilation. *Trends Plant Sci* **8**, 546–553.
Rajasekhar VK and Mulligan RM (1993) RNA editing in plant mitochondria: [alpha]-Phosphate is retained during C-to-U conversion in mRNAs. *Plant Cell* **5**, 1843–1852.
Rapp WD, Lupold DS, Mack S and Stern DB (1993) Architecture of the maize mitochondrial atp1 promoter as determined by linker-scanning and point mutagenesis. *Mol Cell Biol* **13**, 7232–7238.
Rapp WD and Stern DB (1992) A conserved 11 nucleotide sequence contains an essential promoter element of the maize mitochondrial atp1 gene. *EMBO J* **11**, 1065–1073.
Rasmusson AG, Soole KL and Elthon TE (2004) Alternative NAD(P)H dehydrogenases of plant mitochondria. *Annu Rev Plant Biol* **55**, 23–39.
Reed ML, Peeters NM and Hanson MR (2001) A single alteration 20 nt 5′ to an editing target inhibits chloroplast RNA editing *in vivo*. *Nucleic Acids Res* **29**, 1507–1513.
Remacle C and Marechal-Drouard L (1996) Characterization of the potato mitochondrial transcription unit containing 'native' trnS (GCU), trnF (GAA) and trnP (UGG). *Plant Mol Biol* **30**, 553–563.
Richter U, Kiessling J, Hedtke B, *et al.* (2002) Two RpoT genes of *Physcomitrella patens* encode phage-type RNA polymerases with dual targeting to mitochondria and plastids. *Gene* **290**, 95–105.
Roberti M, Polosa PL, Bruni F, Musicco C, Gadaleta MN and Cantatore P (2003) DmTTF, a novel mitochondrial transcription termination factor that recognises two sequences of Drosophila melanogaster mitochondrial DNA. *Nucleic Acids Res* **31**, 1597–1604.
Saalaoui E, Litvak S and Araya A (1990) The apocytochrome b from an alloplasmic line of wheat (*T. aestivum*, cytoplasm-T *Timopheevi*) exists in two differently expressed forms. *Plant Sci* **66**, 237–246.
Salinas T, Schaeffer C, Marechal-Drouard L and Duchene AM (2005) Sequence dependence of tRNA(Gly) import into tobacco mitochondria. *Biochimie* **87**, 863–872.
Sanchez H, Fester T, Kloska S, Schroder W and Schuster W (1996) Transfer of rps19 to the nucleus involves the gain of an RNP-binding motif which may functionally replace RPS13 in *Arabidopsis* mitochondria. *EMBO J* **15**, 2138–2149.

Sarria R, Lyznik A, Vallejos CE and Mackenzie SA (1998) A cytoplasmic male sterility-associated mitochondrial peptide in common bean is post-translationally regulated. *Plant Cell* **10**, 1217–1228.

Satoh M, Kubo T, Nishizawa S, Estiati A, Itchoda N and Mikami T (2004) The cytoplasmic male-sterile type and normal type mitochondrial genomes of sugar beet share the same complement of genes of known function but differ in the content of expressed ORFs. *Mol Genet Genomics* **272**, 247–256.

Schmitz-Linneweber C, Williams-Carrier R and Barkan A (2005) RNA immunoprecipitation and microarray analysis show a chloroplast Pentatricopeptide repeat protein to be associated with the 5′ region of mRNAs whose translation it activates. *Plant Cell* **17**, 2791–2804.

Schock I, Marechal-Drouard L, Marchfelder A and Binder S (1998) Processing of plant mitochondrial tRNAGly and tRNASer(GCU) is independent of RNA editing. *Mol Gen Genet* **257**, 554–560.

Schubot FD, Chen CJ, Rose JP, Dailey TA, Dailey HA and Wang BC (2001) Crystal structure of the transcription factor sc-mtTFB offers insights into mitochondrial transcription. *Protein Sci* **10**, 1980–1988.

Schuster W (1993) Ribosomal protein gene rpl5 is cotranscribed with the nad3 gene in Oenothera mitochondria. *Mol Gen Genet* **240**, 445–449.

Schuster W and Brennicke A (1989) Conserved sequence elements at putative processing sites in plant mitochondria. *Curr Genet* **15**, 187–192.

Siculella L, Pacoda D, Treglia S, Gallerani R and Ceci LR (1996) GTG as translation initiation codon in the apocytochrome b gene of sunflower mitochondria. *DNA Seq* **6**, 365–369.

Singh M and Brown GG (1991) Suppression of cytoplasmic male sterility by nuclear genes alters expression of a novel mitochondrial gene region. *Plant Cell* **3**, 1349–1362.

Singh M, Hamel N, Menassa R, *et al.* (1996) Nuclear genes associated with a single Brassica CMS restorer locus influence transcripts of three different mitochondrial gene regions. *Genetics* **143**, 505–516.

Small ID and Peeters N (2000) The PPR motif – a TPR-related motif prevalent in plant organellar proteins. *Trends Biochem Sci* **25**, 46–47.

Smart CJ, Monéger F and Leaver CJ (1994) Cell-specific regulation of gene expression in mitochondria during anther development in sunflower. *Plant Cell* **6**, 811–825.

Spencer DF, Schnare MN, Coulthart MB and Gray MW (1992) Sequence and organization of a 7.2 kb region of wheat mitochondrial DNA containing the large subunit (26S) rRNA gene. *Plant Mol Biol* **20**, 347–352.

Staudinger M, Bolle N and Kempken F (2005) Mitochondrial electroporation and *in organello* RNA editing of chimeric atp6 transcripts. *Mol Genet Genomics* **273**, 130–136.

Staudinger M and Kempken F (2003) Electroporation of isolated higher-plant mitochondria: transcripts of an introduced cox2 gene, but not an atp6 gene, are edited *in organello*. *Mol Genet Genomics* **269**, 553–561.

Steinhauser S, Beckert S, Capesius I, Malek O and Knoop V (1999) Plant mitochondrial RNA editing. *J Mol Evol* **48**, 303–312.

Sugiyama Y, Watase Y, Nagase M, *et al.* (2005) The complete nucleotide sequence and multipartite organization of the tobacco mitochondrial genome: comparative analysis of mitochondrial genomes in higher plants. *Mol Genet Genomics* **272**, 603–615.

Sutton CA, Conklin PL, Pruitt KD and Hanson MR (1991) Editing of pre-mRNAs can occur before *cis*- and *trans*-splicing in Petunia mitochondria. *Mol Cell Biol* **11**, 4274–4277.

Takenaka M and Brennicke A (2003) *In vitro* RNA editing in pea mitochondria requires NTP or dNTP, suggesting involvement of an RNA helicase. *J Biol Chem* **278**, 47526–47533.

Takenaka M, Neuwirt J and Brennicke A (2004) Complex *cis*-elements determine an RNA editing site in pea mitochondria. *Nucleic Acids Res* **32**, 4137–4144.

Tang HV, Chang R and Pring DR (1998) Cosegregation of single genes associated with fertility restoration and transcript processing of sorghum mitochondrial orf107 and urf209. *Genetics* **150**, 383–391.

Tang HV, Pring DR, Shaw LC, *et al.* (1996) Transcript processing internal to a mitochondrial open reading frame is correlated with fertility restoration in male-sterile sorghum. *Plant J* **10**, 123–133.

Thomson MC, Macfarlane JL, Beagley CT and Wolstenholme DR (1994) RNA editing of mat-r transcripts in maize and soybean increases similarity of the encoded protein to fungal and bryophyte group II intron maturases: evidence that mat-r encodes a functional protein. *Nucleic Acids Res* **22**, 5745–5752.

Tiranti V, Savoia A, Forti F, *et al.* (1997) Identification of the gene encoding the human mitochondrial RNA polymerase (h-mtRPOL) by cyberscreening of the Expressed Sequence Tags database. *Hum Mol Genet* **6**, 615–625.

Topping JF and Leaver CJ (1990) Mitochondrial gene expression during wheat leaf development. *Planta* **182**, 399–407.

Turmel M, Otis C and Lemieux C (2003) The mitochondrial genome of *Chara vulgaris*: insights into the mitochondrial DNA architecture of the last common ancestor of green algae and land plants. *Plant Cell* **15**, 1888–1903.

Unseld M, Marienfeld JR, Brandt P and Brennicke A (1997) The mitochondrial genome of *Arabidopsis* thaliana contains 57 genes in 366,924 nucleotides. *Nat Genet* **15**, 57–61.

Urantowka A, Knorpp C, Olczak T, Kolodziejczak M and Janska H (2005) Plant mitochondria contain at least two i-AAA-like complexes. *Plant Mol Biol* **59**, 239–252.

Van Der Merwe JA, Takenaka M, Neuwirt J, Verbitskiy D and Brennicke A (2006) RNA editing sites in plant mitochondria can share *cis*-elements. *FEBS Lett* **580**, 268–272.

Vermel M, Guermann B, Delage L, Grienenberger JM, Marechal-Drouard L and Gualberto JM (2002) A family of RRM-type RNA-binding proteins specific to plant mitochondria. *Proc Natl Acad Sci USA* **99**, 5866–5871.

Vogel J and Borner T (2002) Lariat formation and a hydrolytic pathway in plant chloroplast group II intron splicing. *EMBO J* **21**, 3794–3803.

Vogel J, Borner T and Hess WR (1999) Comparative analysis of splicing of the complete set of chloroplast group II introns in three higher plant mutants. *Nucleic Acids Res* **27**, 3866–3874.

Vogel J, Hubschmann T, Borner T and Hess WR (1997) Splicing and intron-internal RNA editing of trnK-matK transcripts in barley plastids: support for MatK as an essential splice factor. *J Mol Biol* **270**, 179–187.

Walter M, Kilian J and Kudla J (2002) PNPase activity determines the efficiency of mRNA 3′-end processing, the degradation of tRNA and the extent of polyadenylation in chloroplasts. *EMBO J* **21**, 6905–6914.

Wang Z, Zou Y, Li X, *et al.* (2006) Cytoplasmic male sterility of rice with Boro II cytoplasm is caused by a cytotoxic peptide and is restored by two related PPR motif genes via distinct modes of mRNA silencing. *Plant Cell* **18**, 676–687.

Weihe A (2004) The transcription of plant organelle genomes. In: *Molecular Biology and Biotechnology of Plant Organelles* (eds Daniell H and Chase C), pp. 213–237. Springer, Dordrecht, the Netherlands.

Wen L and Chase CD (1999) Pleiotropic effects of a nuclear restorer-of-fertility locus on mitochondrial transcripts in male-fertile and S male-sterile maize. *Curr Genet* **35**, 521–526.

Williams MA, Johzuka Y and Mulligan RM (2000) Addition of non-genomically encoded nucleotides to the 3′-terminus of maize mitochondrial mRNAs: truncated rps12 mRNAs frequently terminate with CCA. *Nucleic Acids Res* **28**, 4444–4451.

Williams PM and Barkan A (2003) A chloroplast-localized PPR protein required for plastid ribosome accumulation. *Plant J* **36**, 675–686.

Wise RP, Dill CL and Schnable PS (1996) Mutator-induced mutations of the rf1 nuclear fertility restorer of T-cytoplasm maize alter the accumulation of T-urf13 mitochondrial transcripts. *Genetics* **143**, 1383–1394.

Wise RP, Gobelman-Werner K, Pei D, Dill CL and Schnable PS (1999) Mitochondrial transcript processing and restoration of male fertility in T-cytoplasm maize. *J Hered* **90**, 380–385.

Wissinger B, Schuster W and Brennicke A (1991) *Trans* splicing in Oenothera mitochondria: nad1 mRNAs are edited in exon and *trans*-splicing group II intron sequences. *Cell* **65**, 473–482.

Xu F, Morin C, Mitchell G, Ackerley C and Robinson BH (2004) The role of the LRPPRC (leucine-rich pentatricopeptide repeat cassette) gene in cytochrome oxidase assembly: mutation causes lowered levels of COX (cytochrome c oxidase) I and COX III mRNA. *Biochem J* **382**, 331–336.

Yamazaki H, Tasaka M and Shikanai T (2004) PPR motifs of the nucleus-encoded factor, PGR3, function in the selective and distinct steps of chloroplast gene expression in *Arabidopsis*. *Plant J* **38**, 152–163.

Yang AJ and Mulligan RM (1991) RNA editing intermediates of cox2 transcripts in maize mitochondria. *Mol Cell Biol* **11**, 4278–4281.

Yehudai-Resheff S, Portnoy V, Yogev S, Adir N and Schuster G (2003) Domain analysis of the chloroplast polynucleotide phosphorylase reveals discrete functions in RNA degradation, polyadenylation, and sequence homology with exosome proteins. *Plant Cell* **15**, 2003–2019.

Young DA, Allen RL, Harvey AJ and Lonsdale DM (1998) Characterization of a gene encoding a single-subunit bacteriophage-type RNA polymerase from maize which is alternatively spliced. *Mol Gen Genet* **260**, 30–37.

Yu J, Hu S, Wang J, *et al.* (2002) A draft sequence of the rice genome (*Oryza sativa* L. ssp. indica). *Science* **296**, 79–92.

Yu W and Schuster W (1995) Evidence for a site-specific cytidine deamination reaction involved in C to U RNA editing of plant mitochondria. *J Biol Chem* **270**, 18227–18233.

4 Import of nuclear-encoded mitochondrial proteins

Elzbieta Glaser and James Whelan

4.1 Introduction

Mitochondria are believed to originate from an endosymbiotic α-proteobacterial ancestor (Andersson and Kurland, 1999; Gray *et al.*, 1999; Lang *et al.*, 1999). Most of the bacterial DNA has been either lost or transferred to the nucleus during evolution and the residual DNA encodes from 3 to 67 proteins (Gray *et al.*, 1999) in different plant species. In the completely sequenced *Arabidopsis* genome, 25 498 genes were annotated (Arabidopsis Genome Initiative, 2000). Subcellular localization programs, TargetP (cbs.dtu.dk/services/TargetP/), Predotar (genoplante-info.infobiogen.fr/predotar/) and MitoProt (ihg.gsf.de/ihg/mitoprot.html), predict a subset of 5–10% of the *Arabidopsis* proteome to be localized in the mitochondria, but a more stringent analysis including new versions of prediction programs, and analysis of the yeast (Sickmann *et al.*, 2003) and human (Taylor *et al.*, 2003) mitochondrial proteomes, localizes a smaller set of ~1000 proteins to mitochondria. The most thorough proteome analysis of *Arabidopsis* mitochondria identified 416 proteins (Heazlewood *et al.*, 2004). Even though there is considerable difference between the predicted and experimental determined proteome in *Arabidopsis*, it is a conservative estimate that ~1000 mitochondrial proteins are encoded by the nuclear genome, expressed on cytosolic polyribosomes and have to be imported into the organelle.

Most of the nuclear-encoded mitochondrial precursors characterized to date carry an N-terminal extension, the targeting peptide (also called a presequence). Molecular chaperones and a series of other cytosolic factors interact with precursor proteins to mediate and facilitate mitochondrial protein import (Glaser and Soll, 2004). Conventionally, the mitochondrial import process is viewed as posttranslational; however, cotranslational import of mitochondrial precursors mediated by mRNA binding to mitochondria has also been suggested for a subset of proteins (Marc *et al.*, 2002). Mitochondrial precursors, targeted to the organelle surface, are recognized by organellar receptors and transported across the organellar membranes via high molecular mass, oligomeric import complexes, named TOM or TIM, for translocase of the outer, or inner mitochondrial membrane, respectively. After import, targeting peptides are cleaved off by the general mitochondrial processing peptidase (MPP) that is integrated in plants into the cytochrome bc_1 complex of the respiratory chain (Glaser and Dessi, 1999). There are different routes for intraorganellar transport of proteins to the final destination. The mature proteins are assembled with their partner proteins either spontaneously or upon action of molecular chaperones. The cleaved targeting peptides, potentially harmful to biological membranes, are degraded inside the organelle by a newly identified presequence peptide degrading zinc metalloprotease, presequence protease (PreP) (Stahl *et al.*, 2002).

In the last 25 years, over 2000 papers have been published on the process of protein import into mitochondria. Although *Neurospora crassa* was initially a prominent model, budding yeast (*Saccharomyces cerevisiae*) has clearly served as the main model organism, largely due to the relative ease of various genetic approaches that have been used to investigate and characterize this process (Harmey *et al.*, 1977; Hartl *et al.*, 1989; Neupert, 1997; Pfanner and Geissler, 2001). Although it is tempting simply to presume that the components of the import apparatus in plants function in an identical manner to their homologues in yeast, this is clearly not the case. We will make comparisons to yeast and other organisms to highlight similarities or differences. For a thorough review of mitochondrial protein import in yeast and mammals, readers are directed to several excellent reviews of these processes (Neupert, 1997; Pfanner and Geissler, 2001; Hoogenraad *et al.*, 2002; Endo *et al.*, 2003).

4.2 Mitochondrial-targeting signals

Targeting of most mitochondrial proteins is mediated by cleavable N-terminal mitochondrial-targeting peptides (mTPs). mTPs are recognized by mitochondrial import receptors and target precursor proteins into the mitochondrial matrix. In the matrix, mTPs are cleaved off resulting in the production of the mature protein (Sjöling and Glaser, 1998). Some inner membrane proteins require a non-cleavable hydrophobic stop signal sequence located C-terminal to the mTP; the intermembrane space proteins and some other membrane proteins contain an additional cleavable signal peptide located directly down-stream of the mTP, a so-called bipartite-targeting peptide. A second group of mitochondrial-targeting signals has been identified in hydrophobic carrier proteins of the inner membrane, the internal signals (Brix *et al.*, 1999). These internal non-cleavable signals are not as well defined as mTPs, do not require presence of mTPs, are not processed and seem to be distributed throughout the protein. Likewise, outer mitochondrial membrane proteins contain non-cleavable targeting and sorting signal within the protein itself (Rapaport, 2003).

Finally, a third group of proteins exists, which are encoded by single nuclear genes, translated in the cytosol and targeted to both chloroplasts and mitochondria. These proteins contain an N-terminal signal peptide referred to as a dual-targeting peptide (ambiguous targeting signal) that is recognized by both mitochondrial and chloroplast receptors and has capacity to target the precursor protein to both the mitochondrial matrix and chloroplast stroma (Peeters and Small, 2001; Silva-Filho, 2003).

4.2.1 *Features of mTPs*

The features of plant mTPs have been analysed by statistical and structural prediction methods using a data set containing 58 plant mTPs with an experimentally determined cleavage site (Zhang and Glaser, 2002) as well as on 385 proteins (Bhushan *et al.*, 2006) found in the mitochondrial proteome of *Arabidopsis* (Heazlewood *et al.*, 2004). The length of organellar signal peptides varies substantially, from 18 to 136 residues with an average length of 42 amino acids residues. The great majority (>80%) of the mTPs are in the range of 20–60 residues long. Analysis of the *Arabidopsis* mitochondrial proteome with SequenceLogos, a program that calculates how often each residue occurs at each position, in conjunction with the degree of sequence conservation (Schneider and Stephens, 1990), revealed a targeting peptide to consist of 40 amino acids on average. At this selected cut-off, the number of acidic amino acid residues increases significantly, indicating the start of the mature protein. The overall composition of mTPs shows a high content of 33–35% hydrophobic (leucine, alanine, phenalalanine, isoleucine, valine), 22–23% hydroxylated (serine, threonine) and 14–15% positively charged (arginine, lysine) amino acid residues, and a very low abundance of acidic amino acids (Plate 4.1). Furthermore, proline and glycine are well represented and constitute about 11% of the total content. In comparison to mTPs from non-plant sources, plant mTPs are about seven to nine amino-acid residues longer and contain about two- to fivefold more serine residues (Sjöling and Glaser, 1998).

Comparison of the overall amino acid composition between plant mTPs and chloroplast-targeting peptides (cTPs) between a variety of plant species (Zhang and Glaser, 2002) or only in *Arabidopsis* proteomes (Bhushan *et al.*, 2006a) shows a remarkable similarity. The differences in organelle-targeting signals are very small. There are significant differences in arginine and proline content, but they are compensated by lysine and glycine, respectively. Serine and cysteine, however, show significant relative increases of 8 and 23%, respectively, in chloroplast transit peptides. Hydroxylated residues of cTPs contain the potential phosphorylation motif, which has been shown to be important for the interaction of cTPs with the cytosolic 14-3-3 proteins shown to form an import guidance complex with the precursors (May and Soll, 2000). Despite considerable similarities in overall amino acid composition, some differences have been found between mTPs and cTPs when the targeting peptides were analysed using SequenceLogos (Plate 4.1). The SequenceLogos of the first 16 amino acids of the targeting peptides of the mitochondrial and chloroplast proteins found in organellar proteomes (100 N-terminal residues of precursor proteins) show considerable differences. The main difference found is that arginine is greatly overrepresented (+52%) in the N-terminal portion of the mitochondrial presequences. The N-terminal portion of cTPs on the other hand has a significant excess of serine (+28%) and proline (+41%). In contrast to the results obtained in the study with targeting peptides from different plant sources (Zhang and Glaser, 2002), there was no significantly increased leucine content in mTPs compared to cTPs.

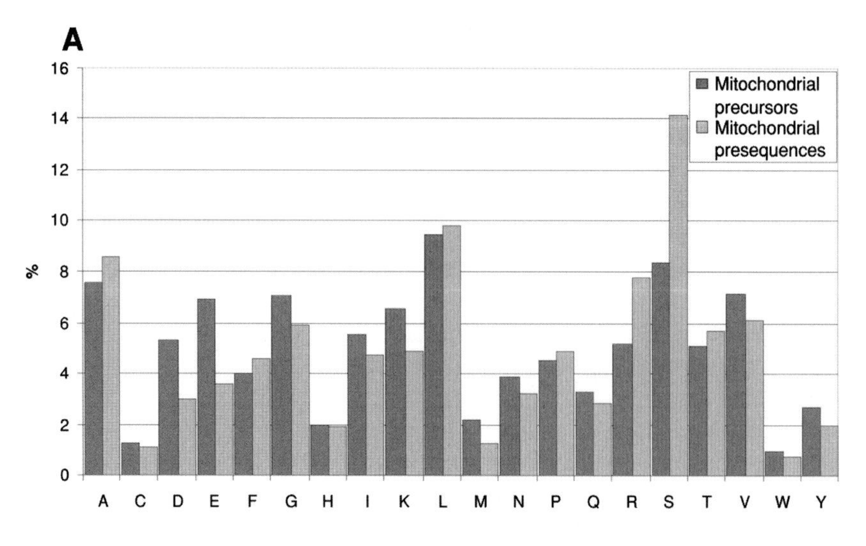

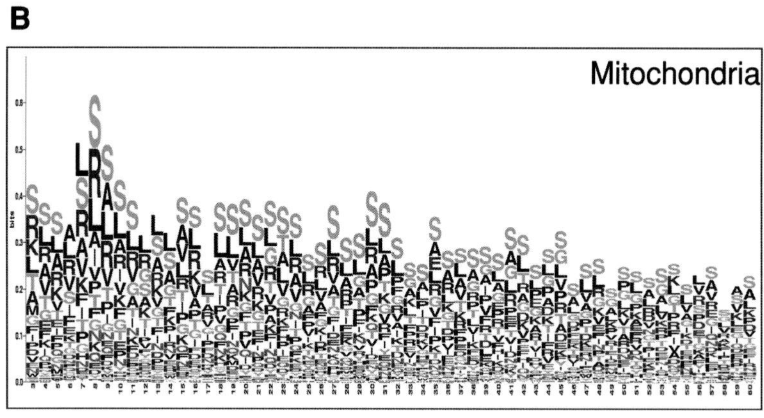

Plate 4.1 Analysis of the amino acid composition of the N-terminal part of proteins from the mitochondrial proteome of *Arabidopsis* containing a data set of 385 proteins. The global amino acid composition of the mitochondrial precursors and presequences (mTPs) (A) derived from SequenceLogos for the mitochondrial (B) proteins. The overall composition of mTPs shows a high content of 33–35% hydrophobic (Leu, Ala, Phe, Ile, Val), 22–23% hydroxylated (Ser, Thr) and 14–15% positively charged (Arg, Lys) amino acid residues and a very low abundance of acidic amino acids. Comparison of SequenceLogos for 385 mitochondrial and 567 chloroplast proteins show a clear difference in arginine content in the first 16 amino acids that is highly enriched in the mitochondrial presequences (not shown). See colour plate section following p. 110 for colour version of this figure.

Although targeting sequences are largely unstructured in aqueous solution, a lipid environment can induce helix elements. mTPs have the potential to form an amphiphilic α-helix (von Heijne, 1986) that is important for import. The predicted overall helical content for mTPs is ~40%. The structures of a few relatively short mTPs, e.g. of aldehyde dehydrogenase (ALDH) (Karslake *et al.*, 1990) and cytochrome c oxidase subunit IV (Chupin *et al.*, 1995), were determined to form amphiphilic α-helices in membrane-like environments. The first NMR structure of a mTP from higher plants, the *Nicotiana plumbaginifolia* F_1 β-ATPase, revealed the presence of two helices: a N-terminal amphipathic α-helix – the putative receptor binding site – and a C-terminal α-helix upstream of the cleavage site that are separated by a largely unstructured long internal domain (Moberg *et al.*, 2004). The NMR structure of the cytosolic domain of rat Tom20 in complex with the presequence of ALDH revealed that the hydrophobic side of the pALDH amphipathic α-helix is located in a hydrophobic groove of the Tom20 receptor (Abe *et al.*, 2000). Also, other mTPs interacted with the same binding site of Tom20 through hydrophobic interactions (Muto *et al.*, 2001). The NMR structure of Tom20 from *Arabidopsis* suggests a similar binding mechanism (Perry *et al.*, 2006). In comparison, cTPs form random coils although recent reports indicate an α-helical structure in a hydrophobic environment (Bruce, 2000). cTPs have regions with the potential to form a helical element(s), and the overall helical content has been calculated to $\sim$20%. The presence of helical elements in cTPs has been confirmed by NMR studies (Wienk *et al.*, 1999).

Nearly all mTPs contain Hsp70 binding motifs, a hydrophobic strech of five amino acid residues flanked by positively charged residues (Zhang and Glaser, 2002). In confirmed plant mTPs, 97% contain Hsp70-binding site(s). Interaction of mTPs with mtHsp70 during the import process has been supported by biochemical studies (Zhang *et al.*, 1999). mtHsp70 has an essential role in the mitochondrial import process (Okamoto *et al.*, 2002). mTPs also contain information determining the specificity for cleavage by MPP. This information is located as a loosely conserved motif at the C-terminus of the peptides, in combination with structural properties upstream of the cleavage site (Sjöling and Glaser, 1998; Zhang and Glaser, 2002; see Section 4.5.1).

4.2.2 *Dual-targeting peptides*

The mitochondrial and chloroplastic protein import systems are usually very specific; however, a subset of proteins has been shown to be dually targeted to both mitochondria and chloroplasts although encoded by a single nuclear gene (Peeters and Small, 2001; Silva-Filho, 2003). Such proteins are referred to as dual-targeted proteins. Since the first report of dual targeting of pea glutathione reductase (GR) in 1995 (Creissen *et al.*, 1995), 31 dual-targeted proteins have been identified and it is expected that this event may be far more common than originally thought. Dual targeting results from 'twin' or 'ambiguous' targeting peptides (Peeters and Small, 2001; Silva-Filho, 2003). Twin targeting signals may arise from multiple transcription initiation sites, alternative pre-mRNA processing or from variable posttranslation modification, all resulting in multiple precursor proteins each possessing different targeting information. For example, translation of protoporphirin oxidase II (protox-II) generates two products using in-frame initiation codons for either chloroplast or mitochondrial import (Watanabe *et al.*, 2001). *Arabidopsis* THI1 is encoded by a single nuclear gene and directed simultaneously to mitochondria and chloroplasts using two in-frame translational start codons (Chabregas *et al.*, 2001). Translation initiation at the first AUG directs translocation of THI1 to chloroplasts, and when translation starts from the second AUG THI1 is addressed to mitochondria.

The ambiguous signal arises from genes encoding single precursors with the same targeting signal being recognized by the import apparatus of both organelles (Small *et al.*, 1998). The majority of proteins that belong to this group are involved in gene expression, e.g. aminoacyl-tRNA synthetases, RNA polymerase, methionine aminopeptidases and a peptidyl deformylase. A recent study reported that among the 24 identified organellar aminoacyl-tRNA synthetases (aaRSs), 15 (and probably 17) are shared between mitochondria and plastids and 5 are shared between cytosol and mitochondria (one of these aaRSs being present also in chloroplasts) (Duchene *et al.*, 2005). The authors concluded that dual targeting of aaRSs appeared to be a general rule in *Arabidopsis*. Other dual-targeted enzymes are related to protection against oxidative stress, e.g. GR and ascorbate peroxidase (Chew *et al.*, 2003a,b), or to cellular protein turnover, such as the targeting peptide-degrading zinc metallopeptidase, PreP (Bhushan *et al.*, 2003; Moberg *et al.*, 2003).

The evidence from *in vitro* work suggests that the import pathway used by those proteins with ambiguous signals is indistinguishable from that taken by other imported precursors. Peeters and Small (2001) compared features of ambiguous targeting peptides with a large set of mTPs and cTPs, concluding that dual-targeting peptides contain classical features of both mTPs and cTPs but they contain fewer alanines and a greater abundance of phenylalanine and leucine, suggesting that they are more hydrophobic. Secondary structure predictions demonstrate a lower abundance of potential N-terminal helices in ambiguous targeting sequences than in classical mitochondrial-targeting sequences. It appears that these sequences are intermediary in character between mitochondrial- and plastid-targeting sequences and contain features from both. SequenceLogos of 45 amino acid residues of all known targeting peptides of the dual-targeted proteins supports this conclusion (Plate 4.2).

A study in which mTPs and cTPs were fused in tandem showed that the position of the targeting sequence is important and the N-terminal portion has the greatest importance for the final location (Silva-Filho *et al.*, 1997). Hedtke *et al.* (2000) reported that the dual-targeted protein, RNA polymerase RpoT;2, may have a domain structure with the N-terminal region being required for chloroplastic import and the C-terminal region being important for mitochondrial targeting. Also, the dual-targeting peptide of PreP reveals a domain structure but with an opposite orientation, the N-terminal region mediates mitochondrial import and the C-terminal region is important for chloroplastic targeting (Bhushan *et al.*, 2003). Furthermore, studies of the dual-targeting peptide of GR suggested a triple domain structure with two domains responsible for the organellar targeting and the third N-terminal domain controlling organellar import efficiency (Rudhe *et al.*, 2002b). Overall single mutations of positive and hydrophobic residues had a greater effect on mitochondrial import in comparison to chloroplast import (Chew *et al.*, 2003b). Investigations of determinants for processing on the GR dual targeting peptide in mitochondria by MPP and in chloroplasts by the stromal processing protease (SPP) (Richter and Lamppa, 1999) revealed that (i) recognition of processing site on a dual-targeted

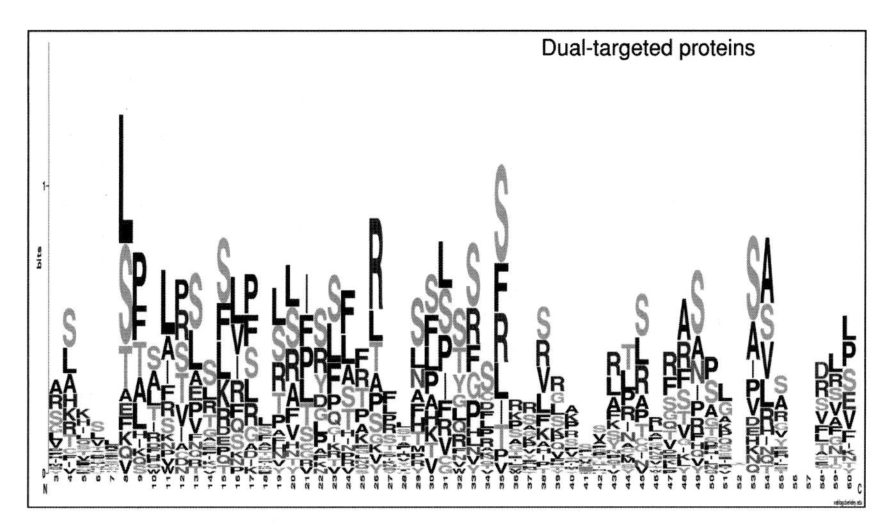

Plate 4.2 SequenceLogos of the 45 N-terminal amino acid residues of a collection of 27 known dual-targeting peptides (Silva-Filho, 2003). See colour plate section following p. 110 for colour version of this figure.

GR precursor differs between MPP and SPP, (ii) the GR targeting signal has similar determinants for processing by MPP as signals targeting only to mitochondria, and (iii) processing by SPP shows a low level of sensitivity to single mutations on targeting peptide and likely involves recognition of the physiochemical properties of the sequence in the vicinity of cleavage rather than a requirement for specific amino acid residues (Rhude *et al.*, 2004). It will be of great interest to perform a high-through-put investigation to identify all proteins that are dual targeted to mitochondria and chloroplasts in the plant cell, their determinants for import and mode of interaction with organellar receptors.

4.3 Cytosolic factors

Plant mitochondrial import shares several conceptual similarities with the chloroplast protein import process. Chaperones of the Hsp70 class have a general role in maintaining nascent proteins in a soluble and import-competent state (Murakami *et al.*, 1988). Hsp70s may operate in concert with specific cytosolic factors enhancing the interaction between mitochondrial receptors and the precursor/chaperone complex. The mammalian mitochondrial import-stimulating factor (MSF), a member of the 14-3-3 protein family (Hachiya *et al.*, 1993), was considered to have an important role in the targeting of a subset of mitochondrial precursors. Precursor-binding factor (PBF), an oligomer of 50-kDa subunits, was reported to bind to the presequence (Murakami *et al.*, 1992) and interact with Hsp70. Targeting Factor (TF), a 200-kDa homo-oligomer made from the 28-kDa subunits, was shown to specifically bind to the presequence of rat ornithine aminotransferase precursor protein

(Ono and Tuboi, 1990). Recently, using two-hybrid screening, Yano *et al.* (2003) identified arylhydrocarbon receptor-interacting protein (AIP) interacting with the mitochondrial receptor Tom20. Binding to Tom20 was mediated by the tetratricopeptide repeats (TPR) of AIP. AIP binds specifically to mitochondrial preproteins and forms a ternary complex with Tom20 and preprotein. AIP-enhanced import of preornithine transcarbamylase and depletion of AIP by RNA interference impaired the import. Furthermore, AIP has a chaperone-like activity and prevents substrate proteins from aggregation. These results suggest that AIP functions as a cytosolic factor that mediates preprotein import into mitochondria. There is an *Arabidopsis* homologue of AIP that shows 45% sequence similarity (27% identity); however, its function has not yet been investigated.

Surprisingly, mitochondrial precursor proteins synthesized in rabbit reticulocyte lysate (RRL) are easily imported into mitochondria, whereas the same precursors synthesized in wheat germ extract (WGE) fail to be imported (Dessi *et al.*, 2003). Only a precursor that does not require addition of extramitochondrial ATP for import, the F_Ad ATP synthase subunit (Dessi *et al.*, 1996; Tanudji *et al.*, 2001), could be imported from WGE. Investigation of chimeric constructs of precursors with switched presequences revealed that the mature domain, not the presequence, of the F_Ad precursor defines the import competence in WGE. The dual-targeted GR precursor synthesized in WGE was imported into chloroplasts but not into mitochondria (Dessi *et al.*, 2003). Investigations on the composition of WGE have been carried out by several laboratories, with regard to how WGE components interact with precursor proteins. Hsp70 in WGE associates with nascent precursors in an ATP-dependent manner (Miernyk *et al.*, 1992). A series of studies using the precursor of aspartate aminotransferase showed that the protein makes a very stable complex with Hsp70, possibly due to a lack or inactivity of cytosolic factors required to release proteins from plant chaperones (Mattingly *et al.*, 1993). Results of import experiments with an overexpressed $F_1\beta$ precursor depleted of any cytosolic factors point to the conclusion that the import incompetence of WGE-synthesized mitochondrial precursors is a result of interaction of WGE inhibitory factors with the mature portion of precursor proteins (Dessi *et al.*, 2003). There also exists a WGE factor that inhibits chloroplast protein import (Schleiff *et al.*, 2002).

Several recent observations support the cotranslational mitochondrial import of some proteins (Lithgow, 2000). In *S. cerevisiae*, cotranslational import is required to transport fumarase into the mitochondria (Stein *et al.*, 1994). The 3′ untranslated region (UTR) of ATM1, a gene encoding an ABC transporter of the mitochondrial inner membrane, is required for the localization of its mRNA to the surface of mitochondria (Corral-Debrinski *et al.*, 2000) and subsequent import. An mRNA-binding protein, protein kinase A anchoring protein (AKAP121) of the mitochondrial outer membrane (Ginsberg *et al.*, 2003), contains an RNA-binding domain that was shown to interact specifically with the 3′UTR of transcripts encoding the manganese superoxide dismutase (Mn-SOD) and the F_O f subunit of the ATP synthase. Yeast mutants of the nascent polypeptide-associated complex (NAC) and the ribosome-associated complex (RAC) display a slow growth phenotype on nonfermentable carbon sources and have defects in protein targeting to mitochondria, suggesting that translation and import processes are associated (George *et al.*, 1998; Funfschilling and Rospert, 1999).

4.4 Sorting of precursors between mitochondria and chloroplasts

The targeting of proteins to mitochondria and chloroplasts can be studied using both *in vitro* or *in vivo* approaches. *In vivo* approaches use an intact cellular system and obviously reflect the *in vivo* targeting capacity of a signal. However, they have several limitations arising from the fact that the investigated proteins are overexpressed at high levels, no kinetics can be assessed and passenger proteins intsead of native mature proteins are used. *In vitro* import approaches overcome these limitations, but they have other disadvantages due to a lack of an intact cellular system, e.g. lack of cytosolic factors and/or competition with other organelles.

Mitochondrial and chloroplast protein import was demonstrated to be highly specific *in vivo* (Boutry *et al.*, 1987; Glaser *et al.*, 1998; Soll, 2002). However, whereas mis-sorting of mitochondrial proteins into chloroplasts has not been reported, several studies have reported mis-sorting of chloroplast proteins into mitochondria *in vitro* using different chloroplast precursors, such as the PsaF protein (Hugosson *et al.*, 1995), triose-3-phosphoglycerate phosphate translocator (Brink *et al.*, 1994; Silva- Filho *et al.*, 1997), plastocyanin, the 33-kDa photosystem II protein (Cleary *et al.*, 2002) and Rubisco small subunit (Lister *et al.*, 2001). Furthermore, mis-sorting was also dependent on the source of plant mitochondria.

In order to overcome the limitations of the *in vitro* import system and ensure the correct specificity of targeting, a novel *in vitro* dual import system for simultaneous targeting of precursor proteins into mitochondria and chloroplasts has been developed (Rudhe *et al.*, 2002b). Purified organelles are mixed, incubated with precursors and repurified after import. This allows the determination of the targeting specificity into either organelle and the use of authentic precursors. By using the dual import system, it has been shown that the chloroplastic precursor of the small subunit of Rubisco was mistargeted to pea mitochondria in a single import system but was imported only into chloroplasts in the dual import system, whereas the dual-targeted GR precursor was targeted to both mitochondria and chloroplasts in both systems (Rudhe *et al.*, 2002a).

There are qualitative and positional differences in the distribution of amino acids in the very N-terminal portion of the mitochondrial and cTPs visualized with SequenceLogos (Plate 4.1B). Furthermore, SequenceLogos revealed that the content and distribution of residual amino acids of chloroplast transit peptides highly resemble a mitochondrial presequence. These observations led to a hypothesis that N-terminally deleted chloroplast transit peptides would direct proteins into mitochondria. However, experimental investigations of this hypothesis demonstrated that the deletion mutants were neither imported into chloroplasts nor mistargeted to mitochondria *in vitro* or *in vivo*, showing that the entire transit peptide is necessary for correct targeting as well as mis-sorting of chloroplast precursors to mitochondria (Bhushan *et al.*, 2006).

4.5 Translocation machinery

The availability of complete *Arabidopsis* and rice genome sequences, combined with the detailed characterization of the mechanisms involved in mitochondrial protein import in yeast, has provided a new platform to allow a dissection of the machinery and processes of protein import in plants (Arabidopsis Genome Initiative, 2000; Goff *et al.*, 2002; Yu *et al.*, 2002; Vij *et al.*, 2006). The primary reason for this is that even in yeast, where mitochondria are the most abundant organelle, it was genetic approaches that proved successful in elucidating the components of the multisubunit protein complexes involved in protein import. This was especially the case with the inner membrane components that are in low abundance compared to the more abundant respiratory chain components and metabolite carriers. Furthermore, a comparison of the import machineries between several organisms where genome sequence is available allows the identification of common or core verse lineage-specific components (Lister *et al.*, 2003; Murcha *et al.*, 2006). The availability of genome sequence also allows the expression, or regulation of expression, of import components to be viewed in a broader cellular context, especially relevant to multicellular organisms where organ- or tissue-specific demands require an adaptable and flexible system to ensure that mitochondria are specialized for the requirements of specific cell types.

A comparison of the genes present in various organisms encoding components of the mitochondrial protein import apparatus clearly indicates that this process is well conserved across diverse phylogenetic groups, with approximately 75% of components characterized in yeast having counterparts in several other organisms (Lister *et al.*, 2003; Murcha *et al.*, 2006). However, a detailed analysis indicates that the presence of a component in two organisms does not mean that they serve the same function and/or are homologous. Examples include Tom20 that appears to have a distinct evolutionary history between fungi/animals and plants (Lister and Whelan, 2006; Perry *et al.*, 2006), Tim17 that links the outer and inner membranes in plants (Murcha *et al.*, 2003, 2005a,b) and a modified role for Tom9 in plants that is equivalent to Tom22 in other organisms but lacks the cytosolic receptor domain (Macasev *et al.*, 2000, 2004).

4.5.1 Translocase of the outer membrane – the discriminating outer membrane

Although potato was initially used to characterize the Tom complex from plants, identification of all the components came from studies in *Arabidopsis* (Heins and Schmitz, 1996; Jansch *et al.*, 1996). A single dynamic Tom complex is present in the outer membrane, which must recognize all mitochondrial proteins but yet reject several thousand proteins that do not belong in mitochondria. Six Tom components could be identified in *Arabidopsis*: Tom 5, 6, 7, 9, 20 and 40 (Werhahn *et al.*, 2001, 2003; Werhahn and Braun, 2002). Of the three small Toms, 5, 6 and 7, only the latter appears to be conserved across all species (Murcha *et al.*, 2006). The basis of the specificity of the plant Tom receptors for mitochondrial proteins is of great interest due to the presence of plastids in plant cells whose targeting signals resemble mitochondrial presequences (See Section 4.2). Comparison of Tom20 between species indicates that the *At*Tom20s are C-terminally anchored compared to the N-terminal-anchored fungal and mammalian counterparts (Pfanner *et al.*, 1996; Werhahn *et al.*, 2001). TPR domains, important in protein–protein interactions, are present in Tom20s from yeast, mammals and *Arabidopsis*. An NMR structure of *At*Tom20 revealed that the binding site is similar to that of rat Tom20 in that they both were TPR proteins and hydrophobic interactions dominated (Abe *et al.*, 2000; Muto *et al.*, 2001; Perry *et al.*, 2006). This is consistent with other studies where mutations in positive charges in the presequence did not affect presequence binding in rat (Abe *et al.*, 2000; Muto *et al.*, 2001). Peptide scan studies with yeast Tom20 also indicate that hydrophobic interactions are important (Brix *et al.*, 1999). In contrast, studies with soluble hTom20 indicated a two-domain structure and positive charge, as well as hydrophobicity, was important for binding. Additionally, these studies suggested that the different domains of human Tom20 allow for interactions with different groups of precursor proteins (Schleiff and Turnbull, 1998; Schleiff *et al.*, 1999).

The structural similarities in binding site between plant and rat Tom20 are hardly surprising as it has been well documented that mitochondrial precursors from one organism are readily imported into mitochondria of another organism (Chaumont *et al.*, 1990; Luzikov *et al.*, 1994). However, detailed structural analysis combined with a wide variety of genome sequence and extensive EST databases from a variety of organisms suggested that Tom20 in plants verse animals/fungi represents a case of convergent evolution (Perry *et al.*, 2006). This is based on the fact that as well as being anchored differently in the membrane, the TPR domain of plant Tom20 is longer and belongs to a different clan to the domain in animal/fungi Tom20s. Thus plant Tom20 contains all the essential features except in reverse, and no mechanism of recombination can be envisaged to explain this. Instead an independent ancestry can readily explain this structure. The similarity in binding site and structure would have been driven by the nature of the protein to be targeted to mitochondria: many of the essential features being already present in the proteins encoded in the genome of bacterial mitochondrial ancestor. Thus this convergent process was driven by the need for higher recognition specificity of a common mitochondrial-targeting signal (Lister and Whelan, 2006).

A similar situation exists with the apparent lack of Tom70 in plants. No homologues can be detected in *Arabidopsis* suggesting that this protein is found only in fungi and animals (Lister *et al.*, 2003; Chan *et al.*, 2006; Murcha *et al.*, 2006). The location of a protein similar to Tom70 in the plant outer mitochondrial membrane, OM64, suggests that the protein may fulfil some of the functions of Tom70 (Chew *et al.*, 2004). Plant OM64 is not orthologous to Tom70 in fungi and animals but does contain some TPR domains in the cytosolic exposed domain, suggesting that it could play a role in the recognition of precursor proteins destined to be imported into plant mitochondria. Surprisingly, OM64 does display a high degree of sequence homology to translocase of the outer chloroplast envelope 64 (Toc64), which has been shown to act as a receptor for proteins delivered to plastids that are bound to HSP90 (Qbadou *et al.*, 2006).

The absence of the N-terminal cytosolic receptor domain of yeast Tom22 in plant Tom9 and the absence of Tom5 in plants suggests that plant mitochondria differ in how precursor proteins are bound. Plant Tom9 represents the functional equivalent of Tom22 in other organisms but the receptor domain has been lost (Macasev *et al.*, 2000, 2004). It is tempting to suggest that this is due to the presence of plastids in a plant cell, and thus receptor specificity for precursor proteins in plant mitochondria is higher than for fungal mitochondria. Several studies using mutagenesis of targeting signals indicate a critical role for hydrophobic residues in mitochondrial-targeting, in agreement with the structural characterization of *At*Tom20 (Perry *et al.*, 2006); one study showed the importance of hydrophobic residues within the predicted α-helix (Duby *et al.*, 2001a). Studies in our laboratory with other precursors have also shown that hydrophobic residues important for import are also close together, i.e. within ten amino acids, further supporting binding to a hydrophobic pocket-like structure (Lee and Whelan, 2004). Studies on plant presequences indicate that positive residues throughout the presequence are important since a combination of positively charged residues distributed at different positions in the preseqeunce had additive positive effects (Tanudji *et al.*, 1999; Zhang *et al.*, 2001; Lee and Whelan, 2004).

The bound preproteins are passed to the pore-forming unit Tom40. *Arabidopsis* Tom40 lacks the conserved NPGT(S) motif present in fungi, *Caenorhabditis elegans*, *Drosophila* and mammals, which is necessary for assembly and stability (Rapaport *et al.*, 2001; Murcha *et al.*, submitted). In yeast, there is a sequential assembly pathway in which a 250-kDa intermediate consisting of yeast Tom40, 7 and 5 interacts with a 100-kDa intermediate of yeast Tom22 and Tom6 to create a core complex of 400 kDa. The primary receptors yeast Tom20 and yeast Tom70 interact with this core to form the holo-complex (Stan *et al.*, 2000; Bruce, 2001). The core complex from *N. crassa* forms a 12 nm $\times$ 7 nm $\times$ 7 nm structure with two pores of 2.2 nm (Ahting *et al.*, 2001). Structural studies of purified *Nc*Tom40 indicate that more than one Tom40 protomer is necessary to form the observed channel (Ahting *et al.*, 2001). Studies with the isolated plant complex show a smaller complex that migrates at 250 kDa and contains the Tom20 receptor, in contrast to the 400-kDa core complex of yeast which does not contain Tom20 (Model *et al.*, 2001, 2002; Werhahn *et al.*, 2001).

In addition to the Tom complex of the outer membrane, an additional protein complex for the insertion and sorting of β-barrel proteins has been characterized in yeast, called the Tob (topogenesis of β-barrel proteins) or Sam (sorting and assembly machinery) complex (Pfanner *et al.*, 2004; Paschen *et al.*, 2005). To date, two core subunits have been defined in yeast, Sam35 (Tob38) (Milenkovic *et al.*, 2004; Waizenegger *et al.*, 2004; Habib *et al.*, 2005) and Sam50 (Tob55) (Paschen *et al.*, 2003, 2005; Wiedemann *et al.*, 2003; Pfanner *et al.*, 2004), together with an additional subunit Sam37 (Tob37/Tom37/Mas37 that is associated with the complex) (Pfanner *et al.*, 2004; Habib *et al.*, 2005; Paschen *et al.*, 2005). Additionally, two subunits Mdm10 and Mim1 (also called Tom13) interact with this complex, or affect this protein import pathway (Paschen *et al.*, 2005). This protein import pathway functions for the import of outer membrane β-barrel proteins that are initially recognized by the receptor components of the Tom complex and then transferred to the Sam complex for assembly into the outer membrane. Small Tim proteins (see below) in the intermembrane space may assist this transfer process. In plants, homologues of the Sam50 component can be clearly identified (Murcha *et al.*, 2006). Sam38, Mdm10 and Min1 appear to be only present in yeast or other fungi. The situation with Sam37 is less clear. It was originally identified in yeast and called Tom37 as it appeared to have some receptor function (Gratzer *et al.*, 1995; Hachiya *et al.*, 1995). It has very low-sequence identity with a mammalian protein called Metaxin, identified independently as affecting the import of proteins into the matrix (Armstrong *et al.*, 1997). A predicted protein in *Arabidopsis* is similar to metaxin (Murcha *et al.*, submitted). The function of Sam37 is unclear; it is not an essential protein like Sam35, or Sam50 in yeast, but yet in Sam37-deleted cells β-barrel proteins accumulate in the Tob complex, suggesting that it is important for release of assembled β-barrel proteins (Pfanner *et al.*, 2004; Paschen *et al.*, 2005). The roles of metaxin in mammals and the related protein in *Arabidopsis* are unknown.

Thus it appears that as with the two outer membrane complexes required for protein import, the channel-forming subunitsTom40 and Sam50, both β-barrel proteins, are well conserved in eukaryotes (Murcha *et al.*, 2006). However, significant differences likely exist with other subunits that appear to have different evolutionary ancestry, but may overlap in function with yeast analogues.

4.5.2 Rediscovered intermembrane space

Arabidopsis ESTs homologous to yTim8, 9, 10 and 13 can be identified. These proteins contain four conserved cysteine residues ($CX_3CX_{14\text{-}17}CX_3C$), which comprise the Zn^{2+} binding motif in yTim9 and 10. Potato Tim9 and 10 have been identified in an intermembrane space fraction from potato and the zinc-dependent import stimulation activity of carrier proteins demonstrated (Lister *et al.*, 2002; Murcha *et al.*, 2005a,b). No homologue of Tim12 could be identified in other species, suggesting that it is restricted to yeast. Two other proteins MIA40 and ERV1 that play an essential role in the import of small intermembrane space proteins are also present in *Arabidopsis* (Murcha *et al.*, submitted); these proteins are involved in a disulphite relay system that is involved in metal addition to these proteins as they are imported into the intermembrane space. No homologue of HOT13p is evident in *Arabidopsis*. However, as this protein has been suggested to be involved in reducing Mia40 and ERV1, it is possible that this role is undertaken by other proteins. Notably, in plants both ascorbate synthesis and gluthione reductase are located in the intermembrane space and could fulfil this role (Millar *et al.*, 2004).

4.5.3 Inner membrane – a tale of two Tims and more

Two multisubunit translocases on the inner membrane are responsible for the import of proteins into or across the inner membrane. Proteins that contain N-terminal targeting signals are imported via the Tim17:23 complex, called the general import pathway, while proteins containing internal targeting signals are imported via the Tim22 complex, called the carrier import pathway.

Homologues to all components of the Tim17:23 translocase can be identified in *Arabidopsis*, Tim21, Tim17, Tim23 and Tim50. Tim17 and 23 are integral membrane proteins; the latter forms the translocation channel required for the insertion of protein into or across the inner membrane (Truscott *et al.*, 2001). Tim21 and Tim50 have been identified more recently and both have large domains exposed to the intermembrane space and play a role linking the translocation of precursor from Tom to Tim (Geissler *et al.*, 2002; Yamamoto *et al.*, 2002; Mokranjac *et al.*, 2003, 2005a; Guo *et al.*, 2004; Chacinska *et al.*, 2005). The *Arabidopsis* homologues identified for Tim17 and 23 differ substantially in predicted structure compared to yeast (and other organisms) (Rassow *et al.*, 1999). *At*Tim17 is the one of the most conserved proteins of the import apparatus in *Arabidopsis* relative to yeast, displaying 44% identity or 52% similarity, just lower than HSP60 and 70 that are well conserved throughout evolution. *At*Tim17 is encoded by three genes on chromosomes 1, 2 and 5 but produces two different proteins. *At*Tim17 on chromosomes 1 and 2 (*At*Tim17-1 and -2) has a long C-terminal extension compared to homologues in other species and is predicted to produce proteins with predicted masses of 23 and 26 kDa, respectively. In contrast, *At*Tim17 encoded on chromosome 5 (*At*Tim17-5) is shorter than other species, with a predicted molecular mass of 14 kDa. It is truncated at the C-terminus and may lack the fourth transmembrane segment typical of this protein from a wide variety of organism (Bauer *et al.*, 1999; Rehling *et al.*, 2004). Several other plant species examined also have *Tim17* genes encoding proteins with a C-terminal extension. Studies in yeast have shown that the four-membrane-spanning protein is necessary for function, so it is difficult to envisage that *At*Tim17-5 can function in a similar manner to four-membrane-spanning Tim17 proteins. *At*Tim23 lacks the first 40 amino acids compared to the yeast counterpart. The N-terminal region of yeast Tim23 has been shown to be exposed at the outer membrane, and this region is necessary for efficient import into mitochondria (Donzeau *et al.*, 2000). Deletion of this region reduces import kinetics and cell growth greater than tenfold. Yeast Tim23 alone has been shown to be a voltage-dependent, cation-selective channel, and selectivity for presequences is mediated by the N-terminal intermembrane space region, much of which appears to be lacking in *At*Tim23. The leucine zipper region present in yeast Tim23 between residues 61 and 83 is not conserved in the *Arabidopsis* (or human) homologues and has been implicated in membrane potential-dependent Tim23 dimer formation (Bauer *et al.*, 1996).

In *Arabidopsis* AtTim17-2 links the outer and inner membranes. The C-terminal extension of the larger *At*Tim17 isoform must be removed before this protein can complement a yeast Tim17 mutant; neither the intact larger isoform nor the small isoform *At*Tim17-5 can complment the yeast mutant. Similarly, *At*Tim23 fails to complement a yeast Tim23 mutant. Furthermore, the link between the inner and outer membrane formed by Tim17-2 in *Arabidopsis* is necessary for efficient protein import into mitochondria when the Tom complex is by-passed, strongly

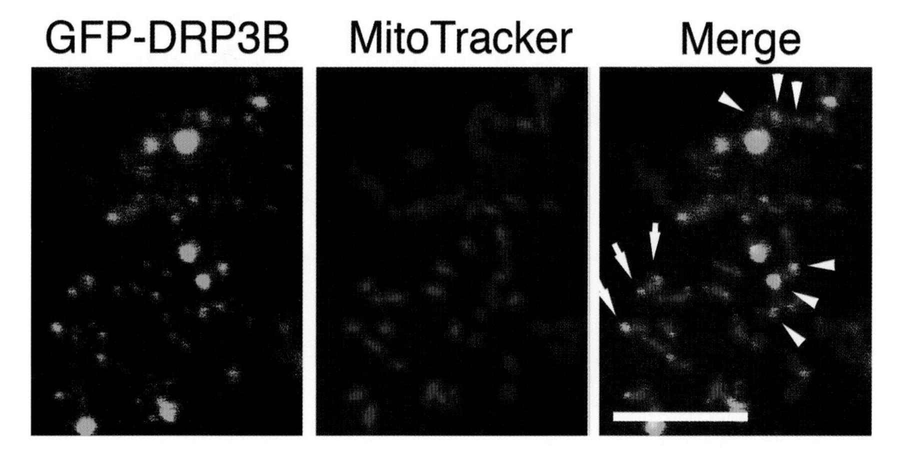

Plate 1.1 Localization of GFP-DRP3B with tips and constriction sites of mitochondria. BY-2 cells transiently expressing GFP-DRP3B, and stained with the mitochondrial marker MitoTracker, were examined by confocal laser scanning microscopy. High-magnification images of a part of a single BY-2 cell are shown. GFP-DRP3B (green) was localized in a punctate distribution and many foci colocalized with the tips (arrows) and constricted sites (arrow-heads) of mitochondria. Scale bar = 5 μm. (Reproduced from Arimura and Tsutsumi, 2002, copyright National Academy of Sciences, USA.)

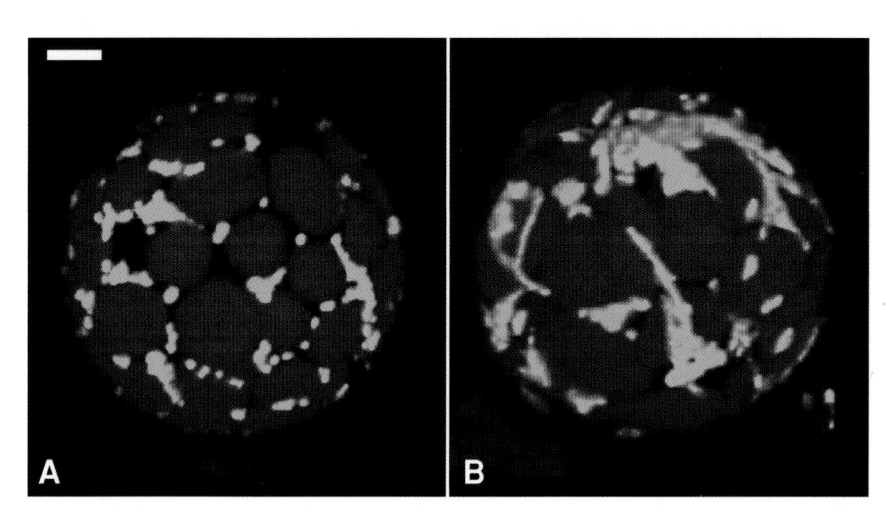

Plate 1.3 Mitochondrial distribution in the wild type and in the *friendly* (*fmt*) mutant of *Arabidopsis*. (A) Protoplast of wild-type *Arabidopsis* expressing mito-GFP, displaying the characteristic even distribution of mitochondria (green) amongst the cortical chloroplasts (red autofluorescence). (B) Large clusters of mitochondria in the *friendly* mutant. Scale bar = 5 μm.

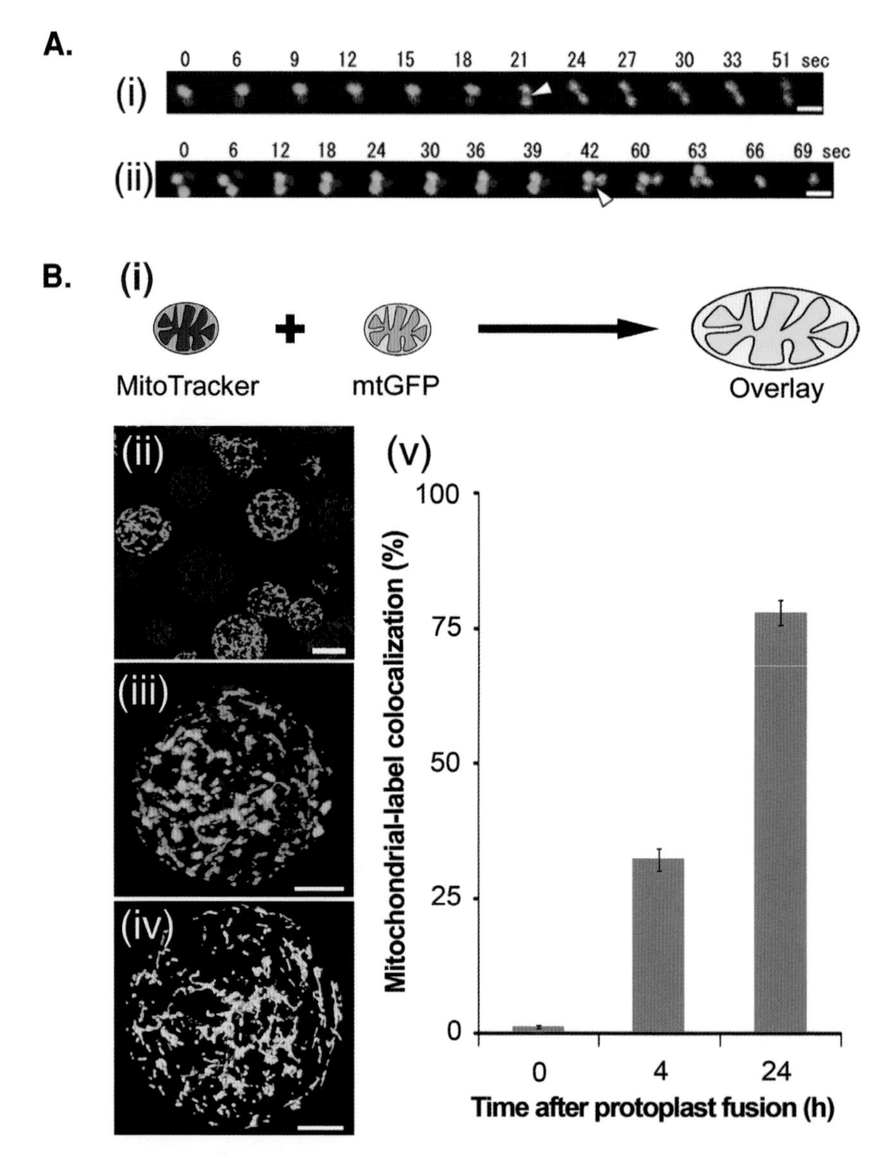

Plate 1.2 Mitochondrial fusion. (A) Mitochondrial fusion in onion bulb epidermal cells. (i) Mitochondrial fusion. After 21 s of observation, the previously physically discrete red or green mitochondria fuse, demonstrated by the appearance of yellow mitochondria. (ii) Mitochondrial fusion and fission. Three mitochondria are shown and fusion between the lower two organelles occurs between 39 and 42 s after the start of observation. This fusion event is only transient and by the 60-s time point the newly fused mitochondria have divided. Arrowheads indicate constrictions at the point of fusion. Scale bars = 2 μm. (Reproduced from Arimura *et al.*, 2004b, copyright National Academy of Sciences, USA.) (B) Mitochondria fuse during early protoplast culture. (i) Schematic representation of mitochondrial fusion as demonstrated by colocalization of green (mtGFP) and red (MitoTracker) mitochondrial labels to produce a yellow signal. (ii) Mixture of MitoTracker-stained (red) and mtGFP-expressing (green) protoplasts. (iii) Limited colocalization of label after 4-h protoplast culture. (iv) Almost complete colocalization after 24 h indicated that fusion occurred throughout the population of mitochondria. (v) Quantification of colocalization revealed that mitochondrial fusion was three-quarters complete in cultured protoplasts 24 h after protoplast fusion. Scale bar = 20 μm in (ii) and 10 μm in (iii) and (iv). (Reproduced from Sheahan *et al.*, 2005, copyright Blackwell Publishing Ltd.)

A

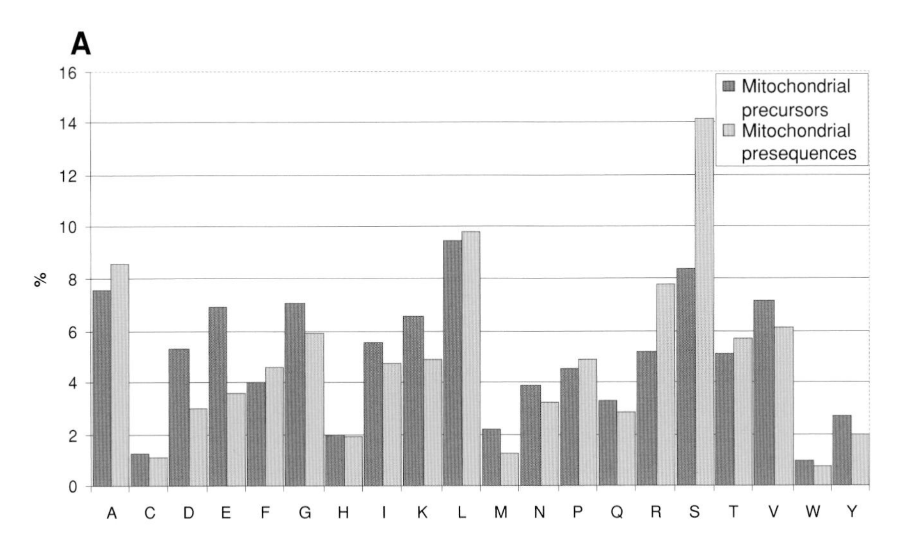

B

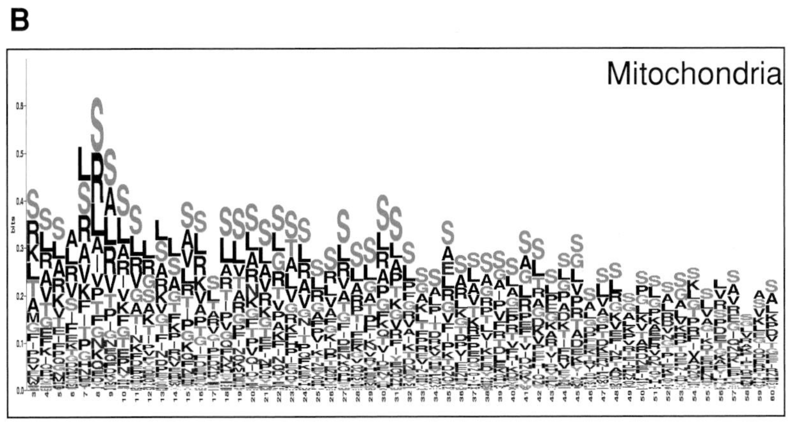

Plate 4.1 Analysis of the amino acid composition of the N-terminal part of proteins from the mitochondrial proteome of *Arabidopsis* containing a data set of 385 proteins. The global amino acid composition of the mitochondrial precursors and presequences (mTPs) (A) derived from SequenceLogos for the mitochondrial (B) proteins. The overall composition of mTPs shows a high content of 33–35% hydrophobic (Leu, Ala, Phe, Ile, Val), 22–23% hydroxylated (Ser, Thr) and 14–15% positively charged (Arg, Lys) amino acid residues and a very low abundance of acidic amino acids. Comparison of SequenceLogos for 385 mitochondrial and 567 chloroplast proteins show a clear difference in arginine content in the first 16 amino acids that is highly enriched in the mitochondrial presequences (not shown).

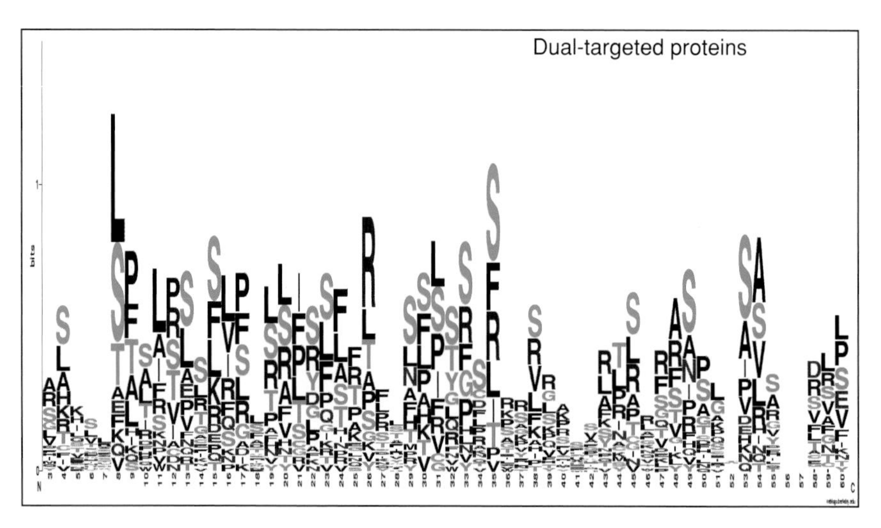

Plate 4.2 SequenceLogos of the 45 N-terminal amino acid residues of a collection of 27 known dual-targeting peptides (Silva-Filho, 2003).

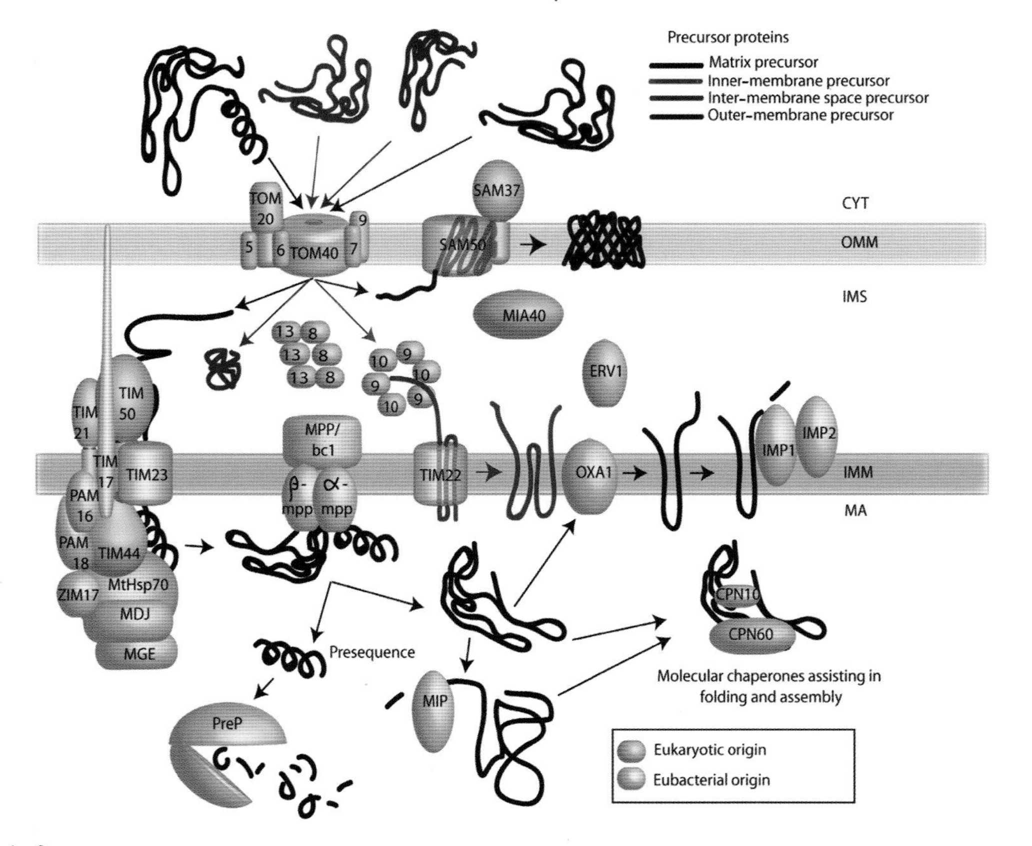

Plate 4.3 (legend overleaf)

Plate 4.3 The plant mitochondrial protein import machinery. Mitochondrial precursor proteins synthesized in the cytosol are specifically recognized by receptors on the TOM complex and translocated through the general import pore. The numerical labels refer to the name of the component and denote the apparent molecular mass of each protein in kilodaltons. Proteins with N-teminal targeting signals are recognized by receptors in the TIM23 complex, and translocated into the matrix. Oxa further sorts a small number of proteins imported into the matrix to the inner membrane. Tim17 component of the TIM23 complex extends from the inner to the outer mitochondrial membrane. Carrier proteins destined for insertion into the inner membrane interact with Tim9-Tim10 chaperone complexes in the intermembrane space, ferrying it from the TOM complex to the TIM22 complex, where the protein is inserted into the membrane. Outer membrane β-barrel proteins are imported through the TOM complex and inserted into the outer membrane by SAM. N-terminal signal peptides are cleaved off by the mitochondrial processing protease, MPP that in plants is integrated into the cytochrome bc_1 of the respiratory chain. The proteins can be further maturated by the MIP protease in the matrix or by IMP1 and IMP2 proteases after export to the intermembrane space or insertion into the inner membrane. The cleaved signal peptides are toxic to the mitochondrial membranes and degraded by the PreP. Mature proteins are folded and assembled to native protein complexes either spontaneously or with assistance of molecular chaperones, CPN60 and CPN10. Import components are coloured to indicate their putative evolutionary origin: 'eubacterial origin' denotes those components for which a likely ancestor in the endosymbiont has been proposed, while 'eukaryotic origin' denotes those components with no relevant similarity to bacterial proteins and which might have developed specifically in the host cell genome during or after the conversion of the endosymbiont to an organelle. Abbreviations: TOM, translocase of the outer membrane; TIM, translocase of the inner membrane; PAM, presequence associated moter; SAM, sorting and assembly machinery.

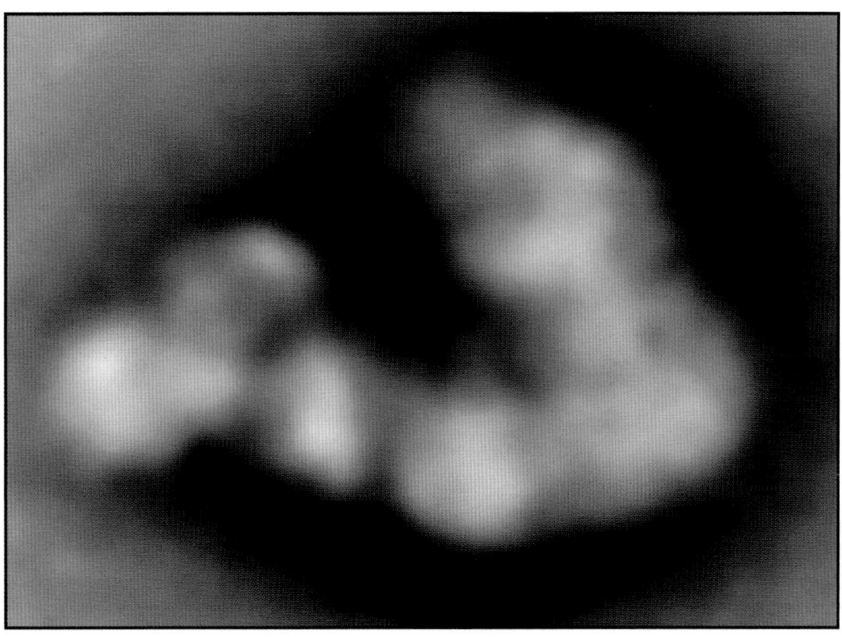

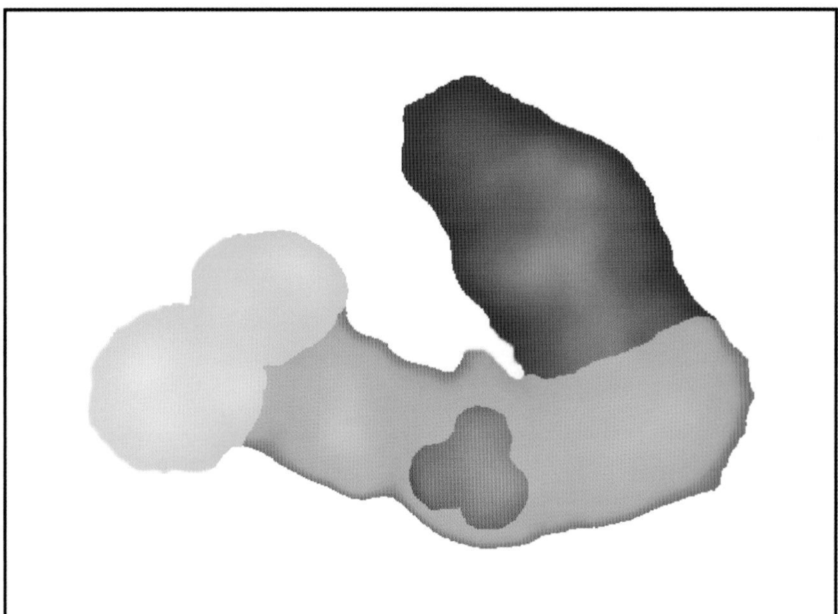

Plate 6.1 Structure of the I + III_2 supercomplex of *Arabidopsis*. Top: electron microscopy projection map obtained by single-particle averaging (see Dudkina *et al*., 2005a). Bottom: model of the supercomplex. Complex I is given in three shades of green (light green: matrix arm [this arm protrudes out of the plane of the image]; middle green: membrane arm; dark green: carbonic anhydrase domain).

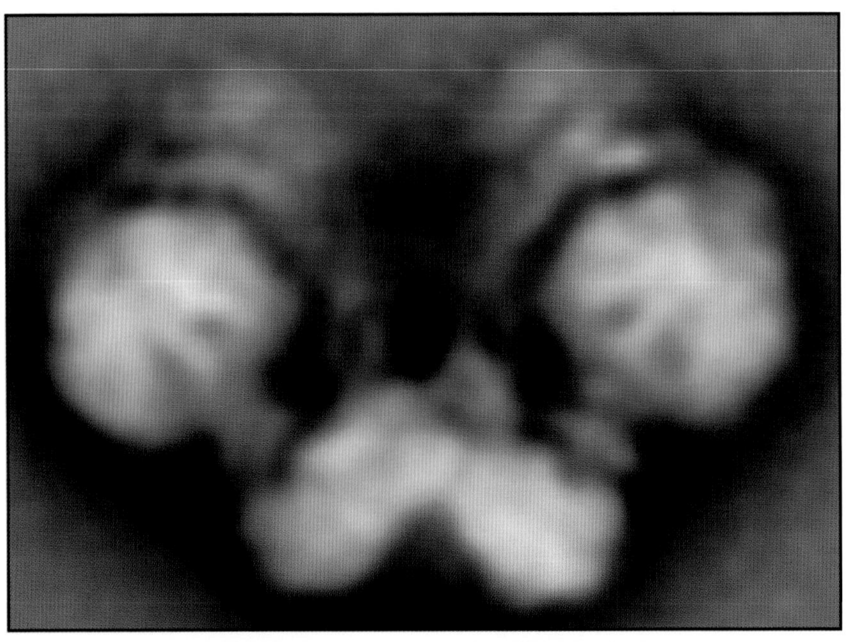

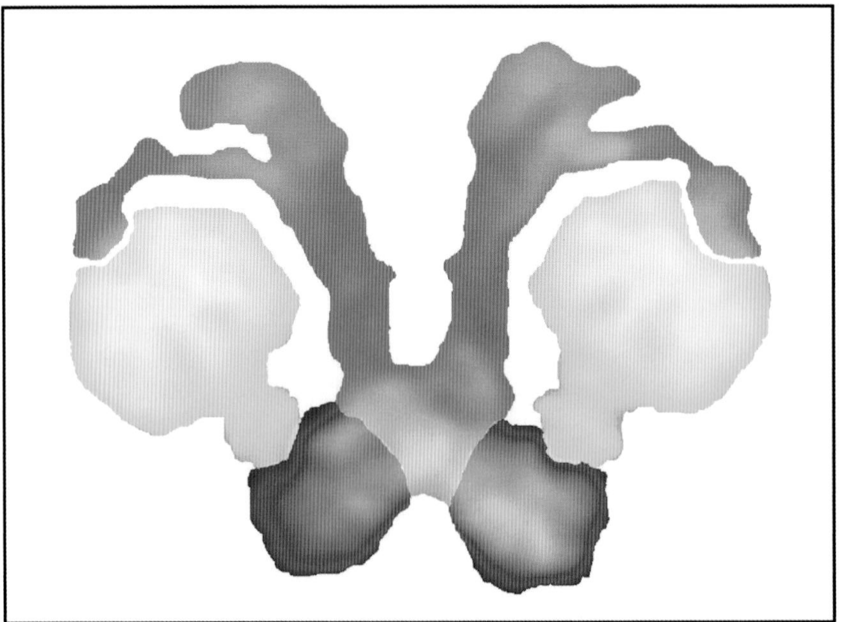

Plate 6.2 Structure of the dimeric ATP synthase supercomplex of *Polytomella*. Top: EM projection map obtained by single-particle averaging (see Dudkina *et al*., 2005b). Bottom: model of the supercomplex. The membrane-bound F_0 parts are given in red, the matrix-exposed F_1 parts and the central stalks in yellow, and the peripheral stalks in orange.

suggesting it plays a role in linking translocation between the Tom and Tim complex in *Arabidopsis* (Murcha *et al.*, 2005a,b). Thus, although the components of the Tim17:23 complex appear to be highly conserved across a range of species at the sequence level (Murcha *et al.*, submitted), structurally and functionally they may perform different roles in different species, and in some cases species-specific roles.

There is direct experimental evidence for the existence of a carrier import pathway in plants. Biochemical reconstitution experiments indicated that carrier proteins can be imported into the inner membrane in outer membrane-ruptured mitochondria and this import can be stimulated by soluble intermembrane space proteins, in particular by fractions enriched with small Tim proteins of the inter-membrane space (Lister *et al.*, 2002; Murcha *et al.*, 2005a,b). In plants, some carrier proteins contain an N-terminal extension (Bathgate *et al.*, 1989; Winning *et al.*, 1992). Both carrier proteins containing N-terminal cleavable extensions, and those without an extension, and proteins containing four transmembrane regions have been shown to be imported via this pathway in plants. Import experiments using this reconstituted system demonstrated that the extensions played a role in increasing the efficiency of insertion into the inner membrane with the aid of soluble intermembrane space components (Murcha *et al.*, 2005a,b). The N-terminal extension of these carrier proteins are removed in a two-step process; the first step carried out by the MPP on the matrix side of the inner membrane and the second by an unknown protease on the intermembrane space side of the inner membrane (Murcha *et al.*, 2004). This results in the N- and C-termini being exposed to the intermembrane space, an arrangment that is typical of carrier proteins.

Despite the abundant evident for a distinct carrier import pathway in plants, the identity of Tim22 in plants cannot be defined with confidence. This is due to the fact that in *Arabidopsis* there are 17 genes that encode proteins of the PRAT (protein and amino acid transporter) family; in rice there are 24 genes in this family (Murcha *et al.*, submitted). Three genes each encode isoforms of Tim17 and 23. Several of the other genes encode PRAT proteins that are targeted to mitochondria or plastids, or both (Murcha *et al.*, submitted). Four proteins that are targeted only to mitochondria do not branch with Tim17 or Tim23 and also do not group with yeast Tim22, so it is unclear which of these proteins represent Tim22 in *Arabidopsis*. Other components of the carrier import pathway in yeast such as Tim12, Tim18 and Tim54 do not appear to have homologues in the *Arabidopsis*, or rice genomes, and appear to be yeast- or fungal-specific components (Lister *et al.*, 2003). Thus it appears that although a carrier-type import pathway exists in plants, it differs considerably to that defined in yeast.

The final component of the import apparatus present in the inner membrane is Oxa1p, originally characterized as necessary for cytochrome oxidase assembly in yeast. Functional data exists for the *Arabidopsis* protein: it can complement a yeast mutant and contains an N-terminal presequence that targets to plant mitochondria (Hamel *et al.*, 1997; Sakamoto *et al.*, 2000). Thus, it can be concluded that an Oxa1p-like translocase exists in plants. Yeast Oxa1p belongs to a family of proteins that mediate insertion and/or assembly of membrane proteins (Yi and Dalbey, 2005). A homologue is present in chloroplasts and called Alb3 due to the mutant albino phenotype. Oxa1p and Alb3 are proposed to be related to YidC in *Escherichia coli*. Although additional components of this translocase have been shown to exist in thylakoids and mitochondria by cross-linking approaches, no information on their identity or role is known. In *E. coli*, YidC has been reported to be necessary for the insertion of all membrane proteins tested (Samuelson *et al.*, 2000). In mitochondria and chloroplasts, the homologues play a more limited role in the insertion of some proteins into the mitochondrial inner membrane, or thylakoid membrane. In yeast, Oxa1p appears to play a role in stabilizing membrane proteins before assembly. In its absence, these proteins are rapidly degraded by membrane-located ATP-dependant proteases (Lemaire *et al.*, 2000).

4.5.4 The matrix – the engine room

The mitochondrial matrix is the site of the import motor that pulls proteins into mitochondria. Given the fundamental nature of this requirement, it is not surprising that these components are generally well conserved across all species. The presequence translocase-associated motor (PAM) contains six subunits: mtHsp70 (Ssc1, DnaK), Tim44, Mge1 (GrpE), PAM18 (Tim14, DnaJ), Pam17 and Pam16 (Tim16) (Truscott *et al.*, 2003; Frazier *et al.*, 2004; Li *et al.*, 2004; Mokranjac *et al.*, 2005b). An additional associated subunit Zim17 (Tim15) has also been identified (Burri *et al.*, 2004). The matrix-located mtHsp70 forms the motor and binds the precursor polypeptide chain. Pam18 contains a J-domain that stimulates the ATPase activity of mtHsp70 similar to the bacterial DnaJ. However, in contrast to its bacterial counterpart, Pam18 is a membrane protein that binds to Tim17. Tim44 anchors (recruits) mtHSP70 to the membrane and the Tim17:23 channel. Pam16 and Pam 18 form a subcomplex that controls the activity of Pam18. Pam 17 is required for the formation of this subcomplex. Mge1 promotes nucleotide exchange, like GrpE in bacteria. Zim17 does not appear to be associated with the PAM complex directly but plays a more general role in preventing aggregation of mtHsp70 (Sanjuan Szklarz *et al.*, 2005). An additional J protein, Mdj2, has also been characterized in yeast and is present in humans. Mdj2 has 55% sequence identity with Pam18, and overexpression of Mdj2 can rescue a Pam18 mutant. Of the six proteins associated with Pam and Zim17, all can be identified in a variety of species except Pam17, that is only present in yeast and *C. elegans*. The HSP70 motor system in mitochondria is the most complex characterized to date compared to other organelles.

4.6 Proteolytic events

Proteolysis is crucial for the biogenesis, morphology and homeostasis of mitochondria. Genome analysis and proteomic data suggest that mitochondria contain about 40 proteases, but only a small number of them have been characterized so far (Esser *et al.*, 2002). A few proteases have been characterized as participating in the processing of precursor proteins, i.e. cleavage of the targeting peptide (Neupert, 1997; Gakh *et al.*, 2002). Furthermore, a novel metalloprotease-degrading targeting peptides after processing has been identified (Stahl *et al.*, 2002). The activities of these proteases are ATP independent. However, there is a group of ATP-dependent proteases that participate in the assembly of mature proteins by regulating the stochiometry of polypeptides in protein complexes; they also mediate the removal of misfolded and damaged organellar proteins. Mitochondria contain a few ATP-dependent proteases including the membrane bound FtsH proteases (also called AAA proteases, ATPases associated with a number of cellular activities) and the soluble Lon and ClpP proteases (Kaser and Langer, 2000; Adam and Clarke, 2002; Urantowka *et al.*, 2005). Beside processing proteases, only a few ATP-independent proteases have been identified in the mitochondria: MOP (metallopeptidase), a thimet oligopeptidase found in the intermembrane space of yeast (Buchler *et al.*, 1994; Kambacheld *et al.*, 2005) that is a functional homologue of the PreP; Oma, a membrane-bound metalloprotease (Kaser *et al.*, 2003) that has proteolytic functions overlapping with the m-AAA protease (and was therefore termed Oma); and a serine rhomboid membrane protease responsible for the degradation of intermembrane space proteins (Van der Bliek and Koehler, 2003). Generally, it can be concluded that ATP-dependent proteases catalyze the initial step of degradation by cleaving proteins into peptides that are subsequently degraded to smaller peptides, or free amino acids by the ATP-independent proteases (Adam and Clarke, 2002; Arnold and Langer, 2002).

4.6.1 Mitochondrial processing peptidase, MPP

MPP is a metalloendopeptidase that cleaves off specifically, in a single cut, presequences from several hundred mitochondrial precursor proteins (for review, see Neupert, 1997; Glaser and Dessi, 1999). MPP is essential in yeast (Baker and Schatz, 1991) and has been purified from *S. cerevisiae* (Yang *et al.*, 1988), *N. crassa* (Hawlitschek *et al.*, 1988), rat liver (Ou *et al.*, 1989) and several plant species (Braun and Schmitz, 1993, 1995b, 1999; Eriksson *et al.*, 1993, 1994; Brumme *et al.*, 1998). MPP consists of two structurally related subunits, α-MPP and β-MPP of $\sim$50 kDa each, which cooperate in processing. In mammals and yeast both MPP subunits are localized to the matrix. In *N. crassa,* 70% of β-MPP was found as a Core 1 protein of the cytochrome bc_1 complex of the respiratory chain (Schulte *et al.*, 1989). MPP contains an inverted zinc-binding motif ($HXXEHX_{74\text{-}76}E$) located on the β-MPP subunit, which classifies the protease to the pitrilysin family (subfamily M16B).

4.6.1.1 Integration of plant MPP into the cytochrome bc_1 complex of the respiratory chain

In contrast to nonplant sources, higher plant MPP activity was shown to reside in the mitochondrial inner membrane (Eriksson and Glaser, 1992; Glaser *et al.*, 1994), and the MPP subunits were shown to be completely integrated into the bc_1 complex of the respiratory chain as Core1 and Core 2 proteins of the complex in several higher plant species such as potato, spinach or wheat (Braun *et al.*, 1992, 1995; Eriksson and Glaser, 1992; Emmermann *et al.*, 1994; Emmermann and Schmitz, 1993; Eriksson *et al.*, 1993, 1994). In addition, the bc_1 complexes from lower plants, the staghorn fern *Platycerium bifurcatum* and the horsetail *Equisetum arvense* contained processing activity (Brumme *et al.*, 1998) indicating that integration of MPP into the bc_1 complex of the respiratory chain is a general feature for plants. Despite the fact that MPP/bc_1 complex in plants is bifunctional, the processing activity, however, is not dependent on the electron transfer within the complex (Emmermann and Schmitz, 1993; Eriksson *et al.*, 1994, 1996). The integration of MPP into a membrane-bound respiratory complex raises another intriguing question: are the presequences cleaved off upon import into mitochondria and does processing occur after translocation? There is no clear answer to that question. It has been shown that an inhibitor of processing – 1,10-phenanthroline, a metal chelator that can cross both mitochondrial membranes – abolished import of mitochondrial precursors, whereas phenanthroline analogues that do not penetrate membrane did not inhibit import (Whelan *et al.*, 1996). This study indicated connection between import and processing. However, experiments with immobilized import intermediates of different lengths (chimeric constructs of the cytochrome b_2 presequence attached to different lengths of the mature protein ranging from 25 to 136 residues, and fused to mouse dihydrofolate reductase), projecting into the matrix, showed that even the longest precursor could not be processed (Dessi *et al.*, 2000). The inability to process translocation intermediates implied that the MPP/bc1 complex in plants is distant from the translocation channel and cannot be directly involved in the translocation mechanism. It was concluded that presequence removal occurs after translocation and the translocation of the precursor and processing catalyzed by the MPP/bc1 complex are independent events in plants (Dessi *et al.*, 2000).

4.6.1.2 Catalysis and substrate specificity of MPP

A striking feature of MPP is that it is a general peptidase, as it acts on so many mitochondrial precursor proteins that show no sequence similarity, yet MPP is specific as it recognizes a distinct cleavage site on the presequences. How does the MPP recognize such a diversity of mitochondrial precursor proteins? What features determine the MPP cleavage?

Although mitochondrial presequences do not share any sequence similarity, there are some amino acids that are more abundant around the cleavage site and hence might be needed for processing. Plant mitochondrial presequences can be divided into three groups (Zhang, 2001). Two major groups representing 38% and 42% of presequences contain an arginine either at position -2 (-2Arg) or -3 (-3Arg) relative to the cleavage site, and have a loosely conserved motif around the cleavage site, Arg-X ↓ Ser-Thr/Ser-Thr, and Arg-X-Phe/Tyr ↓ Ala/Ser-Thr/Ser/Ala, respectively (Sjöling and Glaser, 1998). A third group includes presequences lacking a conserved arginine (no Arg) in the C-terminal portion of the signal sequence (Zhang *et al.*, 2001). In the case of -3Arg, there is an aromatic residue both at the –1 and at the +1 position. An extensive mutagenesis study of the plant mitochondrial presequence of alternative oxidase, which belongs to the -2Arg group, showed that positive residues throughout the presequence (both proximal and distant residues in relation to the processing site) played an important but not essential role in processing and mutation of -2Arg had the greatest inhibitory effect (Tanudji *et al.*, 1999). Furthermore,

Tanudji *et al.* (1999) showed that replacement of the serine residue at the +2 position relative to the cleavage site with a glycine residue, or deletion of gutamate at +1, drastically inhibited processing of soybean alternative oxidase. These results indicated that the processing reaction depends not only on the consensus upstream to the cleavage site but also on the downstream residues. In addition, there is often a proline residue between distal and proximal arginines that may serve as a flexible linker (Niidome *et al.*, 1994; Hammen *et al.*, 1994, 1996). Using intramolecularly quenched fluorescent substrates, Ogishima *et al.* (1995) demonstrated that at least one proline and one glycine residue between the distal and proximal arginine residues in mammalian malate dehydrogenase are important while other connecting sequences were dispensable. This would make it possible for the distal basic residues to vary in distance from the cleavage site for interaction with negative charges on MPP. This could facilitate the formation of a specific structure so that the scissile bond can be correctly presented in the enzyme pocket to an active water molecule on the catalytic metal. This has been suggested to be an 'induced fitting' as the substrate peptides lack secondary structure in an aqueous environment.

Studies of a presequence that does not contain arginine close to the processing site, the F_1-ATPase β-subunit ($F_1\beta$) from *N. plumbaginifolia*, showed that higher order structural properties of the C-terminal portion of the presequence, including the –2 position, even if not an arginine residue, are important determinants for the processing (Zhang *et al.*, 2001). Another studies with the *N. plumbaginifolia* $F_1\beta$ presequence and the 12 first residues of the mature protein fused to GFP (green fluorescent protein) showed that the residues located at position –1 and +2 relative to the cleavage site were also important for processing by MPP (Duby *et al.*, 2001b). The first NMR solution structure of a mitochondrial presequence from higher plants, *N. plumbaginifolia* $F_1\beta$, revealed the presence of two helices, an N-terminal amphipathic α-helix, the putative import receptor-binding site and a C-terminal α-helix upstream of the cleavage site, separated by a largely unstructured long internal domain (Moberg *et al.*, 2004). The amphiphilic α-helix in the N-terminal domain is a common predicted feature for mitochondrial presequences and has been shown to be important for mitochondrial targeting (von Heijne, 1986; Karslake *et al.*, 1990; Thornton *et al.*, 1993; Hammen *et al.*, 1996; Abe *et al.*, 2000). A few studies have also suggested a second α-helical structure in the C-terminal portion of presequences with no Arg as an important recognition determinant for cleavage by MPP (Sjöling *et al.*, 1994; Duby *et al.*, 2001a; Zhang *et al.*, 2001). However, structural studies of both mammalian presequences and the *N. plumbaginifolia* $F_1\beta$ presequence clearly showed a requirement for flexibility in the helical elements of the presequences for efficient processing (Karslake *et al.*, 1990; Thornton *et al.*, 1993; Wang *et al.*, 1993; Hammen *et al.*, 1996; Sjöling *et al.*, 1996; Zhang *et al.*, 2001). The presequences that formed a long continuous helix through the signal sequence were not processed. Mutational insertions, disrupting the helix and making the signal more flexible, plus insertion of a typical –3Arg cleavage site, made two mammalian nonprocessable proteins processable by rat (Waltner and Weiner, 1995) and spinach MPP (Sjöling *et al.*, 1996). However, the same mutations of thiolase did not make it processable, indicating again that a typical cleavage motif, –2Arg or –3Arg, is not the only criterion for efficient processing. Processing studies of plant MPP with mutant peptides derived from the *N. plumbaginifolia* $F_1\beta$ presequence also revealed the importance of structural flexibility (Sjöling *et al.*, 1994).

These results show that higher-order structural elements upstream of the cleavage site are important for processing by MPP. It has been suggested that the α-MPP, which probably recognizes a three-dimensional (3D) motif adopted by the presequence, presents the presequence to β-MPP (Luciano *et al.*, 1997). The structural helix-turn-helix element, adopted by cleavable presequences in a membrane mimetic environment, may be required for processing but is not sufficient for proteolysis. Presequence binding by α-MPP tolerates a high degree of mutations in the presequence. α-MPP may present a degenerate cleavage site motif to β-MPP in an accessible conformation for processing (Luciano *et al.*, 1997).

Secondary structure predictions of plant presequences with confirmed cleavage sites (Schneider *et al.*, 1998; Sjöling and Glaser, 1998), using nnPredict software, show that there is a common secondary structure around the scissile bond. This structure consists of a helix, or a helix followed by an extended stretch, located in the C-terminal region of the presequence in front of the known cleavage site. This predicted motif can be found in at least 50% of the plant presequences (Sjöling and Glaser, 1998).

Crystallization of recombinant yeast MPP in complex with synthetic signal peptides of the –2Arg group demonstrated that signal sequences were bound to MPP in an extended form (Taylor *et al.*, 2001). Furthermore, recognition sites for the arginine at position –2 and +1 aromatic residue were observed. The authors suggested that the presequences adopt context-dependent conformations through mitochondrial import and processing, helical for recognition by mitochondrial import machinery and extended for cleavage by the main processing protease.

In summary, the known determinants for recognition of the processing site include proximal arginines, distal positively charged residues, flexible linkers, residues on the C-terminal side of the cleavage site and secondary structure elements such as a helix followed by an extended conformation, which may facilitate the recognition event.

4.6.2 Mitochondrial intermediate peptidase, MIP

Several reports showed that a few mitochondrial precursor proteins targeted to the mitochondrial matrix in rat or *S. cerevisiae* are processed in two steps: first by MPP and then by the mitochondrial intermediate peptidase (MIP) (for review, see Gakh *et al.*, 2002). These precursors contain a characteristic octapeptide sequence, the MIP cleavage motif, (F/L/I)XX(T/S/G)XXXX downstream of the MPP cleavage site, in the C-terminal portion of the presequence. MIP is a thiol-dependent metallopeptidase that belongs to the thimet (thiol- and metal dependent) oligopeptidase family (Barrett *et al.*, 1995). It has been identified as a monomer of ~75 kDa in the matrix of fungi and mammals (Isaya *et al.*, 1991; Kalousek *et al.*, 1992). MIP contains a metal-binding site, HEXXH, and is enriched in cysteine residues. Single amino acid substitutions in the HEXXH motif resulted in loss of MIP activity, consistent with the metal dependence of MIP *in vitro* (Chew *et al.*, 1996). MIP is not essential for viability in *S. cerevisiae*, but disruption of the *mip1* gene causes loss of respiratory competence indicating involvement of MIP in the maturation of respiratory components. CoxIV, ubiquinol-cytochrome c reductase iron sulphur protein and malate dehydrogenase are among known MIP substrates in *S. cerevisiae* (Branda and Isaya, 1995). However, the sequence analysis data indicates that approximately 65 mitochondrial matrix proteins can be processed by MIP in *S. cerevisiae*. The biological significance of the cleavage reaction catalyzed by MIP is unknown. It has been shown that the N-terminal region of MIP cleaved precursor proteins is structurally incompatible with cleavage by MPP and octapepides may have evolved to overcome this incompatibility (Isaya *et al.*, 1991).

The *Arabidopsis* genome contains an MIP homologue that is predicted to be localized to the mitochondrial matrix. In contrast to nonplant species, *Arabidopsis* MIP harbors an expanded C-terminal domain that is predicted to form a transmembrane segment. *Arabidopsis* mutants homozygous for a T-DNA insertion in the *MIP* gene are viable. We have found that a typical MIP substrate (*S. cerevisiae* CoxIV) could be processed twice: first by the isolated MPP/bc_1 complex from *S. tuberosum* and second by matrix extract from *S. tuberosum* that completed maturation (A. Ståhl and E. Glaser, unpublished results). Investigation of a collection of 58 plant mitochondrial presequences with confirmed cleavage sites identified six potential MIP substrates that contained the –10R (A. Ståhland E. Glaser, unpublished results).

4.6.3 Inner membrane peptidase, IMP

There are a few known nuclear-encoded proteins destined for the intermembrane space that are first imported into the matrix and then exported to the intermembrane space. These proteins are synthesized with a bipartite N-terminal signal sequence consisting of a matrix-targeting signal, which is processed by MPP followed by a hydrophobic sorting signal that is subsequently cleaved by IMP. IMP has been identified in the *Arabidopsis*, *S. cerevisiae* and human genomes. In *S. cerevisiae*, IMP consists of two catalytic subunits, Imp1 and Imp2, that display 31% sequence identity (Schneider *et al.*, 1994), and an additional auxiliary subunit, Som1, that was identified as a suppressor of an Imp1 point mutation (Esser *et al.*, 1996). Som1 interacts physically with Imp1 improving its cleavage efficiency (Jan *et al.*, 2000; Liang *et al.*, 2004). Both Imp1 and Imp2 are anchored to the mitochondrial inner membrane via an N-terminal membrane-spanning domain while the C-terminal domain carrying the catalytic faces the intermembrane space.

Imp1 and Imp2 belong to the type-I signal peptidase (SP) family of proteases, including endoplasmic and eubacterial SPs (Dalbey, 1991). SPs process the signal sequences of membrane and secretory proteins. The catalytic domain of Imp1 and Imp2 is characterized by a conserved Ser-Lys dyad (Chen *et al.*, 1999). In contrast to the functional conservation exhibited by most enzymes in the type-I SP family, the IMP subunits display nonoverlapping substrate specificities (Nunnari *et al.*, 1993). Imp1 has four known substrates – the precursors to NADH-cytochrome b_5 reductase, $cytb_2$, FAD-dependent glycerol-3-phosphate (G-3-P) dehydrogenase and

CoxII – whereas Imp2 has only one known substrate – the precursor to cyt c_1 (Nunnari *et al.*, 1993). Involvement of IMP in processing of electron transport components accounts for the fact that IMP is required for cell respiration in yeast. Type-I SPs recognize, in general, small-uncharged amino acids at –1 position and also to some extent at –3 position in relation to the processing site. However, Imp1 tolerates a variety of structurally distinct residues at –1 position (Chen *et al.*, 1999). Position +1 has been implicated for both Imp1 and Imp2 to be important for substrate specificity; however, in a recent study, Luo *et al.* (2006) have shown that Imp1, but not Imp2, recognizes specific residues in +1 position (acidic residues). In addition, Imp1 showed substrate specificity for –3 position (hydrophobic nonaromatic residues), whereas Imp2 recognized both the –1 (alanine) and the –3 (small uncharged or polar residues) positions. This indicates that Imp2 displays the most similar substrate specificity to the type-I SPs. Interestingly, studies of coexpression of mammalian IMP with mammalian IMP precursors in yeast showed that G-3-P dehydrogenase has switched from Imp1 in yeast to Imp2 in mammals (Lou *et al.*, 2006). Homologues of both Imp1 and Imp2 have been found in *Arabidopsis*, but no homologue of Som1 has been identified.

4.6.4 Mitochondrial peptidasome, the presequence protease, PreP

Free targeting peptides inside mitochondria and chloroplasts are potentially harmful for the integrity and function of the organelles as they can penetrate membranes and dissipate membrane potential. The free targeting peptides must, therefore, be degraded or removed from the organelles. We have recently identified a novel metalloprotease responsible for the degradation of targeting peptides in both mitochondria and chloroplasts called PreP (Stahl *et al.*, 2002; Bhushan *et al.*, 2003, 2005, 2006b; Moberg *et al.*, 2003; Johnson *et al.*, 2006).

4.6.4.1 Identification, intracellular localization and expression of PreP

In order to investigate the fate of the mTPs after import is complete, we have produced a recombinant truncated mitochondrial import-competent precursor protein, a 15-kDa N-terminal fragment of *N. plumbaginifolia* $F_1\beta$ precursor of the ATP synthase, $N_{15}pF_1\beta$ (Pavlov *et al.*, 1999) and studied its *in vitro* import. While processing of $N_{15}pF_1\beta$ with the isolated MPP/bc$_1$ complex resulted in detection of a 54 amino acids-long presequence, the presequence could not be detected after import, indicating *in organello* degradation of the presequence (Stahl *et al.*, 2000). Chromatographic protein purification studies resulted in the isolation of a matrix protease responsible for the presequence degradation and its identification by mass spectrometric analysis. The protease was identified as a metalloendopeptidase belonging to the pitrilysin M16A subfamily. This subfamily contains proteases of ~100 kDa that are also called inverzincins as they contain an inverted zinc-binding motif (HXXEHX$_{74}$E) (Becker and Roth, 1993). *Arabidopsis* contains two PreP paralogues, *At*PreP1-AAL90904 on chromosome 3 and *At*PreP2-AAG13049 on chromosome 1 (Stahl *et al.*, 2002) that display 93% sequence similarity. Both human (hMP1) and yeast (Ydr430cp) PreP homologues have been found in the NCBI database.

Catalysis studies with recombinant *At*PreP proteins showed that PreP was able to degrade not only mTPs but also chloroplastic transit peptides thereby raising the question of the intracellular localization of *At*PreP proteases. In order to study the subcellular localization of *At*PreP1 and *At*PreP2, we have used a wide range of approaches including bioinformatics, immunological studies, substrate specificity studies, mass spectrometry, *in vitro* import and transient *in vivo* expression in tobacco protoplasts and leaves of GFP-fusion constructs containing *At*PreP1 and *At*PreP2 targeting peptides (Stahl *et al.*, 2002; Bhushan *et al.*, 2003, 2005; Moberg *et al.*, 2003). The intracellular location prediction programs, Predotar and TargetP, classified both *At*PreP1 and *At*PreP2 as either mitochondrial or chloroplastic, but there was no consistency in the predicted localization. Immunological studies with anti-*At*PreP1 antibodies localized PreP to both mitochondria and chloroplasts; however, the antibodies did not differentiate between *At*PreP1 and *At*PreP2. Mass spectrometer analysis of *Arabidopsis* chloroplasts provided evidence for the presence of *At*PreP1 and indicated that *At*PreP2 may also be present (Bhushan *et al.*, 2005). As PreP was originally identified as a mitochondrial protein, a possible dual targeting of *At*PreP1 and *At*PreP2 was studied using *in vivo* and *in vitro* protein import approaches. The targeting peptides of *At*PreP1 and *At*PreP2 predicted to be 85-amino acids long and 40-amino acid residues of the mature protein of *At*PreP1 and *At*PreP2 were fused to GFP, and transient expression of these constructs *in vivo* was investigated in tobacco protoplasts and leaves (Bhushan *et al.*, 2003, 2005) infiltrated with *Agrobacterium* carrying the GFP fusion constructs. Confocal microscopy analysis showed that targeting peptides of both *At*PreP1 and *At*PreP2 have dual-targeting properties to both mitochondria and chloroplasts. Targeting peptide of *At*PreP1 was shown to be organized in domains in which the first domain is necessary for the mitochondrial import and the second is sufficient for import into chloroplasts (Bhushan *et al.*, 2003). *In vitro* import experiments of *At*PreP1 and *At*PreP2 using both single and dual import system confirmed the conclusions drawn from the *in vivo* expression of the GFP constructs showing that both *At*PreP1 and *At*PreP2 are dually targeted to both mitochondria and chloroplast and the targeting peptide of *At*PreP1 is organized in domains specific for targeting to a single organelle (Bhushan *et al.*, 2003, 2005). In conclusion, PreP is dually targeted and executing its function not only in mitochondria but also in chloroplasts (Bhushan *et al.*, 2003; Moberg *et al.*, 2003).

Expression analysis of the *AtPreP1* and *AtPreP2* transcripts was studied using semi-quantitative RT-PCR for the quantitative measurements of the transcripts. Both the *AtPreP1* and *AtPreP2* transcripts were detected in young seedlings, however in varying amount. The *AtPreP1* transcript was detected to be present in silique and flower although the transcript level was much higher in flower. In contrast to the *AtPreP1* transcript, the *AtPreP2* transcript was found to be present in leaf, flower and root with no transcript detected in shoot and silique. These results showed that both *AtPreP1* and *AtPreP2* are expressed in an organ-specific manner in *Arabidopsis* plants (Bhushan *et al.*, 2005).

An important question to be addressed was whether or not the PreP protease is responsible for degradation of targeting peptides within both mitochondria and chloroplasts. Immunoinactivation studies with PreP-specific antibodies removed the proteolytic activity against targeting peptides from the mitochondrial matrix as well as from the chloroplast stroma suggesting that PreP is indeed the targeting peptide degrading protease both in mitochondria and in chloroplasts (Stahl *et al.*, 2002; Moberg *et al.*, 2003).

4.6.4.2 Catalysis and substrate specificity of PreP

The PreP protease contains an inverted zinc-binding motif that constitutes residues H77-I78-L79-E80-H81-X_{96}-E177 in *At*PreP1. Three recombinant mutants of *At*PreP1 (H77L, E80Q and H81L) were found to be inactive confirming the importance of these residues in proteolysis (Moberg *et al.*, 2003). However, additional residues contributing to the catalytic site were first shown to be essential when the crystal structure of the enzyme was determined at 2.1 Å resolution (Bhushan *et al.*, 2006b; Johnson *et al.*, 2006). PreP is the first M16 protease to have its crystal structure determnined in the closed, substrate-bound configuration. Proteolysis of the substrate takes place inside a chamber formed by two halves of the enzyme (the N- and the C-terminal parts) connected by a hinge region. The third zinc ligand that was previously unknown has been estimated to be Glu177. The most astonishing discovery was the presence of C-terminal residues at the active site, namely Arg848 and Tyr854, residues separated by almost 800 residues in sequence from the active site. Substituting these residues by Ala and Phe, respectively, resulted in the inactivation of the enzyme (Johnson *et al.*, 2006). The way the substrates are processed inside a chamber is reminiscent of the proteasome, and therefore PreP was named as a peptidasome (Johnson *et al.*, 2006). On the basis of the structure and mutagenesis experiments in which disulphide bridges were introduced to close and open the enzyme by changing redox conditions, a mechanism was proposed in which access to the active site involves hinge-bending motions that cause the peptidasome to open and close in response to substrate binding (Bhushan *et al.*, 2006b; Johnson *et al.*, 2006).

Substrate specificity studies showed that PreP degrades not only mTPs and cTPs but also other unstructured unrelated peptides, such as insulin B-chain and galanin (Moberg *et al.*, 2003). A tightly folded peptide, ala-α_3w, could not be degraded. The maximum length of substrates subjected to proteolysis by PreP was not longer than 70 amino acids when the assay employed peptides derived from the N-terminal part of the $F_1\beta$ precursor protein consisting of 63 or 73 amino acid residues, or when insulin-like growth factor I (IGF-I), a polypeptide of 70 amino acid residues that forms three α-helices, which are connected by a 12-residue linker, or calbindin D_{9K}, a polypeptide of 75 amino acid residues that forms an antiparallel four-helix bundle, where used in the assay. The minimum length for a peptide substrate was longer than seven amino acids. Taken together, these results show that *At*PreP1 and *At*PreP2 have a narrow activity window and only recognize and cleave peptides of approximately 10–65 amino acids in length (Stahl *et al.*, 2005). We have recently shown that human PreP degrades amyloid β-peptides, Aβ(1-40), Aβ(1-42) and

Arctic Aβ(1-40) associated with Alzheimer's disease in human brain mitochondria (Falkevall *et al.*, 2006).

The subsite specificities of *At*PreP1 and *At*PreP2 were analysed by studying their proteolytic activity against the $F_1\beta$ presequence, different peptides and mutants thereof, derived from the presequence as well as nonmitochondrial peptides. These studies indicated that PreP does not recognize amino acid sequence *per se* (Stahl *et al.*, 2005). The degradation products were identified by MALDI-TOF spectrometry and superimposed on the 3D structure of the $F_1\beta$ presequence (Moberg *et al.*, 2004). PreP showed preference for basic amino acids in the P_1' position, and small uncharged amino acids, or serines in the P_1 position. Both PreP paralogues cleaved almost exclusively toward the ends of the α-helical elements of the $F_1\beta$ presequence. However, despite the high sequence identity between *At*PreP1 and *At*PreP2 and similarities in cleavage specificities, cleavage-site recognition differs for both proteases and is context- and structure dependent.

4.6.5 A membrane-bound ATP-dependent protease, FtsH

Following import and maturation, proteins have to be assembled into their native protein complexes and superfluous subunits have to be removed. The FtsH proteases, also called AAA proteases, are membrane-bound ATP-dependent metalloproteases that have been shown to function in degrading the superfluous subunits (Langer, 2000). These proteases have been found in eubacteria and in mitochondria and chloroplasts. Whereas there is a single FtsH protease gene in bacteria and three orthologues in yeast and humans (Arnold and Langer, 2002), the *Arabidopsis* genome contains a total of 12 *FtsH*-like genes (Sokolenko *et al.*, 2002). Mitochondria contain two types of AAA proteases differing in membrane topology, m-AAA (Yta10 and Yta12) facing the matrix and i-AAA (Yta11 or Yme1) facing the intermembrane space. It is assumed that AAA proteases participate in degradation and assembly of membrane proteins. Nonassembled cytochrome oxidase subunit 2 (Pearce and Sherman, 1995) and protein inhibitors of the ATP synthase (Kominsky *et al.*, 2002) have been confirmed as native substrates of the i-AAA protease. The protease is suggested to harbor a chaperone-like activity, and it has been shown that yeasts harboring an inactivated *yme1* gene display several phenotypes associated with mitochondrial dysfunctions (Thorsness *et al.*, 1993). A novel metalloprotease, Oma1, with an activity overlapping with that of the *m*-AAA protease was identified in the inner membrane of *S. cerevisiae* mitochondria (Kaser *et al.*, 2003). It was shown that misfolded Oxa1 can be degraded in an ATP-dependent manner by the *m*-AAA proteases, or in an ATP-independent manner by Oma1.

Of the 12 *Arabidopsis* FtsH-like proteins, 8 were suggested to be targeted to chloroplasts and 4 to mitochondria (Adam and Clarke, 2002). Relatively little is known about FtsH proteases in plant mitochondria. A homologue of m-AAA protease in *Pisum sativum* mitochondria, PsFtsH, has been identified, characterized and suggested to be involved in the assembly of the ATP synthase (Kolodziejczak *et al.*, 2002). Two *Arabidopsis* FtsH isoforms, AtFtsH3 and AtFtsH10, show high sequence identity to PsFtsH and group together close to Yta10 and Yta12 in a phylogenetic analysis, indicating that they are subunits of the m-AAA complex. Two other isoforms, AtFtsH4 and AtFtsH11, encode proteins with a high similarity to Yme1p, a subunit of the i-AAA complex in yeast mitochondria. In addition, phylogenetic analysis groups together the AtFtsH4, AtFtsH11 and Yme1 proteins. By using biochemical and immunological approaches, it has been recently demonstrated that AtFtsH4 and AtFtsH11 are integral proteins of the inner mitochondrial membrane and expose their catalytic sites toward the intermembrane space, thereby corresponding to the i-AAA complex. Furthermore, AtFtsH4 is exclusively a mitochondrial protein while AtFtsH11 is found in both chloroplasts and mitochondria (Urantowka *et al.*, 2005). Database searches revealed that orthologs of AtFtsH4 and AtFtsH11 are present in both monocotyledons and dicotyledons. The two plant i-AAA proteases differ significantly in their termini: the FtsH4 proteins have a characteristic stretch of alanine residues at the C-terminal end, while FtsH11s have long N-terminal extensions. Blue-native gel electrophoresis revealed that AtFtsH4 and AtFtsH11 form at least two complexes with apparent molecular masses of ~1500 kDa. This finding implies that plants, in contrast to fungi and metazoa, have more than one complex with a topology similar to that of yeast i-AAA (Urantowka *et al.*, 2005).

4.6.6 Lon-like ATP-dependent proteases

The ATP-dependent Lon (or La) protease is a serine protease localized in the mitochondrial matrix that plays an important role in mitochondrial biogenesis. The protease has been implicated to have a regulatory role in DNA metabolism, in degradation of misfolded, or abnormal mitochondrial proteins (Suzuki *et al.*, 1994) and in the cellular defence against oxidative stress via the selective removal of oxidation-damaged proteins (Bota and Davies, 2002; Bota *et al.*, 2002). Lon proteases are highly conserved and found in archae, eubacteria and in mitochondria and chloroplasts (Van Dyck and Langer, 1999). *S. cerevisiae* mitochondria contain the Lon homologue, Pim1 (protease in mitochondria 1), that is a homo-oligomer composed of seven subunits, totalling ~800 kDa (Stahlberg *et al.*, 1999). Pim1 consists of two catalytic domains: a chaperone-like ATPase domain and a protease domain. Native substrates of Pim1 in *S. cerevisiae* mitochondria include various nonassembled polypeptides, such as β-MPP, and subunits α, β, γ of the ATP synthase (Kaser and Langer, 2000). Mechanistic studies of ATP-dependent proteolysis demonstrated that substrate unfolding is a prerequisite for processive peptide-bond hydrolysis. Initial cleavage occurred preferentially between hydrophobic amino acids located within highly charged environments at the surface of the folded protein. Subsequent cleavage proceeded sequentially along the primary polypeptide sequence. Lon was proposed to recognize specific surface determinants or folds, initiate proteolysis at solvent-accessible sites and generate unfolded polypeptides that are then processively degraded (Ondrovicova *et al.*, 2005). Recently, it has been shown that downregulation of the human Lon protease results in disruption of mitochondrial structure, loss of function and cell death, with the majority of cells undergoing caspase 3 activation and apoptosis within 4 days (Bota *et al.*, 2005). Overall, the findings show an important role of the Lon proteolytic system for the degradation of oxidized protein within the mitochondrial matrix and for the maintenance of mitochondrial structural and functional integrity (Bulteau *et al.*, 2006).

Four Lon protease isoforms are found in *Arabidopsis*, of which Lon1 is localized to mitochondria (homologue of Pim1) and the remaining three isoforms to the chloroplasts (Adam *et al.*, 2001). Only one of the three chloroplast-targeted gene products is upregulated at high light intensity (Sinvany-Villalobo *et al.*, 2004). Lon transcripts are much less abundant than those of other ATP-dependent proteases. It is also evident that transcripts encoding proteases that are targeted to chloroplasts are more abundant than those destined to mitochondria (Sinvany-Villalobo *et al.*, 2004). A gene encoding Lon protease has been identified in maize (Barakat *et al.*, 1998). Maize Lon1p can replace the Pim1p function in yeast for maintaining mitochondrial DNA integrity. In common bean mitochondria, Lon1 has been identified as at least one protease involved in the degradation of orf239, a cytoplasmic male sterility-associated mitochondrial mutation (Sarria *et al.*, 1998).

4.6.7 ATP-dependent Clp protease

Clp proteases (Caseinolytic protease) are ATP-dependent serine proteases that have both proteolytic and chaperone functions. They have been identified in bacteria, mammalian and plant mitochondria (but not in *S. cerevisiae*) (Corydon *et al.*, 1998; Santagatra *et al.*, 1999; Halperin *et al.*, 2001a,b), and in chloroplasts (Adam and Clarke, 2002; Adam *et al.*, 2006). The Clp protease in *E. coli* forms a multimeric complex consisting of proteolytic (ClpP) and regulatory (ClpA or ClpX) subunits. The ClpP subunit has an active site characteristic of serine proteases: a catalytic triad of Ser-His-Asp. The regulatory subunits have a chaperone-like function. The crystal structure at 2.3 Å resolution of the Clp protease in *E. coli* has been solved (Wang *et al.*, 1997). The proteolytic complex is composed of two central heptameric rings of ClpP that form a hollow chamber of 50 Å, which is flanked by one or two hexameric rings of ClpA or ClpX. The catalytic chamber has two narrow openings at either end of about 10 Å, suggesting that the substrates are unfolded by the chaperone activity of ClpA or ClpX before transfer and presented to the ClpP active site. However, *E. coli* ClpXP can accommodate two to three polypeptides at the same time, indicating a wider opening of about 20–25 Å (Burton *et al.*, 2001). The structural features of ClpP are strikingly similar to those observed in the 20S proteasome of *S. cerevisiae* and *Thermoplasma acidophilum* (Lowe *et al.*, 1995) and *E. coli* HslV (Bochtler *et al.*, 2000). Bacterial Clp proteases are involved in degradation of specific regulatory proteins, in the general removal of aggregated and misfolded proteins as well as nascent peptide chains that are stalled on the cytosolic ribosomes, and proteins involved in the starvation and in oxidative stress responses (Gottesman *et al.*, 1998). It can be concluded that Clp proteases are a part of a cellular quality control system (Adam *et al.*, 2006).

Plants have a far more complex Clp proteolytic system than bacteria. The *Arabidopsis* Clp system includes at least 26 Clp-related nuclear genes with 15 genes encoding plastid-localized proteins: 5 serine-type ClpP proteases, which all possess the catalytic triad motifs, 4 ClpP-related ClpR proteins that are homologous to ClpP but do not contain the conserved proteolytic three residues motif, 3 ClpA homologues (ClpC1, ClpC2 and ClpD) and 3 Clp family members of unknown function (ClpS1, ClpS2 and ClpT) (Halperin *et al.*, 2001a; Peltier *et al.*, 2004; Adam *et al.*, 2006). An additional gene encoding a ClpP-type protein is located on the plastid genome (Kuroda and Maliga, 2003). ClpP2, ClpX1 and ClpX2 were predicted to be localized to mitochondria. Specific polyclonal antibodies confirmed ClpP2 and ClpX localization in plant mitochondria. Transcripts for both *ClpX* and *ClpP2* genes were detected in various tissues and under different growth conditions, with no significant variation in mRNA abundance (Helperin *et al.*, 2001). By using beta-casein as a substrate, plant mitochondria were found to possess an ATP-stimulated, serine-type proteolytic activity that could be strongly inhibited by antibodies specific for ClpX or ClpP2, suggesting an active ClpXP protease.

Plastids contain a tetradecameric Clp protease core complex of ~325 kDa consisting of one to three ClpP and four ClpR proteins. In contrast, mitochondria from potato tuber contained only a single homo-tetra-decameric ClpP2 complex (Peltier *et al.*, 2004). No evidence was found for accumulation of any other Clp proteins predicted to be located in mitochondria. The ClpP2:GFP fusion protein was transiently expressed in tobacco leaves and shown to be imported into mitochondria confirming the mitochondrial localization of ClpP2 (Peltier *et al.*, 2004). Despite the fact that no natural ClpXP substrate was identified in mitochondria the plant Clp protease is likely to play a housekeeping role (Halperin *et al.*, 2001b; Adam *et al.*, 2006).

4.6.8 An integral membrane Rhomboid protease

The Rhomboid protease is an integral membrane protease that can cleave substrates within a membrane-spanning segment. It was first identified as present in the *Drosophila* Golgi apparatus, where it is responsible for the proteolytic activation of epidermal growth factor receptor ligands (Urban *et al.*, 2001) and regulates signal transduction during development. The catalytic triad of the Rhomboid protease, composed of Ser-His-Asn, is located within the transmembrane domains, indicating that the protease is an unusual serine protease with an active site located within the membrane bilayer. As a result, the Rhomboid protease has an important role in the assembly of membrane proteins. Bioinformatic analysis revealed that Rhomboid proteases are ubiquitous, being present in archea, eubacteria and eukaryotes (Koonin, 2003). Whereas most prokaryotes have a single gene copy of the protease, *Drosophila* has seven genes and *Arabidopsis* eight. One of the yeast isoforms of the Rhomboid protease, Rbd1, has been shown to be a mitochondrial protein involved in processing the bipartite signal peptides of cytochrome c peroxidase, Ccp1 and a Dynamin-like GTPase, Mgm1, both located in the intermembrane space of mitochondria (Esser *et al.*, 2002; Herlan *et al.*, 2003, 2004; McQuibban *et al.*, 2003; van der Blick and Koehler, 2003; see Chapter 1). Ccp1 and Mgm1 were shown to be first cleaved by the m-AAA protease and MPP, respectively, and then by Rbd1. It has been suggested that cleavage of Mgm1 by the Rhomboid protease regulates mitochondrial membrane remodeling indicating an important role of the protease in regulation of membrane biogenesis (McQuibban *et al.*, 2003; see Chapter 1). So far there is no information on the role of the Rhomboid protease in plants, but out of eight isoforms identified in *Arabidopsis* at least two are predicted by Predotar and TargetP to reside in mitochondria, and possibly play a role in the processing of the N-terminal extension present on some carrier proteins in plants (Murcha *et al.*, 2004).

4.7 Evolution of protein import components

The endosymbiotic origin of mitochondria immediately provides a clue to the origin of the components of the protein import apparatus (Plate 4.3). Overall, it is apparent that the basic pore-forming subunits, Tom40, Sam50, Tim17, 22, 23 and Oxa1 can all be clearly defined as having bacterial ancestors (Rassow *et al.*, 1999; Herrmann, 2003; Paschen *et al.*, 2003; Dyall *et al.*, 2004; Gentle *et al.*, 2005). Likewise, the PAM complex contains several chaperone-like components that have counterparts in bacteria. Thus, it is not difficult to envisage how genes encoding these components duplicated, and subsequently underwent specialization to play roles in protein import. For other components, a clear evolutionarily picture does not exist as the low levels of sequence homology makes identification of homologues unreliable. In the case of the Tom complex, both Tom9 and Tom7 components can be identified in all organisms examined to date, while for the Tim translocases, the widespread occurrence of Tim8, 9, 10, 13, 21, 44 and 55 suggests that they were derived from bacterial components or arose early in the evolution of the eukaryotic cell. In the case of Tom20 and Tom70, it is now apparent that plants lack a homologue to the latter and plant Tom20 is not homologous to that of fungi or mammals (Chan *et al.*, 2006; Perry *et al.*, 2006). As outlined above, the plant Tom20 receptor represents an elegant example of convergent evolution (Lister and Whelan, 2006; Perry *et al.*, 2006).

Mitochondrial proteases, including processing proteases, MPP, MIP and IMP, the targeting-peptide degrading PreP protease, as well as the ATP-dependent FtsH, Lon and Clp proteases, the Rhomboid protease all have bacterial homologues.

An interesting evolutionary model has been suggested for MPP based on the fact that this protease is localized to different intracellular compartments in different organisms (Braun *et al.*, 1995; Braun and Schmitz, 1995a). It is suggested that early in evolution, MPP evolved from a bacterial protease and became part of the cytochrome c reductase complex. Later in evolution, the two subunits of MPP became detached from the enzyme complex to allow independent regulation of protein processing and respiration in some organisms. The detachment was realized by gene duplications, because the coevolution of cytochrome c reductase and MPP led the two MPP subunits to become indispensable for assembly of this respiratory protein complex. From the perspective of this model, the core subunits are relics of an ancient processing peptidase, and the bifunctional cytochrome c reductase complex in plants represents a situation that was originally present in the mitochondria from all organisms.

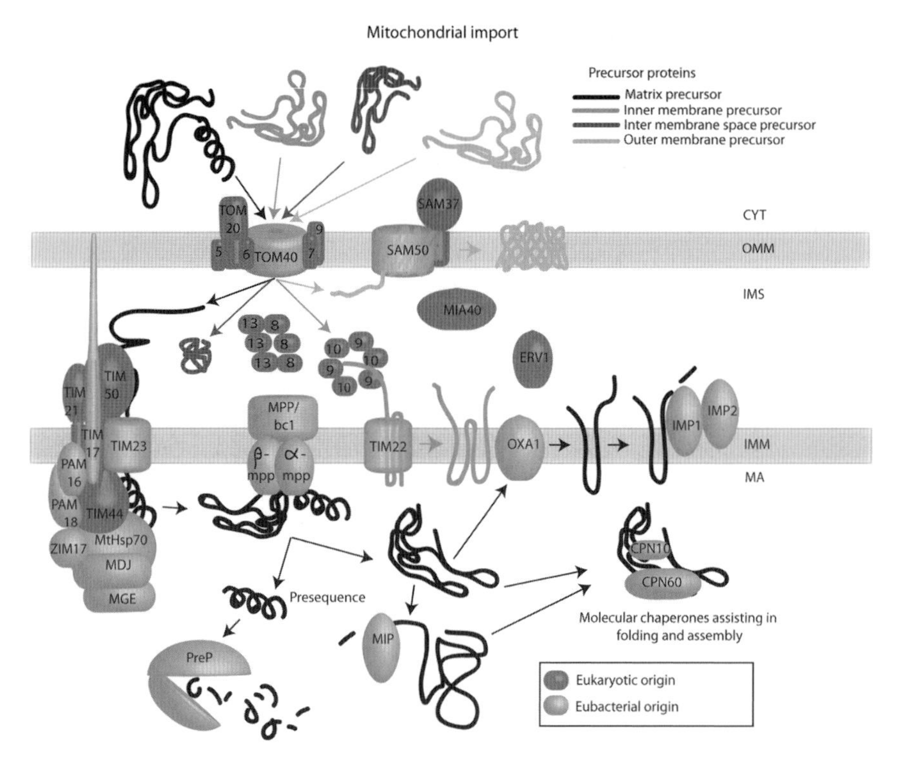

Plate 4.3 The plant mitochondrial protein import machinery. Mitochondrial precursor proteins synthesized in the cytosol are specifically recognized by receptors on the TOM complex and translocated through the general import pore. The numerical labels refer to the name of the component and denote the apparent molecular mass of each protein in kilodaltons. Proteins with N-teminal targeting signals are recognized by receptors in the TIM23 complex, and translocated into the matrix. Oxa further sorts a small number of proteins imported into the matrix to the inner membrane. Tim17 component of the TIM23 complex extends from the inner to the outer mitochondrial membrane. Carrier proteins destined for insertion into the inner membrane interact with Tim9-Tim10 chaperone complexes in the intermembrane space, ferrying it from the TOM complex to the TIM22 complex, where the protein is inserted into the membrane. Outer membrane β-barrel proteins are imported through the TOM complex and inserted into the outer membrane by SAM. N-terminal signal peptides are cleaved of by the mitochondrial processing protease, MPP that in plants is integrated into the cytochrome bc_1 of the respiratory chain. The proteins can be further maturated by the MIP protease in the matrix or by IMP1 and IMP2 proteases after export to the intermembrane space or insertion into the inner membrane. The cleaved signal peptides are toxic to the mitochondrial membranes and degraded by the PreP. Mature proteins are folded and assembled to native protein complexes either spontaneously or with assistance of molecular chaperones, CPN60 and CPN10. Import components are coloured to indicate their putative evolutionary origin: 'eubacterial origin' denotes those components for which a likely ancestor in the endosymbiont has been proposed, while 'eukaryotic origin' denotes those components with no relevant similarity to bacterial proteins and which might have developed specifically in the host cell genome during or after the conversion of the endosymbiont to an organelle. Abbreviations: TOM, translocase of the outer membrane; TIM, translocase of the inner membrane; PAM, presequence associated moter; SAM, sorting and assembly machinery. See colour plate section following p. 110 for colour version of this figure.

4.8 Genomic perspective of mitochondrial protein import components

The availablility of the *Arabidopsis* and rice genome sequences (Arabidopsis Genome Initiative, 2000; Goff *et al.*, 2002; Yu *et al.*, 2002) have not just allowed the painting of a detailed picture of plant mitochondrial protein import as outlined above, but also allows insights into the control of gene expression with respect to tissue and developmental factors, a facet of regulation that is not relevant in yeast. Overall, it is evident that many components of the mitochondrial protein import apparatus are encoded in small gene families in various species (Lister *et al.*, 2003; Murcha *et al.*, 2006). To some degree, plants represent an extreme situation where many components are encoded by more genes than in higher animals. Although studies are limited it is apparent that the expression pattern of these genes differs (Lister *et al.*, 2004), but as yet no difference in function has been reported. Even when comparing *Arabidopsis* and rice, the number of genes encoding a given component varies; for instance, rice appears to have a single gene encoding Tom20 whereas there are 4 in *Arabidopsis*, and also 17 genes encode PRAT-like proteins in *Arabidopsis* compared to 24 in rice (Murcha *et al.*, 2006).

4.9 Concluding remarks

Although the availability of the complete *Arabidopsis* genome sequence has enabled the identification of components of the mitochondrial protein import apparatus in plants, functional genomic approaches now need to be applied to characterize the precise function of the newly identified proteins and the mechanism involved. In contrast to many components of the plastid protein import apparatus, use of the powerful reverse genetic approach has not been reported for any genes encoding mitochondrial protein import components (Soll and Schleiff, 2004; Bedard and Jarvis, 2005; Gutensohn *et al.*, 2006). Thus, although there is a clear *in silico* picture of the protein import apparatus, there is little empirical proof. For instance, apart from the fact that plant mitochondria contain a Tom20 receptor, with a determined NMR structure (Perry *et al.*, 2006), it is unknown if there are other receptors, or if the four genes that encode Tom20 in *Arabidopsis* have undergone functional specialization (Lister and Whelan, 2006). The latter question is relevant for many other components of the protein import apparatus. Even though many components are present in gene families, often the encoded proteins differ considerably in length, or lack domains previously described as essential for function; thus, it appears that some functional diversification has taken place (Murcha *et al.*, 2003, 2004, 2006). The other possibility is that gene families exist for regulatory specialization, raising the possibility that different cell types have a different protein import apparatus – the question is why. Do the Tom and Tim complexes in plants contain a single or multiple isoforms of proteins?

The *in silico* picture is also limited in that only components characterized in yeast will be evident. Given that there appear to be several yeast, or fungal-specific components, it is reasonable to expect that there will be several plant-specific components (Murcha *et al.*, 2006). Another puzzle to solve is the phenomenon of dual targeting of proteins, especially with respect to those targeted by ambiguous signals (Peeters and Small, 2001). How are these signals recognized by mitochondria and plastids while the majority of plastid and mitochondrial proteins are only recognized by a single organelle? This question is further complicated by the fact that components of the plastid and mitochondrial protein import apparatus themselves display unexpected similarity. The Toc64 receptor of plastids and the OM64 protein of mitochondria display ~70% amino acid sequence identity, and PRAT proteins are found in both mitochondria and plastids (Chew *et al.*, 2004; Murcha *et al.*, submitted).

It is also interesting to study novel proteases involved in the processes of maturation, assembly and degradation of specific components. One very interesting and novel protease is the Rhomboid protease that regulates mitochondrial membrane remodeling indicating an important role of this protease in membrane biogenesis (McQuibban *et al.*, 2003; see Section 4.6.8 and Chapter 1). Another example is the newly identified PreP (Stahl *et al.*, 2002; Bhushan *et al.*, 2003, 2005; Moberg *et al.*, 2003) that degrades not only targeting peptides but also other unstructured peptides that might accumulate in the organelles and become toxic to the organellar functions (Moberg *et al.*, 2003; Stahl *et al.*, 2005). Other areas of interest for future research include an investigation of a double knockout mutant of *At*PreP in order to understand the effect of PreP catalyzed proteolysis on organellar activities. Similarly, it will be interesting to determine if the overexpression of PreP either in mitochondria or chloroplasts, or in both organelles, increases efficiency of the organellar functions and whether overexpressors exhibit any effects on plant growth and productivity. Furthermore, PreP's ability to degrade amyloid-β peptides and other potentially harmful peptides in mitochondria (Falkevall *et al.*, 2006) makes the study of PreP interesting also in the medical context.

In conclusion, functional characterization of plant components of the protein import apparatus is at an early but exciting stage. Comparative genomics enable the identification of many of the components, and this in turn will facilitate full characterization of plant-specific components. Reverse genetic approaches to characterize function and genetic screens to identify proteins interacting with known components will provide much-needed functional data about the novel components of the plant mitochondrial protein import apparatus.

Acknowledgements

This study was supported by grants from The Swedish Research Council to E. Glaser and from The Australian Research Council Centre of Excellence Program to J. Whelan. We thank Dr C. Roth for the analysis of targeting peptides derived from the mitochondrial and chloroplast proteomes of *Arabidopsis* and S. Nilsson for the graphical presentation of the plant mitochondrial import machinery.

References

Abe Y, Shodai T, Muto T, *et al.* (2000) Structural basis of presequence recognition by the mitochondrial protein import receptor Tom20. *Cell* **100**, 551–560.

Adam Z, Adamska I, Nakabayashi K, *et al.* (2001) Chloroplast and mitochondrial proteases in *Arabidopsis*. A proposed nomenclature. *Plant Physiol* **125**, 1912–1918.

Adam Z and Clarke AK (2002) Cutting edge of chloroplast proteolysis. *Trends Plant Sci* **7**, 451–456.

Adam Z, Rudella A and van Wijk KJ (2006) Recent advances in the study of Clp, FtsH and other proteases located in chloroplasts. *Curr Opin Plant Biol* **9**, 234–240.

Ahting U, Thieffry M, Engelhardt H, Hegerl R, Neupert W and Nussberger S (2001) Tom40, the pore-forming component of the protein-conducting TOM channel in the outer membrane of mitochondria. *J Cell Biol* **153**, 1151–1160.

Andersson SGE and Kurland CG (1999) Origins of mitochondria and hydrogenosomes. *Curr Opin Microbiol* **2**, 535–541.

Arabidopsis Genome Initiative (2000) Analysis of the genome sequence of the flowering plant *Arabidopsis thaliana*. *Nature* **408**, 796–815.

Armstrong LC, Komiya T, Bergman BE, Mihara K and Bornstein P (1997) Metaxin is a component of a preprotein import complex in the outer membrane of the mammalian mitochondrion. *J Biol Chem* **272**, 6510–6518.

Arnold I and Langer T (2002) Membrane protein degradation by AAA proteases in mitochondria. *Biochim Biophys Acta* **1592**, 89–96.

Baker KP and Schatz G (1991) Mitochondrial proteins essential for viability mediate protein import into yeast mitochondria. *Nature* **349**, 205–208.

Barakat S, Pearce DA, Sherman F and Rapp WD (1998) Maize contains a Lon protease gene that can partially complement a yeast pim1-deletion mutant. *Plant Mol Biol* **37**, 141–154.

Barrett AJ, Brown MA, Dando PM, *et al.* (1995) Thimet oligopeptidase and oligopeptidase M or neurolysin. *Methods Enzymol* **248**, 529–556.

Bathgate B, Baker A and Leaver CJ (1989) Two genes encode the adenine nucleotide translocator of maize mitochondria isolation, characterisation and expression of the structural genes. *Eur J Biochem* **183**, 303–310.

Bauer MF, Gempel K, Reichert AS, *et al.* (1999) Genetic and structural characterization of the human mitochondrial inner membrane translocase. *J Mol Biol* **289**, 69–82.

Bauer MF, Sirrenberg C, Neupert W and Brunner M (1996) Role of Tim23 as voltage sensor and presequence receptor in protein import into mitochondria. *Cell* **87**, 33–41.

Becker AB and Roth RA (1993) Identification of glutamate-169 as the third zinc-binding residue in proteinase III, a member of the family of insulin-degrading enzymes. *Biochem J* **292**, 137–142.

Bedard J and Jarvis P (2005) Recognition and envelope translocation of chloroplast preproteins. *J Exp Bot* **56**, 2287–2320.

Bhushan S, Johnson K, Eneqvist T and Glaser E (2006b) Mechanism of proteolysis of a novel mitochondrial and chloroplastic peptidasome. *Biol Chem* 387, 1087–1090.

Bhushan S, Kuhn C, Berglund AK, Roth C and Glaser E (2006a) The role of the N-terminal domain of chloroplast targeting peptides in organellar protein import and miss-sorting. *FEBS Lett* **580**, 3966–3972.

Bhushan S, Lefebvre B, StÅhl A, *et al.* (2003) Signal peptide degrading Zinc-metalloprotease is dually targeted to both mitochondria and chloroplasts. *EMBO Rep* **4**, 1073–1078.

Bhushan S, Stahl A, Nilsson S, *et al.* (2005) Catalysis, subcellular localization, expression and evolution of the targeting peptides degrading protease, AtPreP2. *Plant Cell Physiol* **46**, 985–996.

Bochtler M, Hartmann C, Song HK, Bourenkov GP, Bartunik HD and Huber R (2000) The structures of HslU and the ATP-dependent protease HslU-HslV. *Nature* **403**, 800–805.

Bota DA and Davies KJ (2002) Lon protease preferentially degrades oxidized mitochondrial aconitase by an ATP-stimulated mechanism. *Nat Cell Biol* **4**, 674–680.

Bota DA, Ngo JK and Davis KJ (2005) Downregulation of the human Lon protease impairs mitochondrial structure and function and causes cell death. *Free Radic Biol Med* **38**, 665–677.

Bota DA, Van Remmen H and Davies KJ (2002) Modulation of Lon protease activity and aconitase turnover during aging and oxidative stress. *FEBS Lett* **532**, 103–106.

Boutry M, Nagy F, Poulsen C, Aoyagi K and Chua NH (1987) Targeting of bacterial chloramphenicol acetyltransferase to mitochondria in transgenic plants. *Nature* **328**, 340–342.

Branda SS and Isaya G (1995) Prediction and identification of new natural substrates of the yeast mitochondrial intermediate peptidase. *J Biol Chem* **270**, 27366–27373.

Braun HP, Emmermann M, Kruft V and Schmitz UK (1992) The general mitochondrial processing peptidase from potato is an integral part of cytochrome c reductase of the respiratory chain. *EMBO J* **11**, 3219–3227.

Braun HP and Schmitz UK (1993) Purification and sequencing of cytochrome b from potato reveal methionine cleavage of a mitochondrially encoded protein. *FEBS Lett* **316**, 128–132.

Braun HP and Schmitz UK (1995a) Are the 'core' proteins of the mitochondrial bc1 complex evolutionary relics of a processing protease? *Trends Biochem Sci* **20**, 171–175.

Braun HP and Schmitz UK (1995b) The bifunctional cytochrome c reductase/processing peptidase complex from plant mitochondria. *J Bioenerg Biomembr* **27**, 423–436.

Braun HP and Schmitz UK (1999) The protein-import apparatus of plant mitochondria. *Planta* **209**, 267–274.

Brink S, Flugge UI, Chaumont F, *et al.* (1994) Preproteins of chloroplast envelope inner membrane contain targeting information for receptor-dependent import into fungal mitochondria. *J Biol Chem* **269**, 16478–16485.

Brix J, Rudiger S, Bukau B, Schneider-Mergener J and Pfanner N (1999) Distribution of binding sequences for the mitochondrial import receptors Tom20, Tom22, and Tom70 in a presequence-carrying preprotein and a non-cleavable preprotein. *J Biol Chem* **274**, 16522–16530.

Bruce BD (2000) Chloroplast transit peptides: structure, function and evolution. *Trends Cell Biol* **10**, 440–447.

Bruce BD (2001) The paradox of plastid transit peptides: conservation of function despite divergence in primary structure. *Biochim Biophys Acta* **1541**, 2–21.

Brumme S, Kruft V, Schmitz UK and Braun HP (1998) New insights into the co-evolution of cytochrome c reductase and the mitochondrial processing peptidase. *J Biol Chem* **273**, 13143–13149.

Buchler M, Tisljar U and Wolf DH (1994) Proteinase yscD (oligopeptidase yscD) structure, function and relationship of the yeast enzyme with mammalian thimet oilgopeptidase (metalloendopeptidase, EP 2415). *Eur J Biochem* **219**, 627–639.

Bulteau A-L, Szweda LI and Friguet B (2006) Mitochondrial protein oxidation and degradation in response to oxidative stress and aging. *Exp Gerontol* **41**, 653–657.

Burri L, Vascotto K, Fredersdorf S, Tiedt R, Hall MN and Lithgow T (2004) Zim17, a novel zinc finger protein essential for protein import into mitochondria. *J Biol Chem* **279**, 50243–50249.

Burton BM, Williams TL and Baker TA (2001) ClpX-mediated remodeling of mu transpososomes: selective unfolding of subunits destabilizes the entire complex. *Mol Cell* **8**, 447–456.

Chabregas SM, Luche DD, van Sluys M-A, Menck CF and Silva-Filho MC (2001) Differential usage of two in-frame translational start codons regulates subcellular localization of *Arabidopsis* THI1. *J Cell Sci* **116**, 285–291.

Chacinska A, Lind M, Frazier AE, *et al.* (2005) Mitochondrial presequence translocase: switching between TOM tethering and motor recruitment involves Tim21 and Tim17. *Cell* **120**, 817–829.

Chan NC, Likic VA, Waller RF, Mulhern TD and Lithgow T (2006) The C-terminal TPR domain of Tom70 defines a family of mitochondrial protein import receptors found only in animals and fungi. *J Mol Biol* **358**, 1010–1022.

Chaumont F, O'Riordan V and Boutry M (1990) Protein transport into mitochondria is conserved between plant and yeast species. *J Biol Chem* **265**, 16856–16862.

Chen X, Van Valkenburgh C, Fang H and Green N (1999) Signal peptides having standard and nonstandard cleavage sites can be processed by Imp1p of the mitochondrial inner membrane protease. *J Biol Chem* **274**, 37750–37754.

Chew O, Lister R, Qbadou S, *et al.* (2004) A plant outer mitochondrial membrane protein with high amino acid sequence identity to a chloroplast protein import receptor. *FEBS Lett* **557**, 109–114.

Chew A, Rollins RA, Sakati WR and Isaya G (1996) Mutations in a putative zinc-binding domain inactivate the mitochondrial intermediate peptidase. *Biochem Biophys Res Commun* **226**, 822–829.

Chew O, Rudhe C, Glaser E and Whelan J (2003a) Characterisation of the targeting signal of dual targeted pea glutathione reductase. *Plant Mol Biol* **53**, 341–356.

Chew O and Whelan J (2004) Just read the message: a model for sorting of proteins between mitochondria and chloroplasts. *Trends Plant Sci* **9**, 318–319.

Chew O, Whelan J and Millar AH (2003b) Molecular definition of the ascorbate-glutathione cycle in *Arabidopsis* mitochondria reveals dual targeting of antioxidant defenses in plants. *J Biol Chem* **278**, 46869–46877.

Chupin V, Leenhouts JM, de Kroon AI and de Kruijff B (1995) Cardiolipin modulates the secondary structure of the presequence peptide of cytochrome oxidase subunit IV: a 2D 1H-NMR study. *FEBS Lett* **373**, 239–244.

Cleary SP, Tan F-C, Nakrieko K-A, *et al.* (2002) Isolated plant mitochondria import chloroplast precursor proteins in vitro with the same efficiency as chloroplasts. *J Biol Chem* **277**, 5562–5569.

Corral-Debrinski M, Blugeon C and Jacq C (2000) In yeast, the 3′ untranslated region or the presequence of ATM1 is required for the exclusive localization of its mRNA to the vicinity of mitochondria. *Mol Cell Biol* **20**, 7881–7892.

Corydon TJ, Bross P, Holst HU, *et al.* (1998) A himan homologue of *Escherichia coli* ClpP caeinolytic preotease: recombinant expression, intracellular processing and subcellualr localisation. *Biochem J* **331**, 309–316.

Creissen G, Reynolds H, Xue Y and Mullineaux P (1995) Simultaneous targeting of pea glutathione reductase and of a bacterial fusion protein to chloroplasts and mitochondria in transgenic tobacco. *Plant J* **8**, 167–175.

Dalbey RE (1991) Leader peptidase. *Mol Microbiol* **5**, 2855–2860.

Dessi P, Pavlov P, WÅllberg F, *et al.* (2003) Investigations on the in vitro import ability of mitochondrial precursor proteins synthesised in wheat germ transcription–translation extract. *Plant Mol Biol* **52**, 259–271.

Dessi P, Rudhe C and Glaser E (2000) Studies on the topology of the protein import channel in relation to the plant mitochondrial processing peptidase integrated into the cytochrome bc1 complex. *Plant J* **24**, 637–644.

Dessi P, Smith MK, Day DA and Whelan J (1996) Characterization of the import pathway of the F(A)d subunit of mitochondrial ATP synthase into isolated plant mitochondria. *Arch Biochem Biophys* **335**, 358–368.

Donzeau M, Kaldi K, Adam A, *et al.* (2000) Tim23 links the inner and outer mitochondrial membranes. *Cell* **101**, 401–412.

Duby G, Degand H and Boutry M (2001b) Structure requirement and identification of a cryptic cleavage site in the mitochondrial processing of a plant F1-ATPase beta-subunit presequence. *FEBS Lett* **505**, 409–413.

Duby G, Oufattole M and Boutry M (2001a) Hydrophobic residues within the predicted *N*-terminal amphiphilic alpha-helix of a plant mitochondrial targeting presequence play a major role in in vivo import. *Plant J* **27**, 539–549.

Duchene AM, Giritch A, Hoffmann B, *et al.* (2005) Dual targeting is the rule for organellar aminoacyl-tRNA synthetases in *Arabidopsis thaliana*. *Proc Natl Acad Sci USA* **102**, 16484–16489.

Dyall SD, Brown MT and Johnson PJ (2004) Ancient invasions: from endosymbionts to organelles. *Science* **304**, 253–257.

Emmermann M, Braun HP and Schmitz UK (1994) The mitochondrial processing peptidase from potato: a self-processing enzyme encoded by two differentially expressed genes. *Mol Gen Genet* **245**, 237–245.

Emmermann M and Schmitz UK (1993) The cytochrome c reductase integrated processing peptidase from potato mitochondria belongs to a new class of metalloendoproteases. *Plant Physiol* **103**, 615–620.

Endo T, Yamamoto H and Esaki M (2003) Functional cooperation and separation of translocators in protein import into mitochondria, the double-membrane bounded organelles. *J Cell Sci* **116**, 3259–3267.

Eriksson AC and Glaser E (1992) Mitochondrial processing protease. A general processing protease of spinach leaf mitochondria is a membrane bound enzyme. *Biochim Biophys Acta* **1140**, 208–214.

Eriksson AC, Sjöling S and Glaser E (1993) A general processing proteinase of spinach leaf mitochondria is associated with the bc1 complex of the respiratory chain. In: *Plant Mitochondria* (eds Brennicke A and Kuch U), pp. 233–241. VCH Verlags-gesellschaft, Weinheim, Germany.

Eriksson AC, Sjöling S and Glaser E (1994) The ubiquinol cytochrome c oxidoreductase complex of spinach leaf mitochondria is involved in both respiration and protein processing. *Biochim Biophys Acta* **1186**, 221–231.

Eriksson AC, Sjöling S and Glaser E (1996) Characterization of the bifunctional mitochondrial processing peptidase (MPP)/bc1 complex in Spinacia oleracea. *J Bioenerg Biomembr* **28**, 285–292.

Esser K, Pratje E and Michaelis G (1996) SOM1, a new gene required for mitochondrial inner membrane peptidase function in *Saccharomyces cerevisiae*. *Mol Gen Genet* **252**, 437–445.

Esser K, Tursun B, Ingenhoven M, Michaelis G and Pratje E (2002) A novel two-step mechanism for removal of a mitochondrial signal sequence involves the mAAA complex and the putative rhomboid protease Pcp1. *J Mol Biol* **323**, 835–843.

Falkevall A, Alikhani N, Bhushan S, *et al.* (2006) Degradation of the amyloid beta-protein by the novel mitochondrial peptidasome, PreP. *J Biol Chem* **281**, 29096–29104.

Frazier AE, Dudek J, Guiard B, *et al.* (2004) Pam16 has an essential role in the mitochondrial protein import motor. *Nat Struct Mol Biol* **11**, 226–233.

Funfschilling U and Rospert S (1999) Nascent polypeptide-associated complex stimulates protein import into yeast mitochondria. *Mol Biol Cell* **10**, 3289–3299.

Gakh O, Cavadini P and Isaya G (2002) Mitochondrial processing peptidases. *Biochim Biophys Acta* **1592**, 63–77.

Geissler A, Chacinska A, Truscott KN, *et al.* (2002) The mitochondrial presequence translocase: an essential role of Tim50 in directing preproteins to the import channel. *Cell* **111**, 507–518.

Gentle IE, Burri L and Lithgow T (2005) Molecular architecture and function of the Omp85 family of proteins. *Mol Microbiol* **58**, 1216–1225.

George R, Beddoe T, Landl K and Lithgow T (1998) The yeast nascent polypeptide-associated complex initiates protein targeting to mitochondria in vivo. *Proc Natl Acad Sci USA* **95**, 2296–2301.

Ginsberg MD, Feliciello A, Jones JK, Avvedimento EV and Gottesman ME (2003) PKA-dependent binding of mRNA to the mitochondrial AKAP121 protein. *J Mol Biol* **327**, 885–897.

Glaser E and Dessi P (1999) Integration of the mitochondrial processing peptidase into the bc1 complex of the respiratory chain in plants. *J Bioenerg Biomembr* **31**, 259–274.

Glaser E, Eriksson AC and Sjöling S (1994) Bifunctional role of the bc1 complex in plants. Mitochondrial bc1 complex catalyses both electron transport and protein processing. *FEBS Lett* **346**, 83–87.

Glaser E, Sj Sjöling S, Tanudji M and Whelan J (1998) Mitochondrial protein import in plants. *Plant Mol Biol* **38**, 311–338.

Glaser E and Soll J (2004) Targeting signals and import machinery of plastids and plant mitochondria. In: *Molecular Biology and Biotechnology of Plant Organelles* (eds Daniell H and Chase C), pp 385–418. Springer, Dordecht.

Goff SA, Ricke D, Lan TH, *et al.* (2002) A draft sequence of the rice genome (*Oryza sativa* L ssp japonica). *Science* **296**, 92–100.

Gottesman S, Roche E, Zhou Y and Sauer RT (1998) The ClpX and ClpAP proteases degrade proteins with carboxy-terminal peptide tails added by the SsrA-tagging system. *Genes Dev* **12**, 1339–1347.

Gratzer S, Lithgow T, Bauer RE, *et al.* (1995) Mas37p, a novel receptor subunit for protein import into mitochondria. *J Cell Biol* **129**, 25–34.

Gray MW, Burger G and Lang BF (1999) Mitochondrial evolution. *Science* **283**, 1476–1481.

Guo Y, Cheong N, Zhang Z, *et al.* (2004) Tim50, a component of the mitochondrial translocator, regulates mitochondrial integrity and cell death. *J Biol Chem* **279**, 24813–24825.

Gutensohn M, Fan E, Frielingsdorf S, *et al.* (2006) Toc, Tic, Tat *et al.*: structure and function of protein transport machineries in chloroplasts. *J Plant Physiol* **163**, 333–347.

Habib SJ, Waizenegger T, Lech M, Neupert W and Rapaport D (2005) Assembly of the TOB complex of mitochondria. *J Biol Chem* **280**, 6434–6440.

Hachiya N, Alam R, Sakasegawa Y, Sakaguchi M, Mihara K and Omura T (1993) A mitochondrial import factor purified from rat liver cytosol is an ATP-dependant conformational modulator for precursor proteins. *EMBO J* **12**, 1579–1586.

Hachiya N, Mihara K, Suda K, Horst M, Schatz G and Lithgow T (1995) Reconstitution of the initial steps of mitochondrial protein import. *Nature* **376**, 705–709.

Halperin T, Ostersetzer O and Adam Z (2001a) ATP-dependent association between subunits of Clp protease in pea chloroplasts. *Planta* **213**, 614–619.

Halperin T, Zheng B, Itzhaki H, Clarke AK and Adam Z (2001b) Plant mitochondria contain proteolytic and regulatory subunits of the ATP-dependent Clp protease. *Plant Mol Biol* **45**, 461–468.

Hamel P, Sakamoto W, Wintz H and Dujardin G (1997) Functional complementation of an oxa1-yeast mutation identifies an *Arabidopsis thaliana* cDNA involved in the assembly of respiratory complexes. *Plant J* **12**, 1319–1327.

Hammen PK, Gorenstein DG and Weiner H (1994) Structure of the signal sequences for two mitochondrial matrix proteins that are not proteolytically processed upon import. *Biochemistry* **33**, 8610–8617.

Hammen PK, Gorenstein DG and Weiner H (1996) Amphiphilicity determines binding properties of three mitochondrial presequences to lipid surfaces. *Biochemistry* **35**, 3772–3781.

Harmey MA, Hallermayer G, Korb H and Neupert W (1977) Transport of cytoplasmically synthesized proteins into the mitochondria in a cell free system from *Neurospora crassa*. *Eur J Biochem* **81**, 533–544.

Hartl FU, Pfanner N, Nicholson DW and Neupert W (1989) Mitochondrial protein import. *Biochim Biophys Acta* **988**, 1–45.

Hawlitschek G, Schneider H, Schmidt B, Tropschug M, Hartl FU and Neupert W (1988) Mitochondrial protein import: identification of processing peptidase and of PEP, a processing enhancing protein. *Cell* **53**, 795–806.

Heazlewood JL, Tonti-Filippini JS, Gout AM, Day DA, Whelan J and Millar AH (2004) Experimental analysis of the *Arabidopsis* mitochondrial proteome highlights signaling and regulatory components, provides assessment of targeting prediction programs, and indicates plant-specific mitochondrial proteins. *Plant Cell* **16**, 241–256.

Hedtke B, Borner T and Weihe A (2000) One RNA polymerase serving two genomes. *EMBO Rep* **1**, 435–440.

Heins L and Schmitz UK (1996) A receptor for protein import into potato mitochondria. *Plant J* **9**, 829–839.

Herlan M, Bornhovd C, Hell K, Neupert W and Reichert AS (2004) Alternative topogenesis of Mgm1 and mitochondrial morphology depend on ATP and a functional import motor. *J Cell Biol* **165**, 167–173.

Herlan M, Vogel F, Bornhovd C, Neupert W and Reichert AS (2003) Processing of Mgm1 by the rhomboid-type protease Pcp1 is required for maintenance of mitochondrial morphology and of mitochondrial DNA. *J Biol Chem* **278**, 27781–27788.

Herrmann JM (2003) Converting bacteria to organelles: evolution of mitochondrial protein sorting. *Trends Microbiol* **11**, 74–79.

Hoogenraad NJ, Ward LA and Ryan MT (2002) Import and assembly of proteins into mitochondria of mammalian cells. *Biochim Biophys Acta* **1592**, 97–105.

Hugosson M, Nurani G, Glaser E and Franzén LG (1995) Peculiar properties of the PsaF photosystem I protein from the green alga *Chlamydomonas reinhardtii*: presequence independent import of the PsaF protein into both chloroplasts and mitochondria. *Plant Mol Biol* **28**, 525–535.

Isaya G, Kalousek F, Fenton WA and Rosenberg LE (1991) Cleavage of precursors by the mitochondrial processing peptidase requires a compatible mature protein or an intermediate octapeptide. *J Cell Biol* **113**, 65–76.

Jan PS, Esser K, Pratje E and Michaelis G (2000) Som1, a third component of the yeast mitochondrial inner membrane peptidase complex that contains Imp1 and Imp2. *Mol Gen Genet* **263**, 483–491.

Jansch L, Kruft V, Schmitz UK and Braun HP (1996) New insights into the composition, molecular mass and stoichiometry of the protein complexes of plant mitochondria. *Plant J* **9**, 357–368.
Johnson KA, Bhushan S, StÅhl A, *et al.* (2006) The closed structure of presequence protease PreP forms a unique 10 000 Å^3 chamber for proteolysis. *EMBO J* **25**, 1977–1986.
Kalousek F, Isaya G and Rosenberg LE (1992) Rat liver mitochondrial intermediate peptidase (MIP): purification and initial characterization. *EMBO J* **11**, 2803–2809.
Kambacheld M, Augustin S, Tatsuta T, Muller S and Langer T (2005) Role of a novel metallopeptidase Mop112 and saccharolysin for the complete degradation of proteins residing in different subcompartmnets of mitochondria. *J Biol Chem* **280**, 20132–20139.
Karslake C, Piotto ME, Pak YK, Weiner H and Gorenstein DG (1990) 2D NMR and structural model for a mitochondrial signal peptide bound to a micelle. *Biochemistry* **29**, 9872–9878.
Kaser M, Kambacheld M, Kisters-Woike B and Langer T (2003) Oma1, a novel membrane-bound metallopeptidase in mitochondria with activities overlapping with the m-AAA protease. *J Biol Chem* **278**, 46414–46423.
Kaser M and Langer T (2000) Protein degradation in mitochondria. *Semin Cell Dev Biol* **11**, 181–190.
Kolodziejczak M, Kolaczkowska A, Szczesny B, *et al.* (2002) A higher plant mitochondrial homologue of the yeast m-AAA protease molecular cloning, localization, and putative function. *J Biol Chem* **277**, 43792–43798.
Kominsky DJ, Brownson MP, Updike DL and Thorsness PE (2002) Genetic and biochemical basis for viability of yeast lacking mitochondrial genomes. *Genetics* **162**, 1595–1604.
Koonin EV (2003) Comparative genomics, minimal gene-sets and the last universal common ancestor. *Nat Rev Microbiol* **1**, 127–136.
Kuroda H and Maliga P (2003) The plastid clpP1 protease gene is essential for plant development. *Nature* **425**, 86–98.
Lang BF, Gray MW and Burger G (1999) Mitochondrial genome evolution and the origin of eukaryotes. *Annu Rev Genet* **33**, 351–397.
Langer T (2000) AAA proteases: cellular machines for degrading membrane proteins. *Trends Biochem Sci* **25**, 247–251.
Lee MN and Whelan J (2004) Identification of signals required for import of the soybean F(A)d subunit of ATP synthase into mitochondria. *Plant Mol Biol* **54**, 193–203.
Lemaire C, Hamel P, Velours J and Dujardin G (2000) Absence of the mitochondrial AAA protease Yme1p restores F0-ATPase subunit accumulation in an oxa1 deletion mutant of *Saccharomyces cerevisiae*. *J Biol Chem* **275**, 23471–23475.
Li Y, Dudek J, Guiard B, Pfanner N, Rehling P and Voos W (2004) The presequence translocase-associated protein import motor of mitochondria: Pam16 functions in an antagonistic manner to Pam18. *J Biol Chem* **279**, 38047–38054.
Liang H, Lou W, Green N and Fang H (2004) Cargo sequences are important for Som1p-dependent signal peptide cleavage in yeast mitochondria. *J Biol Chem* **279**, 4943–4948.
Lister R, Chew O, Lee MN, *et al.* (2004) A transcriptomic and proteomic characterization of the *Arabidopsis* mitochondrial protein import apparatus and its response to mitochondrial dysfunction. *Plant Physiol* **134**, 777–789.
Lister R, Chew O, Lee M and Whelan J (2001) *Arabidopsis thaliana* ferrochelatase-I and -II are not imported into *Arabidopsis* mitochondria. *FEBS Lett* **506**, 291–295.
Lister R, Mowday B, Whelan J and Millar AH (2002) Zinc-dependent intermembrane space proteins stimulate import of carrier proteins into plant mitochondria. *Plant J* **30**, 555–566.
Lister R, Murcha MW and Whelan J (2003) The mitochondrial protein import machinery of plants (MPIMP) database. *Nucleic Acids Res* **31**, 325–327.
Lister R and Whelan J (2006) Mitochondrial protein import: convergent solutions for receptor structure. *Curr Biol* **16**, R197–199.
Lithgow T (2000) Targeting of proteins to mitochondria. *FEBS Lett* **476**, 22–26.
Lowe J, Stock D, Jap B, Zwickl P, Baumeister W and Huber R (1995) Crystal structure of the 20S proteasome from the archaeon *T. acidophilum* at 34 A resolution. *Science* **268**, 533–539.

Luciano P, Geoffroy S, Brandt A, Hernandez JF and Geli V (1997) Functional cooperation of the mitochondrial processing peptidase subunits. *J Mol Biol* **272**, 213–225.

Luo W, Fang H and Green N (2006) Substrate specificity of inner membrane peptidase in yeast mitochondria. *Mol Gen Genomics* **275**, 431–436.

Luzikov VN, Novikova LA, Whelan J, Hugosson M and Glaser E (1994) Import of the mammalian cytochrome P450 (scc) precursor into plant mitochondria. *Biochem Biophys Res Commun* **199**, 33–36.

Macasev D, Newbigin E, Whelan J and Lithgow T (2000) How do plant mitochondria avoid importing chloroplast proteins? Components of the import apparatus Tom20 and Tom22 from *Arabidopsis* differ from their fungal counterparts. *Plant Physiol* **123**, 811–816.

Macasev D, Whelan J, Newbigin E, Silva-Filho MC, Mulhern TD and Lithgow T (2004) Tom22′, an 8-kDa trans-site receptor in plants and protozoans, is a conserved feature of the TOM complex that appeared early in the evolution of eukaryotes. *Mol Biol Evol* **21**, 1557–1564.

Marc P, Margeot A, Devaux F, Blugeon C, Corral-Debrinski M and Jacq C (2002) Genome-wide analysis of mRNAs targeted to yeast mitochondria. *EMBO Rep* **3**, 159–164.

Mattingly JR,Youssef J, Iriarte A and Martinez-Carrion M (1993) Protein folding in a cell-free translation system the fate of the precursor to mitochondrial aspartate aminotransferase. *J Biol Chem* **268**, 3925–3937.

May T and Soll J (2000) 14-3-3 proteins form a guidance complex with chloroplast precursor proteins in plants. *The Plant Cell* **12**, 53–63.

McQuibban GA, Saurya S and Freeman M (2003) Mitochondrial membrane remodelling regulated by a conserved rhomboid protease. *Nature* **423**, 537–541.

Miernyk J, Duck N, Shatters RJ and Folk W (1992) The 70-kilodalton heat shock cognate can act as a molecular chaperone during the membrane translocation of a plant secretory protein precursor. *Plant Cell* **4**, 821–829.

Milenkovic D, Kozjak V, Wiedemann N, *et al.* (2004) Sam35 of the mitochondrial protein sorting and assembly machinery is a peripheral outer membrane protein essential for cell viability. *J Biol Chem* **279**, 22781–22785.

Millar AH, Day DA and Whelan J (2004) Mitochondrial biogenesis and function in *Arabidopsis*. In: *The Arabidopsis Book* (eds Somerville CR and Meyerowitz EM), pp. 1–36. American Society of Plant Biologists, Rockville, MD. Available at: www.aspborg/publications/arabidopsis.

Moberg P, Nilsson S, Stahl A, Eriksson AC, Glaser E and Maler L (2004) NMR solution structure of the mitochondrial F1β presequence from *Nicotiana plumbaginifolia*. *J Mol Biol* **336**, 1129–1140.

Moberg P, Stahl A, Bhushan S, *et al.* (2003) Characterization of a novel zinc metalloprotease involved in degrading signal peptides in mitochondria and chloroplasts. *Plant J* **36**, 616–628.

Model K, Meisinger C, Prinz T, *et al.* (2001) Multistep assembly of the protein import channel of the mitochondrial outer membrane. *Nat Struct Biol* **8**, 361–370.

Model K, Prinz T, Ruiz T, *et al.* (2002) Protein translocase of the outer mitochondrial membrane: role of import receptors in the structural organization of the TOM complex. *J Mol Biol* **316**, 657–666.

Mokranjac D, Paschen SA, Kozany C, *et al.* (2003) Tim50, a novel component of the TIM23 preprotein translocase of mitochondria. *EMBO J* **22**, 816–825.

Mokranjac D, Popov-Celeketic D, Hell K and Neupert W (2005a) Role of Tim21 in mitochondrial translocation contact sites. *J Biol Chem* **280**, 23437–23440.

Mokranjac D, Sichting M, Popov-Celeketic D, Berg A, Hell K and Neupert W (2005b) The import motor of the yeast mitochondrial TIM23 preprotein translocase contains two different J proteins, Tim14 and Mdj2. *J Biol Chem* **280**, 31608–31614.

Murakami H, Pain D and Blobel G (1988) 70-kD heat shock-related protein is one of at least two distinct cytosolic factors stimulating protein import into mitochondria. *J Cell Biol* **107**, 2051–2057.

Murakami K, Tanase S, Morino Y and Mori M (1992) Presequence binding factor-dependent and -independent import of proteins into mitochondria. *J Biol Chem* **267**, 13119–13122.

Murcha MW, Elhafez D, Lister R, Tonti-Filippini J, Baumgartner M, Philippar K, Carrie C, Mokranjac D, Soll J, and Whelan J (2006) Characterisation of the preprotein and amino acid transporter family in *Arabidopsis*. Plant Physiol (in Press) DOI 10.1104/pp. 106.090688.
Murcha MW, Elhafez D, Millar AH and Whelan J (2004) The *N*-terminal extension of plant mitochondrial carrier proteins is removed by two-step processing: the first cleavage is by the mitochondrial processing peptidase. *J Mol Biol* **344**, 443–454.
Murcha MW, Elhafez D, Millar AH and Whelan J (2005a) The C-terminal region of TIM17 links the outer and inner mitochondrial membranes in *Arabidopsis* and is essential for protein import. *J Biol Chem* **280**, 16476–16483.
Murcha MW, Lister R, Ho AY and Whelan J (2003) Identification, expression, and import of components 17 and 23 of the inner mitochondrial membrane translocase from *Arabidopsis*. *Plant Physiol* **131**, 1737–1747.
Murcha MW, Millar AH and Whelan J (2005b) The *N*-terminal cleavable extension of plant carrier proteins is responsible for efficient insertion into the inner mitochondrial membrane. *J Mol Biol* **351**, 16–25.
Muto T, Obita T, Abe Y, Shodai T, Endo T and Kohda D (2001) NMR identification of the Tom20 binding segment in mitochondrial presequences. *J Mol Biol* **306**, 137–143.
Neupert W (1997) Protein import into mitochondria. *Annu Rev Biochem* **66**, 863–917.
Niidome T, Kitada S, Shimokata K, Ogishima T and Ito A (1994) Arginine residues in the extension peptide are required for cleavage of a precursor by mitochondrial processing peptidase. Demonstration using synthetic peptide as a substrate. *J Biol Chem* **269**, 24719–24722.
Nunnari J, Fox TD and Walter P (1993) A mitochondrial protease with two catalytic subunits of nonoverlapping specificities. *Science* **262**, 1997–2004.
Ogishima T, Niidome T, Shimokata K, Kitada S and Ito A (1995) Analysis of elements in the substrate required for processing by mitochondrial processing peptidase. *J Biol Chem* **270**, 30322–30326.
Okamoto K, Brinker A, Paschen SA, *et al.* (2002) The protein import motor of mitochondria: a targeted molecular ratchet driving unfolding and translocation. *EMBO J* **21**, 3659–3671.
Ondrovicova G, Liu T, Singh K, *et al.* (2005) Cleavage site selection within a folded substrate by the ATP-dependent lon protease. *J Biol Chem* **280**, 25103–25110.
Ono H and Tuboi S (1990) Purification and identification of a cytosolic factor required for import of precursors of mitochondrial proteins into mitochondria. *Arch Biochem Biophys* **280**, 299–304.
Ou W, Ito A, Okazaki H and Omura T (1989) Purification and characterization of a processing protease from rat liver mitochondria. *EMBO J* **8**, 2605–2612.
Paschen SA, Neupert W and Rapaport D (2005) Biogenesis of beta-barrel membrane proteins of mitochondria. *Trends Biochem Sci* **30**, 575–582.
Paschen SA, Waizenegger T, Stan T, *et al.* (2003) Evolutionary conservation of biogenesis of beta-barrel membrane proteins. *Nature* **426**, 862–866.
Pavlov P, Moberg P, Zhang X-P and Glaser E (1999) CNBr chemical cleavage of the overexpressed mitochondrial F1b precursor. A new strategy to construct an import competent preprotein. *Biochem J* **341**, 95–103.
Pearce DA and Sherman F (1995) Degradation of cytochrome oxidase subunits in mutants of yeast lacking cytochrome c and suppression of the degradation by mutation of yme1. *J Biol Chem* **270**, 20879–20882.
Peeters N and Small I (2001) Dual targeting to mitochondria and chloroplasts. *Biochim Biophys Acta* **1541**, 54–63.
Peltier JB, Ripoll DR, Friso G, *et al.* (2004) Clp protease complexes from photosynthetic and non-photosynthetic plastids and mitochondria of plants, their predicted three-dimensional structures, and functional implications. *J Biol Chem* **279**, 4768–4781.
Perry AJ, Hulett JM, Likic VA, Lithgow T and Gooley PR (2006) Convergent evolution of receptors for protein import into mitochondria. *Curr Biol* **16**, 221–229.
Pfanner N, Douglas MG, Endo T, *et al.* (1996) Uniform nomenclature for the protein transport machinery of the mitochondrial membranes. *Trends Biochem Sci* **21**, 51–52.

Pfanner N and Geissler A (2001) Versatility of the mitochondrial protein import machinery. *Nat Rev Mol Cell Biol* **2**, 339–349.

Pfanner N, Wiedemann N, Meisinger C and Lithgow T (2004) Assembling the mitochondrial outer membrane. *Nat Struct Mol Biol* **11**, 1044–1048.

Qbadou S, Becker T, Mirus O, Tews I, Soll J and Schleiff E (2006) The molecular chaperone Hsp90 delivers precursor proteins to the chloroplast import receptor Toc64. *EMBO J* **25**, 1836–1847.

Rapaport D (2003) Finding the right organelle targeting signals in mitochondrial outer-membrane proteins. *EMBO Rep* **4**, 948–952.

Rapaport D, Taylor RD, Kaser M, Langer T, Neupert W and Nargang FE (2001) Structural requirements of Tom40 for assembly into preexisting TOM complexes of mitochondria. *Mol Biol Cell* **12**, 1189–1198.

Rassow J, Dekker PJ, van Wilpe S, Meijer M and Soll J (1999) The preprotein translocase of the mitochondrial inner membrane: function and evolution. *J Mol Biol* **286**, 105–120.

Rehling P, Brandner K and Pfanner N (2004) Mitochondrial import and the twin-pore translocase. *Nat Rev Mol Cell Biol* **5**, 519–530.

Richter S and Lamppa GK (1999) Stromal processing peptidase binds transit peptides and initiates their ATP-dependent turnover in chloroplasts. *J Cell Biol* **147**, 33–44.

Rudhe C, Chew O, Whelan J and Glaser E (2002a) A novel in vitro system for simultanous import of precursor proteins into chloroplast and mitochondria. *Plant J* **30**, 213–220.

Rudhe C, Clifton R, Chew O, *et al.* (2004) Processing of the dual targeted precursor protein of glutathione reductase in mitochondria and chloroplasts. *J Mol Biol* **343**, 639–647.

Rudhe C, Clifton R, Whelan J and Glaser E (2002b) *N*-terminal domain of the dual targeted pea glutathione reductase signal peptide controls organellar targeting efficiency. *J Mol Biol* **324**, 577–585.

Sakamoto W, Spielewoy N, Bonnard G, Murata M and Wintz H (2000) Mitochondrial localization of AtOXA1, an *arabidopsis* homologue of yeast Oxa1p involved in the insertion and assembly of protein complexes in mitochondrial inner membrane. *Plant Cell Physiol* **41**, 1157–1163.

Samuelson JC, Chen M, Jiang F, *et al.* (2000) YidC mediates membrane protein insertion in bacteria. *Nature* **406**, 637–641.

Sanjuan Szklarz LK, Guiard B, Rissler M, *et al.* (2005) Inactivation of the mitochondrial heat shock protein zim17 leads to aggregation of matrix hsp70s followed by pleiotropic effects on morphology and protein biogenesis. *J Mol Biol* **351**, 206–218.

Santagatra S, Bhattacharyya D, Wang FH, Singha N, Hodstev A and Spanopoulou E (1999) Molecular cloning and characterization of a mouse homolog of bacterial ClpX, a novel mammalian classs II member of the Hsp100/Clp chaperone family. *J Biol Chem* **274**, 16311–16319.

Sarria R, Lyznik A, Vallejos CE and Mackenzie SA (1998) A cytoplasmic male sterility-associated mitochondrial peptide in common bean is post-translationally regulated. *Plant Cell* **10**, 1217–1228.

Schleiff E, Heard TS and Weiner H (1999) Positively charged residues, the helical conformation and the structural flexibility of the leader sequence of pALDH are important for recognition by hTom20. *FEBS Lett* **461**, 9–12.

Schleiff E, Motzkus M and Soll J (2002) Chloroplast protein import inhibition by soluble factor from wheat germ lysate. *Plant Mol Biol* **50**, 177–185.

Schleiff E and Turnbull JL (1998) Characterization of the *N*-terminal targeting signal binding domain of the mitochondrial outer membrane receptor, Tom20. *Biochemistry* **37**, 13052–13058.

Schneider A, Oppliger W and Jeno P (1994) Purified inner membrane protease I of yeast mitochondria is a heterodimer. *J Biol Chem* **269**, 8635–8638.

Schneider G, Sjöling S, Wallin E, Wrede P, Glaser E and von Heijne G (1998) Feature-extraction from endopeptidase cleavage sites in mitochondrial targeting sequences. *Proteins: Struct Funct Genet* **30**, 49–60.

Schneider TD and Stephens RM (1990) Sequence logos: a new way to display consensus sequences. *Nucleic Acids Res* **18**, 6097–6100.

Schulte U, Arretz M, Schneider H, *et al.* (1989) A family of mitochondrial proteins involved in bioenergetics and biogenesis. *Nature* **339**, 147–149.
Sickmann A, Reinders J, Wagner Y, *et al.* (2003) The proteome of *Saccharomyces cerevisiae* mitochondria. *Proc Natl Acad Sci USA* **100**, 13207–13212.
Silva-Filho MC (2003) One ticket for multiple destinations: dual targeting of proteins to distinct subcellular locations. *Curr Opin Plant Biol* **6**, 589–595.
Silva-Filho MDC, Wieers M-C, Flugge U-I, Chaumont F and Boutry M (1997) Different in vitro and in vivo targeting properties of the transit peptide of a chloroplast envelope inner membrane protein. *J Biol Chem* **272**, 15264–15269.
Sinvany-Villalobo G, Davydov O, Ben-Ari G, Zaltsman A, Raskind A and Adam Z (2004) Expression in multigene families. Analysis of chloroplast and mitochondrial proteases. *Plant Physiol* **135**, 1336–1345.
Sjöling S, Eriksson AC and Glaser E (1994) A helical element in the C-terminal domain of the *N. plumbaginifolia* F1 beta presequence is important for recognition by the mitochondrial processing peptidase. *J Biol Chem* **269**, 32059–32062.
Sjöling S and Glaser E (1998) Mitochondrial targeting peptides in plants. *Trends Plant Sci* **3**, 136–140.
Sjöling S, Waltner M, Kalousek F, Glaser E and Weiner H (1996) Studies on protein processing for membrane-bound spinach leaf mitochondrial processing peptidase integrated into the cytochrome bc1 complex and the soluble rat liver matrix mitochondrial processing peptidase. *Eur J Biochem* **242**, 114–121.
Small I, Wintz H, Akashi K and Mireau H (1998) Two birds with one stone: genes that encode products targeted to two or more compartments. *Plant Mol Biol* **38**, 265–277.
Sokolenko A, Pojidaeva E, Zinchenko V, *et al.* (2002) The gene complement for proteolysis in the cyanobacterium Synechocystis sp. PCC 6803 and *Arabidopsis thaliana* chloroplasts. *Curr Genet* **41**, 291–310.
Soll J (2002) Protein import into chloroplasts. *Curr Opin Plant Biol* **5**, 529–535.
Soll J and Schleiff E (2004) Protein import into chloroplasts. *Nat Rev Mol Cell Biol* **5**, 198–208.
Stahl A, Moberg P, Ytterberg J, *et al.* (2002) Isolation and identification of a novel mitochondrial metalloprotease (PreP) that degrades targeting presequences. *J Biol Chem* **277**, 41931–41939.
Stahl A, Nilsson S, Lundberg P, *et al.* (2005) Two novel targeting peptide degrading proteases, PrePs, in mitochondria and chloroplasts, so similar and still different. *J Mol Biol* **349**, 847–860.
Stahl A, Pavlov PF, Szigyarto C and Glaser E (2000) Rapid degradation of the presequence of the F1 beta precursors of the ATP synthase inside mitochondria. *Biochem J* **349**, 703–707.
Stahlberg H, Kutejova E, Suda K, *et al.* (1999) Mitochondrial Lon of *Saccharomyces cerevisiae* is a ring-shaped protease with seven flexible subunits. *Proc Natl Acad Sci USA* **96**, 6787–6790.
Stan T, Ahting U, Dembowski M, *et al.* (2000) Recognition of preproteins by the isolated TOM complex of mitochondria. *EMBO J* **19**, 4895–4902.
Stein I, Peleg Y, Even-Ram S and Pines O (1994) The single translation product of the FUM1 gene (fumarase) is processed in mitochondria before being distributed between the cytosol and mitochondria in *Saccharomyces cerevisiae*. *Mol Cell Biol* **14**, 4770–4778.
Suzuki CK, Suda K, Wang N and Schatz G (1994) Requirement for the yeast gene LON in intramitochondrial proteolysis and maintenance of respiration. *Science* **264**, 891.
Tanudji M, Dessi P, Murcha M and Whelan J (2001) Protein import into plant mitochondria: precursor proteins differ in ATP and membrane potential requirements. *Plant Mol Biol* **45**, 317–325.
Tanudji M, Sjöling S, Glaser E and Whelan J (1999) Signals required for the import and processing of the alternative oxidase into mitochondria. *J Biol Chem* **274**, 1286–1293.
Taylor AB, Smith BS, Kitada S, *et al.* (2001) Crystal structures of mitochondrial processing peptidase reveal the mode for specific cleavage of import signal sequences. *Structure (Camb)* **9**, 615–625.
Taylor SW, Fahy E, Zhang B, *et al.* (2003) Characterization of the human heart mitochondrial proteome. *Nat Biotechnol* **21**, 281–286.
Thornton K, Wang Y, Weiner H and Gorenstein DG (1993) Import, processing, and two-dimensional NMR structure of a linker-deleted signal peptide of rat liver mitochondrial aldehyde dehydrogenase. *J Biol Chem* **268**, 19906–19914.

Thorsness PE, White KH and Fox TD (1993) Inactivation of YME1, a member of the ftsH-SEC18-PAS1-CDC48 family of putative ATPase-encoding genes, causes increased escape of DNA from mitochondria in *Saccharomyces cerevisiae*. *Mol Cell Biol* **13**, 5418–5426.

Truscott KN, Kovermann P, Geissler A, *et al.* (2001) A presequence- and voltage-sensitive channel of the mitochondrial preprotein translocase formed by Tim23. *Nat Struct Biol* **8**, 1074–1082.

Truscott KN, Voos W, Frazier AE, *et al.* (2003) A J-protein is an essential subunit of the presequence translocase-associated protein import motor of mitochondria. *J Cell Biol* **163**, 707–713.

Urantowka A, Knorpp C, Olczak T, Kolodziejczak M and Janska H (2005) Plant mitochondria contain at least two i-AAA-like complexes. *Plant Mol Biol* **59**, 239–252.

Urban S, Lee JR and Freeman M (2001) Drosophila rhomboid-1 defines a family of putative intramembrane serine proteases. *Cell* **107**, 173–182.

Van Der Bliek AM and Koehler CM (2003) A mitochondrial rhomboid protease. *Dev Cell* **4**, 769–770.

van Dyck L and Langer T (1999) ATP-dependent proteases controlling mitochondrial function in the yeast *Saccharomyces cerevisiae*. *Cell Mol Life Sci* **56**, 825–842.

Vij S, Gupta V, Kumar D, *et al.* (2006) Decoding the rice genome. *Bioessays* **28**, 421–432.

von Heijne G (1986) Mitochondrial targeting sequences may form amphiphilic helices. *EMBO J* **5**, 1335–1342.

Waizenegger T, Habib SJ, Lech M, *et al.* (2004) Tob38, a novel essential component in the biogenesis of beta-barrel proteins of mitochondria. *EMBO Rep* **5**, 704–709.

Waltner M and Weiner H (1995) Conversion of a nonprocessed mitochondrial precursor protein into one that is processed by the mitochondrial processing peptidase. *J Biol Chem* **270**, 26311–26317.

Wang N, Gottesman S, Willingham MC, Gottesman MM and Maurizi MR (1993) A human mitochondrial ATP-dependent protease that is highly homologous to bacterial Lon protease. *Proc Natl Acad Sci USA* **90**, 11247–11251.

Wang J, Hartling JA and Flanagan JM (1997) The structure of ClpP at 23 a resolution suggests a model for ATP-dependent proteolysis. *Cell* **91**, 447–456.

Watanabe N, Che FS, Iwano M, Takayama S, Yoshida S and Isogai A (2001) Dual targeting of spinach protoporphyrinogen oxidase II to mitochondria and chloroplasts by alternative use of two in-frame initiation codons. *J Biol Chem* **276**, 20474–20481.

Werhahn W and Braun HP (2002) Biochemical dissection of the mitochondrial proteome from *Arabidopsis thaliana* by three-dimensional gel electrophoresis. *Electrophoresis* **23**, 640–646.

Werhahn W, Jansch L and Braun HP (2003) Identification of novel subunits of the TOM complex from *Arabidopsis thaliana*. *Plant Physiol Biochem* **41**, 407–416.

Werhahn W, Niemeyer A, Jansch L, Kruft V, Schmitz UK and Braun HP (2001) Purification and characterization of the preprotein translocase of the outer mitochondrial membrane from *Arabidopsis* Identification of multiple forms of TOM20. *Plant Physiol* **125**, 943–954.

Whelan J, Tanudji MR, Smith MK and Day DA (1996) Evidence for a link between translocation and processing during protein import into soybean mitochondria. *Biochim Biophys Acta* **1312**, 48–54.

Wiedemann N, Kozjak V, Chacinska A, *et al.* (2003) Machinery for protein sorting and assembly in the mitochondrial outer membrane. *Nature* **424**, 565–571.

Wienk HL, Czisch M and de Kruijff B (1999) The structural flexibility of the preferredoxin transit peptide. *FEBS Lett* **453**, 318–326.

Winning BM, Sarah CJ, Purdue PE, Day CD and Leaver CJ (1992) The adenine nucleotide translocator of higher plants is synthesized as a large precursor that is processed upon import into mitochondria. *Plant J* **2**, 763–773.

Yamamoto H, Esaki M, Kanamori T, Tamura Y, Nishikawa S and Endo T (2002) Tim50 is a subunit of the TIM23 complex that links protein translocation across the outer and inner mitochondrial membranes. *Cell* **111**, 519–528.

Yang M, Jensen RE, Yaffe MP, Opplinger W and Schatz G (1988) Import of proteins into yeast mitochondria: the purified matrix processing protease contains two subunits which are encoded by the nuclear MAS1 and MAS2 genes. *EMBO J* **7**, 3857–3862.

Yano M, Terada K and Mori M (2003) AIP is a mitochondrial import mediator that binds to both import receptor Tom20 and preproteins. *J Cell Biol* **163**, 45–56.

Yi L and Dalbey RE (2005) Oxa1/Alb3/YidC system for insertion of membrane proteins in mitochondria, chloroplasts and bacteria. *Mol Membr Biol* **22**, 101–111.

Yu J, Hu S, Wang J, *et al.* (2002) A draft sequence of the rice genome (Oryza sativa L ssp indica). *Science* **296**, 79–92.

Zhang XP (2001) *Structure and Function of Mitochondrial and Chloroplast Signal Peptides*. PhD thesis, Department of Biochemistry, Stockholm University, Stockholm.

Zhang XP, Elofsson A, Andreu D and Glaser E (1999) Interaction of mitochondrial presequences with DnaK and mitochondrial Hsp70. *J Mol Biol* **288**, 177–190.

Zhang XP and Glaser E (2002) Interaction of plant mitochondrial and chloroplast signal peptides with Hsp70 molecular chaperone. *Trends Plant Sci* **7**, 14–21.

Zhang XP, Sj Söling S, Tanudji M, *et al.* (2001) Mutagenesis and computer modelling approach to study determinants for recognition of signal peptides by the mitochondrial processing peptidase. *Plant J* **27**, 427–438.

5 Mitochondrial respiratory complex biogenesis: communication, gene expression and assembly

Philippe Giegé

5.1 Introduction

Mitochondria and chloroplasts are the power stations in the plant cell. They are often referred to as semi-autonomous organelles because they have retained a genome after the symbiosis of their ancestor with its host, and throughout evolution. As a power station, one of the most important functions of mitochondria is to perform oxidative phosphorylation. This pathway, operated by respiratory complexes I to IV, generates a proton gradient between the mitochondrial matrix and inter-membrane space. Protons re-enter the matrix through the ATP synthase complex, also termed complex V, thus generating ATP. However, alternative pathways also exist in plant mitochondria, with the occurrence of alternative dehydrogenases and an alternative oxidase (see Chapters 6 and 7).

The plant mitochondrial genome encodes subunits of the respiratory complexes, proteins involved in cytochrome *c* maturation, an essential process to obtain functional cytochrome *c* and cytochrome c_1, a subunit of respiratory complex III (see below), and proteins, and RNAs necessary for the translation of mitochondrial genes, i.e. ribosomal proteins, rRNAs and tRNAs (Unseld *et al.*, 1997). Consequently, it seems that the sole purpose of the mitochondrial genome, directly or indirectly, is to express respiratory proteins.

However, the mitochondrial genome alone is far from being able to express all the required proteins. Of the estimated 75 proteins composing the functional respiratory complexes, only about 25% are encoded in the mitochondrial genome (e.g. 18 proteins in *Arabidopsis*). The remaining proteins are encoded in the nucleus, expressed in the cytosol, imported into mitochondria (see Chapter 4) and assembled into their respective respiratory complexes. Therefore, various and precise communication mechanisms seem to be necessary for the biogenesis of respiratory complexes, and especially for the modulation of this biogenesis. A number of studies have established that mitochondrial respiration can be modulated in the plant cell in response to environmental stimuli, at some particular developmental stages, or in response to stress (Wood *et al.*, 1996; Svensson and Rasmusson, 2001; Giegé *et al.*, 2005; Ribas-Carbo *et al.*, 2005). If this modulation of respiration is due to changes in the number of respiratory complexes per cell, one of the conclusions that can be drawn is that the biogenesis of respiratory complexes might also be adjustable.

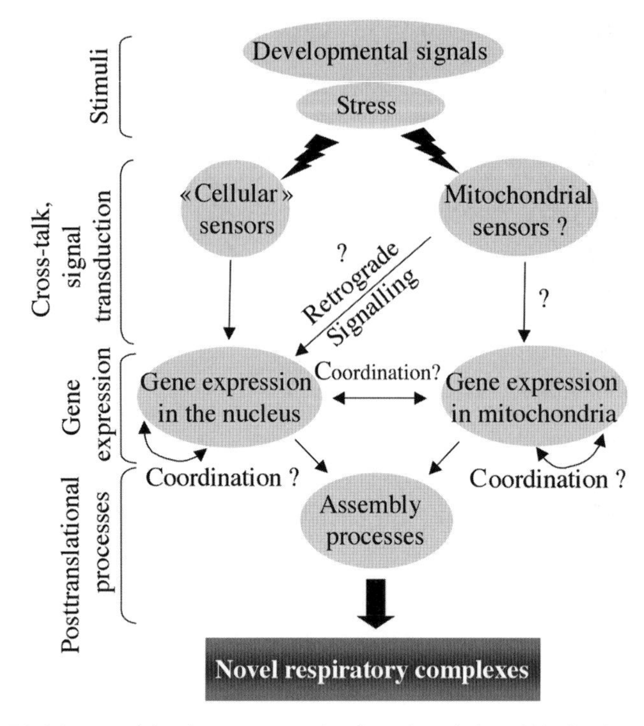

Figure 5.1 Model summarising the processes and envisaged regulation within the plant cell necessary for the biogenesis of plant mitochondrial respiratory complexes composed of subunits encoded in both the nuclear and the mitochondrial genomes. *Cellular sensors* refer to all signal-sensing components in the plant cell not localised to mitochondria.

For this adjustment, the plant cell would first need to sense the need for a modulation of respiratory complexes biogenesis (Figure 5.1). This could be achieved at different levels. First, gene expression of mitochondrial proteins could respond to precise developmental signals. This is very likely since transcripts for many mitochondrial proteins were found to be differentially expressed in different organs, and at different developmental stages (Smart *et al.*, 1994; Ribichich *et al.*, 2001). Second, in the case of stresses linked to mitochondrial function, such as oxidative stress, retrograde signalling to the nucleus appears to be elicited (Wagner, 1995; Maxwell *et al.*, 1999; Moller and Kristensen, 2004). Putative signals, in this case maybe reactive oxygen species (ROS), have to be sensed and the message transferred to the nucleus where gene expression is modulated. Most of the available clues suggesting the occurrence of retrograde signalling for plant mitochondria describe the induction of alternative oxidase (AOX). However other clues that will be discussed here suggest that mitochondrial retrograde pathways could also control the expression of genes necessary for the biogenesis of respiratory complexes.

After sensing and signalling, the biogenesis of respiratory complexes depends on precise gene expression in the nucleus, and mitochondria. Different kinds of gene expression co-ordination are possible (Figure 5.1). First, a co-ordination of gene expression might exist between mitochondrial-encoded genes, and nuclear genes both encoding subunits of the same respiratory complex. Second, a co-ordination could exist at the level of gene expression of nuclear-encoded components of respiratory complexes, i.e. different nuclear genes encoding components of the same complex might be under the control of promoters with conserved elements, possibly with specific transcription factors that would control the expression of functional gene families. This could be true as well for the expression of mitochondrial-encoded genes (see Chapter 3). This chapter will review and discuss what has been documented about the co-ordination, or lack of co-ordination, in the gene expression patterns between mitochondria and the nucleus, and within the nucleus. The nature of the proposed specific promoters for nuclear gene expression and the corresponding trans acting factors will also be discussed.

The biogenesis of respiratory complexes next depends on the synthesis of essential sub-components, or prosthetic groups, found in the respiratory chain such as Fe–S clusters or hemes. Thereafter, biogenesis depends on the fixation to individual subunits of the prosthetic groups and other co-factors such as metals. Ultimately, the construction of multi-subunit respiratory complexes depends on precise assembly processes mediated by a number of dedicated proteins. These proteins interact transiently with the respiratory subunits and are typically present in much smaller amounts than the core subunits of the functional assembled complexes. The information available about these assembly processes in plant mitochondria and the nature of the assembly proteins will be discussed and put in perspective with the more abundant data already existing for other model organisms.

5.2 Biogenesis of prosthetic groups essential for respiration

5.2.1 Mitochondrial Fe–S clusters biogenesis

As part of the mitochondrial respiratory chain, Fe–S clusters are essential prostethic groups that can transfer electrons via the redox state (Fe^{2+} or Fe^{3+}) of their iron atoms. These clusters are found in complex I, in the second subunit of complex II and attached to the Rieske protein in complex III. However, the ubiquitous Fe–S clusters are not only found in the respiratory chain. They are also co-factors of other enzymes in mitochondria such as aconitase, adrenoxin, biotin synthase and lipoate synthase (Beinert, 2000). Moreover, in plants, Fe–S clusters are also found in chloroplasts, the cytosol and the nucleus (Johnson *et al.*, 2005). Nonetheless, Fe–S clusters, and consequently their biogenesis, are crucial to the biogenesis of respiratory complexes. The biogenesis of these clusters has been much studied in yeast and bacteria (Muhlenhoff and Lill, 2000; Lill and Muhlenhoff, 2005). Sulphur and iron atoms are assembled at the level of scaffold proteins. Among them, ISU-like and SUF-like proteins have been described (Garland *et al.*, 1999; Loiseau *et al.*, 2003).

In plants, recent studies (reviewed by Balk and Lobreaux, 2005) strongly suggest that plant mitochondria possess all the proteins necessary for the biogenesis of their own Fe–S clusters. Similar to cytochrome *c* maturation (see below), plant cells possess at least two independent pathways for Fe–S biogenesis. A simple possibility would be that chloroplasts possess a pathway that uses SUF proteins, whereas the mitochondrial one uses ISU proteins. However, this is too simplistic since, for example, AtSUFE was found recently to be essential not only for plastid but also for mitochondrial Fe–S cluster biogenesis (Xu and Moller, 2006). Although independent, i.e. most of the proteins they use are different, both systems show the same core functions, and appear to be prokaryotic in origin (Muhlenhoff and Lill, 2000).

A summarised model for the biogenesis of Fe–S clusters in plant mitochondria is shown in Figure 5.2. Briefly, a desulfurase, probably NFS1 (Picciocchi *et al.*, 2003), catalyses the release of sulphurs from cysteine residues. Iron atoms donated by frataxin (Busi *et al.*, 2004) are assembled with sulphurs at the level of scaffold proteins. Plant mitochondria contain many scaffold proteins including ISU1, ISU2,

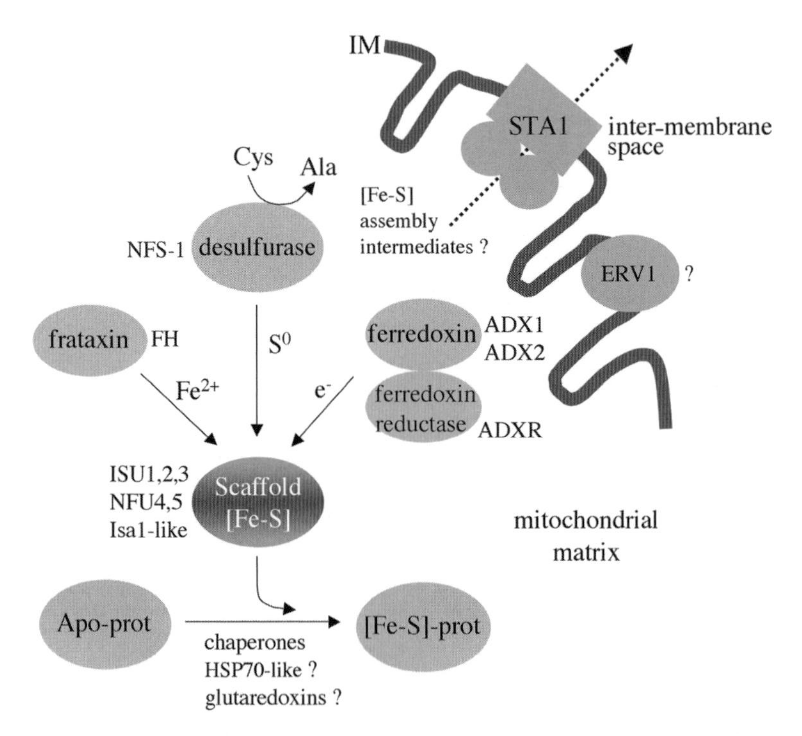

Figure 5.2 Model for the biogenesis of Fe–S clusters in plant mitochondria. The different functions of this pathway are represented by the name of the corresponding proteins identified in plant mitochondria as reviewed by Balk and Lobréaux (2005).

ISU3 (Tone *et al.*, 2004; Leon *et al.*, 2005), NFU4 and NFU5 (Leon *et al.*, 2003; Yabe *et al.*, 2004). Two Isa1-like scaffold proteins are also predicted to be localised in mitochondria (Abdel-Ghany *et al.*, 2005). The precise biological explanation for the presence of the different scaffold proteins is unknown. A ferredoxin, ADX1 and/or ADX2, together with a ferredoxin reductase ADXR (Takubo *et al.*, 2003), produce the reducing equivalents necessary for the assembly. After this process, the Fe–S clusters have to be inserted in their respective target proteins. In yeast and bacteria, glutaredoxins and chaperones are predicted to be involved in this process. In plant mitochondria, glutaredoxins and HSP70-like chaperones that resemble the yeast and bacterial proteins are also present (Heazlewood *et al.*, 2004; Rouhier *et al.*, 2004) and could be involved in the biogenesis of Fe–S proteins. Finally, in plant mitochondria, STA1, the *Starik* ABC transporter (Kushnir *et al.*, 2001) and possibly also ERV1, a sulfhydril oxidase, are predicted, as in yeast (Levitan *et al.*, 2004), to be specifically required for the assembly of Fe–S clusters in the cytosol. Thus it is predicted that mitochondria also provide the cytosol with intermediates for Fe–S clusters biogenesis.

5.2.2 Late steps of heme B synthesis

Similar to Fe–S clusters, heme B are essential prosthetic groups of the mitochondrial respiratory chain. They are found associated with cytochromes where they transfer electrons via the redox state of their iron atoms. However, they are also found in other enzymes such as catalase, peroxidase and nitrate reductase. In higher plants, heme biosynthesis must be envisaged within the overall context of the synthesis of tetrapyrroles. This crucial metabolic pathway produces compounds such as siroheme, heme, phytochromobilin and chlorophyll (for review, see Moulin and Smith, 2005). Seven steps of tetrapyrrole biosynthesis generate protoporphyrin IX from glutamyl tRNA and are common to the synthesis of heme B and chlorophyll. Protogen oxidase catalyses the change of protoporphyrinogen IX to protoporphyrin IX. This compound marks the branch point between heme B and chloropyll synthesis. Heme B is then generated by the insertion of Fe^{2+} into protoporphyrin IX, catalysed by ferrochelatase (Figure 5.3). The enzymes involved in these heme and chlorophyll syntheses are also present in chloroplasts. However, there is an ongoing debate on the presence or not of the last two enzymes of heme B synthesis, namely protogen oxidase and ferrochelatase in plant mitochondria.

Ferrochelatase activity has been described as being associated with mitochondria in potato (Porra and Lascelles, 1968) and in barley (Little and Jones, 1976). A recent study using pea quantified the ferrochelatase activity as being 90% plastidic and 10% mitochondrial in origin (Cornah *et al.*, 2002). Moreover, a single precursor of ferrochelatase could be imported *in vitro* to both chloroplasts and mitochondria (Chow *et al.*, 1997). Similarly, protogen oxidase activity has been described for both plastids and mitochondria (Smith *et al.*, 1993). This enzyme was identified as dual targeted to mitochondria and chloroplasts in spinach via an alternate use of in-frame translation initiation codons that generate two forms of the protein, one being chloroplastic and the other mitochondrial (Watanabe *et al.*, 2001).

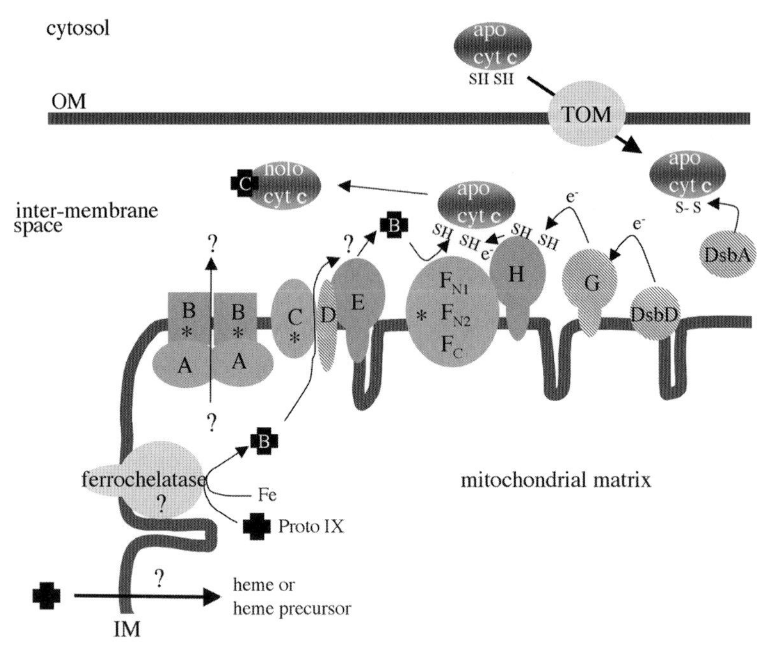

Figure 5.3 Model for the assembly of *c*-type cytochromes in plant mitochondria. A to H stands for CCMA to CCMH. Hatched proteins exist in bacteria but have yet to be identified in plants. Proteins marked by an asterisk (*) are encoded in plant mitochondrial genomes. Black crosses represent heme or its precursor. Heme B is generated by the attachment of iron to protoporphryn IX, catalysed by ferrochelatase. After covalent attachment to cysteine residues of cytochrome *c*, the heme is termed heme C. Abbreviation: TOM, translocase of the outer membrane complex.

However, other studies have shown that the products of the two *Arabidopsis* ferrochelatase genes could not be imported *in vitro* into mitochondria but could be imported into chloroplasts (Lister *et al.*, 2001). Similarly, another study using *in situ* green fluorescent protein (GFP)-coupled transient expression assays showed that the two ferrochelatases of cucumber were both plastidic (Masuda *et al.*, 2003). In chlamydomonas, where one gene exists for protogen oxidase and one for ferrochelatase, both gene products were exclusively found in plastids, but not in mitochondria, thereby suggesting that the absence of heme synthesis, and consequently the requirement for heme import into mitochondria, is a common feature of photosynthetic eukaryotes (van Lis *et al.*, 2005).

It is possible, however, that GFP and the *in vitro* import methods used in the studies mentioned above were not suited to the detection of such low-abundance proteins. Indeed, in all cases, the two final enzymes of heme B synthesis appear to be present at low levels in plant mitochondria relative to chloroplasts. The debate on the localisation of these two proteins remains unsettled. New answers might be provided by an investigation of the transport mechanisms for heme and/or protoporphyrinogen IX in the plant cell.

5.2.3 Heme A synthesis

Heme a is a modified form of heme B (the product of ferrochelatase, see above). As part of the respiratory chain, this type of heme is found in complex IV, cytochrome oxidase. Two such hemes are attached to COX1, one of the core subunits of complex IV. One heme transfers electrons and the other one interacts with copper to form a hetero-metallic active site (Carr and Winge, 2003). In yeast and animal mitochondria, heme A is synthesised from heme b as follows. In the first reaction, a hydroxyethylfarnesyl group from heme B is substituted to a vinyl group by COX10, a farnesyl transferase (Nobrega *et al.*, 1990) to generate heme O. In the second reaction, COX15, the heme A synthase, catalyses the oxidation of a pyrrolmethyl group to form a formyl residue, and thus the final heme A (Glerum *et al.*, 1997). This second step is performed in conjunction with a ferredoxin and a ferredoxin reductase (Barros *et al.*, 2002). In plants, COX10 and COX15 have clear homologues in the genome of *Arabidopsis* (Arabidopsis genome initiative, 2000; Welchen and Gonzalez, 2005). This suggests that the pathway used by plants to synthesise heme A should be similar to the one described in yeast and animals (Figure 5.4).

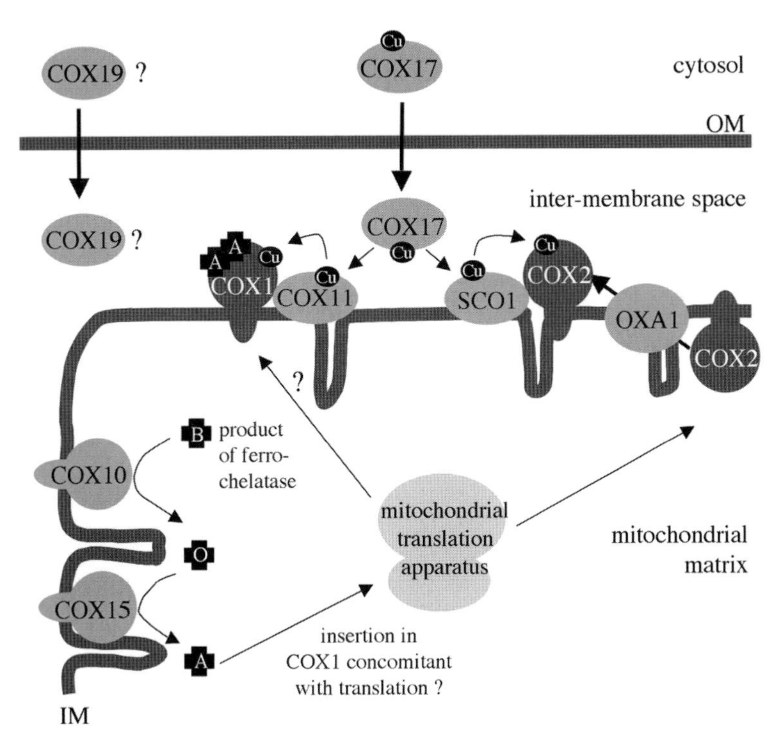

Figure 5.4 Model for some aspects of the assembly of complex IV in plant mitochondria, as derived from yeast complex IV assembly mechanisms reviewed by Carr and Winge (2003). Black crosses represent the various forms of heme.

5.3 Mitochondrial respiratory complexes assembly

All the plant mitochondrial respiratory complexes, with the exception of complex II, are composed of subunits encoded in both the mitochondrial and the nuclear genome. Complex I, NADH dehydrogenase, is thought to be composed of 30 proteins (Heazlewood *et al.*, 2003). Among them, nine are encoded by the mitochondrial genome in *Arabidopsis*. Complex II, succinate dehydrogenase, is composed of eight nuclear-encoded proteins (Millar *et al.*, 2004). Sequences for a complex II subunit are also found in mitochondrial genomes of higher plants, but only as pseudo-genes (Giegé *et al.*, 1998). Complex III, cytochrome *c* reductase, is composed of 9 nuclear-encoded proteins and 1 mitochondrial-encoded protein (Braun *et al.*, 1994), while complex IV, cytochrome *c* oxidase, comprises 11 nuclear-encoded and 3 mitochondrial-encoded proteins (Millar *et al.*, 2004). Finally, complex V, the ATP synthase has 5 mitochondrial-encoded proteins out of a total of 11 in *Arabidopsis* (Heazlewood *et al.*, 2003). Downstream of gene expression in the two compartments, and import in the case of nuclear-encoded subunits, proteins must be properly folded, co-factors added and mature subunits assembled into the correct respiratory complexes. Most of these assembly processes are still poorly understood in plants. However, some preliminary information is available, particularly concerning the assembly of co-factors essential for respiration.

5.3.1 Complex I assembly

The NADH ubiquinone oxidoreductase, commonly know as complex I, is the first enzyme of the respiratory chain. It is L shaped with a membrane-localised arm and a peripheral arm (see Chapter 6). The complex is conserved from bacteria to animals, with the notable exception of yeast (Guenebaut *et al.*, 1998). In plant mitochondria, although the core structure of the complex is maintained, some specific subunits were found, in particular ferripyochelin-binding proteins (Heazlewood *et al.*, 2003). Recently, the same proteins were described in complex I as gamma carbonic anhydrases (Perales *et al.*, 2004). One of them was found to be essential for the stability of the complex (Perales *et al.*, 2005).

Complex I assembly has been studied extensively in the model fungus *Neurospora crassa* (for review, see Schulte, 2001). Several mutant studies in *Neurospora* have revealed that assembly of the full complex is preceded by the assembly of three sub-complexes: the peripheral arm, a large and a small membrane sub-complex. The two membrane-bound sub-complexes are brought together before being attached to the peripheral arm. Recently, it has been shown that human mitochondrial complex I follows the same assembly pattern (for review, see Ugalde *et al.*, 2004). Assembly of the large membrane arm in both model organisms requires chaperones such as complex I intermediate associated (CIA) proteins. CIA30 is found in both model organisms, and CIA84 is found only in fungi (Schulte, 2001; Vogel *et al.*, 2005). Other proteins found to be important for the biogenesis of complex I in *Neurospora* are mitochondrial acyl carrier proteins (mtACP) (Schneider *et al.*, 1997). It is believed that in all eukaryotes that possess complex I, one mtACP is free in the matrix, and involved in lipoic acid biosynthesis, whereas another mtACP is integrated into complex I, where it is possibly involved in the myristoylation of NAD5 (Plesofsky *et al.*, 2000). Finally, in yeast and animals it appears that prohibitins (Phb proteins) are essential for the assembly of complex I, although maybe not specifically for this complex. Phb1 and Phb2 were found to form a large molecular weight complex in the mitochondrial membrane that binds newly synthesised mitochondrial-encoded proteins and protects them from degradation from membrane-bound metalloproteases (Nijtmans *et al.*, 2002; Tatsuta *et al.*, 2005). Recently, a direct interaction between prohibitins and a sub-complex of complex I was detected (Bourges *et al.*, 2004), thus further strengthening the hypothesis of their involvement in the assembly of complex I.

In plants, a clear homologue of CIA30 is found in the *Arabidopsis* genome. Moreover, three other proteins resemble CIA30 but with an additional N-terminal extension exhibiting a typical NAD(P)H binding site. However, no clear homologues are found for CIA84 (Schulte, 2001). In *Arabidopsis,* three mtACPs share higher sequence identity with the ACPs found in *Neurospora* complex I than with plastidic ACPs (Chuman and Brody, 1989; Shintani and Ohlrogge, 1994; Koo *et al.*, 2005). Being an ubiquitous class of chaperones, prohibitins are also identified in plant mitochondria (Millar *et al.*, 2001). They were also detected in complexes of the size of complex I, or of slightly higher molecular weight (Giegé *et al.*, 2003; Heazlewood *et al.*, 2003). In conclusion, the assembly of complex I is only partially understood. However, the discovery in plants of sequences homologous to those of assembly factors in other model organisms indicates conservation among eukaryotes of aspects of complex I assembly.

5.3.2 Cytochrome c maturation is required for the biogenesis of complex III

One of the crucial steps for the biogenesis of complex III, also termed the bc_1 complex, is the maturation of *c*-type cytochromes. This class of cytochromes comprises cytochrome *c,* a soluble protein that shuttles electrons from complex III to complex IV during respiration, and cytochrome c_1, a membrane protein localised in complex III. These cytochromes are essential electron transporters of the respiratory chain. They distinguish themselves from other cytochromes by the covalent attachment of their prosthetic group, a heme C, to two apocytochrome cysteine residues. The term *cytochrome c maturation* describes the mechanisms leading to this covalent attachment.

Different pathways have been described for cytochrome *c* maturation (Kranz *et al.*, 1998). Yeast and animals have evolved a comparatively simple mechanism relying on the cytochrome *c* heme lyases (CCHL) (Dumont *et al.*, 1987) and CC_1HL (Nicholson *et al.*, 1989; Zollner *et al.*, 1992), and on the redox protein Cyc2 (Bernard *et al.*, 2005). Surprisingly, in plant cells, mitochondria and chloroplasts have each retained a distinct system derived from their respective prokaryotic ancestors. The mitochondrial system resembles the one found in some Gram-negative bacteria (Thöny-Meyer, 1997). Similar to the respiratory complexes themselves, the maturation system involves proteins encoded in both the plant nuclear and mitochondrial genomes. The involvement of plant proteins in this process was initially deduced from sequence similarities with bacterial proteins shown to be involved in cytochrome *c* maturation by mutant studies.

The model for the maturation process in plants is shown in Figure 5.3. Briefly, a heme delivery pathway shuttles heme from the matrix, where their synthesis is completed (although, this is a matter of debate, see above), to the inter-membrane space. It involves CcmC (Bonnard and Grienenberger, 1995) and the heme chaperone CCME (Spielewoy *et al.*, 2001) (CcmD is also involved in this delivery in bacteria but no clear homologue of this protein could be found in plants) (Ahuja and Thony-Meyer, 2005). The two cysteines on apocytochrome *c* present in the inter-membrane space have to be reduced for the covalent ligation with heme to occur. This reduction appears to be performed by CCMH a thiol disulfide oxidoreductase (Meyer *et al.*, 2005). Three other proteins forming an entire reduction cascade in bacteria have not been identified in plants (Thony-Meyer, 2002). The attachment of hemes to apocytochromes is thought to be catalysed by $CcmF_{N1}$, F_{N2} and F_C (Gonzalez *et al.*, 1993; Giegé *et al.*, 2004), three mitochondrial-encoded proteins homologous to three domains of the bacterial protein CcmF (Thöny-Meyer *et al.*, 1995). Finally, an ABC transporter is predicted to be essential for cytochrome *c* maturation (Thöny-Meyer *et al.*, 1995). Its exact function in the process and the nature of its substrate are still unknown. This transporter is very interesting because it represents the only example of an ABC transporter composed of two domains encoded by separate genomes, i.e. its trans-membrane domain, named CcmB (Faivre-Nitschke *et al.*, 2001), is encoded in mitochondria and its ATP-binding domain, namely CCMA, is encoded in the nucleus (N. Rayapuram, unpublished results).

In addition to cytochrome *c* maturation, other processes are needed for the assembly of complex III. Unfortunately, nothing is known about these processes in plants. In yeast however, Bcs1 was found to be essential for the assembly of the Rieske Fe/S protein (Cruciat *et al.*, 1999). Similarly, Cbp3 and Cbp4 were identified as complex III assembly factors; they are specific chaperones necessary for its assembly (Wu and Tzagoloff, 1989; Crivellone, 1994). Assembly appears to be a sequential process that requires the preliminary formation of subcomplexes (Zara *et al.*, 2004); in mutants of Cbp3 or Cbp4 an accumulation of intermediate-sized forms of complex III was observed (Kronekova and Rodel, 2005).

5.3.3 Complex IV assembly

Cytochrome oxidase, known as complex IV, is the terminal complex of the mitochondrial electron transport chain, reducing molecular oxygen to water. In plants, the three core subunits of the complex COX1, COX2 and COX3 are encoded in the mitochondrial genome (Unseld *et al.*, 1997; Adams and Palmer, 2003), and assembled with about 10 smaller nuclear-encoded proteins. The biogenesis of this complex is an intricate process because of the number and variety of co-factors attached to it, including two heme A, three Cu, one Zn, one Mg and one Na ion, (Carr and Winge, 2003). In plants, few investigations have been performed into the assembly of this complex. However, the discovery of proteins homologous to yeast or animal proteins and data from yeast mutant complementation assays strongly suggest that the biogenesis of complex IV in plant mitochondria occurs in a similar fashion to that in yeast and animal mitochondria (Figure 5.4).

In yeast and animal mitochondria, nascent COX2, translated in the mitochondrial matrix, is anchored in the inner-membrane by an N-terminal helix, and translocated to the inter-membrane space with the help of OXA1 (Altamura *et al.*, 1996; Green-Willms *et al.*, 2001). OXA1, which is essential for the assembly of complex IV, is however not specific for COX2 (Kermorgant *et al.*, 1997; Kuhn *et al.*, 2003). Other proteins belonging to the OXA1 family were found to be involved in protein translocation as well. Among them, OXA2 was found in *Neurospora* (Funes *et al.*, 2004). In yeast and humans, COX18, a protein similar to OXA2, was found to be involved in the translocation of the C-terminal domain of COX2 to the inter-membrane space (Saracco and Fox, 2002; Sacconi *et al.*, 2005). After the translocation, metals have to be added to the individual subunits. Little is known concerning the delivery of Zn or Mg. However a number of proteins, namely COX17, found in the cytosol and mitochondrial inter-membrane space, as well as COX11 and SCO1, found in the mitochondrial inner membrane, were found to be essential for the delivery of Cu to complex IV (Harrison *et al.*, 1999). COX17 is proposed to be a Cu chaperone that binds Cu through a CCxC motif and is thought to deliver Cu to the inter-membrane space where assembly is performed (Horng *et al.*, 2004; Palumaa *et al.*, 2004). COX11 and SCO1 are believed to be metal chaperones that are involved in the later steps of Cu transfer to complex IV (Carr and Winge, 2003). SCO1 mediates the transfer of Cu to COX2 via a conserved CxxxCH motif (Schulze and Rodel, 1988), whereas COX11 is important for the transfer of Cu to COX1 (Carr *et al.*, 2002). COX19 is similar to COX17; it is essential for respiration. However, it lacks the CCxC conserved motif, and its function appears to be independent from Cu binding. However, its dual localisation, similar to that of COX17, and the presence of four conserved cysteines suggest that COX19 could be a metal transporter as well (Nobrega *et al.*, 2002). Finally, during, or after the attachment of the co-factors, the subunits have to be assembled into a fully functional complex. In yeast, a model has recently been proposed where COX1 and COX2, the two largest subunits of complex IV, are first equipped with their respective co-factors on independent assembly lines before being assembled into a full complex IV. These two sub-complexes are assembled with the help of specific chaperones, i.e. Mss51 for COX1 and COX20 for COX2 (for review, see Herrmann and Funes, 2005).

In plants, OXA1, COX17, COX19, COX11 and SCO1 all have homologues in the *Arabidopsis* genome (Arabidopsis genome initiative, 2000; Welchen and Gonzalez, 2005). Moreover, the product of an *Arabidopsis OXA1* gene that could complement yeast *oxa1* mutants (Hamel *et al.*, 1997; Cardazzo *et al.*, 1998) was found localised in the inner membrane of mitochondria (Sakamoto *et al.*, 2000). Similarly, *Arabidopsis COX17* was able to complement yeast *cox17* mutants (Balandin and Castresana, 2002), thus further strengthening the hypothesis that plants assemble complex IV in a similar manner to yeast and animals.

5.3.4 Complex V assembly

The respiratory chain generates a proton gradient between the matrix and the mitochondrial inter-membrane space. Protons re-enter the matrix dynamically through the ATP synthase, also commonly known as complex V, thus catalysing the formation of ATP. Complex V is composed of the F_1 catalytic domain, extrinsically attached to the mitochondrial inner-membrane and the F_O domain embedded in this membrane.

Similar to most of the other complexes described above, the assembly of complex V has not been investigated in plants. In yeast however, the study of respiratory deficient mutants has revealed that several chaperone-like proteins are essential for the biogenesis of the ATP synthase. Among them, ATP11 and ATP12 are important for the assembly of the F_1 domain. The core of the F_1 sub-complex is composed of an $\alpha_3\beta_3$ hexamer. ATP11 and ATP12 are involved in the formation of this hexamer (Ackerman and Tzagoloff, 1990; Ackerman, 2002). ATP11 binds ATP2 (the β subunit) and ATP12 binds ATP1 (the α subunit). Three of each of these duplexes are joined to form the core $\alpha_3\beta_3$, with the subsequent release of ATP11 and ATP12. In *atp11* and *atp12* mutants, ATP1 and ATP2 form unspecific aggregates (Lefebvre-Legendre *et al.*, 2005). In yeast, FMC1 was also found to be essential for complex V assembly (Lefebvre-Legendre *et al.*, 2001). Similar to ATP11 and ATP12, FMC1 appears to be involved in the formation of the F_1 sub-complex. In *fmc1* mutants, the same unspecific ATP1/ATP2 aggregates are detected, however only when mutants are grown at 37°C. This indicates that FMC1 is involved in a heat-sensitive assembly step (Lefebvre-Legendre *et al.*, 2001). ATP10 (Ackerman and Tzagoloff, 1990) and ATP22 (Helfenbein *et al.*, 2003) appear to be involved in the formation of the membrane-embedded F_0 sub-complex. ATP22 is a protein of the mitochondrial inner membrane that acts at the post-translational level (Helfenbein *et al.*, 2003), while ATP10 seems to be an ATP6-specific chaperone (Tzagoloff *et al.*, 2004). It should be noted that all these proteins are unrelated to the major chaperone families such as HSP60 or HSP70 that have many different substrates. These chaperone proteins seem to be specific for the formation of complex V (Ackerman, 2002).

A recent phylogenetic study that investigated all the complete genome sequences available revealed that ATP11 and ATP12 have conserved homologues in plants (Pickova *et al.*, 2005). The same is true for ATP10. However, FCM1 and ATP22 are not found in plants; they only have clear homologues in fungal genomes (Pickova *et al.*, 2005). Altogether, these findings suggest that some key aspects of the biogenesis of complex V are conserved from yeast to plants. However, it also appears that some specific, as-yet-unidentified, mechanisms for the biogenesis of the ATP synthase have evolved in plant mitochondria, or have been retained from their prokaryotic ancestors.

5.4 Gene expression for the biogenesis of mitochondrial respiratory complexes

Before the assembly processes discussed above can take place, the biogenesis of mitochondrial respiratory complexes requires precise gene expression from both the mitochondrial and the nuclear genome. Genes in the nucleus that encode components of respiratory complexes might be regulated specifically for this biogenesis to occur, i.e. under the control of specific promoters recognised by dedicated transcription factors. In mitochondria, the control of gene expression entirely governed by nuclear-encoded factors is crucial for the biogenesis of respiratory complexes. Mitochondrial gene expression and its control are discussed in Chapter 3. The precise stoichiometry of nuclear- and mitochondrial-encoded subunits within the respiratory complexes infers that different types of gene-expression co-ordination mechanisms could be operative between mitochondrial genes, nuclear genes and/or between mitochondrial and nuclear genes. In some cases, this seems to be achieved directly at the level of transcription, whereas in other cases, this appears to be achieved at later stages, i.e. during complex assembly.

5.4.1 Nuclear co-ordination of the expression of electron transport chain genes

In plants, the identification of nuclear-encoded components of the mitochondrial respiratory complexes and the investigation of their expression patterns have suggested that some co-ordinated gene expression could exist for nuclear genes. In particular, for complex I genes, transcripts of the 22-kDa PSST subunit, 55-kDa NADH binding subunit and 28-kDa TYKY subunit were all found to be upregulated six- to tenfold in flowers (Grohmann *et al.*, 1996; Heiser *et al.*, 1996; Schmidt-Bleek *et al.*, 1997). Similarly, another nuclear-encoded respiratory chain gene encoding the Rieske Fe–S subunit of complex III was also found highly expressed in flowers relative to other organs (Huang *et al.*, 1994). On the basis of the work on complex I genes, Zabaleta *et al.* (1998) searched for conserved promoter sequences for these three genes. They identified a seven-nucleotide-long conserved motif (TGTGGTT) in the 5′ promoter regions that was involved in the specific expression of these nuclear-encoded components of complex I in anthers and pollen, but not in other flower tissues. Loss-of-function as well as gain-of-function experiments showed that this regulation, this co-ordinated gene expression, was achieved at the transcript level (Zabaleta *et al.*, 1998). However, similar so-called pollen box sequences were also found for other genes, i.e. not only for genes encoding complex I or respiratory genes (Twell *et al.*, 1991; Eyal *et al.*, 1995). The results of this study nevertheless showed that co-ordinated gene expression in the nucleus could be driven by conserved *cis*-acting regulatory elements, even if in this case, the motifs identified were organ specific, and not necessarily specific for a functional family of genes such as complex I genes.

If specific *cis*-acting elements exist for the expression of complex I genes, then specific *trans*-acting factors should exist as well. Recently, a protein described as a gamma carbonic anhydrase has been identified as an interactor with complex I, and as essential for complex I biogenesis (see above) (Perales *et al.*, 2004). This protein had been identified several times in complexes of the size of complex I (Eubel *et al.*, 2003; Giegé *et al.*, 2003; Heazlewood *et al.*, 2003). This protein was also described as a putative transcription factor that could be responsible for specific gene expression of complex I subunits in anthers (Parisi *et al.*, 2004; E. Zabaleta, unpublished results). If this role as a transcription factor is real, then it provides an interesting perspective on complex I gene expression in plant mitochondria.

Recently, comparison of promoter regions for two of the three genes encoding the second subunit of complex II, namely SDH2-1 and SDH2-2, was reported (Elorza *et al.*, 2004). Two promoters were shown to induce enhanced expression in anthers and pollen. However, only the promoter for SDH2-2 induced significant expression in root tips. Surprisingly, promoter mapping revealed that the region driving the specific expression in anthers was located at different positions in the two promoters. In this case, co-ordinated expression could not be attributed to a conserved *cis*-element (Elorza *et al.*, 2004). The absence of co-ordination here may not be surprising given the functional redundancy of the two genes.

In other recent studies, the promoter regions of the nuclear genes encoding cytochrome *c*, and complex IV subunits, were investigated. Specifically, promoter elements were characterised for *Cytc-1* and *Cytc-2* (Welchen and Gonzalez, 2005), *COX5b-1* (Welchen *et al.*, 2004), *COX5c-1* and *COX5c-2* (Curi *et al.*, 2005). Different expression patterns were observed for the various genes, with a constant being the detection of high gene expression in flowers during early development. A closer investigation of the *Cytc-1* promoter revealed that gene expression in root, shoot meristems and anthers was entirely dependent on very small motifs, namely two site II motifs (TGGGCC/T), and a telomeric repeat motif found downstream. However, mutant analysis demonstrated that site II elements were of predominant importance because the deletion of a single one led to the total abolition of gene expression (Welchen and Gonzalez, 2005). These promoter elements had already been described as implicated in the expression of genes in proliferating cells, and as localised roughly 100–200 nucleotides upstream of the initiation codon (Kosugi and Ohashi, 1997; Tremousaygue *et al.*, 2003). The most interesting extrapolation from the cytochrome *c* promoter analysis was the discovery that site II elements are found in predicted promoters for most of the nuclear-encoded respiratory proteins (Welchen and Gonzalez, 2006). A bioinformatic analysis of 103 predicted promoters of respiratory chain genes in *Arabidopsis* revealed that over 80% of them have site II elements. The same analysis was performed on the rice genome and gave the same result. Moreover, these elements cluster in a region upstream of the initiation codons of the respective genes, a localisation compatible with their predicted role as conserved *cis*-acting elements (Figure 5.5). Note that genes encoding members of the alternative pathways, such as alternative NAD(P)H dehydrogenases, or alternative oxidases typically lack site II motifs in their promoter regions (Welchen and Gonzalez, 2006). Though, as appealing as it seems, site II elements alone cannot define respiratory chain genes because they are found in the promoter region

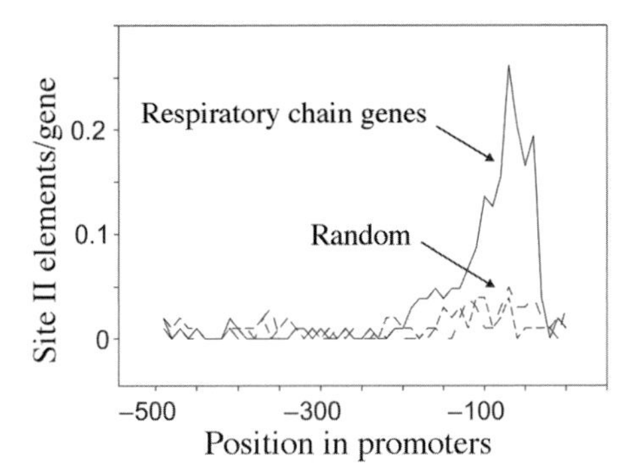

Figure 5.5 Site II elements (TGGGCC/T) frequency (number per gene) and position in 103 promoter regions of *Arabidopsis* nuclear genes encoding components of the respiratory chain (solid line) compared to their distribution in three random sets of *Arabidopsis* nuclear genes (dotted lines) (Welchen and Gonzalez, 2006).

of over 5% of the genes in *Arabidopsis* genome (Welchen and Gonzalez, 2006). However, site II elements together with other, yet unidentified, motifs might very well be involved in the co-ordinated gene expression of nuclear-encoded respiratory proteins.

Previous studies have shown that TCP-domain transcription factors can bind site II motifs (Kosugi and Ohashi, 1997; Tremousaygue *et al.*, 2003). In the case of *Cytc-1*, Welchen and Gonzalez (2005) demonstrated that the exemplary chosen AtTCP20 could bind their promoter region *in vitro*. This is not surprising since all TCP transcription factors are predicted to bind site II motifs. The TCP transcription factors family has over 20 members in *Arabidopsis*. It is thus possible that some of them could be specific for the expression of nuclear-encoded respiratory-chain genes. Other promoter mutation studies as well as systematic approaches such as one hybrid screens and *Chromatin Immunoprecipitation-Chips* (van Steensel, 2005) should help in identifying potential conserved promoter elements and the *trans*-acting factors that bind them. This will possibly confirm the site II/TCP transcription factors hypothesis and should help in determining whether co-ordination really exists for the expression of nuclear genes encoding components of mitochondrial respiratory complexes and, if it does, its extent and possible role in the modulation of respiratory complex biogenesis.

5.4.2 Co-ordination between mitochondrial and nuclear gene expression

It has been long speculated as to whether or not co-ordinated expression of mitochondrial and nuclear genes exists in plants, and if it exists, whether or not this is a house-keeping or default phenomenon, or a much rarer occurrence that would take place in response to very particular situations, during development, or in response to stresses.

In favour of the co-ordinated expression hypothesis, Ribichich *et al.* (2001) found an induction of cytochrome *c* during flower development in sunflower, whereas Smart *et al.* (1994) observed an induction of mitochondrial gene expression at the same developmental stage, also in sunflower. This enabled Ribichich *et al.* (2001) to suggest that co-ordinated gene expression could indeed exist between mitochondria and the nucleus for the biogenesis of respiratory complexes. Similarly, a recent study using rice germination as a model showed that the expression of some mitochondrial- and nuclear-encoded transcripts were both induced during the early steps of rice germination (Howell *et al.*, 2006). Finally, data mining in the publicly available microarray databases and in published results reveals that certain mitochondrial genes are sometimes up- or downregulated when nuclear-encoded genes from the same complexes are regulated in the same direction (Clifton *et al.*, 2005). The relevance of such microarray results is unclear, especially when the gene expression data concerns single genes encoding subunits of respiratory complexes, and not whole functional families of genes. However, researchers have so far failed to describe a co-ordination mechanism between gene expression in mitochondria and the nucleus of plant cells (Mackenzie and McIntosh, 1999).

In light of the absence of a general model for gene expression co-ordination between mitochondria and the nucleus, an investigation on a global scale was initiated by Giegé *et al.* (2005) to determine whether or not gene expression co-ordination between mitochondria and the nucleus was taking place at the transcript level, and if not, at which level it would ultimately be achieved. For this purpose, sucrose starvation imposed on an *Arabidopsis* cell suspension culture was used as a model system. The sucrose starvation treatment had the effect of reducing the volume and/or number of mitochondria per cell, resulting in decreased respiration. This could be explained by a decrease in the biogenesis of respiratory complexes. A custom microarray was designed where the entire mitochondrial genome gene content was represented together with nuclear-encoded genes encoding proteins found in the respiratory complexes. The microarray analysis clearly showed an increase in the steady-state abundances of mitochondrial transcripts, whereas nuclear gene expression was downregulated. These results were obtained repeatedly and could be reversed when sucrose was added back to the cell suspension. These data demonstrate that in this case no gene expression co-ordination was taking place at the transcript level between mitochondria and the nucleus. Further studies at the protein level, including measurements of the steady-state abundances of proteins, protein abundance in assembled complexes, measurements of *in organello* protein synthesis and the capacity to import proteins, revealed that the effect of sucrose starvation on nuclear or mitochondrial gene, i.e. the production of an excess of mitochondrial-encoded subunits compared to their nuclear-encoded counterparts, would most probably only be sorted out at the level of complex assembly (Figure 5.6). No co-ordination of the genomes expression had been achieved in response to sucrose starvation.

The results of this analysis have to be tempered by the fact that they arise from a specific treatment applied to cell suspensions, and not to plants with differentiated tissues. It is, therefore, not possible to extrapolate too much from these results. It is

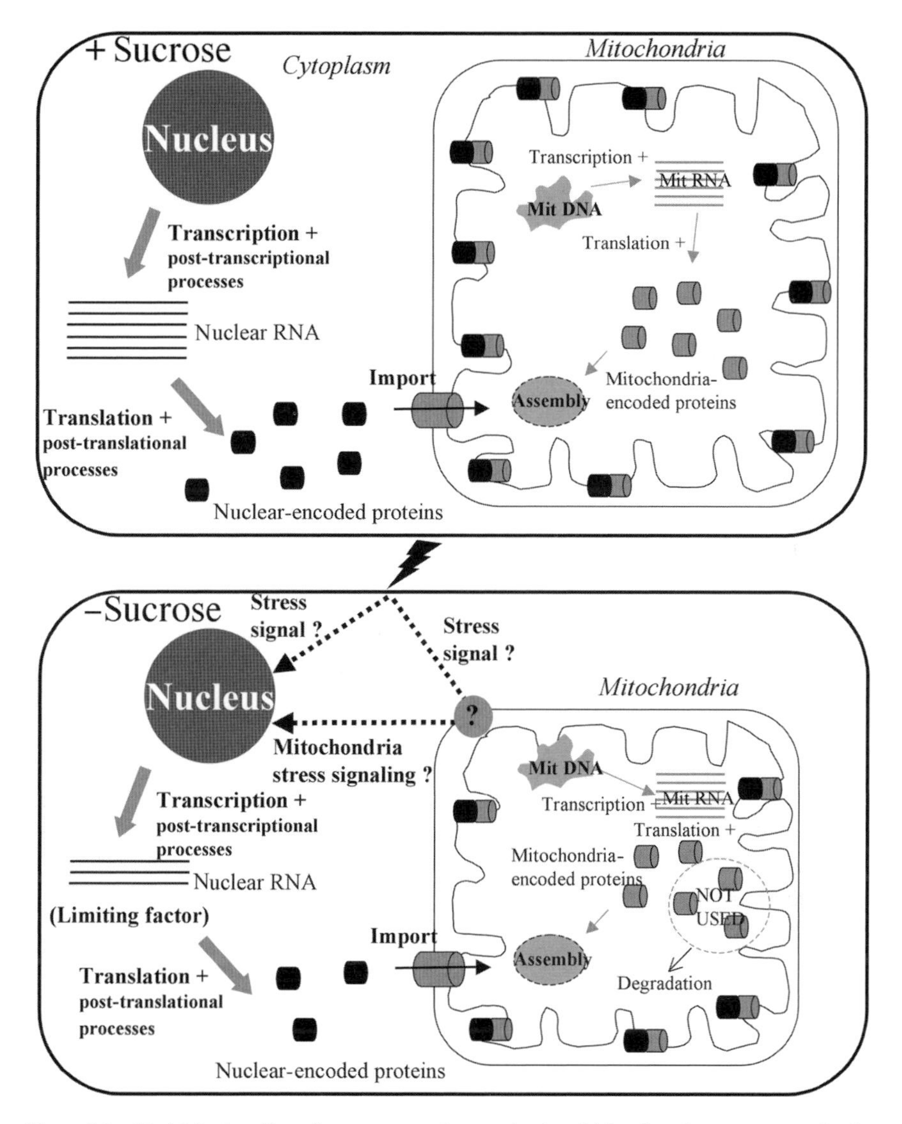

Figure 5.6 Model for the effect of sucrose starvation on mitochondrial and nuclear gene expression for the biogenesis of respiratory complexes in *Arabidopsis* mitochondria as described by Giegé *et al.* (2005). Nucleus-encoded proteins (black cylinders) and mitochondrial-encoded proteins (grey cylinders) are assembled in multi-subunits respiratory complexes (black/grey cylinders).

possible that some co-ordination mechanisms are operative for specific cell types, and/or in response to particular stresses, or developmental stages. However, even if this is happening in some cases, the results published to date tend to show that an absence of co-ordination between gene expression in mitochondria and the nucleus could be the normal situation in the plant cell.

5.5 Mitochondria–nucleus cross talk

As should be clear from the previous sections, the biogenesis of mitochondrial respiratory complexes depends entirely on gene expression in the nucleus. The assembly factors essential for the construction of complexes and their co-factors, the proteins controlling gene expression in the nucleus, and also in mitochondria, are all encoded in the nucleus. Thus, since mitochondria are unable to directly control their own genetic system and modulate the biogenesis of their respiratory complexes, some communication mechanisms must be operative between mitochondria and the nucleus. In some cases, mitochondria sense dysfunctions, or stresses imposed on them, or a specific developmental stage and elicit a signal that will be transferred to the nucleus where specific gene-expression changes will respond to the mitochondrial situation. This is referred to as retrograde signalling. This type of signalling mechanism is well established and much studied for yeast and animal mitochondria (Butow and Avadhani, 2004; Moye-Rowley, 2005). In both yeast and animal models, retrograde signalling is linked to the target of rapamycin signalling pathway (Jacinto and Hall, 2003). In mammals, mitochondrial retrograde signalling involves Ca^{2+} dynamics and proteins such as NFκB, NFAT and ATF (for review, see Butow and Avadhani, 2004).

In plants, a precise model for retrograde signalling between organelles and the nucleus remains to be established (Leister, 2005). However, studies on plastid to nucleus signalling using genetic and biochemical approaches have revealed the existence of at least three partially overlapping retrograde-signalling pathways (for review, see Nott *et al.*, 2006). The most studied pathway uses Mg-protoporphyrinogen IX as a signal molecule (Strand *et al.*, 2003) and involves genomes uncoupled proteins (Susek *et al.*, 1993). A second pathway requires chloroplast gene expression, while a third pathway is linked to the redox state of the photosynthetic electron transport chain. ROS generated by chloroplasts were found to play a role in nuclear gene expression (Karpinski *et al.*, 1999). ROS appear to initiate a plastid-signalling pathway (op den Camp *et al.*, 2003) involving the EXECUTER 1 protein (Wagner *et al.*, 2004).

Such signalling pathways have not yet been described for mitochondria. A growing number of examples tend to show that multiple retrograde-signalling pathways could be operative between plant mitochondria and the nucleus (Gray *et al.*, 2004a,b; Noctor *et al.*, 2004) and it could respond to situations where the biogenesis of respiratory complexes is affected.

5.5.1 Clues for plant mitochondrial retrograde signalling

In plant mitochondria, many of the attempts to identify mitochondrial retrograde regulation have focused on describing the induction of components of the alternative pathways, such as AOXs in response to mitochondrial functions or stresses (for review, see Juszczuk and Rychter, 2003). Interesting examples come from maize non-chromosomal stripe mutants (NCS) where the mitochondrial genome is mutated or rearranged. In the *NCS2* mutant, rearranged mitochondrial DNA has created a chimeric *nad4-nad7* gene, ultimately resulting in decreased complex I function (Marienfeld and Newton, 1994). In *NCS3* and *NCS4* mutants, genes for ribosomal proteins are affected (Newton *et al.*, 1996). Finally, *NCS5* and *NCS6* mutants have deletions in the *cox2* gene (Newton *et al.*, 1990). Each *NCS* mutant exhibits high-level expression of a specific *aox* gene in all non-photosynthetic tissues tested, and the expression pattern is specific for each type of mitochondrial lesion. For example, the NADH dehydrogenase–defective *NCS2* mutant has high expression of *aox2*, whereas the cytochrome oxidase–defective *NCS6* mutant predominantly expresses *aox3* (Karpova *et al.*, 2002). This could be a clue to the existence of retrograde signalling in response to defective respiration in mitochondria.

Recently, a study using a novel reporter-gene system, where luciferase is under the control of the promoter region of AtAOX1a, suggested that the induction of AtAOX1a in response to different mitochondrial perturbations, e.g. inhibition of the respiratory chain, occurs by distinct, yet overlapping signalling pathways. These pathways appear to be tissue specific (Zarkovic *et al.*, 2005). Moreover, another study using the same system and promoter deletion analyses identified a region in the AtAOX1a promoter to be essential for the induced expression of AtAOX1a in response to mitochondrial retrograde regulation (Dojcinovic *et al.*, 2005).

Retrograde signalling is not, however, believed to be involved only in the control of components of alternative pathways. In some situations, a few of them described below, results indicate that retrograde signalling could also operate to modulate the expression of nuclear genes encoding genes essential for the biogenesis of the mitochondrial respiratory chain.

In the *Starik* mutant, where STA1, a gene encoding a mitochondrial ABC transporter possibly involved in the export of Fe–S clusters precursors to the cytosol (see above), is disrupted, increased levels of free iron were observed in the mitochondrial matrix (Kushnir *et al.*, 2001). Nuclear transcript abundance analysis in this mutant revealed an induction of expression of genes encoding DNA repair proteins, ROS scavenging enzymes and a putative mitochondrial L-cysteine desulfurase (see above). The latter gene expression modification could reflect an attempt by the plant cell to compensate for the disruption in the pathways for Fe–S clusters synthesis induced by the mutation. In this case, it appears that a particular mitochondrial dysfunction has induced very specific gene expression modifications in the nucleus (Kushnir *et al.*, 2001), suggesting that retrograde signalling could be operating here. If this was the case, the primary signal could be the levels of iron and/or of ROS in mitochondria. In another example, transgenic lines of *Arabidopsis* and tobacco were generated where unedited versions of ATP9 were expressed (Hernould *et al.*, 1998; Gomez-Casati *et al.*, 2002). These lines resulted as expected in dysfunctional mitochondria with reduced respiration rates and, in particular, a male sterile phenotype. Expression of the three, exemplary chosen, PSST, TYKY and NADH, binding subunits of complex I were induced in the mutants (Gomez-Casati *et al.*, 2002). It is unclear, however, if these changes were specific to respiratory complexes genes since a global transcriptome analysis was not performed. The authors attribute these gene expression modifications to a compensatory response of the nucleus that would sense the mitochondrial dysfunction and react at the level of the nucleus by increasing the biogenesis of mitochondrial respiratory complexes. Finally, when sugar starvation was imposed on *Arabidopsis* cells, a decrease in the amount of respiratory complexes was induced. This resulted in decreased transcript abundance for nuclear-encoded respiratory protein genes (Giegé *et al.*, 2005). In this case as well, retrograde signalling could be operating. However, similar to the examples given above, it is unclear whether perception of stress or mitochondrial dysfunction, i.e. the deprivation of sucrose in this case, was taking place directly at the mitochondrial level or elsewhere in the plant cell. The direct involvement of plant mitochondria in this particular stress response has not been established.

5.5.2 Potential components of retrograde-signalling pathways

ROS (see Chapter 7) are obvious candidates for involvement in mitochondrial retrograde signalling. These active molecules are produced in many circumstances, in particular as by-products of mitochondrial function or stresses (Neill *et al.*, 2002a; Vranova *et al.*, 2002). ROS are therefore often regarded as markers of mitochondrial oxidative stress (Smirnoff, 1998). On the other hand, ROS are known to be general signal molecules for plant responses (Neill *et al.*, 2002b). Therefore, it is possible that mitochondrial ROS, similar to plastid-derived ROS (Nott *et al.*, 2006), has a signalling function. Sensors, maybe ROS-scavenging proteins, could participate in the early steps of a signalling pathway that would ultimately lead to gene expression changes in the nucleus (Gray *et al.*, 2004a).

Retrograde signalling for plant mitochondria could also involve protein phosphorylation and dephosphorylation events through, for example, MAP kinases that are known to be involved in stress responses, in particular in oxidative stress response in plants (Agrawal *et al.*, 2003). Plant mitochondrial retrograde signalling could also use calcium signalling in a similar fashion to animals (Butow and Avadhani, 2004). However, the occurrence of these specific types of signalling pathway between mitochondria and the nucleus in plants is still speculative.

Other potential participants in mitochondrial retrograde regulation might be found among proteins involved in two-component pathways. This type of signalling pathway, which was until recently thought to be exclusively prokaryotic, has been found in plants (for review, see Grefen and Harter, 2004). It involves at least two proteins: a signal-sensing histidine kinase and a response regulator that elicits the output response. In multi-step two-component pathways, the two elements are linked by histidine-containing phosphotransfer (HPt) domain proteins. These proteins are linked by His- to Asp phospho-relays (Figure 5.7). In plants, these types of proteins were found to be involved in organelle signalling. ARR4, a response regulator protein, was found to modulate red light signalling at the level of phytochrome B photoreceptors (Sweere *et al.*, 2001). ARR2, another response regulator protein predominantly expressed in pollen, is localised in the nucleus where it acts as a transcription factor. ARR2 binds *in vivo* to the promoter region of the PSST-subunit gene of complex I. It also binds *in vitro* two other promoters of complex I subunit genes, i.e. the TYKY and 55 kDa subunits (Lohrmann *et al.*, 2001). These three genes were found to be upregulated in pollen (see above). However, ARR2 is not

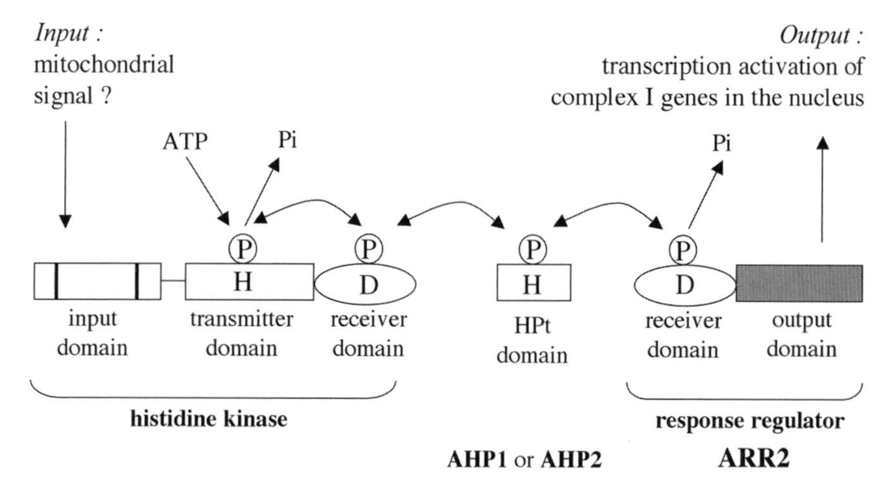

Figure 5.7 Model for a multi-step two-component pathway as described by Lohrmann *et al.* (2001), potentially involved in mitochondria to nucleus signalling. After auto-phosphorylation of a histidine kinase at a conserved histidine (H), the signal is transmitted by phosphorylations at conserved aspartic acid residues (D) and histidines on HPt domain proteins and on the response regulator, in this case ARR2 that acts as a transcription factor in *Arabidopsis* nucleus (derived from Grefen and Harter, 2004).

only an effector of a signalling pathway involved in the control of mitochondrial respiratory genes expression. A recent study shows that ARR2 is also involved in ethylene signalling. It appears to be central in the fine tuning and cross talk of a number of signalling pathways in higher plants (Hass *et al.*, 2004).

5.6 Concluding remarks

In the plant cell, the biogenesis of mitochondrial respiratory complexes and its regulation are intricate processes. They involve mitochondria, the nucleus and chloroplasts where essential co-factors, like heme, or their precursors are synthesised. The control of respiratory complex biogenesis could involve retrograde signalling. Signals elicited by mitochondria might be perceived by the nucleus, where gene expression could be consequently altered. Nuclear genes encoding subunits of respiratory complexes are possibly expressed in a coordinated fashion. Other nuclear genes encoding specific respiratory complex assembly factors, enzymes for the synthesis of co-factors such as Fe–S clusters and proteins necessary for the expression of the mitochondrial genome, i.e. its transcription, transcript processing and translation, are all expressed, translated in the cytosol and their products imported into mitochondria. In mitochondria, on the other hand, gene expression, although under the control of nuclear-encoded proteins, appears to take place independently of the cellular environment. Transcript abundance of mitochondrial-encoded subunits of respiratory complexes generally does not vary co-ordinately with transcript abundance of the nuclear-encoded subunits in response to stresses or developmental stage. Thus, it appears that before the start of the assembly processes the pools of mitochondrial- and nuclear-encoded subunits are not present in the correct ratios required by the stoichiometry of the respiratory complexes. Newly synthesised subunits encoded in the mitochondrial and the nuclear genomes are equipped with metals and other co-factors, before and/or during assembly, and are brought together by chaperones and other specific assembly factors before finally being inserted as newly synthesised respiratory complexes in the mitochondrial inner membrane. Excess mitochondrial- or nuclear-encoded subunits leftover from the assembly processes appears to be degraded by a clean-up system involving as-yet-unidentified proteases. PreP (Stahl *et al.*, 2005) or AAA metalloproteases identified in plant mitochondria (Urantowka *et al.*, 2005) may be involved in this process.

Acknowledgements

My apologies to all colleagues working on the various aspects of plant mitochondrial biogenesis discussed here whose work could not be cited in this chapter because of space limitations. Thanks to Dr. D. Gonzalez (Santa Fe, Argentina) for critical reading and discussion, and for welcoming me in his laboratory during the time when this chapter was written. Thanks as well to Dr. G. Bonnard, Dr. D. Gagliardi and Dr. H. Lange for critical reading.

References

Abdel-Ghany SE, Ye H, Garifullina GF, Zhang L, Pilon-Smits EA and Pilon M (2005) Iron–sulfur cluster biogenesis in chloroplasts. Involvement of the scaffold protein CpIscA. *Plant Physiol* **138**, 161–172.

Ackerman SH (2002) Atp11p and Atp12p are chaperones for F(1)-ATPase biogenesis in mitochondria. *Biochim Biophys Acta* **1555**, 101–105.

Ackerman SH and Tzagoloff A (1990) Identification of two nuclear genes (ATP11, ATP12) required for assembly of the yeast F1-ATPase. *Proc Natl Acad Sci USA* **87**, 4986–4990.

Adams KL and Palmer JD (2003) Evolution of mitochondrial gene content: gene loss and transfer to the nucleus. *Mol Phylogenet Evol* **29**, 380–395.

Agrawal GK, Agrawal SK, Shibato J, Iwahashi H and Rakwal R (2003) Novel rice MAP kinases OsMSRMK3 and OsWJUMK1 involved in encountering diverse environmental stresses and developmental regulation. *Biochem Biophys Res Commun* **300**, 775–783.

Ahuja U, Thony-Meyer L (2005) CcmD is involved in complex formation between CcmC and the heme chaperone CcmE during cytochrome *c* maturation. *J Biol Chem* **280**, 236–243.

Altamura N, Capitanio N, Bonnefoy N, Papa S and Dujardin G (1996) The *Saccharomyces cerevisiae* OXA1 gene is required for the correct assembly of cytochrome *c* oxidase and oligomycin-sensitive ATP synthase. *FEBS Lett* **382**, 111–115.

Arabidopsis Genome Initiative (2000) Analysis of the genome sequence of the flowering plant *Arabidopsis thaliana*. *Nature* **408**, 796–815.

Balandin T and Castresana C (2002) AtCOX17, an *Arabidopsis* homolog of the yeast copper chaperone COX17. *Plant Physiol* **129**, 1852–1857.

Balk J and Lobreaux S (2005) Biogenesis of iron-sulfur proteins in plants. *Trends Plant Sci* **10**, 324–331.

Barros MH, Nobrega FG and Tzagoloff A (2002) Mitochondrial ferredoxin is required for heme A synthesis in *Saccharomyces cerevisiae*. *J Biol Chem* **277**, 9997–10002.

Beinert H (2000) Iron–sulfur proteins: ancient structures, still full of surprises. *J Biol Inorg Chem* **5**, 2–15.

Bernard DG, Quevillon-Cheruel S, Merchant S, Guiard B and Hamel PP (2005) Cyc2p, a membrane-bound flavoprotein involved in the maturation of mitochondrial *c*-type cytochromes. *J Biol Chem* **280**, 39852–39859.

Bonnard G and Grienenberger JM (1995) A gene proposed to encode a transmembrane domain of an ABC transporter is expressed in wheat mitochondria. *Mol Gen Genet* **246**, 81–99.

Bourges I, Ramus C, Mousson de Camaret B, *et al.* (2004) Structural organization of mitochondrial human complex I: role of the ND4 and ND5 mitochondria-encoded subunits and interaction with prohibitin. *Biochem J* **383**, 491–499.

Braun HP, Kruft V and Schmitz UK (1994) Molecular identification of the ten subunits of cytochrome-*c* reductase from potato mitochondria. *Planta* **193**, 99–106.

Busi MV, Zabaleta EJ, Araya A and Gomez-Casati DF (2004) Functional and molecular characterization of the frataxin homolog from *Arabidopsis thaliana*. *FEBS Lett* **576**, 141–144.

Butow RA and Avadhani NG (2004) Mitochondrial signaling: the retrograde response. *Mol Cell* **14**, 1–15.

Cardazzo B, Hamel P, Sakamoto W, Wintz H and Dujardin G (1998) Isolation of an Arabidopsis thaliana cDNA by complementation of a yeast abc1 deletion mutant deficient in complex III respiratory activity. *Gene* **221**, 117–125.

Carr HS, George GN and Winge DR (2002) Yeast Cox11, a protein essential for cytochrome *c* oxidase assembly, is a Cu(I)-binding protein. *J Biol Chem* **277**, 31237–31242.

Carr HS and Winge DR (2003) Assembly of cytochrome *c* oxidase within the mitochondrion. *Acc Chem Res* **36**, 309–316.

Chow KS, Singh DP, Roper JM and Smith AG (1997) A single precursor protein for ferrochelatase-I from *Arabidopsis* is imported *in vitro* into both chloroplasts and mitochondria. *J Biol Chem* **272**, 27565–27571.

Chuman L and Brody S (1989) Acyl carrier protein is present in the mitochondria of plants and eucaryotic micro-organisms. *Eur J Biochem* **184**, 643–649.

Clifton R, Lister R, Parker KL, *et al.* (2005) Stress-induced co-expression of alternative respiratory chain components in *Arabidopsis thaliana*. *Plant Mol Biol* **58**, 193–212.

Cornah JE, Roper JM, Pal Singh D and Smith AG (2002) Measurement of ferrochelatase activity using a novel assay suggests that plastids are the major site of haem biosynthesis in both photosynthetic and non-photosynthetic cells of pea (Pisum sativum L.). *Biochem J* **362**, 423–432.

Crivellone MD (1994) Characterization of CBP4, a new gene essential for the expression of ubiquinol-cytochrome *c* reductase in *Saccharomyces cerevisiae*. *J Biol Chem* **269**, 21284–21292.

Cruciat CM, Hell K, Folsch H, Neupert W and Stuart RA (1999) Bcs1p, an AAA-family member, is a chaperone for the assembly of the cytochrome bc(1) complex. *EMBO J* **18**, 5226–5233.

Curi GC, Chan RL and Gonzalez DH (2005) The leader intron of *Arabidopsis thaliana* genes encoding cytochrome *c* oxidase subunit 5c promotes high-level expression by increasing transcript abundance and translation efficiency. *J Exp Bot* **56**, 2563–2571.

Dojcinovic D, Krosting J, Harris AJ, Wagner DJ and Rhoads DM (2005) Identification of a region of the *Arabidopsis* AtAOX1a promoter necessary for mitochondrial retrograde regulation of expression. *Plant Mol Biol* **58**, 159–175.

Dumont ME, Ernst JF, Hampsey DM and Sherman F (1987) Identification and sequence of the gene encoding cytochrome *c* heme lyase in the yeast *Saccharomyces cerevisiae*. *EMBO J* **6**, 235–241.

Elorza A, Leon G, Gomez I, *et al.* (2004) Nuclear SDH2-1 and SDH2-2 genes, encoding the iron–sulfur subunit of mitochondrial complex II in *Arabidopsis*, have distinct cell-specific expression patterns and promoter activities. *Plant Physiol* **136**, 4072–4087.

Eubel H, Jansch L and Braun HP (2003) New insights into the respiratory chain of plant mitochondria. Supercomplexes and a unique composition of complex II. *Plant Physiol* **133**, 274–286.

Eyal Y, Curie C and McCormick S (1995) Pollen specificity elements reside in 30 bp of the proximal promoters of two pollen-expressed genes. *Plant Cell* **7**, 373–384.

Faivre-Nitschke E, Nazoa P, Gualberto JM, Grienenberger JM and Bonnard G (2001) Wheat mitochondria ccmB encodes the membrane domain of a putative ABC transporter involved in cytochromes *c* biogenesis. *Biochim Biophys Acta* **1519**, 199–208.

Funes S, Nargang FE, Neupert W and Herrmann JM (2004) The Oxa2 protein of Neurospora crassa plays a critical role in the biogenesis of cytochrome oxidase and defines a ubiquitous subbranch of the Oxa1/YidC/Alb3 protein family. *Mol Biol Cell* **15**, 1853–1861.

Garland SA, Hoff K, Vickery LE and Culotta VC (1999) *Saccharomyces cerevisiae* ISU1 and ISU2: members of a well-conserved gene family for iron-sulfur cluster assembly. *J Mol Biol* **294**, 897–907.

Giegé P, Knoop V and Brennicke A (1998) Complex II subunit 4 (sdh4) homologous sequences in plant mitochondrial genomes. *Curr Genet* **34**, 313–317.

Giegé P, Rayapuram N, Meyer EH, Grienenberger JM and Bonnard G (2004) CcmF(C) involved in cytochrome *c* maturation is present in a large sized complex in wheat mitochondria. *FEBS Lett* **563**, 165–169.

Giegé P, Sweetlove L and Leaver C (2003) Identification of mitochondrial protein complexes in *Arabidopsis* using two-dimensional Blue-Native polyacrylamide gel electrophoresis. *Plant Mol Biol Reporter* **21**, 133–144.

Giegé P, Sweetlove LJ, Cognat V and Leaver CJ (2005) Coordination of nuclear and mitochondrial genome expression during mitochondrial biogenesis in *Arabidopsis*. *Plant Cell* **17**, 1497–1512.

Glerum DM, Muroff I, Jin C and Tzagoloff A (1997) COX15 codes for a mitochondrial protein essential for the assembly of yeast cytochrome oxidase. *J Biol Chem* **272**, 19088–19094.

Gomez-Casati DF, Busi MV, Gonzalez-Schain N, Mouras A, Zabaleta EJ and Araya A (2002) A mitochondrial dysfunction induces the expression of nuclear-encoded complex I genes in engineered male sterile *Arabidopsis thaliana*. *FEBS Lett* **532**, 70–74.

Gonzalez DH, Bonnard G and Grienenberger JM (1993) A gene involved in the biogenesis of *c*-type cytochromes is co-transcribed with a ribosomal protein gene in wheat mitochondria. *Curr Genet* **21**, 248–255.

Gray GR, Maxwell DP, Villarimo AR and McIntosh L (2004a) Mitochondria/nuclear signaling of alternative oxidase gene expression occurs through distinct pathways involving organic acids and reactive oxygen species. *Plant Cell Rep* **23**, 497–503.

Gray GR, Villarimo AR, Whitehead CL and McIntosh L (2004b) Transgenic tobacco (*Nicotiana tabacum* L.) plants with increased expression levels of mitochondrial NADP+-dependent isocitrate dehydrogenase: evidence implicating this enzyme in the redox activation of the alternative oxidase. *Plant Cell Physiol* **45**, 1413–1425.

Green-Willms NS, Butler CA, Dunstan HM and Fox TD (2001) Pet111p, an inner membrane-bound translational activator that limits expression of the *Saccharomyces cerevisiae* mitochondrial gene COX2. *J Biol Chem* **276**, 6392–6397.

Grefen C and Harter K (2004) Plant two-component systems: principles, functions, complexity and cross talk. *Planta* **219**, 733–742.

Grohmann L, Rasmusson AG, Heiser V, Thieck O and Brennicke A (1996) The NADH-binding subunit of respiratory chain complex I is nuclear-encoded in plants and identified only in mitochondria. *Plant J* **10**, 793–803.

Guenebaut V, Schlitt A, Weiss H, Leonard K and Friedrich T (1998) Consistent structure between bacterial and mitochondrial NADH:ubiquinone oxidoreductase (complex I). *J Mol Biol* **276**, 105–112.

Hamel P, Sakamoto W, Wintz H and Dujardin G (1997) Functional complementation of an oxa1-yeast mutation identifies an *Arabidopsis thaliana* cDNA involved in the assembly of respiratory complexes. *Plant J* **12**, 1319–1327.

Harrison MD, Jones CE and Dameron CT (1999) Copper chaperones: function, structure and copper-binding properties. *J Biol Inorg Chem* **4**, 145–153.

Hass C, Lohrmann J, Albrecht V, *et al.* (2004) The response regulator 2 mediates ethylene signalling and hormone signal integration in Arabidopsis. *EMBO J* **23**, 3290–3302.

Heazlewood JL, Howell KA and Millar AH (2003) Mitochondrial complex I from *Arabidopsis* and rice: orthologs of mammalian and fungal components coupled with plant-specific subunits. *Biochim Biophys Acta* **1604**, 159–169.

Heazlewood JL, Tonti-Filippini JS, Gout AM, Day DA, Whelan J and Millar AH (2004) Experimental analysis of the *Arabidopsis* mitochondrial proteome highlights signaling and regulatory components, provides assessment of targeting prediction programs, and indicates plant-specific mitochondrial proteins. *Plant Cell* **16**, 241–256.

Heazlewood JL, Whelan J and Millar AH (2003) The products of the mitochondrial orf25 and orfB genes are FO components in the plant F1FO ATP synthase. *FEBS Lett* **540**, 201–205.

Heiser V, Brennicke A and Grohmann L (1996) The plant mitochondrial 22 kDa (PSST) subunit of respiratory chain complex I is encoded by a nuclear gene with enhanced transcript levels in flowers. *Plant Mol Biol* **31**, 1195–1204.

Helfenbein KG, Ellis TP, Dieckmann CL and Tzagoloff A (2003) ATP22, a nuclear gene required for expression of the F0 sector of mitochondrial ATPase in *Saccharomyces cerevisiae*. *J Biol Chem* **278**, 19751–19756.

Hernould M, Suharsono S, Zabaleta E, *et al.* (1998) Impairment of tapetum and mitochondria in engineered male-sterile tobacco plants. *Plant Mol Biol* **36**, 499–508.

Herrmann JM and Funes S (2005) Biogenesis of cytochrome oxidase-sophisticated assembly lines in the mitochondrial inner membrane. *Gene* **354**, 43–52.

Horng YC, Cobine PA, Maxfield AB, Carr HS and Winge DR (2004) Specific copper transfer from the Cox17 metallochaperone to both Sco1 and Cox11 in the assembly of yeast cytochrome C oxidase. *J Biol Chem* **279**, 35334–35340.

Howell KA, Millar AH and Whelan J (2006) Ordered assembly of mitochondria during rice germination begins with promitochondrial structures rich in components of the protein import apparatus. *Plant Mol Biol* **60**, 201–223.

Huang J, Struck F, Matzinger DF and Levings CS (1994) Flower-enhanced expression of a nuclear-encoded mitochondrial respiratory protein is associated with changes in mitochondrion number. *Plant Cell* **6**, 439–448.

Jacinto E and Hall MN (2003) Tor signalling in bugs, brain and brawn. *Nat Rev Mol Cell Biol* **4**, 117–126.

Johnson DC, Dean DR, Smith AD and Johnson MK (2005) Structure, function, and formation of biological iron–sulfur clusters. *Annu Rev Biochem* **74:** 247–281.

Juszczuk IM and Rychter AM (2003) Alternative oxidase in higher plants. *Acta Biochim Pol* **50**, 1257–1271.

Karpinski S, Reynolds H, Karpinska B, Wingsle G, Creissen G and Mullineaux P (1999) Systemic signaling and acclimation in response to excess excitation energy in *Arabidopsis*. *Science* **284**, 654–657.

Karpova OV, Kuzmin EV, Elthon TE and Newton KJ (2002) Differential expression of alternative oxidase genes in maize mitochondrial mutants. *Plant Cell* **14**, 3271–3284.

Kermorgant M, Bonnefoy N and Dujardin G (1997) Oxa1p, which is required for cytochrome *c* oxidase and ATP synthase complex formation, is embedded in the mitochondrial inner membrane. *Curr Genet* **31**, 302–307.

Koo AJ, Fulda M, Browse J and Ohlrogge JB (2005) Identification of a plastid acyl-acyl carrier protein synthetase in *Arabidopsis* and its role in the activation and elongation of exogenous fatty acids. *Plant J* **44**, 620–632.

Kosugi S and Ohashi Y (1997) PCF1 and PCF2 specifically bind to cis elements in the rice proliferating cell nuclear antigen gene. *Plant Cell* **9**, 1607–1619.

Kranz R, Lill R, Goldman B, Bonnard G and Merchant S (1998) Molecular mechanisms of cytochrome *c* biogenesis: three distinct systems. *Mol Microbiol* **29**, 383–396.

Kronekova Z and Rodel G (2005) Organisation of assembly factors Cbp3p and Cbp4p and their effect on bc(1) complex assembly in *Saccharomyces cerevisiae*. *Curr Genet* **47**, 203–212.

Kuhn A, Stuart R, Henry R and Dalbey RE (2003) The Alb3/Oxa1/YidC protein family: membrane-localized chaperones facilitating membrane protein insertion? *Trends Cell Biol* **13**, 510–516.

Kushnir S, Babiychuk E, Storozhenko S, *et al.* (2001) A mutation of the mitochondrial ABC transporter Sta1 leads to dwarfism and chlorosis in the *Arabidopsis* mutant starik. *Plant Cell* **13**, 89–100.

Lefebvre-Legendre L, Salin B, Schaeffer J, *et al.* (2005) Failure to assemble the alpha 3 beta 3 subcomplex of the ATP synthase leads to accumulation of the alpha and beta subunits within inclusion bodies and the loss of mitochondrial cristae in *Saccharomyces cerevisiae*. *J Biol Chem* **280**, 18386–18392.

Lefebvre-Legendre L, Vaillier J, Benabdelhak H, Velours J, Slonimski PP and di Rago JP (2001) Identification of a nuclear gene (FMC1) required for the assembly/stability of yeast mitochondrial F(1)-ATPase in heat stress conditions. *J Biol Chem* **276**, 6789–6796.

Leister D (2005) Genomics-based dissection of the cross-talk of chloroplasts with the nucleus and mitochondria in *Arabidopsis*. *Gene* **354**, 110–116.

Leon S, Touraine B, Briat JF and Lobreaux S (2005) Mitochondrial localization of *Arabidopsis thaliana* Isu Fe-S scaffold proteins. *FEBS Lett* **579**, 1930–1934.

Leon S, Touraine B, Ribot C, Briat JF and Lobreaux S (2003) Iron–sulphur cluster assembly in plants: distinct NFU proteins in mitochondria and plastids from *Arabidopsis thaliana*. *Biochem J* **371**, 823–830.

Levitan A, Danon A and Lisowsky T (2004) Unique features of plant mitochondrial sulfhydryl oxidase. *J Biol Chem* **279**, 20002–20008.

Lill R and Muhlenhoff U (2005) Iron–sulfur–protein biogenesis in eukaryotes. *Trends Biochem Sci* **30**, 133–141.

Lister R, Chew O, Rudhe C, Lee MN and Whelan J (2001) *Arabidopsis thaliana* ferrochelatase-I and -II are not imported into *Arabidopsis* mitochondria. *FEBS Lett* **506**, 291–295.

Little HN and Jones OT (1976) The subcellular loclization and properties of the ferrochelatase of etiolated barley. *Biochem J* **156**, 309–314.

Lohrmann J, Sweere U, Zabaleta E, *et al.* (2001) The response regulator ARR2: a pollen-specific transcription factor involved in the expression of nuclear genes for components of mitochondrial complex I in *Arabidopsis*. *Mol Genet Genomics* **265**, 2–13.

Loiseau L, Ollagnier-de-Choudens S, Nachin L, Fontecave M and Barras F (2003) Biogenesis of Fe–S cluster by the bacterial Suf system: SufS and SufE form a new type of cysteine desulfurase. *J Biol Chem* **278**, 38352–38359.

Mackenzie SA and McIntosh L (1999) Higher plant mitochondria. *Plant Cell* **11**, 571–585.

Marienfeld JR and Newton KJ (1994) The maize NCS2 abnormal growth mutant has a chimeric nad4-nad7 mitochondrial gene and is associated with reduced complex I function. *Genetics* **138**, 855–863.

Masuda T, Suzuki T, Shimada H, Ohta H and Takamiya K (2003) Subcellular localization of two types of ferrochelatase in cucumber. *Planta* **217**, 602–609.

Maxwell DP, Wang Y and McIntosh L (1999) The alternative oxidase lowers mitochondrial reactive oxygen production in plant cells. *Proc Natl Acad Sci USA* **96**, 8271–8276.

Meyer EH, Giegé P, Gelhaye E, *et al.* (2005) AtCCMH, an essential component of the *c*-type cytochrome maturation pathway in *Arabidopsis* mitochondria, interacts with apocytochrome *c*. *Proc Natl Acad Sci USA* **102**, 16113–16118.

Millar AH, Eubel H, Jansch L, Kruft V, Heazlewood JL and Braun HP (2004) Mitochondrial cytochrome *c* oxidase and succinate dehydrogenase complexes contain plant specific subunits. *Plant Mol Biol* **56**, 77–90.

Millar AH, Sweetlove LJ, Giegé P and Leaver CJ (2001) Analysis of the *Arabidopsis* mitochondrial proteome. *Plant Physiol* **127**, 1711–1727.

Moller IM and Kristensen BK (2004) Protein oxidation in plant mitochondria as a stress indicator. *Photochem Photobiol Sci* **3**, 730–735.

Moulin M and Smith AG (2005) Regulation of tetrapyrrole biosynthesis in higher plants. *Biochem Soc Trans* **33**, 737–742.

Moye-Rowley WS (2005) Retrograde regulation of multidrug resistance in *Saccharomyces cerevisiae*. *Gene* **354**, 15–21.

Muhlenhoff U and Lill R (2000) Biogenesis of iron–sulfur proteins in eukaryotes: a novel task of mitochondria that is inherited from bacteria. *Biochim Biophys Acta* **1459**, 370–382.

Neill S, Desikan R and Hancock J (2002a) Hydrogen peroxide signalling. *Curr Opin Plant Biol* **5**, 388–395.

Neill SJ, Desikan R, Clarke A, Hurst RD and Hancock JT (2002b) Hydrogen peroxide and nitric oxide as signalling molecules in plants. *J Exp Bot* **53**, 1237–1247.

Newton KJ, Knudsen C, Gabay-Laughnan S and Laughnan JR (1990) An abnormal growth mutant in maize has a defective mitochondrial cytochrome oxidase gene. *Plant Cell* **2**, 107–113.

Newton KJ, Mariano JM, Gibson CM, Kuzmin E and Gabay-Laughnan S (1996) Involvement of S2 episomal sequences in the generation of NCS4 deletion mutation in maize mitochondria. *Dev Genet* **19**, 277–286.

Nicholson DW, Stuart RA and Neupert W (1989) Biogenesis of cytochrome c_1. Role of cytochrome *c*1 heme lyase and of the two proteolytic processing steps during import into mitochondria. *J Biol Chem* **264**, 10156–10168.

Nijtmans LG, Artal SM, Grivell LA and Coates PJ (2002) The mitochondrial PHB complex: roles in mitochondrial respiratory complex assembly, ageing and degenerative disease. *Cell Mol Life Sci* **59**, 143–155.

Nobrega MP, Bandeira SC, Beers J and Tzagoloff A (2002) Characterization of COX19, a widely distributed gene required for expression of mitochondrial cytochrome oxidase. *J Biol Chem* **277**, 40206–40211.

Nobrega MP, Nobrega FG and Tzagoloff A (1990) COX10 codes for a protein homologous to the ORF1 product of *Paracoccus denitrificans* and is required for the synthesis of yeast cytochrome oxidase. *J Biol Chem* **265**, 14220–14226.

Noctor G, Dutilleul C, De Paepe R and Foyer CH (2004) Use of mitochondrial electron transport mutants to evaluate the effects of redox state on photosynthesis, stress tolerance and the integration of carbon/nitrogen metabolism. *J Exp Bot* **55**, 49–57.

Nott A, Jung HS, Koussevitzky S and Chory J (2006) Plastid-to-nucleus retrograde signaling. *Annu Rev Plant Biol* **57**, 739–759.

op denCamp RG, Przybyla D, Ochsenbein C, *et al.* (2003) Rapid induction of distinct stress responses after the release of singlet oxygen in *Arabidopsis*. *Plant Cell* **15**, 2320–2332.

Palumaa P, Kangur L, Voronova A and Sillard R (2004) Metal-binding mechanism of Cox17, a copper chaperone for cytochrome *c* oxidase. *Biochem J* **382**, 307–314.

Parisi G, Perales M, Fornasari MS, *et al.* (2004) Gamma carbonic anhydrases in plant mitochondria. *Plant Mol Biol* **55**, 193–207.

Perales M, Eubel H, Heinemeyer J, Colaneri A, Zabaleta E and Braun HP (2005) Disruption of a nuclear gene encoding a mitochondrial gamma carbonic anhydrase reduces complex I and supercomplex I +III2 levels and alters mitochondrial physiology in *Arabidopsis*. *J Mol Biol* **350**, 263–277.

Perales M, Parisi G, Fornasari MS, *et al.* (2004) Gamma carbonic anhydrase like complex interact with plant mitochondrial complex I. *Plant Mol Biol* **56**, 947–957.

Picciocchi A, Douce R and Alban C (2003) The plant biotin synthase reaction. Identification and characterization of essential mitochondrial accessory protein components. *J Biol Chem* **278**, 24966–24975.

Pickova A, Potocky M and Houstek J (2005) Assembly factors of F1FO-ATP synthase across genomes. *Proteins* **59**, 393–402.

Plesofsky N, Gardner N, Videira A and Brambl R (2000) NADH dehydrogenase in *Neurospora crassa* contains myristic acid covalently linked to the ND5 subunit peptide. *Biochim Biophys Acta* **1495**, 223–230.

Porra RJ and Lascelles J (1968) Studies on ferrochelatase. The enzymic formation of haem in proplastids, chloroplasts and plant mitochondria. *Biochem J* **108**, 343–348.

Ribas-Carbo M, Taylor NL, Giles L, *et al.* (2005) Effects of water stress on respiration in soybean leaves. *Plant Physiol* **139**, 466–473.

Ribichich KF, Tioni MF, Chan RL and Gonzalez DH (2001) Cell-type-specific expression of plant cytochrome *c* mRNA in developing flowers and roots. *Plant Physiol* **125**, 1603–1610.

Rouhier N, Gelhaye E and Jacquot JP (2004) Plant glutaredoxins: still mysterious reducing systems. *Cell Mol Life Sci* **61**, 1266–1277.

Sacconi S, Trevisson E, Pistollato F, *et al.* (2005) hCOX18 and hCOX19: two human genes involved in cytochrome *c* oxidase assembly. *Biochem Biophys Res Commun* **337**, 832–839.
Sakamoto W, Spielewoy N, Bonnard G, Murata M and Wintz H (2000) Mitochondrial localization of AtOXA1, an *Arabidopsis* homologue of yeast Oxa1p involved in the insertion and assembly of protein complexes in mitochondrial inner membrane. *Plant Cell Physiol* **41**, 1157–1163.
Saracco SA and Fox TD (2002) Cox18p is required for export of the mitochondrially encoded *Saccharomyces cerevisiae* Cox2p C-tail and interacts with Pnt1p and Mss2p in the inner membrane. *Mol Biol Cell* **13**, 1122–1131.
Schmidt-Bleek K, Heiser V, Thieck O, Brennicke A and Grohmann L (1997) The 28.5-kDa iron–sulfur protein of mitochondrial complex I is encoded in the nucleus in plants. *Mol Gen Genet* **253**, 448–454.
Schneider R, Brors B, Massow M and Weiss H (1997) Mitochondrial fatty acid synthesis: a relic of endosymbiontic origin and a specialized means for respiration. *FEBS Lett* **407**, 249–252.
Schulte U (2001) Biogenesis of respiratory complex I. *J Bioenerg Biomembr* **33**, 205–212.
Schulze M and Rodel G (1988) SCO1, a yeast nuclear gene essential for accumulation of mitochondrial cytochrome *c* oxidase subunit II. *Mol Gen Genet* **211**, 492–498.
Shintani DK and Ohlrogge JB (1994) The characterization of a mitochondrial acyl carrier protein isoform isolated from *Arabidopsis thaliana*. *Plant Physiol* **104**, 1221–1229.
Smart CJ, Moneger F and Leaver CJ (1994) Cell-specific regulation of gene expression in mitochondria during another development in sunflower. *Plant Cell* **6**, 811–825.
Smirnoff N (1998) Plant resistance to environmental stress. *Curr Opin Biotechnol* **9**, 214–219.
Smith AG, Marsh O and Elder GH (1993) Investigation of the subcellular location of the tetrapyrrole-biosynthesis enzyme coproporphyrinogen oxidase in higher plants. *Biochem J* **292**, 503–508.
Spielewoy N, Schulz H, Grienenberger JM, Thony-Meyer L and Bonnard G (2001) CCME, a nuclear-encoded heme-binding protein involved in cytochrome *c* maturation in plant mitochondria. *J Biol Chem* **276**, 5491–5497.
Stahl A, Nilsson S, Lundberg P, *et al.* (2005) Two novel targeting peptide degrading proteases, PrePs, in mitochondria and chloroplasts, so similar and still different. *J Mol Biol* **349**, 847–860.
Strand A, Asami T, Alonso J, Ecker JR and Chory J (2003) Chloroplast to nucleus communication triggered by accumulation of Mg-protoporphyrinIX. *Nature* **421**, 79–83.
Susek RE, Ausubel FM and Chory J (1993) Signal transduction mutants of *Arabidopsis* uncouple nuclear CAB and RBCS gene expression from chloroplast development. *Cell* **74**, 787–799.
Svensson AS and Rasmusson AG (2001) Light-dependent gene expression for proteins in the respiratory chain of potato leaves. *Plant J* **28**, 73–82.
Sweere U, Eichenberg K, Lohrmann J, *et al.* (2001) Interaction of the response regulator ARR4 with phytochrome B in modulating red light signaling. *Science* **294**, 1108–1111.
Takubo K, Morikawa T, Nonaka Y, *et al.* (2003) Identification and molecular characterization of mitochondrial ferredoxins and ferredoxin reductase from *Arabidopsis*. *Plant Mol Biol* **52**, 817–830.
Tatsuta T, Model K and Langer T (2005) Formation of membrane-bound ring complexes by prohibitins in mitochondria. *Mol Biol Cell* **16**, 248–259.
Thony-Meyer L (2002) Cytochrome *c* maturation: a complex pathway for a simple task? *Biochem Soc Trans* **30**, 633–638.
Thöny-Meyer L (1997) Biogenesis of respiratory cytochromes in bacteria. *Microbiol Mol Biol Rev* **61**, 337–376.
Thöny-Meyer L, Fischer F, Kunzler P, Ritz D and Hennecke H (1995) *Escherichia coli* genes required for cytochrome *c* maturation. *J Bacteriol* **177**, 4321–4326.
Tone Y, Kawai-Yamada M and Uchimiya H (2004) Isolation and characterization of *Arabidopsis thaliana* ISU1 gene. *Biochim Biophys Acta* **1680**, 171–175.
Tremousaygue D, Garnier L, Bardet C, Dabos P, Herve C and Lescure B (2003) Internal telomeric repeats and 'TCP domain' protein-binding sites co-operate to regulate gene expression in *Arabidopsis thaliana* cycling cells. *Plant J* **33**, 957–966.

Twell D, Yamaguchi J, Wing RA, Ushiba J and McCormick S (1991) Promoter analysis of genes that are coordinately expressed during pollen development reveals pollen-specific enhancer sequences and shared regulatory elements. *Genes Dev* **5**, 496–507.

Tzagoloff A, Barrientos A, Neupert W and Herrmann JM (2004) Atp10p assists assembly of Atp6p into the F0 unit of the yeast mitochondrial ATPase. *J Biol Chem* **279**, 19775–19780.

Ugalde C, Vogel R, Huijbens R, Van Den Heuvel B, Smeitink J and Nijtmans L (2004) Human mitochondrial complex I assembles through the combination of evolutionary conserved modules: a framework to interpret complex I deficiencies. *Hum Mol Genet* **13**, 2461–2472.

Unseld M, Marienfeld JR, Brandt P and Brennicke A (1997) The mitochondrial genome of *Arabidopsis thaliana* contains 57 genes in 366,924 nucleotides. *Nature Genet* **15**, 57–61.

Urantowka A, Knorpp C, Olczak T, Kolodziejczak M and Janska H (2005) Plant mitochondria contain at least two i-AAA-like complexes. *Plant Mol Biol* **59**, 239–252.

van Lis R, Atteia A, Nogaj LA and Beale SI (2005) Subcellular localization and light-regulated expression of protoporphyrinogen IX oxidase and ferrochelatase in *Chlamydomonas reinhardtii*. *Plant Physiol* **139**, 1946–1958.

van Steensel B (2005) Mapping of genetic and epigenetic regulatory networks using microarrays. *Nat Genet* **37** (Suppl.), S18–S24.

Vogel RO, Janssen RJ, Ugalde C, *et al.* (2005) Human mitochondrial complex I assembly is mediated by NDUFAF1. *Febs J* **272**, 5317–5326.

Vranova E, Inze D and Van Breusegem F (2002) Signal transduction during oxidative stress. *J Exp Bot* **53**, 1227–1236.

Wagner AM (1995) A role for active oxygen species as second messengers in the induction of alternative oxidase gene expression in Petunia hybrida cells. *FEBS Lett* **368**, 339–342.

Wagner D, Przybyla D, Op den Camp R, *et al.* (2004) The genetic basis of singlet oxygen-induced stress responses of *Arabidopsis thaliana*. *Science* **306**, 1183–1185.

Watanabe N, Che FS, Iwano M, Takayama S, Yoshida S and Isogai A (2001) Dual targeting of spinach protoporphyrinogen oxidase II to mitochondria and chloroplasts by alternative use of two in-frame initiation codons. *J Biol Chem* **276**, 20474–20481.

Welchen E, Chan RL and Gonzalez DH (2004) The promoter of the *Arabidopsis* nuclear gene COX5b-1, encoding subunit 5b of the mitochondrial cytochrome *c* oxidase, directs tissue-specific expression by a combination of positive and negative regulatory elements. *J Exp Bot* **55**, 1997–2004.

Welchen E and Gonzalez DH (2005) Differential expression of the Arabidopsis cytochrome *c* genes Cytc-1 and Cytc-2. Evidence for the involvement of TCP-domain protein-binding elements in anther- and meristem-specific expression of the Cytc-1 gene. *Plant Physiol* **139**, 88–100.

Welchen E and Gonzalez DH (2006) Overexpression of elements recognised by TCP-domain transcription factors in the upstream regions of nuclear genes encoding components of the mitochondrial oxidative phosphorylation machinery. *Plant Physiol* **141**, 540–545.

Wood CK, Dudley P, Albury MS, *et al.* (1996) Developmental regulation of respiratory activity and protein import in plant mitochondria. *Biochem Soc Trans* **24**, 746–749.

Wu M and Tzagoloff A (1989) Identification and characterization of a new gene (CBP3) required for the expression of yeast coenzyme QH2-cytochrome *c* reductase. *J Biol Chem* **264**, 11122–11130.

Xu XM and Moller SG (2006) AtSufE is an essential activator of plastidic and mitochondrial desulfurases in *Arabidopsis*. *EMBO J* **25**, 900–909.

Yabe T, Morimoto K, Kikuchi S, Nishio K, Terashima I and Nakai M (2004) The *Arabidopsis* chloroplastic NifU-like protein CnfU, which can act as an iron–sulfur cluster scaffold protein, is required for biogenesis of ferredoxin and photosystem I. *Plant Cell* **16**, 993–1007.

Zabaleta E, Heiser V, Grohmann L and Brennicke A (1998) Promoters of nuclear-encoded respiratory chain complex I genes from *Arabidopsis thaliana* contain a region essential for anther/pollen-specific expression. *Plant J* **15**, 49–59.

Zara V, Palmisano I, Conte L and Trumpower BL (2004) Further insights into the assembly of the yeast cytochrome bc1 complex based on analysis of single and double deletion mutants lacking supernumerary subunits and cytochrome b. *Eur J Biochem* **271**, 1209–1218.

Zarkovic J, Anderson SL and Rhoads DM (2005) A reporter gene system used to study developmental expression of alternative oxidase and isolate mitochondrial retrograde regulation mutants in *Arabidopsis*. *Plant Mol Biol* **57**, 871–888.

Zollner A, Rodel G and Haid A (1992) Molecular cloning and characterisation of *S. cerevisiae* CYT2 gene encoding cytochrome-*c*1-heme lyase. *Eur J Biochem* **207**, 1093–1100.

6 Supramolecular structure of the oxidative phosphorylation system in plants

Jesco Heinemeyer, Natalya V. Dudkina, Egbert J. Boekema and Hans-Peter Braun

6.1 Introduction

The oxidative phosphorylation (OXPHOS) system is localized in the inner mitochondrial membrane (IMM) and consists of various oxidoreductases and the ATP synthase complex. By the combined action of the oxidoreductases, electrons are transferred from metabolites (mainly NADH and $FADH_2$) within the mitochondrial matrix, or the intermembrane space to the terminal electron acceptor O_2. Four multisubunit complexes are of central importance for this so-called respiratory electron transport, the NADH dehydrogenase complex (complex I), the succinate dehydrogenase (complex II), the cytochrome *c* reductase (complex III) and the cytochrome *c* oxidase (complex IV). The lipid ubiquinone mediates transfer of electrons from the dehydrogenases to complex III, and the monomeric protein cytochrome *c* mediates transfer of electrons from complex III to complex IV. Complexes I, III and IV couple electron transfer to proton translocation across the IMM, causing the generation of a chemiosmotic gradient. The adenosine triphosphate (ATP) synthase complex, which is also designated complex V, finally uses this gradient to catalyse the formation of ATP by adenosine diphsophate phosphorylation at the matrix-exposed side of the IMM. Besides the classical oxidoreductase complexes of the respiratory chain, some organisms have further so-called alternative oxidoreductases. As a result, respiratory electron transport is branched. Numerous alternative oxidoreductases occur in plants, including the four distinct alternative NAD(P)H dehydrogenases, and one alternative terminal oxidase (see Chapter 7).

6.2 Structure and function of OXPHOS complexes I–V

Complexes I–V were discovered more than 40 years ago (for review, see Hatefi, 1985). The structure and function of these complexes have since been studied extensively by biochemical procedures in combination with site-directed mutagenesis, electron microscopy and X-ray crystallography. In particular, the detailed structures of the mammalian and yeast OXPHOS complexes have been determined.

6.2.1 *Complex I*

Complex I has a molecular mass of about 1 MDa and is composed of 40–45 distinct protein types (for review, see Friedrich and Böttcher, 2004). At least 10 cofactors are attached to this complex (one flavin mononucleotide and nine Fe–S clusters). Two large functional domains can be defined: an elongated membrane domain (membrane arm) involved in proton translocation, and a matrix-exposed domain (matrix arm) attached to one end of the membrane arm, which is responsible for NADH oxidation. Overall, the enzyme has an L-like shape. In plants, at least 10 of the approximately 40 subunits do not exhibit sequence similarity to subunits of complex I from heterotrophic eukaryotes (Heazlewood *et al.*, 2003a; Cardol *et al.*, 2004). One of these plant-specific subunits is an L-galactono-1,4-lactone dehydrogenase, which represents the terminal enzyme of the mitochondrial ascorbic acid biosynthesis pathway (Millar *et al.*, 2003); five other subunits exhibit sequence homology to an archaebacterial γ-type carbonic anhydrase (Parisi *et al.*, 2004; Perales *et al.*, 2004, 2005). The carbonic anhydrase subunits form an extra matrix-exposed domain, which, as revealed by single-particle electron microscopy, is attached to the central part of the membrane arm of complex I in plants (Sunderhaus *et al.*, 2006). Furthermore, an acyl-carrier protein for mitochondrial fatty acid biosynthesis forms part of complex I, as also reported for yeast, and bovine heart mitochondria (Runswick *et al.*, 1991; Sackmann *et al.*, 1991; Heazlewood *et al.*, 2003a). Plant complex I is therefore a multifunctional enzyme complex.

6.2.2 *Complex II*

Complex II is the smallest OXPHOS complex (for review, see Horsefield *et al.*, 2004). In most organisms it includes four types of subunits, and five cofactors (one flavin adenine dinucleotide, three Fe–S clusters and one heme b). *In vivo*, it most likely has a dimeric or trimeric structure (Yankovskaya *et al.*, 2003). Complex II from plants has at least four additional subunits of unknown function (Eubel *et al.*, 2003; Millar *et al.*, 2004), and these might bring secondary activities to this OXPHOS complex.

6.2.3 *Complex III*

Complex III is a functional dimer of about 500 kDa. Each monomer is composed of 10–11 proteins and 4 cofactors (three hemes and one Fe–S cluster; reviewed in Braun and Schmitz, 1995a; Berry *et al.*, 2000; Hunte *et al.*, 2000). Its overall structure is very similar in *Arabidopsis*, yeast and bovine heart (Dudkina *et al.*, 2005a). The two largest subunits of complex III are termed core proteins (core I and core II), because they were originally thought to form the center of the complex (Silman *et al.*, 1967). However, it is now known that they protrude into the mitochondrial matrix. The core proteins exhibit sequence similarity to the two subunits of the mitochondrial processing peptidase (MPP), which removes the presequences from nuclear-encoded mitochondrial proteins after import. In most heterotrophic organisms, the MPP subunits are localized in the mitochondrial matrix. In contrast, in plants the core subunits of complex III represent the MPP subunits (Braun *et al.*, 1992, 1995; Eriksson *et al.*, 1994). Structural prerequisites for MPP activity are a zinc-binding and a substrate-binding domain, which are completely conserved in core subunits of plant complex III, but incomplete in most core proteins from heterotrophic organisms (Braun and Schmitz, 1995b). Isolated complex III from plants was shown to efficiently remove presequences from mitochondrial precursor proteins. Thus, complex III is a bifunctional enzyme in plants.

6.2.4 *Complex IV*

Complex IV of mammals has a molecular mass of about 210 kDa and includes 13 subunits, some of which are very hydrophobic (Richter and Ludwig, 2003). Four cofactors are attached to the complex (two heme a and two Cu^{2+}). X-ray crystallography revealed that bovine complex IV is a dimer (Tsukihara *et al.*, 1996). However, the supramolecular structure of complex IV is still a matter of debate (Lee *et al.*, 2001). *Arabidopsis* complex IV has a similar number of subunits as the mammalian complex (Millar *et al.*, 2004). Some of the smaller subunits seem to be plant specific. It is possible that plant complex IV has subsidiary activities, but this remains to be confirmed.

6.2.5 *Complex V*

Complex V has a molecular mass of about 600 kDa (for review, see Stock *et al.*, 2000). It is composed of two domains, one within the IMM termed F_0 and the other within the mitochondrial matrix termed F_1, which are linked by a central and a peripheral stalk. Complex V catalysis involves rotation of the central stalk assembly together with an oligomeric ring of c subunits within F_0. Five different subunits form F_1 (α, β, γ, δ and ε) and about ten subunits form F_0 (a, b, c and several additional small subunits which are designated differentially in different organisms). The structure and composition of complex V of plants is very similar to that of heterotrophic eukaryotes as revealed by single-particle electron microscopy (Dudkina *et al.*, 2005b). The number of mitochondrially encoded subunits is especially high in plants (Heazlewood *et al.*, 2003b).

6.3 Supramolecular organization of the OXPHOS system

6.3.1 *Solid-state versus fluid-state model*

Historically, there has been some controversy regarding the supramolecular organization of the OXPHOS system. Two extreme models were proposed (for review, see Rich, 1984; Hackenbrock *et al.*, 1986; Lenaz, 2001; Schägger, 2001a, 2002). According to the solid-state model, OXPHOS complexes stably interact and form supramolecular structures called respiratory supercomplexes. In contrast, the fluid-state model posits that OXPHOS complexes can diffuse freely in a lateral direction within the membrane. Based on this model, electron transfer is believed to take place by random collisions between components (also referred to as the random-collision model; Hackenbrock *et al.*, 1986).

The fluid-state model was based originally on the finding that physiologically active portions of the OXPHOS system could be biochemically purified (for review, see Hatefi, 1985). However, purification protocols applied for the isolation of these portions often resulted in copurification of more than one OXPHOS complex (Fowler and Hatefi, 1961; Hatefi *et al.*, 1961, 1962b; Fowler and Richardson, 1963; Hatefi and Rieske, 1967). Furthermore, isolated OXPHOS complexes were found to assemble in specific stoichiometric ratios (Hatefi *et al.*, 1962a; Fowler and Richardson, 1963; Ragan and Heron, 1978), and the combined activity of the assembled complexes could not be increased upon addition of any of the individual purified OXPHOS complexes (Ragan and Heron, 1978). Together, these data were interpreted in favor of the solid-state model.

In contrast, results of lipid dilution experiments were interpreted in support of the fluid-state model (Schneider *et al.*, 1980a,b). Fusion of IMMs with exogenous lipids was found to cause an increase in the average distance between integral membrane proteins, and at the same time a decrease in respiration rates. Also, measurement of rotation of complexes III and IV reconstituted in artificial liposomes indicated the independent presence of these components (Kawato *et al.*, 1981). However, based on recent findings on the essential role of specific lipids for the structure, and activity of membrane-bound protein complexes (Lange *et al.*, 2001; Domonkos *et al.*, 2004; Fyfe and Jones, 2005), these results may have to be reinterpreted. For instance, the mitochondrial-specific lipid cardiolipin was found to be important for the formation of respiratory supercomplexes in yeast (Zhang *et al.*, 2002; Pfeiffer *et al.*, 2003; Zhang *et al.*, 2005). Dilution of mitochondrial membranes with exogenous phospholipids or reconstitution of OXPHOS complexes into artificial vesicles alters the natural lipid composition surrounding the OXPHOS complexes that might cause destabilization of supramolecular structures.

Theoretical considerations, based on the relationship between the measured activity rates for OXPHOS complexes and their diffusion rates within the IMM under *in vitro* and *in vivo* conditions, did not quite resolve the questions concerning the supramolecular organization of the OXPHOS complexes, but were interpreted in favor of a rather dynamic system (Rich, 1984).

Evidence in support of respiratory supercomplexes came from genetic investigations in mouse, humans and nematodes. Mutations affecting subunits of individual OXPHOS complexes were found to have specific effects on the stability of other OXPHOS complexes (Acin-Perez *et al.*, 2004; Grad and Lemire, 2004; Schägger *et al.*, 2004; Ugalde *et al.*, 2004). However, other data are contradictory. Mutations affecting complex I subunits in tobacco, *Arabidopsis* and trypanosomes were recently shown not to affect complex III abundance or stability (Horváth *et al.*, 2005; Perales *et al.*, 2005; Pineau *et al.*, 2005). Crosswise stabilization between different OXPHOS complex types therefore appears to occur only in some organisms. Evidence in favor of defined associations of OXPHOS complexes also comes from investigations on the homologous electron transfer system in bacteria. Stable respiratory supercomplexes were discovered in several microorganisms (Berry and Trumpower, 1985; Sone *et al.*, 1987; Iwasaki *et al.*, 1995; Niebisch and Bott, 2003; Stroh *et al.*, 2004). A functional fusion of respiratory complexes III and IV, devoid of cytochrome *c* protein, was reported for a thermoacidophilic archaeon (Iwasaki *et al.*, 1995).

Some years ago, a novel experimental strategy to characterize the OXPHOS system was introduced by Hermann Schägger, based on the mild solubilization of mitochondrial membranes with nonionic detergents and subsequent separation of the solubilized protein complexes by blue-native polyacrylamide gel electrophoresis (PAGE) (Arnold *et al.*, 1998; Schägger and Pfeiffer, 2000; Schägger, 2001b). Distinct respiratory supercomplexes could be described by this procedure, such as a dimeric ATP synthase, a $I + III_2$ supercomplex, and even larger supercomplexes including OXPHOS complexes I, III and IV. It was suggested that the larger structures should be termed respirasomes, because they can perform respiration autonomously in the presence of ubiquinol and cytochrome *c* (Schägger and Pfeiffer, 2000). Detergent-solubilized OXPHOS supercomplexes also proved to be stable during protein separations by sucrose-gradient ultracentrifugation (Dudkina *et al.*, 2005a,b). Of course, detergent treatment of biological membranes could also lead to artificial associations of membrane–protein complexes on the basis of random hydrophobic interaction. However, so far, there is no clear evidence that detergent treatment generates artifacts upon solubilization of mitochondrial membranes. Data based on detergent-treated mitochondrial membrane fractions correspond nicely with biochemical data obtained by other procedures, e.g. cross-linking experiments. In addition, the observed interactions between OXPHOS complexes within supramolecular structures on blue-native gels always make sense with respect to the known physiological context, e.g. complex III associates with complex IV in yeast (see below), or complex I with dimeric complex III in bovine and *Arabidopsis* mitochondria (see below).

Recently, detergent-solubilized supercomplexes purified by sucrose-gradient ultracentrifugaion were analyzed by electron microscopy in combination with single-particle analysis (Dudkina *et al.*, 2005a,b). The results clearly exclude the possibility that OXPHOS complexes interact on the basis of nonspecific hydrophobic interaction. All supercomplexes investigated revealed highly specific associations of OXPHOS complexes, e.g. within the $I + III_2$ supercomplex of *Arabidopsis*, or the V_2 supercomplex of the nonphotosynthetic alga *Polytomella*.

Very convincing evidence supporting defined associations of OXPHOS complexes also comes from flux-control experiments in combination with inhibitor titrations. The data obtained reveal that the yeast respiratory chain behaves as a single functional unit (Boumans *et al.*, 1998). Furthermore, inhibitor titrations indicate an interaction between bovine complexes I and III_2 (Bianchi *et al.*, 2003, 2004; Genova *et al.*, 2003). Flux-control analyses are of special value, because they can be carried out *in organello* in the absence of detergent.

In summary, the fluid-state model and the random-collision model cannot explain many of the reported results obtained for the structure of the mitochondrial OXPHOS system. On the other hand, not all OXPHOS complexes can form respiratory supercomplexes at the same time, since their abundances vary within the IMM (Schägger and Pfeiffer, 2001). Most likely, monomeric complexes and supramolecular assemblies coexist under *in vivo* conditions (Lenaz, 2001).

6.3.2 Composition of OXPHOS supercomplexes in plants

Respiratory supercomplexes of plant mitochondria were first described by blue-native/sodium dodecyl sulfate (SDS) PAGE (Eubel *et al.*, 2003). Using membrane solubilization with dodecymaltoside, Triton X-100, or digitonin, a 1500-kDa supercomplex, composed of the subunits of monomeric complex I and dimeric complex III, is visible on the two-dimensional gels (Figure 6.1). The supercomplex does not include proteins that are not components of complexes I or III_2. It is best stabilized in the presence of digitonin. Under these conditions, 50–90% of complex I is associated with dimeric complex III in *Arabidopsis*, potato, bean and barley (Eubel *et al.*, 2003). A small percentage of complex I forms part of an even larger particle of 3000 kDa, which most likely has I_2III_4 composition. Recently, the 1500-kDa $I + III_2$ supercomplex was described for spinach, pea, tobacco and asparagus by blue-native PAGE (Krause *et al.*, 2004; Pineau *et al.*, 2005; Taylor *et al.*, 2005; Dudkina

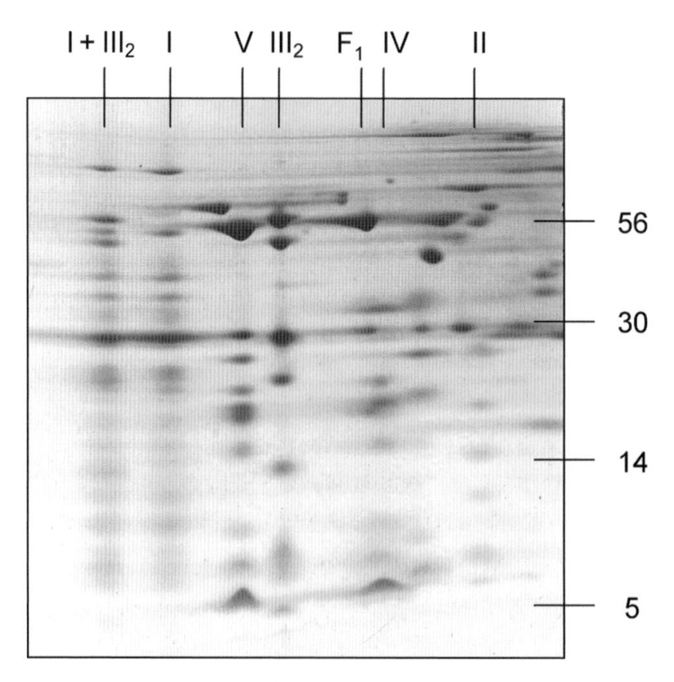

Figure 6.1 Two-dimensional resolution of OXPHOS complexes of *Arabidopsis* by blue-native/SDS PAGE. Mitochondrial membranes were solubilized by digitonin (5 g/g protein). The molecular masses of standard proteins are given to the right and the identities of the protein complexes above the gel. $I + III_2$: supercomplex formed of complexes I and dimeric complex III, I: complex I, V: ATP synthase; III_2: dimeric complex III, F_1: F_1 part of ATP synthase, IV: complex IV, II: complex II.

et al., 2006). Some larger supercomplexes of 1700–3000 kDa, which additionally include one or multiple copies of complex IV, were described for potato, spinach and sunflower (Eubel *et al.*, 2004a; Krause *et al.*, 2004; Sabar *et al.*, 2005). However, compared to mammalian mitochondria, these respirasomelike structures are of comparatively low abundance or stability in plants. Only a very small percentage of complex IV forms part of respiratory supercomplexes under all conditions tested. In contrast, the $I + III_2$ supercomplex proved to be of extraordinary stability in plants, and therefore was subject to single-particle electron microscopy in order to obtain structural information on supercomplex architecture (see below).

Very low detergent concentration during mitochondrial membrane solubilization enables the visualization of a dimeric ATP synthase supercomplex (Arnold *et al.*, 1998; Schägger and Pfeiffer, 2000). This supercomplex was first described for yeast and mammals using blue-native PAGE (Arnold *et al.*, 1998). In yeast, the supercomplex includes three dimer-specific proteins termed subunit e, g and k. If the gene for subunit g is deleted in yeast, the supercomplex is not formed. Furthermore, the characteristic foldings of the inner membrane do not develop in the mutant yeast line, leading to speculation that dimerization of ATP synthase is important for cristae formation (Giraud *et al.*, 2002; Paumard *et al.*, 2002). A similar ultrastructural phenotype was obtained upon *in vivo* cross-linking of the matrix-exposed F_1 parts of ATP synthase, indicating an important role of the F_0 membrane domain of ATP synthase during dimer formation (Gavin *et al.*, 2004).

A dimeric ATP synthase supercomplex could also be described for plant mitochondria upon solubilization of mitochondrial membranes by low concentrations of nonionic detergent in combination with blue-native PAGE (Eubel *et al.*, 2003, 2004b). The supercomplex is best stabilized by very low Triton X-100 concentrations (<0.5 g/g protein). However, overall this supercomplex seems to be less stable/abundant in higher plants. In contrast, the algae *Chlamydomonas* and *Polytomella* were shown to have an exceptionally stable ATP synthase supercomplex (Atteia *et al.*, 2003; van Lis *et al.*, 2003, 2005). In these algae, ATP synthase is exclusively dimeric on blue-native gels, independent of the type and concentration of nonionic detergent used for membrane solubilization (J. Heinemeyer, unpublished results). Compared with other organisms, the algal ATP synthase supercomplex includes an additional 60-kDa subunit termed *m*itochondrial *A*TP *s*ynthase *a*ssociated *p*rotein (MASAP). This protein is speculated to be responsible for dimer stability (van Lis *et al.*, 2003).

6.3.3 Structure of OXPHOS supercomplexes in plants

The $I + III_2$ supercomplex of higher plants and the dimeric ATP synthase supercomplex from green algae proved to be extraordinary stable particles. Therefore, they were selected for structural investigations using electron microscopy in combination with single-particle analyses. In order to omit binding of Coomassie blue during protein purification, which confers negative charge to protein surfaces and, therefore, possibly interferes with native protein structures, protein complexes and supercomplexes were separated by sucrose-gradient ultracentrifugation. Fractions of these gradients were directly analyzed by electron microscopy. Single-particle analyses revealed a highly defined interaction of monomeric complex I and dimeric complex III within the I + III_2 supercomplex of *Arabidopsis* (Plate 6.1, facing p. 110). The complex III dimer is laterally attached to the membrane arm of complex I. At the region of interaction, the membrane arm is slightly bent around the complex III dimer. The overall length of the membrane arm of plant complex I, which exceeds that in mammals, is most likely responsible for the enhanced stability of the I + III_2 supercomplex.

Analysis of the dimeric ATP synthase supercomplex of *Polytomella* also revealed a highly defined interaction between ATP synthase monomers within the V_2 supercomplex (Dudkina *et al.*, 2005b). Interaction takes place exclusively between the membrane-bound F_0 parts of the monomeric complex (Plate 6.2). The long axis of the two ATP synthase monomers forms an angle of about 70°, causing a local bending of the IMM in the region of the dimeric supercomplex. These data nicely support the hypothesized role for ATP synthase dimerization in the formation of cristae (see above). A parallel investigation of bovine ATP synthase supercomplexes by single-particle electron microscopy revealed an angle of 30° between the monomers (Minauro-Sanmiguel *et al.*, 2005). Furthermore, monomers within yeast ATP synthase dimers were recently found to have angles of either 40° or 90° (Dudkina *et al.*, 2006). Therefore, angles between ATP synthase monomers of ATP synthase supercomplexes seem to vary in different organisms. On the other hand, angles fall into two classes, a narrow-angle (30–40°) and a wide-angle (70–90°) class. It is possible that these classes represent two alternative binding arrangements for ATP synthase monomers in oligomeric ATP synthase structures, as described previously based on the results of rapid-freeze deep-etch electron microscopy (Allen *et al.*, 1989; Allen, 1995) or blue-native PAGE (Paumard *et al.*, 2002; Arselin *et al.*, 2003; Krause *et al.*, 2005; Wittig and Schägger, 2005). Recently, it was suggested that the wide-angle ATP synthase dimers are true dimers that represent the building blocks of ATP synthase oligomerization, whereas the narrow-angle dimers are articifial dimers which consist of two monomeric ATP synthase complexes of two neighboring ATP synthase dimers within the oligomers (Dudkina *et al.*, 2006). It has been suggested that the latter dimers are artificially generated as a result of the breakdown of ATP synthase oligomers during the solubilization step. However, it should be pointed out that, according to this hypothesis, both dimer types represent specific interactions of functional relevance. The presence of two different classes of ATP synthase dimers in yeast is also supported by biochemical analyses. Using fluorescence resonance energy transfer, b subunits of different ATP synthase monomers were shown to interact physically (Gavin *et al.*, 2005). Therefore, yeast ATP synthase dimers might be formed by the interaction of either g subunits or b subunits.

6.3.4 *Function of OXPHOS supercomplexes in plants*

The function of the OXPHOS supercomplexes is only partially understood. ATP synthase supercomplexes probably have an important role in the formation of IMM ultrastructure in mitochondria. This might also be the case for other OXPHOS supercomplexes, such as the I + III_2 supercomplex. However, membrane bending due to the binding of monomeric complex I and dimeric complex III is less evident. Dimerization of ATP synthase complexes has been suggested to have a positive effect on the fixation of the nonrotating parts of the ATP synthase monomers because they are attached to each other back-to-back in the region of the peripheral stalks, thereby causing torsional forces of opposite orientation (Rexroth *et al.*, 2004).

Respiratory supercomplexes might allow an overall increase in electron transfer rates as a result of substrate channeling between different OXPHOS components. This has been demonstrated for the NADH-cytochrome *c* oxidoreductase activity of the bovine I + III_2 supercomplex under *in vitro* conditions (Schägger and Pfeiffer, 2001). However, the I + III_2 supercomplex structure, as revealed by single-particle electron microscopy, does not support ubiquinone/ubiquinol channeling between complex I and complex III_2, because the location of the ubiquinone pocket of complex I is not in the region of the complex I–complex III_2 interface (Sazanov and Hinchliffe, 2006). The role of the membrane arm of complex I in proton translocation is still not known. It is possible that there are secondary electron transport chains present within this part of complex I, which could interact with electron transport in dimeric complex III.

Supercomplexes are likely also important for the regulation of respiration. The I + III_2 supercomplex was suggested to be involved in the regulation of the alternative oxidase in plants, because it might reduce access of this enzyme to its substrate, ubiquinol (Eubel *et al.*, 2003). However, as stated above, ubiquinone/ubiquinol channeling between complexes I and III_2 has not been proven. The alternative oxidoreductases of plant mitochondria were not found to form part of the respiratory supercomplexes (Eubel *et al.*, 2003), which might be an important prerequisite for the independent regulation of alternative electron transport pathways.

6.4 Concluding remarks

The OXPHOS supercomplexes may have functions in addition to electron transport and ATP generation, such as providing structural stabilization of membrane–protein complexes or increasing the protein-insertion capacity of the IMM. At present, the physiological roles of the respiratory supercomplexes are poorly understood, and there is much more to learn about supercomplex structure. Unfortunately, this quinternary level of protein structure is fragile, making investigations technically challenging; it is likely that X-ray crystallography will not be applicable for the resolution of supercomplex structure, at least in the near future. However, alternative, noninvasive procedures to investigate the three-dimensional interaction network of proteins and protein complexes should generate much useful data. What is certain is that investigation of protein supercomplexes and protein interaction networks will be crucial for the understanding of the molecular basis of life.

Acknowledgements

Research in our laboratories is supported by the Deutsche Forschungsgemeinschaft (DFG) and the Dutch science foundation, NOW.

References

Acin-Perez R, Bayona-Bafaluy MP, Fernandez-Silva P, *et al.* (2004) Respiratory complex III is required to maintain complex I in mammalian mitochondria. *Mol Cell* **13**, 805–815.

Allen RD (1995) Membrane tubulation and proton pumps. *Protoplasma* **189**, 1–8.

Allen RD, Schroeder CC and Fok AK (1989) An Investigation of mitochondrial inner membranes by rapid-freeze deep-etch techniques. *J Cell Biol* **108**, 2233–2240.

Arnold I, Pfeiffer K, Neupert W, Stuart RA and Schägger H (1998) Yeast mitochondrial F1FO-ATP synthase exists as a dimer: identification of three dimer-specific subunits. *EMBO J* **17**, 7170–7178.

Arselin G, Giraud MF, Dautant A, *et al.* (2003) The GxxxG motif of the transmembrane domain of subunit e is involved in the dimerization/oligomerization of the yeast ATP synthase complex in the mitochondrial membrane. *Eur J Biochem* **170**, 1875–1884.

Atteia A, van Lis R, Mendoza-Hernandez G, *et al.* (2003) Bifunctional aldehyde/alcohol dehydrogenase (ADHE) in chlorophyte algal mitochondria. *Plant Mol Biol* **53**, 175–188.

Berry EA, Guergova-Kuras M, Huang LS and Crofts AR (2000) Structure and function of cytochrome bc complexes. *Annu Rev Biochem* **69**, 1005–1075.

Berry EA and Trumpower BL (1985) Isolation of ubiquinol oxidase from *Paracoccus denitrificans* and resolution into cytochrome bc_1 and cytochrome *c*-aa_3 complexes. *J Biol Chem* **260**, 2458–2467.

Bianchi C, Fato R, Genova ML, Castelli GP and Lenaz G (2003) Structural and functional organization of complex I in the mitochondrial respiratory chain. *Biofactors* **18**, 3–9.

Bianchi C, Genova ML, Castelli GP and Lenaz G (2004) The mitochondrial respiratory chain is partially organized in a supercomplex assembly: kinetic evidence using flux control analysis. *J Biol Chem* **279**, 36562–36569.

Boumans H, Grivell LA and Berden JA (1998) The respiratory chain in yeast behaves as a single functional unit. *J Biol Chem* **273**, 4872–4877.

Braun HP, Emmermann M, Kruft V, Bödicker M and Schmitz UK (1995) The general mitochondrial processing peptidase from wheat is integrated into the cytochrome bc_1 complex of the respiratory chain. *Planta* **195**, 396–402.

Braun HP, Emmermann M, Kruft V and Schmitz UK (1992) The general mitochondrial processing peptidase from potato is an integral part of cytochrome *c* reductase of the respiratory chain. *EMBO J* **11**, 3219–3227.

Braun HP and Schmitz UK (1995a) The bifunctional cytochrome *c* reductase/processing peptidase complex from plant mitochondria. *J Bioenerg Biomembr* **27**, 423–436.

Braun HP and Schmitz UK (1995b) Are the 'core' proteins of the mitochondrial bc_1 complex evolutionary relics of a processing peptidase? *Trends Biochem Sci* **20**, 171–175.

Cardol P, Vanrobaeys F, Devreese B, van Beeuman J, Matagne RF and Remacle C (2004) Higher plant-like subunit composition of mitochondrial complex I from *Chlamydomonas reinhardtii*: 31 conserved components among eukaryotes. *Biochim Biophys Acta* **1658**, 212–224.

Domonkos I, Malec P, Sallai A, *et al.* (2004) Phosphatidylglycerol is essential for oligomerization of photosystem I reaction center. *Plant Physiol* **134**, 1471–1478.

Dudkina NV, Eubel H, Keegstra W, Boekema EJ and Braun HP (2005a) Structure of a mitochondrial supercomplex formed by respiratory chain complexes I and III. *Proc Natl Acad Sci USA* **102**, 3225–3229.

Dudkina NV, Heinemeyer J, Keegstra W, Boekema EJ and Braun HP (2005b) Structure of dimeric ATP synthase from mitochondria: an angular association of monomers induces the strong curvature of the inner membrane. *FEBS Lett* **579**, 5769–5772.

Dudkina NV, Sunderhaus S, Braun HP and Boekema EJ (2006) Characterization of dimeric ATP synthase and cristae membrane ultrastructure from Saccharomyces and Polytomella mitochondria. *FEBS Lett* **580**, 3427–3432.

Eriksson AC, Sjöling S and Glaser E (1994) The ubiquinol cytochrome *c* oxido-reductase complex of spinach leaf mitochondria is involved in both respiration and protein processing. *Biochim Biophys Acta* **1186**, 221–231.

Eubel H, Heinemeyer J and Braun HP (2004a) Identification and characterization of respirasomes in potato mitochondria. *Plant Physiol* **134**, 1450–1459.

Eubel H, Heinemeyer J, Sunderhaus S and Braun HP (2004b) Respiratory chain supercomplexes in plant mitochondria. *Plant Physiol Biochem* **42**, 937–942.

Eubel H, Jänsch L and Braun HP (2003) New insights into the respiratory chain of plant mitochondria: supercomplexes and a unique composition of complex II. *Plant Physiol* **133**, 274–286.

Fowler LR and Hatefi Y (1961) Reconstitution of the electron transport system III. Reconstitution of DPNH oxidase, succinic oxidase, and DPNH succinic oxidase. *Biochem Biophys Res Commun* **5**, 203–208.

Fowler LR and Richardson HS (1963) Studies on the electron transfer system. *J Biol Chem* **238**, 456–463.

Friedrich T and Böttcher B (2004) The gross structure of the respiratory complex I: a Lego system. *Biochim Biophys Acta* **1608**, 1–9.

Fyfe PK and Jones MR (2005) Lipids in and around photosynthetic reaction centres. *Biochem Soc Trans* **33**, 924–930.

Gavin PD, Prescott M and Devenish RJ (2005) Yeast F_1F_0-ATP synthase complex interactions *in vivo* can occur in the absence of the dimer specific subunit e. *J Bioenerg Biomembr* **37**, 55–66.

Gavin PD, Prescott M, Luff SE and Devenish RJ (2004) Cross-linking ATP synthase complexes *in vivo* eliminates mitochondrial cristae. *J Cell Sci* **117**, 2233–2243.

Genova ML, Bianchi C and Lenaz G (2003) Structural organization of the mitochondrial respiratory chain. *Ital J Biochem* **52**, 58–61.

Giraud MF, Paumard P, Soubannier V, *et al.* (2002) Is there a relationship between the supramolecular organization of the mitochondrial ATP synthase and the formation of cristae? *Biochim Biophys Acta* **1555**, 174–180.

Grad LI and Lemire BD (2004) Mitochondrial complex I mutations in *Caenorhabditis elegans* produce cytochrome *c* oxidase deficiency, oxidative stress and vitamin-responsive lactic acidosis. *Hum Mol Genet* **13**, 303–314.

Hackenbrock CR, Chazotte B and Gupta SS (1986) The random collision model and a critical assessment of the diffusion and collision in mitochondrial electron transport. *J Bioenerg Biomembr* **18**, 331–368.

Hatefi Y (1985) The mitochondrial electron transport and oxidative phosphorylation system. *Ann Rev Biochem* **54**, 1015–1069.

Hatefi Y, Haavik AG, Fowler LR and Griffiths DE (1962a) Studies on the electron transfer system. Reconstitution of the electron transfer system. *J Biol Chem* **237**, 2661–2669.

Hatefi Y, Haavik AG and Griffiths E (1962b) Studies on the electron transfer system XL: preparation and properties of mitochondrial DPNH-Coenzyme Q reductase. *J Biol Chem* **237**, 1676–1680.

Hatefi Y, Haavik AG and Jurtshuk P (1961) Studies on the electron transport system XXX. DPNH-cytochrome *c* reductase I. *Biochim Biophys Acta* **52**, 106–118.

Hatefi Y and Rieske JS (1967) The preparation and properties of DPNH-cytochrome *c* reductase (complex I–III of the respiratory chain). *Meth Enzymol* **10**, 225–231.

Heazlewood JL, Howell KA and Millar AH (2003a) Mitochondrial complex I from *Arabidopsis* and rice: orthologs of mammalian and yeast components coupled to plant-specific subunits. *Biochim Biophys Acta* **1604**, 159–169.

Heazlewood JL, Whelan J and Millar AH (2003b) The products of the mitochondrial ORF25 and ORFB genes are F_O components of the plant F_1F_O ATP synthase. *FEBS Lett* **540**, 201–205.

Horsefield R, Iwata S and Byrne B (2004) Complex II from a structural perspective. *Curr Protein Pept Sci* **5**, 107–118.

Horváth A, Horáková E, Dunajcikova P, *et al.* (2005) Downregulation of the nuclear-encoded subunits of the complexes III and IV disrupts their respective complexes but not complex I in procyclic *Trypanosoma brucei*. *Mol Microbiol* **58**, 116–130.

Hunte C, Koepke J, Lange C, Rossmanith T and Michel H (2000) Structure at 2.3 Å resolution of the cytochrome bc_1 complex from the yeast *Saccharomyces cerevisiae* co-crystallized with an antibody fragment. *Structure* **8**, 669–684.

Iwasaki T, Matsuura K and Oshima T (1995) Resolution of the aerobic respiratory system of the thermoacidophilic archaeon, *Sulfolobus* sp. strain 7. I: The archael terminal oxidase supercomplex is a functional fusion of respiratory complexes III and IV with no c-type cytochromes. *J Biol Chem* **270**, 30881–30892.

Kawato S, Sigel E, Carafoli E and Cherry RJ (1981) Rotation of cytochrome oxidase in phospholipid vesicles. *J Biol Chem* **256**, 7518–7527.

Krause F, Reifschneider NF, Goto S and Dencher NA (2005) Active oligomeric ATP synthases in mammalian mitochondria. *Biochem Biophys Res Commun* **329**, 583–590.

Krause F, Reifschneider NH, Vocke D, Seelert H, Rexroth S and Dencher NA (2004) 'Respirasome'-like supercomplexes in green leaf mitochondria of spinach. *J Biol Chem* **279**, 48369–48375.

Lange C, Nett JH, Trumpower BL and Hunte C (2001) Specific roles of protein–phospholipid interactions in the yeast cytochrome bc_1 complex. *EMBO J* **20**, 6591–6600.

Lee SJ, Yamashita E, Abe T, *et al.* (2001) Intermonomer interactions in dimer of bovine heart cytochrome *c* oxidase. *Acta Crystallogr D Biol Crystallogr* **57**, 941–947.

Lenaz G (2001) A critical appraisal of the mitochondrial coenzyme Q pool. *FEBS Lett* **509**, 151–155.

Millar AH, Eubel H, Jänsch L, Kruft V, Heazlewood JL and Braun HP (2004) Mitochondrial cytochrome *c* oxidase and succinate dehydrogenase contain plant-specific subunits. *Plant Mol Biol* **56**, 77–89.

Millar AH, Mittova V, Kiddle G, *et al.* (2003) Control of ascorbate synthesis by respiration and its implications for stress responses. *Plant Physiol* **133**, 443–447.

Minauro-Sanmiguel F, Wilkens S and Garcia JJ (2005) Structure of dimeric mitochondrial ATP synthase: novel F_0 bridging features and the structural basis of mitochondrial cristae biogenesis. *Proc Natl Acad Sci USA* **102**, 12356–12358.

Niebisch A and Bott M (2003) Purification of a cytochrome bc_1-aa_3 supercomplex with quinol oxidase activity from *Corynebacterium glutamicum*. Identification of a fourth subunit of cytochrome aa_3 oxidase and mutational analysis of diheme cytochrome c_1. *J Biol Chem* **278**, 4339–4346.

Parisi G, Perales M, Fornasari MS, *et al.* (2004) Gamma carbonic anhydrases in plant mitochondria. *Plant Mol Biol* **55**, 193–207.

Paumard P, Vaillier J, Coulary B, *et al.* (2002) The ATP synthase is involved in generating mitochondrial cristae morphology. *EMBO J* **21**, 221–230.

Pfeiffer K, Gohil V, Stuart RA, *et al.* (2003) Cardiolipin stabilizes respiratory chain supercomplexes. *J Biol Chem* **278**, 52873–52880.

Perales M, Eubel H, Heinemeyer J, Colaneri A, Zabaleta E and Braun HP (2005) Disruption of a nuclear gene encoding a mitochondrial gamma carbonic anhydrase reduces complex I and supercomplex $I + III_2$ levels and alters mitochondrial physiology in *Arabidopsis*. *J Mol Biol* **350**, 263–277.

Perales M, Parisi G, Fornasari MS, *et al.* (2004) Gamma carbonic anhydrase subunits physically interact within complex I of plant mitochondria. *Plant Mol Biol* **56**, 947–957.

Pineau B, Mathieu C, Gerard-Hirne C, De Paepe R and Chetrit P (2005) Targeting the NAD7 subunit to mitochondria restores a functional complex I and a wild type phenotype in the *Nicotiana sylvestris* CMS II mutant lacking nad7. *J Biol Chem* **280**, 25994–26001.

Ragan CI and Heron C (1978) The interaction between mitochondrial NADH-ubiquinone oxidoreductase and ubiquinol-cytochrome *c* oxidoreductase. *Biochem J* **174**, 783–790.

Rexroth S, Meyer zu Tittingdorf JM, Schwassmann HJ, Krause F, Seelert H and Dencher NA (2004) Dimeric H^+-ATP synthase in the chloroplast of *Chlamydomonas reinhardtii*. *Biochim Biophys Acta* **1658**, 202–211.

Rich PR (1984) Electron and proton transfer through quinones and cytochrome bc complexes. *Biochim Biophys Acta* **768**, 53–79.

Richter OM and Ludwig B (2003) Cytochrome *c* oxidase – structure, function, and physiology of a redox-driven molecular machine. *Rev Physiol Biochem Pharmacol* **147**, 47–74.

Runswick MJ, Fearnley IM, Skehel JM and Walker JE (1991) Presence of an acyl carrier protein in NADH: ubiquinone oxidoreductase from bovine heart mitochondria. *FEBS Lett* **286**, 121–124.

Sabar M, Balk J and Leaver CJ (2005) Histochemical staining and quantification of plant mitochondrial respiratory chain complexes using blue-native polyacrylamide gel electrophoresis. *Plant J* **44**, 893–901.

Sackmann U, Zensen R, Röhlen D, Jahnke U and Weiss H (1991) The acyl-carrier protein in *Neurospora crassa* mitochondria is a subunit of NADH: ubiquinone reductase (complex I). *Eur J Biochem* **200**, 463–469.

Sazanov LA and Hinchliffe P (2006) Structure of the hydrophilic domain of respiratory complex I from *Thermus thermophilus*. *Science* **311**, 1430–1436.

Schägger H (2001a) Respiratory chain supercomplexes. *IUBMB Life* **52**, 119–128.

Schägger H (2001b) Blue-native gels to isolate protein complexes from mitochondria. *Methods Cell Biol* **65**, 231–244.

Schägger H (2002) Respiratory supercomplexes of mitochondria and bacteria. *Biochim Biophys Acta* **1555**, 154–159.

Schägger H, De Coo R, Bauer MF, Hofmann S, Godinot C and Brandt U (2004). Significance of respirasomes for the assembly/stability of human respiratory chain complex I. *J Biol Chem* **279**, 36349–36353.

Schägger H and Pfeiffer K (2000) Supercomplexes in the respiratory chains of yeast and mammalian mitochondria. *EMBO J* **19**, 1777–1783.

Schägger H and Pfeiffer K (2001) The ratio of oxidative phosphorylation complexes I–V in bovine heart mitochondria and the composition of respiratory chain supercomplexes. *J Biol Chem* **276**, 37861–37867.

Schneider H, Lemasters JL, Höchli M and Hackenbrock CR (1980a) Fusion of liposomes with mitochondrial inner membranes. *Proc Natl Acad Sci USA* **77**, 442–446.

Schneider H, Lemasters JL, Höchli M and Hackenbrock CR (1980b) Liposome-mitochondrial inner membrane fusions. *J Biol Chem* **255**, 3748–3756.

Silman HI, Rieske JS, Lipton SH and Baum H (1967) A new protein component of complex III of the mitochondrial electron transfer chain. *J Biol Chem* **242**, 4867–4875.

Sone N, Sekimachi M and Kutoh E (1987) Identification and properties of a quinol oxidase supercomplex composed of a bc_1 complex and cytochrome oxidase in the thermophilic bacterium PS3. *J Biol Chem* **262**, 15386–15391.

Stock D, Gibbons C, Arechaga I, Leslie AG and Walker JE (2000) Rotary mechanism of ATP synthase. *Curr Opin Struct Biol* **10**, 672–679.

Stroh A, Anderka O, Pfeiffer K, *et al.* (2004) Assembly of respiratory complexes I, III and IV into NADH oxidase supercomplexes stabilizes complex I in *Paracoccus denitrificans*. *J Biol Chem* **279**, 5000–5007.

Sunderhaus S, Dudkina NV, Jänsch L, *et al.* (2006) Carbonic anhydrase subunits form a matrix-exposed domain attached to the membrane arm of mitochondrial complex I in plants. *J Biol Chem* **281**, 6482–6488.

Taylor NL, Heazlewood JL, Day DA and Millar AH (2005) Differential impact of environmental stresses on the pea mitochondrial proteome. *Mol Cell Proteomics* **4**, 1122–1133.

Tsukihara T, Aoyama H, Yamashita E, *et al.* (1996) The whole structure of the 13-subunit oxidized cytochrome *c* oxidase at 2.8 Å. *Science* **272**, 1136–1144.

Ugalde C, Janssen RJ, Van Den Heuvel LP, Smeitink JA and Nijtmans LG (2004) Differences in assembly or stability of complex I and other mitochondrial OXPHOS complexes in inherited complex I deficiency. *Hum Mol Genet* **13**, 659–667.

Van Lis R, Atteia A, Mendoza-Hernandez G and Gonzalez-Halphen D (2003) Identification of novel mitochondrial protein components of *Chlamydomonas reinhardtii*. A proteomic approach. *Plant Physiol* **132**, 318–330.

Van Lis R, Mendoza-Hernandez G and Gonzalez-Halphen D (2005) Divergence of the mitochondrial electron transport chains from the green alga *Chlamydomonas reinhardtii* and its colorless close relative *Polytomella* sp. *Biochim Biophys Acta* **1708**, 23–34.

Wittig I and Schägger H (2005) Advantages and limitations of clear native polyacrylamide gel electrophoresis. *Proteomics* **5**, 4338–4346.

Yankovskaya V, Horsefield R, Tornroth S, *et al.* (2003) Architecture of succinate dehydrogenase and reactive oxygen species generation. *Science* **299**, 700–704.

Zhang M, Mileykovskaya E and Dowhan W (2002) Gluing the respiratory chain together. Cardiolipin is required for supercomplex formation in the IMM. *J Biol Chem* **277**, 43553–43556.

Zhang M, Mileykovskaya E and Dowhan W (2005) Cardiolipin is essential for the organization of complexes III and IV into a supercomplex in intact yeast mitochondria. *J Biol Chem* **280**, 29403–29408.

7 Mitochondrial electron transport and oxidative stress

Ian M. Møller

7.1 Introduction

One of the primary functions of mitochondria in the plant cell is to produce ATP, which is mainly exported and used as an energy source throughout the cell. The ATP is produced through the process of oxidative phosphorylation. In this process the oxidation of NADH and $FADH_2$ produced by the Krebs cycle causes the release of chemical energy, which is converted into an electrochemical proton gradient. The energy stored in this gradient is then used to produce ATP from ADP and inorganic phosphate (Pi). An excellent general introduction into the bioenergetics of these processes is found in Nicholls and Ferguson (2002).

The oxidation of NADH and $FADH_2$ takes place in the inner mitochondrial membrane (IMM) of virtually all eukaryotes, and is catalysed by the electron transport chain (ETC). The ultimate electron acceptor is oxygen and it is reduced to water by the terminal oxidase, cytochrome *c* oxidase (CCO). However, oxygen can also accept electrons at other sites, and when that happens, superoxide and other reactive oxygen species (ROS) are formed. Although ROS have signalling functions in the cell (Apel and Hirt, 2004), they also cause damage, and their production is therefore strictly regulated. It is the aim of this chapter to present an overview of the ETC in plant mitochondria, to consider the turnover of ROS and to look at the damage caused by ROS to DNA, lipids and proteins, and the ways in which the mitochondria repair this damage.

7.2 The standard electron transport chain

7.2.1 The standard ETC complexes

The ETC comprises four multisubunit complexes, respiratory complexes I–IV, assisted by ubiquinone and cytochrome *c*. Three of these complexes pump protons from the matrix into the intermembrane space (IMS) to create an electrical gradient, negative inside, and a smaller proton gradient, basic inside (Figure 7.1) (Nicholls and Ferguson, 2002). The complexes are similar in all eukaryote mitochondria, except that budding yeast, *Saccharomyces cerevisiae*, lacks complex I, and are reviewed in Chapters 5 and 6. Here we will only briefly consider their properties.

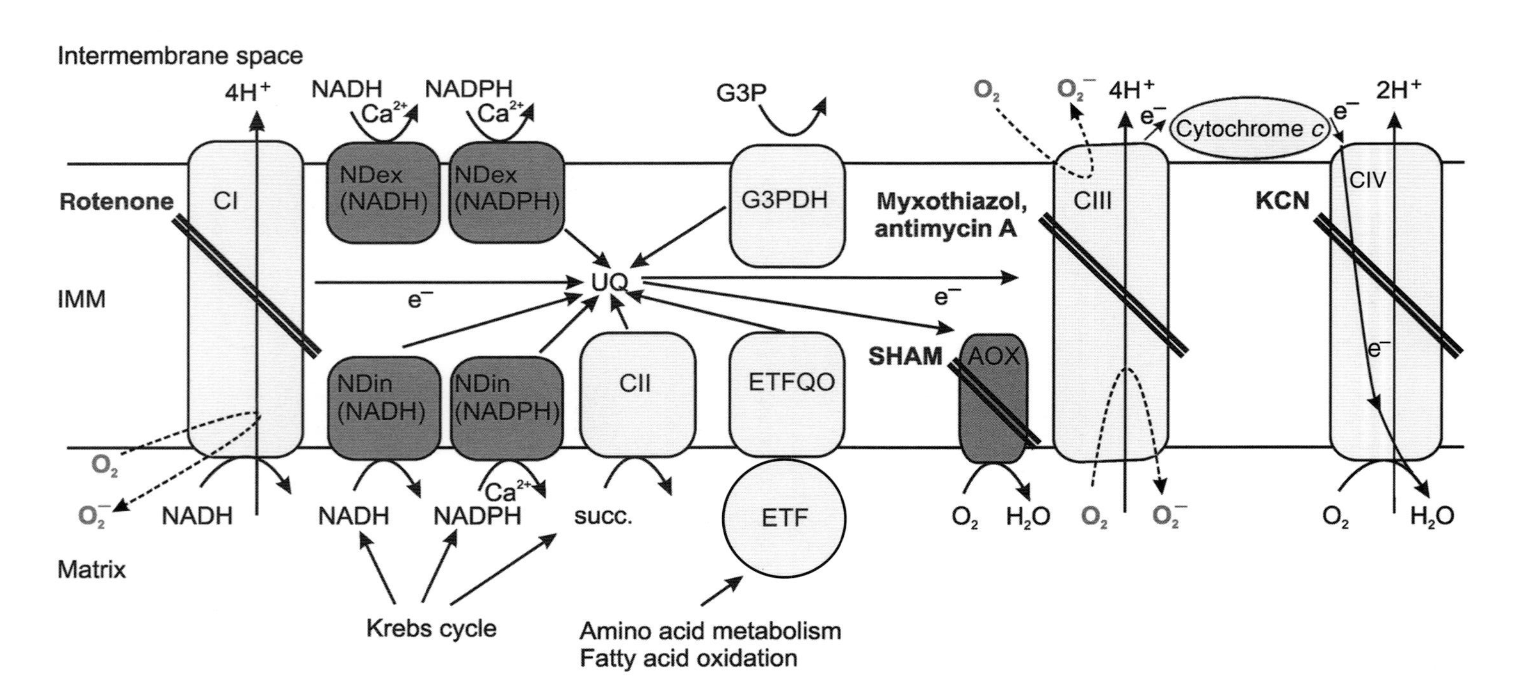

Figure 7.1 The electron transport chain in mitochondria showing both the standard and the plant-specific components. The standard components are in light grey, the plant-specific components are in darker grey and ROS formation is shown with dashed lines. CI–CIV: respiratory complexes I–IV; ETF: electron-transfer flavoprotein; ETFQO: electron-transfer flavoprotein:ubiquinone oxidoreductase; G3P: glycerol-3-phosphate.

The conversion of the electrochemical proton gradient into ATP is catalysed by the ATP synthase, often called complex V, which, like complexes I–IV, is located in the IMM (Figure 7.1). Note that although this is *respiratory complex* V, it is not part of the ETC, as it does not transfer electrons. The total cost of converting one ADP plus one Pi into one exported ATP is four H^+ passing from the IMS into the matrix (Brand, 1994).

7.2.1.1 Complex I

Complex I, NADH dehydrogenase, is a type I NADH dehydrogenase characterised by being multisubunit, proton pumping and inhibited by rotenone. It oxidises NADH with high affinity (low K_m) at an active site containing a flavine mononucleotide (FMN) prosthetic group. The electrons are ultimately used to reduce ubiquinone to ubiquinol. Four protons are pumped for each pair of electrons passing through the complex. Rotenone and piericidin are specific and diagnostic inhibitors of this complex.

7.2.1.2 Complex II

Complex II, succinate dehydrogenase, oxidises succinate at an active site containing FAD. The electrons are ultimately used to reduce ubiquinone to ubiquinol. No protons are pumped by this complex.

7.2.1.3 Complex III

Complex III, ubiquinol-cytochrome *c* reductase, oxidises ubiquinol and reduces cytochrome *c*. Four protons are pumped for each pair of electrons passing through the complex. Antimycin A and myxothiazol are specific and diagnostic inhibitors of this complex.

7.2.1.4 Complex IV

Complex IV, CCO, receives electrons from reduced cytochrome *c* and passes them on to oxygen via cytochromes *a* and a_3. Four electrons are transferred per oxygen molecule, giving water as the product. Two protons are pumped for each pair of electrons passing through the complex. Cyanide, azide and carbon monoxide inhibit this complex, but also other heme-containing enzymes.

7.2.2 Other "standard" ETC enzymes

In addition to the four complexes, two other redox enzymes appear to be located in the IMM and be connected with the ETC (Figure 7.1).

The IMM-localised enzyme, electron-transfer flavoprotein:ubiquinone oxidoreductase, accepts electrons from the matrix enzyme, the electron-transfer flavoprotein. The latter is the physiological electron acceptor for a number of matrix flavoprotein dehydrogenases, e.g. enzymes involved in the oxidation of amino acids. Both enzymes were recently identified in *Arabidopsis*, where they appear to be involved in amino acid degradation and also appear to have a role in chlorophyll degradation (Ishizaki *et al.*, 2005).

Glycerol-3-phosphate dehydrogenase is a flavoprotein located on the outer surface of the IMM. In mammalian mitochondria it oxidises glycerol-3-phosphate to dihydroxyacetone phosphate in a Ca^{2+}-dependent manner and the electrons enter the ETC at ubiquinone (Hansford, 1991). This enzyme has now been found in *Arabidopsis* where it does not appear to be Ca^{2+} dependent. It is highly expressed in the seeds where it is probably involved in metabolising glycerol derived from the breakdown of storage lipids. Its expression is increased by treatment of *Arabidopsis* seedlings with abscisic acid, or in response to drought (Shen *et al.*, 2003).

In addition to these standard components, the plant mitochondrial ETC contains a further five enzymes not found in mammalian mitochondria, the alternative oxidase (AOX) and four rotenone-insensitive NAD(P)H dehydrogenases. None of these enzymes pump protons, making them energy wasteful to the extent that they lead to electron transport bypassing the proton pumping sites in complexes I, III and IV.

7.3 The alternative oxidase

AOX is a 32–36-kDa diiron enzyme located in the inner leaflet of the IMM (Berthold and Stenmark, 2003) (Figure 7.1), which produces water as the final product just like complex IV. It does not pump protons. In *Arabidopsis*, five nuclear genes encode AOXs. AOX is inhibited by salicylhydroxamic acid and pyrogallate, but these are not reliable diagnostic inhibitors as they affect other processes in intact tissues (Møller *et al.*, 1988, and references therein).

AOX works as a dimer. The enzyme is activated by reduction of the disulphide bridge linking the homodimer, probably by the thioredoxin system (Gelhaye *et al.*, 2004). This is followed by binding of the allosteric activator pyruvate, which lowers the K_m for ubiquinol, rather than changing the V_{max}. However, the physiological relevance of these activation mechanisms has been questioned (Millenaar and Lambers, 2003).

Various aspects of AOX properties and function have been reviewed in detail by Berthold and Stenmark (2003), Juszczuk and Rychter (2003), Millenaar and Lambers (2003) and Finnegan *et al.* (2004).

7.4 NAD(P)H dehydrogenases

7.4.1 *Basic properties*

Plant mitochondria contain four different type II NAD(P)H dehydrogenases, which contain only one polypeptide, do not pump protons and are not inhibited by rotenone (Kerscher, 2000) (Figure 7.1). Two of these dehydrogenases are on the *ex*ternal surface of the IMM, ND*ex*(NADH) and ND*ex*(NADPH) (Roberts *et al.*, 1995), and two are on the *in*ternal surface of the IMM, ND*in*(NADH) and ND*in*(NADPH) (Melo *et al.*, 1996). All four enzymes probably contain FAD, typical for type II

Table 7.1 NAD(P)H dehydrogenase genes in the *Arabidopsis* nuclear genome, and the enzymes that they encode

Gene	Locus	Enzyme encoded[a]	Reference
nda1	At1g07180	NDin(NADH)	Rasmusson *et al.*, 1999 Svensson and Rasmusson, 2001 Svensson *et al.*, 2002
nda2	At2g29990	NDin(NADH)?	Michalecka *et al.*, 2003 Elhafez *et al.*, 2006
ndb1	At4g28220	NDex(NADPH)	Michalecka *et al.*, 2004
ndb2	At4g05020	NDex(NADH)?	*b*
ndb3	At4g21490	NDex(NADH)?	*b*
ndb4	At2g20800	NDex(NADH)?	*b*
ndc1	At5g08740	NDin(NADPH)?	Michalecka *et al.*, 2004 Elhafez *et al.*, 2006

[a] Names followed by a question mark are suggestions rather than definite identifications.
[b] All the *ndb* genes have an insert with an EF-hand motif strongly indicating that they bind Ca^{2+}. While NDB1 has an uncharged amino acid at the end of the second β-sheet, NDB2–4 all have a charged amino acid indicating that they will not bind NADPH (Michalecka *et al.*, 2003). NDB2 and NDB4 are external (Elhafez *et al.*, 2006) and, by inference, so is NDB3. So we have a Ca^{2+}-dependent, external enzyme not oxidising NADPH, in other words NDex(NADH).

dehydrogenases, although this has not yet been demonstrated for any of the plant proteins. Their properties have been described in several reviews (Kerscher, 2000; Møller, 2001a, 2002; Rasmusson *et al.*, 2004).

Initially, two genes encoding an NDex (NDB) and an NDin (NDA) were sequenced and characterised in potatoes (Rasmusson *et al.*, 1999). Subsequent analyses of the *Arabidopsis* genome showed that there are seven nuclear genes encoding these enzymes that group into three families: NDA with two members, NDB with four members and NDC with only one member (Michalecka *et al.*, 2003) (Table 7.1).

Judging from localisation experiments and import studies, it appears that the NDB proteins are probably all NDex (Rasmusson *et al.*, 1999; Michalecka *et al.*, 2003; Elhafez *et al.*, 2006). They contain an insert with an EF-hand motif for Ca^{2+} binding, implying that they are regulated by Ca^{2+} (Michalecka *et al.*, 2003). Consistent with this, overexpression of potato NDB1 in tobacco demonstrated that this is the Ca^{2+}-dependent NDex(NADPH) (Michalecka *et al.*, 2004). The related NDE1 in *Neurospora crassa* mitochondria is also an external Ca^{2+}-dependent NDex(NADPH) (Melo *et al.*, 2001). An analysis of the NAD(P)H-binding domain of the NDB proteins indicates that the NADPH specificity of NDB1 and NDE1 is linked to the presence of an uncharged amino acid at the end of the second β-sheet. In contrast, NDB2, NDB3 and NDB4 all have a charged amino acid in that position, and are probably specific for NADH (Michalecka *et al.*, 2004).

Unlike NDB family members, the NDA proteins do not contain a Ca^{2+}-binding domain, but they are imported and contain a charged amino acid in the position critical for substrate selectivity (Rasmusson *et al.*, 1999; Michalecka *et al.*, 2003; Elhafez *et al.*, 2006). This makes them good candidates for NDin(NADH).

Consistent with this, both a cold-induced downregulation of the *nda1* gene in potatoes and a mutation in the equivalent gene in *Arabidopsis* were accompanied by a decrease in NDin (NADH) activity (Svensson *et al.*, 2002; Moore *et al.*, 2003).

The NDC protein has an uncharged amino acid in the critical position of the NAD(P)H-binding domain (Michalecka *et al.*, 2003), and it is imported across the IMM (Elhafez *et al.*, 2006). It is, therefore, a candidate for the enigmatic NDin(NADPH), but since it does not contain an EF-hand motif, some other explanation for the Ca^{2+} dependence of NDin(NADPH) must be found.

7.4.2 Physiology of the NAD(P)H dehydrogenases

The cytoplasmic male-sterile (CMS) tobacco line CMSII has a large deletion in the mitochondrial DNA (mtDNA)-encoded subunit NAD7 and does not assemble complex I. In response to the almost total absence of complex I activity, the plants show a marked upregulation of the activities of AOX and both internal and external alternative dehydrogenases. This indicates that the alternative dehydrogenases can replace complex I to maintain a viable plant (apart from the male-sterile phenotype) (Sabar *et al.*, 2000).

NDA1 is probably involved in photorespiration by helping re-oxidise the NADH produced by glycine oxidation. The expression of the *nda1* gene is light dependent, and the steady-state transcript abundance shows a diurnal cycle with a maximum just after dawn and a minimum later in the day, which is similar to the expression pattern observed with genes encoding photorespiratory enzymes (Svensson and Rasmusson, 2001; Michalecka *et al.*, 2003; Escobar *et al.*, 2004; Elhafez *et al.*, 2006). In addition, NDB2 (Elhafez *et al.*, 2006) and NDC1 (Escobar *et al.*, 2004) may be light regulated.

When nitrogen-deficient *Arabidopsis* plants were transferred to a nitrate-containing medium, two AOX genes and one NDA gene were downregulated. In contrast, transfer to an ammonium-containing medium caused the upregulation of two NDB genes, as well as three members of the AOX gene family. Apparently, the NDex(NADH) encoded by the NDB genes is important for re-oxidation of cytosolic NADH when not needed for the reduction of nitrate. On the other hand, the down-regulation of NDA under nitrate indicates that a lowered capacity to oxidise matrix NADH might allow an increased export of reducing equivalents to the cytosol to be used in nitrate reduction (Escobar *et al.*, 2006).

NDB2 and AOX1a are co-expressed in *Arabidopsis* under a number of conditions suggesting a common role in cellular metabolism (Clifton *et al.*, 2005; Escobar *et al.*, 2006).

7.4.3 Alamethicin is a useful tool to study respiration in intact cells and isolated mitochondria

Alamethicin is a peptide ionophore that inserts into membranes that are negative inside. Thus, alamethicin inserts into the IMM in isolated mitochondria (F. I. Johansson *et al.*, 2004) and into the plasma membrane and the IMM, but not into the tonoplast, in intact cells (Matic *et al.*, 2005). Once in the membrane alamethicin forms channels that permit anions of at least 1–2 kDa to pass through. As most co-enzymes and many intermediary metabolites are anions, it is possible to assay for enzymes inside the plasma membrane and the IMM by adding the appropriate substrates to the medium. This method has been used to measure directly the capacities of complex I and NDin(NADH) in intact potato tuber mitochondria, something that is otherwise quite difficult to do (F. I. Johansson *et al.*, 2004).

7.5 Substrates for the NAD(P)H dehydrogenases

The main source of electrons for the ETC is NAD(P)H either from the cytosol, where it is oxidised by NDex, or from the matrix, where it is oxidised by complex I and the two NDin. In addition, NADPH is also used for the removal of ROS, as we shall see in Section 7.6.2. We, therefore, need to consider the sources and turnover of matrix NAD(P).

7.5.1 NAD(P) transport, synthesis and degradation

The total NAD content is typically 1–5 nmol/mg protein in isolated plant mitochondria, which translates into 1–5 mM in the matrix assuming a matrix volume of 1 μl/mg protein. Generally, the NADP content is about ten times lower (Agius *et al.*, 2001; Igamberdiev and Gardeström, 2003).

NAD^+ appears to be synthesised outside the mitochondria. The NAD^+ transporter has recently been identified in yeast, and it is a member of the IMM carrier family (Todisco *et al.*, 2006). In plant mitochondria, NAD^+ uptake has been well characterised (Tobin *et al.*, 1980; Neuburger *et al.*, 1985), but the transporter has not yet been identified. NADP can be synthesised by phosphorylation, either by an NAD^+ kinase or by an NADH kinase (Turner *et al.*, 2005). Neither has been identified in plant mitochondria to date, but an NADH kinase is responsible for NADPH synthesis in yeast mitochondria (Outten and Culotta, 2003). As an alternative to synthesis inside the mitochondria, $NADP^+$ may be taken up by plant mitochondria (Bykova and Møller, 2001). NAD^+ is also degraded in plant mitochondria, mainly by an NAD glycohydrolase located in the outer mitochondrial membrane (OMM), to yield nicotinamide mononucleotide and adenosine monophosphate (Agius *et al.*, 2001).

7.5.2 The reduction level of NAD(P)

The reduction level is usually monitored either by NAD(P)H fluorescence (Neuburger *et al.*, 1984) or by extraction (Agius *et al.*, 2001), and it is very dependent on the metabolic conditions. In isolated mitochondria, it is typically high in state 2 (substrate added) and state 4 (substrate plus ATP, but no ADP) respiration when the ATP synthase is inactive and the membrane potential is relatively high. The reduction level is low in state 3 (substrate plus ADP), when the membrane potential is smaller. Inhibition of the ETC gives the maximal reduction level, often as high as 80–90% (Agius *et al.*, 2001).

When the NADH reduction level is determined in protoplasts, it is generally fairly low but much higher in the light under low CO_2 concentrations (active photorespiration) than in the light under high CO_2 concentrations, or in darkness (Igamberdiev and Gardeström, 2003). The relatively low *in vivo* values indicate that the mitochondria are closest to state 3 *in situ*.

7.5.3 Free versus bound NADH

The discussion in the previous section was based on total extractable levels of the co-enzymes. However, most of the NADH is bound, probably to enzymes using NAD as co-enzyme, but possibly also to other proteins where it would act as an inactive reserve pool (Hagedorn *et al.*, 2004; Kasimova *et al.*, 2006). In fact, the concentration of free NADH is kept surprisingly constant, and does not decrease upon ADP addition (transfer between the relatively reduced state 4 to the oxidised state 3) where there is a large decrease in the total NADH concentration (Kasimova *et al.*, 2006). It is this free NADH that is available to the NADH dehydrogenases in the IMM. Unfortunately, we do not have similar data for NAD^+, so it is only possible to calculate the minimum reduction level for free NAD, namely free NADH/(free NADH plus total NAD^+), which is around 0.1. Depending on the proportion of NAD^+ bound, the reduction level could be significantly higher. Even at 0.1, it is two orders of magnitude higher than the reduction level estimated for NAD in the cytosol of plant cells (Heineke *et al.*, 1991; Krömer and Heldt, 1991). This has important implications for the transport processes between the cytosol and the mitochondrial matrix. For example, it means that the malate/oxaloacetate transporter can work only to export reducing equivalents, which is thought to be important under photorespiratory conditions (Krömer and Heldt, 1991).

Similar data are not available for NADPH, but there is no reason to think that the situation is substantially different.

7.5.4 Transhydrogenase

In mammalian mitochondria, an energy-linked membrane-bound transhydrogenase catalyses the reaction:

$$\mathrm{NADH} + \mathrm{NADP}^+ + \mathrm{H}^+(\mathrm{out}) \rightarrow \mathrm{NAD}^+ + \mathrm{NADPH} + \mathrm{H}^+(\mathrm{in})$$

In this way the proton gradient across the IMM (acidic outside) is used to keep the NADP pool much more reduced than the NAD pool (Sazanov and Jackson, 1994). Plant mitochondria contain plenty of transhydrogenase activity, (Bykova *et al.*, 1999) and the NADP pool was reported to be much more reduced than the NAD pool in pea leaf protoplast mitochondria (Igamberdiev and Gardeström, 2003), which could be an indication of energy-linked transhydrogenase activity.

However, the *Arabidopsis* genome does not contain a gene encoding an energy-linked transhydrogenase (Arkblad *et al.*, 2001).

Evidence for the participation of the transhydrogenase activities in the metabolism of plant mitochondria is only indirect at present. Glycine decarboxylase is strictly NAD^+ specific, yet rotenone-insensitive glycine oxidation by pea leaf mitochondria is substantially inhibited by very low concentrations of diphenylene iodonium (Bykova and Møller, 2001), so low that only NDin(NADPH) should be inhibited (Agius *et al.*, 1998). This indicates that NADH produced by glycine decarboxylase is converted by a transhydrogenase into NADPH, which is re-oxidised by NDin(NADPH) (Bykova and Møller, 2001).

7.6 Oxidative stress – ROS turnover

The formation of ROS and, as a result, a range of other reactive compounds is an unavoidable consequence of aerobic metabolism. The ETC in the mitochondria is the major site of ROS formation in mammalian cells (Turrens, 2003) and in non-photosynthesising plant cells (Maxwell *et al.*, 1999; Møller, 2001a; Foyer and Noctor, 2003). These reactive compounds can react with and modify proteins, lipids and DNA. However, they can also be signalling molecules (Apel and Hirt, 2004). Under unstressed conditions, a low steady-state ROS concentration is maintained by ROS-removal systems leading to a low level of damage, most of which is repaired. However, under stressed conditions the rate of ROS production exceeds the capacity of the ROS-removal systems. This results in increasing concentrations of ROS that cause greatly increased damage, thereby overpowering the repair systems. Damage accumulates and eventually leads to cell death (Figure 7.2; see Chapter 10).

In the following few sections, I will briefly summarise our knowledge about the three basic strategies the mitochondria employ to regulate ROS production, and limit oxidative damage: avoidance, removal/detoxification and repair/replacement.

7.6.1 Minimising ROS production

The main sites of ROS production in the ETC of mammalian mitochondria are complexes I and III (Turrens, 2003), and this is probably also true of plant mitochondria (Møller, 2001a). As described above, the ETC of plant mitochondria contains several enzymes not present in the mammalian ETC, and the possibility that they contribute to ROS production should be considered. However, AOX reduces oxygen to water and the four NAD(P)H dehydrogenases are all assumed to contain FAD rather than FMN (Kercher, 2000). Since FAD cannot form a semiquinone, and since it is probably the semiquinone in complexes I and III that is responsible for reacting with oxygen, it follows that NDin and NDex are not likely to contribute to ROS formation, although evidence for such a contribution has been presented (Davidson and Schiestl, 2001). Additionally, some matrix enzymes also produce ROS (see later).

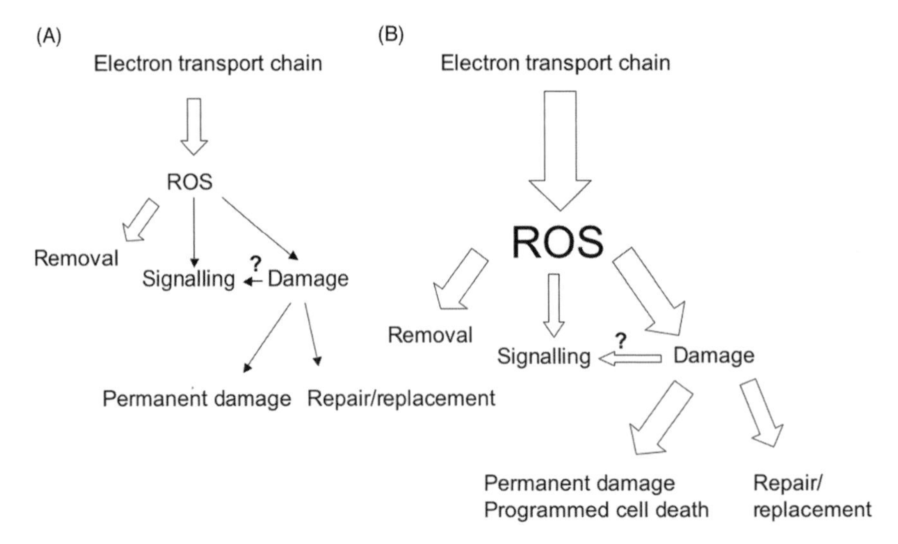

Figure 7.2 The relationship between ROS production, removal, signalling and damage in plant mitochondria under (A) unstressed and (B) stressed conditions. The question marks indicate that some of the so-called damaged molecules may, in fact, be secondary signal molecules.

ROS production can be limited by preventing ETC overreduction. Two means to achieve this are provided by AOX and uncoupling proteins (UCPs). AOX activation, or overexpression, helps minimise ROS production in isolated mitochondria, cell cultures and intact tissues probably by preventing or limiting the overreduction of complex III (Maxwell *et al.*, 1999; Møller, 2001a; Umbach *et al.*, 2005). The UCPs, encoded by a small nuclear gene family in plants (Vercesi *et al.*, 2006), allow protons to flow through the IMM back into the matrix without coupling this flow to ATP production. This lowers the membrane potential and uncouples electron flow from ATP synthesis. UCP activation leads to lower ROS production by isolated mitochondria (for review, see Møller, 2001a). An interesting feedback mechanism for limiting ROS formation is provided by superoxide, which stimulates proton leakage through UCP (Considine *et al.*, 2003).

Since Fe^{2+} reacts with H_2O_2 in the Fenton reaction to produce the hydroxyl radical ($HO^\bullet$), a very reactive ROS, it is essential to keep the concentration of free Fe very low. This might be achieved through the presence of ferritin, a large protein complex able to sequester Fe atoms, and recently reported to be present in plant mitochondria (Zancani *et al.*, 2004). One potential source of free Fe^{2+} is aconitase damaged by ROS (Ilangovan *et al.*, 2005; see also below).

Nitric oxide ($NO^\bullet$) is formed from arginine by the action of nitric oxide synthase in the mitochondria (Guo and Crawford, 2005), but also by nitrite reduction under low-oxygen tensions (Gupta *et al.*, 2005). NO can react with and inhibit CCO (Millar and Day, 1996), and one would expect this to cause increased ROS production because the ETC becomes more reduced. However, NO appears to be able to act as both pro- and anti-oxidant (Guo and Crawford, 2005 and references therein).

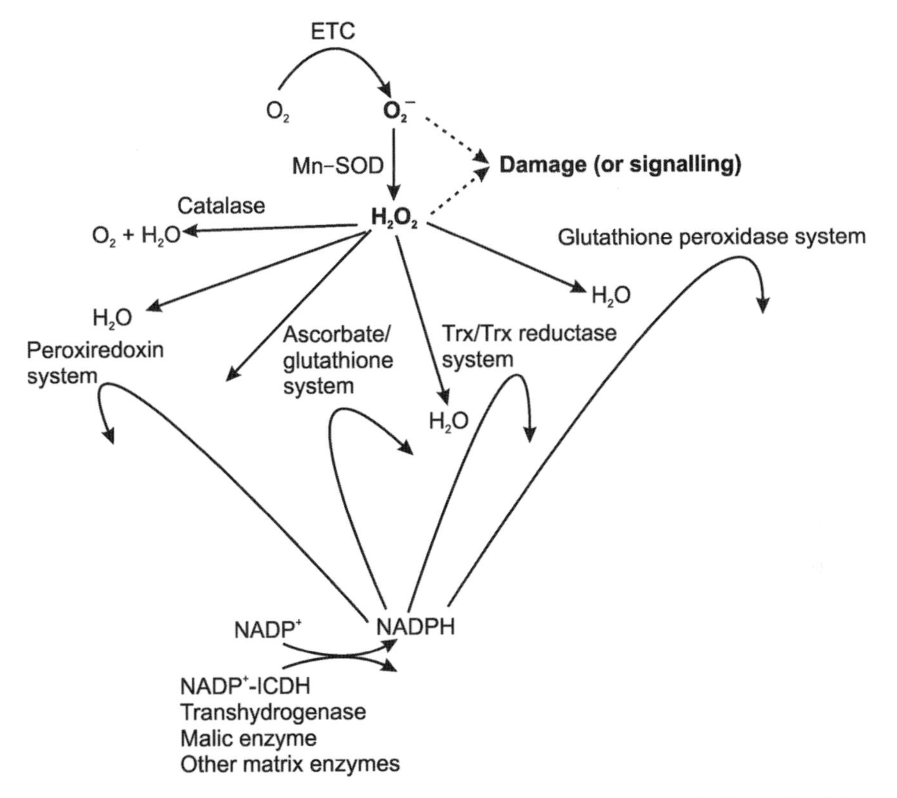

Figure 7.3 Summary of the enzyme systems possibly involved in ROS removal in the matrix of plant mitochondria. ICDH: isocitrate dehydrogenase; Trx: thioredoxin.

7.6.2 *Removal or detoxification of ROS*

A number of enzymes and enzyme systems have been reported to be able to remove superoxide ($O_2^{-\bullet}$) and H_2O_2 from the mitochondrial matrix (Figure 7.3). These systems were reviewed by Møller (2001a), and the reader is referred to that paper for the basic references. Here mainly more recent evidence will be cited. An excellent review is also found in Navrot *et al.* (2007).

7.6.2.1 *ROS removal by SOD and (possibly) catalase*

Superoxide is the ROS initially formed by the ETC, and it is converted into H_2O_2 by Mn-superoxide dismutase (SOD, EC 1.15.1.1) in the matrix of both plant and animal mitochondria. If superoxide is formed outside the IMM, at the external site in complex III (see Figure 7.1), it can be converted into H_2O_2 by CuZn-SOD either in the IMS or in the cytosol (O'Brien *et al.*, 2004). While superoxide is not particularly reactive, H_2O_2 can give rise to the highly reactive $HO^\bullet$ through the Fenton reaction (Halliwell and Gutteridge, 1999), and the concentration of H_2O_2 must therefore be maintained at a low level.

Catalase converts H_2O_2 into oxygen and water. In yeast one of the catalase isoforms is dually targeted to peroxisomes and mitochondria in spite of the fact that it has no classical mitochondrial presequence (Petrova *et al.*, 2004). Whether plant mitochondria contain catalase is still unclear; however, there are several alternative solutions in plant mitochondria: (i) there are several other matrix enzymes that can remove H_2O_2 (see below); (ii) H_2O_2 is quite mobile and it may suffice that it is detoxified elsewhere in the cell; e.g. H_2O_2 produced in the matrix might pass through aquaporins in the IMM to be detoxified in the cytosol (Bienert *et al.*, 2006) and (iii) H_2O_2 may be used as an indicator of oxidative stress, a signalling molecule, to activate nuclear genes involved in the stress response, so a certain low concentration should be maintained.

7.6.2.2 The ascorbate/glutathione cycle

The ascorbate/glutathione cycle is the main ROS-removing system in the chloroplast. It uses four enzymes, ascorbate peroxidase (EC 1.11.1.11), dehydroascorbate reductase (EC 1.8.5.1), monodehydroascorbate reductase (EC 1.6.5.4) and glutathione reductase (EC 1.6.4.2), as well as two low-molecular-mass molecules, ascorbate and glutathione, to remove H_2O_2. All the requisite enzyme activities of the ascorbate/glutathione cycle were found in pea leaf mitochondria (Jimenez *et al.*, 1997), and all four enzymes are imported into *Arabidopsis* mitochondria (Chew *et al.*, 2003).

The levels of glutathione and ascorbate are about 6 and 24 nmol/mg protein in pea leaf mitochondria (Jimenez *et al.*, 1997), and 0.4–1.6 and 7–12 nmol/mg protein in tomato mitochondria (Mittova *et al.*, 2003). This is equivalent to about 0.4–6 and 7–24 mM in the matrix (assuming 1 μl/mg protein). The ascorbate levels are more than 10 times the matrix concentration of NADP, and somewhat above that of NAD (see above). The plant cell can synthesise ascorbate via several routes, but the main route includes a mitochondrial enzyme in the last step (Wheeler *et al.*, 1998). Here the flavoprotein L-galactono-γ-lactone dehydrogenase, an intrinsic protein in the inner membrane with the active site facing the IMS, converts L-galactono-γ-lactone into ascorbate and donates electrons to the ETC between complexes III and IV (Bartoli *et al.*, 2000; Millar *et al.*, 2003). Since the ascorbate is formed in the IMS, it must be transported into the matrix, e.g. by the low-affinity transporter in the IMM described by Szarka *et al.* (2004).

In mammalian mitochondria, glutathione synthesis takes place in the cytosol and reduced glutathione (*N*-(*N*-L-gamma-glutamyl-L-cysteinyl)glycine, GSH) is taken up across the inner membrane via a special transporter (Meister, 1995). Nothing is known about the uptake characteristics of glutathione across the inner membrane of plant mitochondria.

7.6.2.3 Glutathione peroxidases

Glutathione peroxidases (EC 1.11.1.9) comprise another family of enzymes using GSH to reduce H_2O_2, lipid hydroperoxides and other hydroperoxides (Halliwell and Gutteridge, 1999). Glutathione peroxidase is the main enzyme for removing H_2O_2 in mammalian mitochondria, and a member of the plant gene family is predicted to be localised to the mitochondria (Rodriguez Milla *et al.*, 2003). One member of the glutathione peroxidase superfamily is the phospholipid-hydroperoxide glutathione peroxidase (EC 1.11.1.12), which can act directly on lipid hydroperoxide without the need for release of the hydroperoxy fatty acid. Thus, glutathione peroxidases can also contribute to repair – the third line of defence against ROS damage.

The regeneration of GSH from oxidised glutathione (glutathione disulphide, GSSG), whether produced in the ascorbate/glutathione cycle or by glutathione peroxidase, is catalysed by the NADPH-specific glutathione reductase in the mitochondrial matrix. It may well be the main NADPH-consuming enzyme under metabolic conditions when ROS production and glutathione turnover are high. It has been suggested that GSSG inhibits NADPH-consuming enzymes other than those involved in ROS detoxification to conserve NADPH (Del Corso *et al.*, 1994). If this were true in plant mitochondria, NDin(NADPH) would be predicted to be inhibited by GSSG.

7.6.2.4 The thioredoxin and thioredoxin reductase system

Thioredoxin is a small protein (12–14 kDa) with two cysteines in its active site that, upon oxidation, can form an internal dithiol. Together with the enzyme thioredoxin reductase (EC 1.6.4.5), which uses NADPH to convert oxidised into reduced thioredoxin, thioredoxin is involved in the regulation of enzyme activities as well as in scavenging hydroperoxides and H_2O_2 (Halliwell and Gutteridge, 1999). Thioredoxin has long been known to regulate the activities of a number of Calvin cycle enzymes in the chloroplast stroma (Buchanan and Balmer, 2005).

Plant mitochondria contain a fully functional thioredoxin system that may be involved in stress responses (Laloi *et al.*, 2001). Moreover, Balmer *et al.* (2004) identified around 50 enzymes and other proteins including Krebs cycle enzymes, and subunits of respiratory complexes that are potentially regulated by thioredoxin. AOX is also regulated by this reversible mechanism (Gelhaye *et al.*, 2004). Therefore, it is likely that the thioredoxin system will turn out to be a very important mechanism for regulating mitochondrial metabolism.

7.6.2.5 Peroxiredoxin

One of the targets of the thioredoxin reductase system is a peroxiredoxin. Peroxiredoxins form a family of small proteins (15–25 kDa) able to use internal thiol groups to reduce H_2O_2 and alkylperoxides. The thiol groups can be regenerated by reduction with thioredoxin (Finkemeier *et al.*, 2005).

7.7 Mitochondrial DNA and lipid oxidation

Both here and in the next section on protein oxidation, we have to keep in mind that some of the oxidative modifications described may be part of as-yet-unrecognised signal transduction pathways.

7.7.1 DNA oxidation

One of the dominant theories of ageing in mammals is that mitochondrial mutations accumulate during life as a side effect of respiration (Harman, 1972). In fact, not only do mitochondrial mutations accumulate with age, but also mitochondrial protein damage, and both are somehow caused by exposure to ROS (Raha and Robinson, 2000).

In mammalian mitochondria, aconitase is essential for DNA maintenance independent of its catalytic activity (Chen *et al.*, 2005, and references therein). Since aconitase appears to be very susceptible to oxidative modifications (see below), this might be an indirect way for oxidative stress to affect mtDNA.

To the best of my knowledge, no studies of oxidative modifications of plant mtDNA have been published.

7.7.2 Lipid oxidation

Polyunsaturated fatty acids (PUFA) are particularly susceptible to attack by $HO^{\bullet}$, the hydroxyl radical, formed in the Fenton reaction and HO_2, protonated superoxide. Abstraction of a hydrogen atom and reaction with O_2 lead to the formation of a peroxy radical, which can start a chain reaction. The decomposition of lipid peroxides yields, among many other products, the reactive malondialdehyde (MDA) and 4-hydroxy-2-*trans*-nonenal (HNE) (Halliwell and Gutteridge, 1999). These breakdown products can form conjugates with proteins (see below) and may be involved in cytoplasmic male sterility (Liu *et al.*, 2001; Møller, 2001b).

Mitochondria contain about 20% (w/w) lipids, and the majority are phospholipids. Particularly in plant mitochondria, these phospholipids contain a high PUFA percentage (Daum, 1985). This is especially true for diphosphatidylglycerol, also called cardiolipin, which can contain more than 80 mol% 18:2 and 18:3 fatty acids (Bligny and Douce, 1980; Caiveau *et al.*, 2001). When inner membranes of bovine heart mitochondria were exposed to strong oxidative stress *in vitro*, a major decrease in cardiolipin was observed probably due to peroxidation of its PUFA. In parallel, a strong decrease was observed in the activity of CCO which appears to require cardiolipin for activity. The CCO activity completely recovered when the damaged cardiolipin was replaced with intact cardiolipin (Paradies *et al.*, 1998). Thus, PUFA may protect proteins against oxidative damage by reacting with ROS. On the other hand, the aldehydes formed as a result of PUFA peroxidation can react with proteins and create carbonyl groups (Dean *et al.*, 1997). It is easy to imagine that such an indirect side-chain modification could be just as damaging to protein function as direct side-chain oxidation. In the above case, however, CCO recovered completely, so either the enzyme did not sustain such a modification or the modification was not harmful. It is interesting to note that the function of UCP3 may be to rid the matrix of fatty acid peroxides (Goglia and Skulachev, 2003).

Isolated epicotyl mitochondria, where 87% of the fatty acids in the phospholipids are 18:2 and 18:3, contained three times more PUFA breakdown products than seed mitochondria with only 68% PUFA, although it is difficult to know whether

Table 7.2 Common ROS-induced modifications of protein (from Halliwell and Gutteridge, 1999)

Modification	Product	Comment
Met oxidation	Methionine sulphoxide	Reversible
Cys oxidation	Disulphide bond formed	Reversible
Trp oxidation	*N*-formylkynurenine and kynurenine	Conveniently detected by tandem MS (Taylor *et al.*, 2003; Møller and Kristensen, 2006)
Arg, Lys, Pro and Thr	Carbonyl group formed	Can be detected using DNP and anti-DNP antibodies (e.g. Kristensen *et al.*, 2004)
Conjugation with lipid breakdown products	4-Hydroxy-2-nonenal forms carbonyl groups by reacting with Lys, Cys and His	Indirect modification caused by oxidation of PUFA
Chain breakage	Might form carbonyl end groups	

these breakdown products reflected the *in vivo* situation. Likewise, incubation of the isolated epicotyl mitochondria with Fe^{2+} led to the formation of twice as many MDA equivalents than for seed mitochondria. It was suggested that the low PUFA content limits oxidative cellular damage, and in this way contributes to the longevity of pea seeds (Stupnikova *et al.*, 2006).

7.8 Mitochondrial protein oxidation

7.8.1 Reaction with ROS

Proteins can react with ROS in a number of ways and the most important are given in Table 7.2. Although this has not been well studied, different ROS species have different reactivity; e.g. histidine has a high reaction rate with singlet oxygen, whereas tryptophan is sensitive to damage by $HO^{\bullet}$ (Halliwell and Gutteridge, 1999). By determining the specific damage to specific proteins, it is therefore possible to predict what ROS species was involved and where in the cell/organelle the reaction occurred, tracking the footprints of ROS as it were. The different footprints of different ROS species might also impart specificity to ROS signalling.

7.8.2 Identification of oxidised mitochondrial proteins

Consistent with the high level of ROS production in plant mitochondria, protein carbonylation was reported to be higher in the mitochondria of wheat leaves than in chloroplasts and peroxisomes (Bartoli *et al.*, 2004).

Several studies have addressed the question of what mitochondrial proteins are oxidised/affected by oxidative stress:

1. Sweetlove *et al.* (2002) treated *Arabidopsis* cell cultures for 16 h with H_2O_2 or menadione, which generates intracellular superoxide, and identified

mitochondrial protein spots increasing or decreasing in abundance by two-dimensional electrophoresis followed by mass spectrometry. Smaller break-down products of 11 proteins were identified and a further 12 proteins decreased in abundance in response to the oxidative treatment of the cell cultures.

2. Karpova *et al.* (2002) compared carbonylated proteins in mitochondria from wild-type and non-chromosomal stripe CMS-mutant maize seedlings by dinitrophenylhydrazine (DNP) tagging followed by probing one-dimensional Western blots with anti-DNP-specific antibodies. A number of distinct bands were seen in mitochondria from the normal maize lines, consistent with the idea that a pool of oxidised isoform(s) exists for some proteins *in vivo*. None of these oxidised proteins were identified. Surprisingly, there was no increase in the number or intensity of DNP-tagged bands in mitochondria from non-chromosomal stripe mutants. These mutants are defective in the biosynthesis of the respiratory complexes, and an increased ROS production with accompanying protein oxidation might therefore have been expected.
3. Taylor *et al.* (2003) found 39 mitochondrial proteins, containing oxidised tryptophan in non-stressed human cardiac muscle by looking for the signature of tryptophan oxidation in mass spectra of mitochondrial proteins. The authors concluded that they had been oxidised *in vivo*. Many of these proteins were subunits of the respiratory complexes, notably nine subunits of complex I, two subunits of complex III and three of the membrane-bound part of complex V. However, a number of soluble proteins also contained oxidised tryptophan. These included five Krebs cycle enzymes, and several enzymes involved in fatty acid oxidation and other redox processes.
4. In a similar study, proteins from rice leaf mitochondria and potato tuber mitochondria were separated by two-dimensional gel electrophoresis and identified by mass spectrometry. *N*-formylkynurenine was detected in 29 peptides, representing 17 different proteins including subunits of respiratory complexes I, III, IV and V. With one exception, the oxidation-sensitive aconitase, all of these proteins were either redox active themselves or subunits in redox-active enzyme complexes. The results indicate that tryptophan oxidation is a selective process (Møller and Kristensen, 2006).
5. Kristensen *et al.* (2004) identified oxidised (carbonylated) proteins in a soluble protein fraction from green rice leaf mitochondria by DNP-labelling, anti-DNP-antibody immunoprecipitation, trypsin treatment and two-dimensional liquid chromatography followed by mass spectrometry. Twenty oxidised proteins were found, and it was concluded that they were oxidised *in vivo*. After mild oxidation of the matrix sample, which roughly doubled the overall oxidation level, a further 31 oxidised proteins were identified. These 31 proteins represent a group of particularly oxidation-prone proteins, including several Krebs cycle enzymes, all four enzymes involved in glycine decarboxylation, and several other redox enzymes, heat shock proteins, Mn-SOD and porin. Seven of the 11 proteins identified by Sweetlove *et al.*, (2002) as

smaller breakdown products were identified as being carbonylated, so it is possible that chain breakage is responsible for at least part of the carbonylation detected by 2D-LC-MS/MS. Four of the 12 mitochondrial proteins that decreased in amount in response to oxidative treatment of cell cultures (Sweetlove *et al.*, 2002) were also carbonylated in rice leaves.

6. O'Brien *et al.* (2004) compared protein oxidation in mitochondria from paraquat-stressed wild-type cells to that of mutant cell lines lacking the CuZn-SOD and/or the Mn-SOD. They found 14 proteins in which the level of carbonylation was at least twofold higher in the mutant lines. Half of these proteins were also found to be carbonylated by Kristensen *et al.* (2004) in rice leaf mitochondria.
7. Sieger *et al.* (2005) identified 40 mitochondrial proteins that decreased in abundance in phosphorous-deficient tobacco cells lacking AOX, and therefore severely stressed, relative to phosphorous-deficient wild-type cells. It is not known whether increased breakdown or decreased synthesis was responsible for the decrease in abundance.

It is clear from the above and from the comparison in Table 7.3 that many of the same proteins are oxidised in mitochondria from plants, yeast and mammals. This indicates that these proteins are either more susceptible to oxidation or more exposed to ROS, or both. However, most of these proteins are quite abundant, and it is possible that this increases the risk of collision with ROS. The high abundance might also facilitate the detection of the oxidised form.

Since most of mitochondrial ROS is produced in complexes I and III located in the IMM, one might expect most of the oxidatively damaged proteins to be at least membrane-associated if not subunits of these two respiratory complexes. The results of several investigations indicate that such damage is observed. However, even under unstressed conditions, soluble proteins such as the Krebs cycle enzymes, which may not be physically close to the site of ROS synthesis, are affected (Sweetlove *et al.*, 2002; Taylor *et al.*, 2003; E. Johansson *et al.*, 2004; Kristensen *et al.*, 2004; Møller and Kristensen, 2006).

Mitochondrial aconitase appears to be particularly susceptible to oxidative damage. Aconitase contains oxidised tryptophan *in vivo* (Taylor *et al.*, 2003; Møller and Kristensen, 2006), and carbonylation (Kristensen *et al.*, 2004) and fragmentation products (Sweetlove *et al.*, 2002) are detected after *in vitro* and *in vivo* oxidative stress, respectively. Damage to aconitase may lead to the release of iron and this has been suggested to lead to further ROS production via the Fenton reaction, i.e. a vicious circle (Ilangovan *et al.*, 2005).

Interestingly, both 2-oxoglutarate dehydrogenase and pyruvate dehydrogenase produce ROS, especially under high NAD(P)H/NAD(P) ratios, probably via the flavin in the lipoamide hydrogenase (Chinopoulos and Adam-Vizi, 2006). By analogy, it is likely that the glycine decarboxylase complex also produces ROS. Oxidised subunits, including lipoamide dehydrogenase, probably deriving from *in vivo* oxidation events, have been found for all three complexes (Kristensen *et al.*, 2004; Møller and Kristensen, 2006), and the oxidation could have been caused by locally

Table 7.3 A comparison of oxidised mitochondrial proteins identified in plant, yeast and mammalian mitochondria

Protein	Rice leaves matrix fraction[a]	Rice leaf and potato tubers[b]	Human heart[c]	Yeast[d]
ETC				
Complex I[e]	—	3 subunits	9 subunits	—
Complex III[f]	—	2 subunits	2 subunits	1 subunit
Complex IV	—	1 subunit	1 subunit	—
Complex V	2 subunits	—	3 subunits	1 subunit
Krebs cycle and associated enzymes				
Pyruvate DH[g]	+	—	—	+
Dihydrolipoamide DH	+	—	—	+
Aconitase	+	+	+	+
Isocitrate DH	+[h]	—	+[i]	+[j]
ROS-removing enzymes				
SOD	+	+	—	—
Others				
Aldehyde DH	+	+	—	—
Formate DH	+	+	—	—
Glycine decarboxylase complex P-protein	+	+	—	—
Porin (OMM)	+	—	+	+

Proteins in intact mitochondria were analysed except for the data from rice leaves (Kristensen *et al.*, 2004). Only proteins identified in at least two investigations are included.
[a]Carbonylated proteins (Kristensen *et al.*, 2004).
[b]Oxidised Trp (Møller and Kristensen, 2006).
[c]Oxidised Trp (Taylor *et al.*, 2003).
[d]Carbonylated proteins (O'Brien *et al.*, 2004).
[e]Many of the identified subunits are involved in electron transfer. Yeast does not contain complex I.
[f]All the identified subunits are core subunits.
[g]DH, dehydrogenase.
[h]NAD specific.
[i]NADP specific.
[j]Specificity unknown.

produced ROS. Detection of these oxidation products might provide a sensitive indicator of oxidative stress.

7.8.3 Enzymes interacting with ROS as part of their function

Two groups of proteins must be able to function in the presence of ROS – ROS-detoxifying enzymes and enzymes that produce ROS as part of their normal catalytic cycle. SOD qualifies on both of these counts. This raises several questions:

- Are these enzymes more resistant to ROS? And if yes, how is this resistance achieved?
- Does oxidation of these enzymes lead to activation like increased V_{max} or lowered K_m?
- Do these enzymes have a more rapid turnover than other proteins?

CuZn-SOD is known to be inactivated by oxidation of the Cu^{2+} histidine ligand in the catalytic site (Kurahashi *et al.*, 2001). However, this modification was not observed in a study of the role of SOD in neurogenerative diseases. Of four spots identified as CuZn-SOD, one was carbonylated and one contained a cysteine sulphonic acid; no oxidative post-translational modifications were identified in the last two spots (Choi *et al.*, 2005). It is possible that histidine oxidation in the active site leads to immediate degradation. Since a histidine is also the metal ligand in the catalytic site of bacterial Mn-SOD (Whittaker and Whittaker, 1998), it is likely that mitochondrial Mn-SOD, in which this histidine is conserved, is sensitive to oxidative inactivation via the formation of oxo-histidine. The inactivation of Mn-SOD by nitration, caused by oxidation by a reactive nitrogen species, of a tyrosine near the Mn in the active site was recently reported (Yamakura *et al.*, 1998).

What makes a protein susceptible to oxidative damage? The method of directed evolution can provide some answers to this question. Cherry *et al.* (1999) subjected a mushroom peroxidase to multiple rounds of directed evolution and obtained an enzyme that had 100 times the oxidative stability (measured as the half-life at 40°C and pH 10.5 in the presence of 0.2 mM H_2O_2) of the wild-type enzyme. Changing three oxidisable amino acids near the active site heme group improved the stability fivefold only. The other changes required to increase the stability a further 20-fold changed the overall structure of the active site and probably reduced the accessibility of H_2O_2 to susceptible groups in the interior of the protein (Cherry *et al.*, 1999).

7.8.4 Conjugation with products of PUFA oxidation

In vitro treatment of isolated mitochondria or enzyme complexes with HNE, one of the products of PUFA peroxidation, causes an inhibition of the decarboxylating enzyme complexes, 2-oxoglutarate dehydrogenase, pyruvate dehydrogenase and glycine decarboxylase, probably by interaction with their lipoamide moity (Millar and Leaver, 2000; Taylor *et al.*, 2002). The AOX is also inhibited (Winger *et al.*, 2005).

Paraquat treatment of *Arabidopsis* cell cultures or pea plants increases the amount of PUFA breakdown as measured by MDA formation and, at the same time, the amount and number of proteins conjugated with HNE increases dramatically while glycine oxidation is inhibited (Taylor *et al.*, 2002; Winger *et al.*, 2005).

7.9 Fate of oxidised proteins

It is generally thought that oxidised proteins are degraded due to having a more open conformation, which is recognised by proteases. At the same time, massive protein damage can lead to the formation of protein aggregates that apparently cannot be degraded (Davies and Shringarpure, 2006).

The fate of oxidised proteins can be graded by the energetic cost to the cell. The cheapest response is to reverse the damage, i.e. reduce the oxidised amino acid side chain back to its original form. This is what happens to sulphur-containing amino acids, whether or not the oxidation can be categorised as metabolic regulation, signalling or oxidative damage. At the second level, the oxidatively modified polypeptide is removed, degraded and replaced by a new polypeptide. Presumably, the degradation continues to the amino acid level; while the vast majority of undamaged amino acids are conserved, the oxidatively damaged amino acids are further metabolised. If a damaged polypeptide is part of a protein complex, the complex might dissociate, allowing replacement of the damaged peptide followed by complex reassembly. The most expensive solution is obviously to remove the entire organelle by mitophagy (Lemasters, 2005).

The half-life of a number of mitochondrial proteins in mammalian cells has been estimated to range from around 20 to more than 100 h (Hare and Hodges, 1982a,b). This could be a relative measure of the rate at which the proteins absorb damage. However, that does not explain why the half-life of four of the eight subunits of complex III is significantly shorter than that of the other subunits (Hare and Hodges, 1982b). As expressed by Rep and Grivell (1996), 'Presumably, the subunits with short half-lives are replaced by newly synthesised subunits after the complex has fallen apart. The question remains whether disassembly of the complex simply represents a random process or whether (damaged?) complexes are actively disassembled by quality control factors.' Interestingly, conjugation with co-enzyme A is one mechanism that makes mitochondrial proteins less accessible to ATP-dependent protease (Huth *et al.*, 2002), which would extend the lifetime of the targeted proteins.

What is the fate of to oxidatively modified proteins? Are they degraded by specific proteases? Degradation by specific proteases probably does not occur, since there are so many types of oxidative modification. It is more likely that the proteases recognise proteins with a more open conformation. However, an unfolded conformation does not appear to be a requirement for degradation (Röttgers *et al.*, 2002).

Plant mitochondria contain several different proteases with different properties (see Adam *et al.*, 2001; Møller and Kristensen, 2004; Janska, 2005, and references therein) but also peptidases (Kambacheld *et al.*, 2005). The membrane-bound FtsH proteases may be involved in degrading oxidised proteins, but probably mainly membrane-associated ones (Lindahl *et al.*, 2000; Ostersetzer and Adam, 2003). To what extent the other proteases are involved in degrading damaged proteins and misfolded proteins and/or removing unassembled subunits is not known. We need to know much more about the function and specificity of these proteases before we can understand the turnover of oxidised proteins.

7.10 How to prevent the transfer of oxidative damage to the next generation?

It is important to prevent, or at least limit, the transfer of damaged components to the next generation, especially DNA and DNA-containing organelles. For example, biochemical activity is very high in the stamen where pollen are produced and the mitochondria appear to be working close to their maximal capacity since most mutations in genes encoding mitochondrial proteins lead to dramatically lowered pollen production (Vedel *et al.*, 1999; Sabar *et al.*, 2000). High activity generally means high ROS production, although direct evidence for this in stamen is lacking. However, mitochondria (and plastids) are generally maternally inherited in plants. The organelles therefore derive from the relatively quiescent egg cell, thus minimising oxidative damage to the fertilised egg cell (Allen, 1996). Another possible strategy has been proposed for mammalian cells where a small, biochemically relatively inactive subpopulation of 'breeding' mitochondria in each cell is dividing, while the active, and therefore relatively damaged, mitochondria do not divide (Lemasters, 2005). An alternative strategy exists in budding yeast where carbonylated proteins, mitochondrial and otherwise, are prevented from entering the daughter cell by an as-yet-unidentified mechanism (Aguilaniu *et al.*, 2003).

7.11 Concluding remarks

The ETC of plant mitochondria contains four unique NAD(P)H dehydrogenases, as well as an additional oxidase. These enzymes allow mitochondria to adjust their electron transport under stress conditions to avoid overreduction and, in that way, minimise the production of ROS. Although ROS can act as signalling molecules, the ROS level in the matrix is maintained at a low level by several ROS-removing enzymes. When the ROS production overloads the capacity of these enzymes, increasing damage occurs to proteins, lipids and DNA. Virtually nothing is known about damage to lipids and DNA in plant mitochondria. Although the presence of a number of oxidised mitochondrial proteins under unstressed as well as stressed conditions is well established, the consequences for the protein have not yet been determined in any one case.

We need to identify more sites of modification of specific proteins, lipids and DNA. We need to quantify the extent of these modifications and the mechanism(s) by which they arise. We need to study the effect of these modifications on each of the target molecules. And we need to understand the fate of the modified molecules. Finally, we need to consider whether some of the modified molecules have a role as secondary signalling molecules.

Acknowledgements

I am grateful to Drs. Poul Erik Jensen and Andreas Hansson for stimulating discussions and critical comments on the manuscript.

References

Adam Z, Adamska I, Nakabayashi K, *et al.* (2001) Chloroplast and mitochondrial proteases in *Arabidopsis*. A proposed nomenclature. *Plant Physiol* **125**, 1912–1918.

Agius SC, Bykova NV, Igamberdiev AU and Møller IM (1998) The internal otenone-insensitive NADPH dehydrogenase contributes to malate oxidation by potato tuber and pea leaf mitochondria. *Physiol Plant* **104**, 329–336.

Agius SC, Rasmusson AG and Møller IM (2001) NAD(P) turnover in plant mitochondria. *Aust J Plant Physiol* **28**, 461–470.

Aguilaniu H, Gustafsson L, Rigoulet M and Nyström T (2003) Asymmetric inheritance of oxidatively damaged proteins during cytokinesis. *Science* **299**, 1751–1753.

Allen JF (1996) Separate sexes and the mitochondrial theory of ageing. *J Theor Biol* **180**, 135–140.

Apel K and Hirt H (2004) Reactive oxygen species: metabolism, oxidative stress, and signal transduction. *Annu Rev Plant Biol* **55**, 373–399.

Arkblad EL, Betsholtz C, Mandoli D and Rydström J (2001) Characterization of a nicotinamide nucleotide transhydrogenase gene from the green alga Acetabulatia actabulum and comparison of its structure with those of the corrsponding genes in mouse and Chenorhabditis elegans. *Biochim Biophys Acta* **1520**, 115–123.

Balmer Y, Vensel WH, Tanaka CK, *et al.* (2004) Thioredoxin links redox to the regulation of fundamental processes of plant mitochondria. *Proc Natl Acad Sci USA* **101**, 2642–2647.

Bartoli CG, Gomez F, Martinez DE and Guiamet JJ (2004) Mitochondria are the main target for oxidative damage in leaves of wheat (*Tricicum aestivum* L.). *J Exp Bot* **55**, 1663–1669.

Bartoli CG, Pastori GM and Foyer CH (2000) Ascorbate biosynthesis in mitochondria is linked to the electron transport chain between complexes III and IV. *Plant Physiol* **123**, 335–343.

Berthold DA and Stenmark P (2003) Membrane-bound diiron carboxylate proteins. *Annu Rev Plant Biol* **54**, 497–517.

Bienert GP, Schjoerring JK and Jahn TP (2006) Membrane transport of hydrogen peroxide. *Biochim Biophys Acta* **1758**, 994–1003.

Bligny R and Douce R (1980) Precise localization of cardiolipin in plant-cells. *Biochim Biophys Acta* **617**, 254–263.

Brand MD (1994) The stoichiometry of proton pumping and ATP synthesis in mitochondria. *Biochemist* **16** (4), 20–24.

Buchanan BB and Balmer Y (2005) Redox regulation: a broadening horizon. *Annu Rev Plant Biol* **56**, 187–220.

Bykova NV and Møller IM (2001) Involvement of matrix NADP turnover in the oxidation of NAD^+-linked substrates by pea leaf mitochondria. *Physiol Plant* **111**, 448–456.

Bykova NV, Rasmusson AG, Igamberdiev AU, Gardeström P and Møller IM (1999) Two separate transhydrogenase activities are present in plant mitochondria. *Biochem Biophys Res Commun* **265**, 106–111.

Caiveau O, Fortune D, Cantrel C, Zachowski A and Moreau F (2001) Consequences of omega-6-oleate desaturase deficiency on lipid dynamics and functional properties of mitochondrial membranes of *Arabidopsis thaliana*. *J Biol Chem* **276**, 5788–5794.

Chen XJ, Wang X, Kaufman BA and Butow RA (2005) Aconitase couples metabolic regulation to mitochondrial DNA maintenance. *Science* **307**, 714–717.

Cherry JR, Lamsa MH, Schneider P, *et al.* (1999) Directed evolution of a fungal peroxidase. *Nat Biotechnol* **17**, 379–384.

Chew O, Whelan J and Millar AH (2003) Molecular definition of the ascorbate-glutathione cycle in *Arabidopsis* mitochondria reveals dual targeting of antioxidant defenses in plants. *J Biol Chem* **278**, 46869–46877.

Chinopoulos C and Adam-Vizi V (2006) Calcium, mitochondria and oxidative stress in neuronal pathology. Novel aspects of an enduring theme. *FEBS J* **273**, 433–450.

Choi J, Rees HD, Weintraub ST, Levey AI, Chin L-S and Li L (2005) Oxidative modifications and aggregation of Cu,Zn-superoxide dismutase associated with Alzheimer and Parkinson diseases. *J Biol Chem* **280**, 11648–11655.

Clifton R, Lister R, Parker KL, *et al.* (2005) Stress-induced co-expression of alternative respiratory chain components in *Arabidopsis thaliana*. *Plant Mol Biol* **58**, 193–212.

Considine MJ, Goodman M, Echtay KS, *et al.* (2003) Superoxide stimulates a proton leak in potato mitochondria that is related to the activity of the uncoupling protein. *J Biol Chem* **278**, 22298–22302.

Daum G (1985) Lipids of mitochondria. *Biochim Biophys Acta* **822**, 1–42.

Davidson JF and Schiestl RH (2001) Mitochondrial respiratory electron carriers are involved in oxidative stress during heat stress in *Saccharomyces cerevisiae*. *Mol Cell Biol* **21**, 8483–8489.

Davies KJA and Shringarpure R (2006) Preferential degradation of oxidized proteins by the 20S proteasome may be inhibited in aging and in inflammatory neuromuscular diseases. *Neurology* **66**, S93–S96.

Dean RT, Fu SL, Stocker R and Davies MJ (1997) Biochemistry and pathology of radical-mediated protein oxidation. *Biochem J* **324**, 1–18.

Del Corso A, Cappiello M and Mura U (1994) Thiol dependent oxidation of enzymes: the last chance against oxidative stress. *Int J Biochem* **26**, 745–750.

Elhafez D, Murcha MW, Clifton R, Soole KL, Day DA and Whelan J (2006) Characterization of mitochondrial alternative NAD(P)H dehydrogenases in *Arabidopsis*: intraorganelle location and expression. *Plant Cell Physiol* **47**, 43–54.

Escobar MA, Franklin KA, Svensson ÅS, Salter MG, Whitelam GC and Rasmusson AG (2004) Light regulation of the *Arabidopsis* respiratory chain. Multiple discrete photoreceptor responses contribute to induction of type II NAD(P)H dehydrogenase genes. *Plant Physiol* **136**, 2710–2721.

Escobar MA, Geisler DA and Rasmusson AG (2006) Reorganization of the alternative pathways of the *Arabidopsis* respiratory chain by nitrogen supply: opposing effects of ammonium and nitrate. *Plant J* **45**, 775–788.

Finkemeier I, Goodman M, Lamkemeyer P, Kandlbinder A, Sweetlove LJ and Dietz K-J (2005) The mitochondrial type II peroxiredoxin F is essential for redox homeostasis and root growth of *Arabidopsis thaliana* under stress. *J Biol Chem* **280**, 12168–12180.

Finnegan PM, Soole KL and Umbach AL (2004) Alternative mitochondrial electron transport proteins in higher plants. In: *Plant Mitochondria: From Genome to Function, Vol 17, Advances in Photosynthesis and Respiration* (eds Day DA, Millar AH and Whelan J), pp. 163–230. Kluwer Academic Press, Dordrecht, The Netherlands.

Foyer CH and Noctor G (2003) Redox sensing and signalling associated with reactive oxygen in chloroplasts, peroxisomes and mitochondria. *Physiol Plant* **119**, 355–364.

Gelhaye E, Rouhier N, Gerard J, *et al.* (2004) A specific form of thioredoxin h occurs in plant mitochondria and regulates the alternative oxidase. *Proc Natl Acad Sci USA* **101**, 14545–14550.

Goglia F and Skulachev VP (2003) A function for novel uncoupling proteins: antioxidant defense of mitochondrial matrix by translocating fatty acid peroxides from the inner to the outer membrane leaflet. *FASEB J* **17**, 1585–1591.

Guo FQ and Crawford NM (2005) *Arabidopsis* nitric oxide synthase 1 is targeted to mitochondria and protects against oxidative damage and dark-induced senescence. *Plant Cell* **17**, 3436–3450.

Gupta KJ, Stoimenova M and Kaiser WM (2005) In higher plants, only root mitochondria, but not leaf mitochondria reduce nitrite to NO, *in vitro* and *in vivo*. *J Exp Bot* **56**, 2601–2609.

Hagedorn PH, Flyvbjerg H and Møller IM (2004) Modeling NADH turnover in plant mitochondria. *Physiol Plant* **120**, 370–385.

Halliwell B and Gutteridge JMC (1999) *Free Radicals in Biology and Medicine*, 3rd edn. Oxford University Press, Oxford.

Hansford RG (1991) Dehydrogenase activation by Ca^{2+} in cells and tissues. *J Bioenerg Biomembr* **23**, 823–854.

Hare JF and Hodges R (1982a) Turnover of mitochondrial matrix polypeptides in hepatoma monolayer-cultures. *J Biol Chem* **257**, 2950–2953.

Hare JF and Hodges R (1982b) Turnover of mitochondrial inner membrane-proteins in hepatoma monolayer-cultures. *J Biol Chem* **257**, 3575–3580.

Harman D (1972) The biologic clock: the mitochondria? *J Am Geriatr Soc* **20**, 145–147.

Heineke D, Riens B, Grosse H, *et al.* (1991) Redox transfer across the inner chloroplast envelope. *Plant Physiol* **95**, 1131–1137.

Huth W, Rolle S and Wunderlich I (2002) Turnover of matrix proteins in mammalian mitochondria. *Biochem J* **364**, 275–284.

Igamberdiev AU and Gardeström P (2003) Regulation of NAD- and NADP-dependent isocitrate dehydrogenases by reduction levels of pyridine nucleotides in mitochondria and cytosol of pea leaves. *Biochim Biophys Acta* **1606**, 117–125.

Ilangovan G, Venkatakrishnan CD, Bratasz A, *et al.* (2005) Heat shock-induced attenuation of hydroxyl radical generation and mitochondrial aconitase activity in cardiac H9c2 cells. *Am J Physiol Cell Physiol* **290**, C313–C324.

Ishizaki K, Larson TR, Schauer N, Fernie AR, Graham IA and Leaver CJ (2005) The critical role of *Arabidopsis* electron-transfer flavoprotein:ubiquinone oxidoreductase during dark-induced starvation. *Plant Cell* **17**, 2587–2600.

Janska H (2005) ATP-dependent proteases in plant mitochondria: what do we know about them today? *Physiol Plant* **123**, 399–405.

Jimenez A, Hernández JA, del Rio LA and Sevilla F (1997) Evidence for the presence of the ascorbate-glutathione cycle in mitochondria and peroxisomes of pea leaves. *Plant Physiol* **114**, 275–284.

Johansson E, Olsson O and Nyström T (2004) Progression and specificity of protein oxidation in the life cycle of *Arabidopsis thaliana*. *J Biol Chem* **279**, 22204–22208.

Johansson FI, Michalecka AM, Møller IM and Rasmusson AG (2004) Oxidation and reduction of pyridine nucleotides in alamethicin-permeabilised plant mitochondria. *Biochem J* **380**, 193–202.

Juszczuk IM and Rychter AM (2003) Alternative oxidase in higher plants. *Acta Biochim Pol* **50**, 1257–1271.

Kambacheld M, Augustin S, Tatsuta T, Müller S and Langer T (2005) Role of the novel metallopeptidase MoP112 and saccharolysin for the complete degradation of proteins residing in different subcompartments of mitochondria. *J Biol Chem* **280**, 20132–20129.

Karpova OV, Kuzmin EV, Elthon TE and Newton KJ (2002) Differential expression of alternative oxidase genes in maize mitochondrial mutants. *Plant Cell* **14**, 3271–3284.

Kasimova MA, Krab K, Grigiene J, *et al.* (2006) The free NADH concentration is kept constant in plant mitochondria under different metabolic conditions. *Plant Cell* **18**, 688–698.

Kerscher SJ (2000) Diversity and origin of alternative NADH:ubiquinone oxidoreductases. *Biochim Biophys Acta* **1459**, 274–283.

Kristensen BK, Askerlund P, Bykova NV, Egsgaard H and Møller IM (2004) Identification of oxidised proteins in the matrix of rice leaf mitochondria by immunoprecipitation and two-dimensional liquid chromatography-tandem mass spectrometry. *Phytochemistry* **65**, 1839–1851.

Krömer S and Heldt HW (1991) Respiration of pea leaf mitochondria and redox transfer between the mitochondrial and extramitochondrial compartment. *Biochim Biophys Acta* **1057**, 42–50.

Kurahashi T, Miyazaki A, Suwan S and Isobe M (2001) Extensive investigations on oxidized amino acid residues in H_2O_2-treated Cu,Zn-SOD protein with LC-ESI-Q-TOF-MS, MS/MS for the determination of the copper-binding site. *J Am Chem Soc* **123**, 9268–9278.

Laloi C, Rayapuram N, Chartier Y, Grienenberger J-M, Bonnard G and Meyer Y (2001) Identification and characterization of a mitochondrial thioredoxin system in plants. *Proc Natl Acad Sci USA* **98**, 14144–14149.

Lemasters JJ (2005) Selective mitochondrial autophagy, or mitophagy, as a targeted defense against oxidative stress, mitochondrial dysfunction, and aging. *Rejuvenation Res* **8**, 3–5.

Lindahl M, Spetea C, Hundal T, Oppenheim A, Adam Z and Andersson B (2000) The thylakoid FtsH protease plays a role in the light-induced turnover of the photosystem II D1 protein. *Plant Cell* **12**, 419–431.

Liu F, Cui XQ, Horner HT, Weiner H and Schnable PS (2001) Mitochondrial aldehyde dehydrogenase activity is required for male fertility in maize. *Plant Cell* **12**, 1063–1078.

Matic S, Geisler D, Møller IM, Widell S and Rasmusson AG (2005) Alamethicin permeabilises plasma membrane and mitochondria but not tonoplast in tobacco (*Nicotiana tabacum*) suspension cells. *Biochem J* **389**, 695–704.

Maxwell DP, Wang Y and McIntosh L (1999) The alternative oxidase lowers mitochondrial reactive oxygen production in plant cells. *Proc Natl Acad Sci USA* **96**, 8271–8276.

Meister A (1995) Mitochondrial changes associated with glutathione deficiency. *Biochim Biophys Acta* **1271**, 35–42.

Melo AMP, Duarte M, Møller IM, *et al.* (2001) NDE1 is the external calcium-dependent NADPH dehydrogenase in *Neurospora crassa* mitochondria. *J Biol Chem* **276**, 3947–3951.

Melo AMP, Roberts TH and Møller IM (1996) Evidence for the presence of two rotenone-insensitive NAD(P)H dehydrogenases on the inner surface of the inner membrane of potato tuber mitochondria. *Biochim Biophys Acta* **1276**, 133–139.

Michalecka AM, Agius SC, Møller IM and Rasmusson AG (2004) Identification of a mitochondrial external NADPH dehydrogenase by overexpression in transgenic *Nicotiana sylvestris*. *Plant J* **37**, 415–425.

Michalecka AM, Svensson ÅS, Johansson FI, *et al.* (2003) *Arabidopsis* genes encoding mitochondrial type II NAD(P)H dehydrogenases have different evolutionary origin and show distinct responses to light. *Plant Physiol* **133**, 642–652.

Millar AH and Day DA (1996) Nitric oxide inhibits the cytochrome oxidase but not the alternative oxidase of plant mitochondria. *FEBS Lett* **398**, 155–158.

Millar AH and Leaver CJ (2000) The cytotoxic lipid peroxidation product, 4-hydroxy-2-nonenal, specifically inhibits decarboxylating dehydrogenases in the matrix of plant mitochondria. *FEBS Lett* **481**, 117–121.

Millar AH, Mittova V, Kiddle G, *et al.* (2003) Control of ascorbate synthesis by respiration and its implications for stress responses. *Plant Physiol* **133**, 443–447.

Millenaar FF and Lambers H (2003) The alternative oxidase: *in vivo* regulation and function. *Plant Biol* **5**, 2–15.

Mittova V, Tal M, Volokita M and Guy M (2003) Up-regulation of the leaf mitochondrial and peroxisomal antioxidative systems in response to salt-induced oxidative stress in the wild salt-tolerant tomato species *Lycopersicon pennillii*. *Plant Cell Environ* **26**, 845–856.

Møller IM (2001a) Plant mitochondria and oxidative stress. Electron transport, NADPH turnover and metabolism of reactive oxygen species. *Annu Rev Plant Physiol Plant Mol Biol* **52**, 561–591.

Møller IM (2001b) A more general mechanism of cytoplasmic male sterility? *Trends Plant Sci* **6**, 560.

Møller IM (2002) A new dawn for mitochondrial NAD(P)H dehydrogenases. *Trends Plant Sci* **7**, 235–237.

Møller IM, Bérczi A, Van Der Plas LHW and Lambers H (1988) Measurement of the activity and capacity of the alternative pathway in intact plant tissues: identification of problems and possible solutions. *Physiol Plant* **72**, 642–649.

Møller IM and Kristensen BK (2004) Protein oxidation in plant mitochondria as a stress indicator. *Photochem Photobiol Sci* **3**, 730–735.

Møller IM and Kristensen BK (2006) Protein oxidation in plant mitochondria detected as oxidized tryptophan. *Free Radic Biol Med* **40**, 430–435.

Moore CS, Cook-Johnson RJ, Rudhe C, *et al.* (2003) Identification of AtNDI1, an internal non-phosphorylating NAD(P)H dehydrogenase in *Arabidopsis* mitochondria. *Plant Physiol* **133**, 1968–1978.

Navrot N, Rouhier N, Gelhaye E and Jacquot J-P (in press) ROS generation and antioxidant systems in plant mitochondria. *Physiol Plant.*

Neuburger M, Day DA and Douce R (1984) The regulation of malate oxidation in plant mitochondria by the redox state of endogenous pyridine nucleotides. *Physiol Vég* **22**, 571–580.

Neuburger M, Day DA and Douce R (1985) Transport of NAD^+ in Percoll-purified potato tuber mitochondria. *Plant Physiol* **78**, 405–410.

Nicholls DG and Ferguson SJ (2002) *Bioenergetics 3*. Academic Press, Amsterdam.

O'Brien KM, Dirmeier R, Engle M and Poyton RO (2004) Mitochondrial protein oxidation in yeast mutants lacking manganese- (MnSOD) or copper- and zinc-containing superoxide dismutase (CuZnSOD). *J Biol Chem* **279**, 51817–51827.

Ostersetzer O and Adam Z (2003) Light-stimulated degradation of an unassembled Rieske FeS protein by a thylakoid-bound protease: the possible role of the FstH protease. *Plant Cell* **9**, 957–965.

Outten CE and Culotta VC (2003) A novel NADH kinase is the mitochondrial source of NADPH in *Saccharomyces cerevisiae*. *EMBO J* **22**, 2015–2024.

Paradies G, Ruggiero FM, Petrosillo G and Quagliariello E (1998) Peroxidative damage to cardiac mitochondria: cytochrome oxidase and cardiolipin alterations. *FEBS Lett* **424**, 155–158.

Petrova VY, Drescher D, Kujumdzieva AV and Schmitt MJ (2004) Dual targeting of yeast catalase A to peroxisomes and mitochondria. *Biochem J* **380**, 393–400.

Raha S and Robinson BH (2000) Mitochondria, oxygen free radicals, disease and ageing. *Trends Biochem Sci* **25**, 502–508.

Rasmusson AG, Soole KL and Elthon TE (2004) Alternative NAD(P)H dehydrogenases of plant mitochondria. *Annu Rev Plant Biol* **55**, 23–39.

Rasmusson AG, Svensson AS, Knoop V, Grohmann L and Brennicke A (1999) Homologues of yeast and bacterial rotenone insensitive NADH dehydrogenases in higher eukaryotes: two enzymes are present in potato mitochondria. *Plant J* **20**, 79–88.

Rep M and Grivell LA (1996) The role of protein degradation in mitochondrial function and biogenesis. *Curr Genet* **30**, 367–380.

Roberts TH, Fredlund KM and Møller IM (1995) Direct evidence for the presence of two external NAD(P)H dehydrogenases coupled to the electron transport chain in plant mitochondria. *FEBS Lett* **373**, 307–309.

Rodriguez Milla MA, Maurer A, Huete AR and Gustafson JP (2003) Glutathione peroxidase genes in *Arabidopsis* are ubiquitous and regulated by abiotic stresses through diverse signaling pathways. *Plant J* **36**, 602–615.

Röttgers K, Zufall N, Guiard B and Voos W (2002) The ClpB homolog Hsp78 is required for the efficient degradation of proteins in the mitochondrial matrix. *J Biol Chem* **277**, 45829–45837.

Sabar M, de Paepe R and de Kouchkovsky Y (2000) Complex I impairment, respiratory compensations, and photosynthetic decrease in nuclear and mitochondrial male sterile mutants of *Nicotiana sylvestris*. *Plant Physiol* **124**, 1239–1249.

Sazanov LA and Jackson JB (1994) Proton-translocating transhydrogenase and NAD- and NADP-linked isocitrate dehydrogenases operate in a substrate cycle which contributes to fine regulation of the tricarboxylic acid cycle activity in mitochondria. *FEBS Lett* **344**, 109–116.

Shen WY, Wei YD, Dauk M, Zheng ZF and Zou JT (2003) Identification of a mitochondrial glycerol-3-phosphate dehydrogenase from *Arabidopsis thaliana*: evidence for a mitochondrial glycerol-3-phosphate shuttle in plants. *FEBS Lett* **536**, 92–96.

Sieger SM, Kristensen BK, Robson CA, *et al.* (2005) The role of alternative oxidase in modulating carbon use efficiency and growth during macronutrient stress in tobacco cells. *J Exp Bot* **56**, 1499–1515.

Stupnikova I, Benamar A, Tolleter D, *et al.* (2006) Pea seed mitochondria are endowed with a remarkable tolerance to extreme physiological temperatures. *Plant Physiol* **140**, 326–335.

Svensson ÅS, Johansson FI, Møller IM and Rasmusson AG (2002) Cold stress decreases the capacity for respiratory NADH oxidation in potato leaves. *FEBS Lett* **517**, 79–82.

Svensson AS and Rasmusson AG (2001) Light-dependent gene expression for proteins in the respiratory chain of potato leaves. *Plant J* **28**, 73–82.

Sweetlove LJ, Heazlewood JL, Herald V, *et al.* (2002) The impact of oxidative stress on *Arabidopsis* mitochondria. *Plant J* **32**, 891–904.

Szarka A, Horemans N, Bánhegyi G and Asard H (2004) Facilitated glucose and dehydroascorbate transport in plant mitochondria. *Arch Biochem Biophys* **428**, 73–80.

Taylor NL, Day DA and Millar AH (2002) Environmental stress causes oxidative damage to plant mitochondria leading to inhibition of glycine decarboxylase. *J Biol Chem* **277**, 42663–42668.

Taylor SW, Fahy E, Murray J, Capaldi RA and Ghosh SS (2003) Oxidative post-translational modification of tryptophan residues in cardiac mitochondrial proteins. *J Biol Chem* **278**, 19587–19590.

Tobin A, Djerdjour B, Journet E, Neuburger M and Douce R (1980) Effect of NAD^+ on malate oxidation in intact plant mitochondria. *Plant Physiol* **66**, 225–229.

Todisco S, Agrimi G, Castegna A and Palmieri F (2006) Identification of the mitochondrial NAD^+ transporter in *Saccharomyces cerevisiae*. *J Biol Chem* **281**, 1524–1531.

Turner WL, Waller JC and Snedden WA (2005) Identification, molecular cloning and functional characterization of a novel NADH kinase from *Arabidopsis thaliana* (thale cress). *Biochem J* **385**, 217–223.

Turrens JF (2003) Mitochondrial formation of reactive oxygen species. *J Physiol* **552**, 335–344.

Umbach AL, Fiorani F and Siedow JN (2005) Characterization of transformed *Arabidopsis* with altered alternative oxidase levels and analysis of effects on reactive oxygen species in tissue. *Plant Physiol* **139**, 1806–1820.

Vedel F, Lalanne E, Sabar M, Chetrit P and de Paepe R (1999) The mitochondrial respiratory chain and ATP synthase complexes: composition, structure and mutational studies. *Plant Physiol Biochem* **37**, 629–643.

Vercesi AE, Borecky J, Maia IG, Arruda P, Cuccovia IM and Chaimovich H (2006) Plant uncoupling mitochondrial proteins. *Annu Rev Plant Biol* **57**, 383–404.

Wheeler GL, Jones MA and Smirnoff N (1998) The biosynthetic pathway of vitamin C in higher plants. *Nature* **393**, 365–369.

Whittaker M and Whittaker J (1998) A glutamate bridge is essential for dimer stability and metal selectivity in manganese superoxide dismutase. *J Biol Chem* **273**, 22188–22193.

Winger AM, Millar AH and Day DA (2005) Sensitivity of plant mitochondrial terminal oxidases to the lipid peroxidation product 4-hydroxy-2-nonenal (HNE). *Biochem J* **387**, 865–870.

Yamakura F, Taka H, Fujimura T and Murayama K (1998) Inactivation of human manganese-superoxide dismutase by peroxynitrite is caused by exclusive nitration of tyrosine 34 to 3-nitrotyrosine. *J Biol Chem* **273**, 14085–14089.

Zancani M, Peresson C, Biroccio A, *et al.* (2004) Evidence for the presence of ferritin in plant mitochondria. *Eur J Biochem* **271**, 3657–3664.

8 Mitochondrial metabolism

Adriano Nunes-Nesi and Alisdair R. Fernie

8.1 Introduction

Mitochondria are responsible for a wide range of metabolic processes in addition to the synthesis of ATP. This is particularly the case in autotrophic organisms such as plants, in which mitochondria provide precursors for a number of essential biosynthetic processes such as nitrogen fixation and the biosynthesis of amino acids, tetrapyrroles and vitamin co-factors (Douce, 1985; Douce and Neuberger, 1989; MacKenzie and McIntosh, 1999). The advent of genomic and post-genomic technologies has dramatically improved our understanding of the compartmentalisation and functioning of mitochondrial metabolism (Sweetlove *et al.*, 2002; Heazlewood *et al.*, 2004; Dutilleul *et al.*, 2005; Urbanczyk-Wochniak *et al.*, 2006). This is particularly true in the case of proteomics, which is well on the way to providing a complete catalogue of the mitochondrial proteome (Kruft *et al.*, 2001; Millar *et al.*, 2001; Heazlewood *et al.*, 2004), as well as underpinning investigations into the structural organisation of respiratory pathways (Eubel *et al.*, 2003; Sabar *et al.*, 2005). The current widespread availability of T-DNA insertional mutants in *Arabidopsis* is facilitating studies relating to metabolically associated proteins resident in the mitochondria (e.g. Ishizaki *et al.*, 2005). Such approaches will be vital in understanding the functional importance of metabolism in the plant cell. The approaches described above reveal that the unique demands placed on plant mitochondria are reflected by the presence of additional respiratory chain components that are not found in animal mitochondria (Moller, 1986; Vanlerberghe and McIntosh, 1997; Moller, 2001; see Chapters 5 and 7), as well as a number of plant-specific metabolite exchanges between the mitochondria and the cytosol (Douce and Neuberger, 1989; Picault *et al.*, 2002). Plant mitochondria are sites of synthesis for fatty acids (Gueguen *et al.*, 2000) and vitamin co-factors such as folate (Mouillon *et al.*, 2002), and ascorbate (Bartoli *et al.*, 2000), and have been implicated in the synthesis and export of Fe–S clusters (Kushnir *et al.*, 2001), and in the degradation of branched-chain amino acids and phytol (Ishizaki *et al.*, 2005), and fatty acids (Baker *et al.*, 2006). The pivotal importance of mitochondrial function in the physiology and development of higher plants is demonstrated by the fact that mutations of the mitochondrial genome frequently lead to cytoplasmic male sterility (CMS) (Hanson, 1991; Schnable and Wise, 1998) and by the recent implication of plant mitochondria in programmed cell death (Balk and Leaver, 2001; Robson and Vanlerberghe, 2002; Yao *et al.*, 2004; see Chapter 10). However, plant mitochondrial metabolism is dominated by reactions comprising of the TCA cycle, in addition to anabolic and catabolic processes interconnected to it, the vast majority of which are ubiquitous to all kingdoms of life. For this reason, the vast majority of our chapter will focus on understanding the structure and regulation of this essential pathway, as well as the process of oxidative phosphorylation. Our chapter will be split into two major subsections, focussing first on structural and latter on mechanistic aspects of mitochondrial metabolism. To achieve this, we will first adopt a reductionist approach, whereby we characterise current understanding of the function of individual proteins before defining their roles within a broader context. Attention will also be paid to the roles of transgenesis and high-throughput phenotyping in charaterising the response of mitochondrial metabolism to stress, and to the definition of its role in developmental processes. In addition, the clear integration of mitochondrial metabolism within cellular metabolism, and other cellular processes, will be described, and the major hurdles to advancing our understanding of mitochondrial metabolism will be discussed.

8.2 Metabolic pathways of the mitochondria

As stated above, mitochondria harbour multiple metabolic pathways; however, the metabolic activity of this organelle is dominated by the flux of carbon through the TCA cycle. Following the groundbreaking elucidation of the cycle in pigeon muscle (Krebs and Johnson, 1937), considerable evidence accumulated that the same reactions occur in plant cells (Beevers, 1961); namely, the progressive oxidation of pyruvate to carbon dioxide, and the concomitant transfer of electrons to redox co-factors, reducing four molecules of NAD^+ to NADH, and one molecule of FAD to $FADH_2$. This process takes place entirely within the mitochondrial matrix, and synthesises one molecule of ATP directly from ADP and P_i. Despite the fact that the structure of the cycle has been known for many years, and much is known about many of the functions of the cycle, little insight has been obtained into the mechanisms which regulate flux through the cycle. This is possibly due to the extensive subcellular compartmentalisation within the plant cell (ap Rees, 1987) and difficulties in extracting and measuring metabolites, and enzymes, from plant cells due to their large complement of secondary metabolites (Fiehn, 2002). Notwithstanding these technical difficulties, our understanding of the TCA cycle, and of many other mitochondrial biochemical pathways, has developed significantly over recent years as a result of the reductionist approach of characterising the properties of the pathways constituent enzymes.

8.2.1 Enzymes of the TCA cycle

The reactions of the TCA cycle are illustrated in Figure 8.1. For the purposes of this discussion, we have also included two reactions that are not strictly involved in the TCA cycle but are, however, intimately associated with it. Pyruvate dehydrogenase complex (PDC) is the main source of acetyl CoA required for TCA cycle operation and is classically regarded as the entry point of glycolytically derived carbon into the cycle. NAD-malic enzyme may also be a significant route for carbon entry into

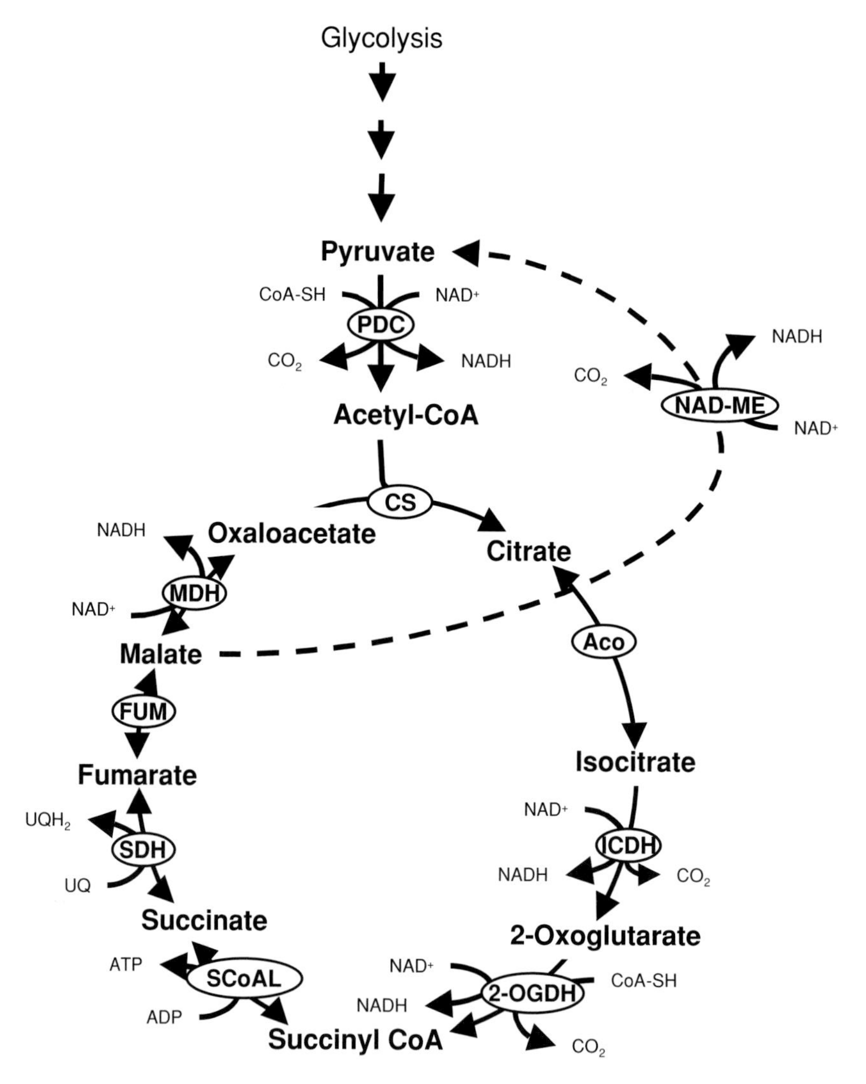

Figure 8.1 Reactions of the TCA cycle in plant mitochondria. Abbreviations: PDC, pyruvate dehydrogenase complex; CS, citrate synthase; Aco, aconitase; ICDH, isocitrate dehydrogenase; 2-OGDH, 2-oxoglutarate dehydrogenase complex; SCoAL, succinyl CoA ligase; SDH, succinate dehydrogenase; FUM, fumarase; MDH, malate dehydrogenase; NAD-ME, NAD-malic enzyme.

the cycle, particularly when mitochondrial pyruvate uptake rates are insufficient to support respiration, or when malate is the primary respiratory substrate.

8.2.1.1 Pyruvate dehydrogenase complex

Entry of carbon into the cycle is catalysed by PDC, a large multi-enzyme complex that catalyses the oxidative descarboxylation of pyruvate:

$$\text{Pyruvate} + \text{NAD}^+ + \text{CoA} \rightarrow \text{acetyl CoA} + \text{NADH} + \text{H}^+ + \text{CO}_2$$

The core of the complex consists of three enzymes. Pyruvate dehydrogenase (E1, EC1.2.4.1) catalyses the decarboxylation of pyruvate, and the transfer of the remaining acetyl moiety to the co-factor thiamine pyrophosphate (TPP). This component consists of two subunits, E1α and E1β. The second component, lipoamide acetyltransferase (E2, EC2.3.1.12), catalyses the transfer of acetyl moiety from TPP to co-enzyme A, with the concomitant reduction of a covalently bound lipoamide co-factor. The final component, lipoamide dehydrogenase (E3, EC1.8.1.4), reoxidises the lipoamide residue and converts NAD^+ to its reduced form, via a bound flavin adenine nucleotide (FAD) co-factor. The PDC reaction scheme is illustrated in Figure 8.2.

Our knowledge of the molecular organisation of mitochondrial PDC in plants is gained from analogy with PDC from other eukaryotes. The bovine complex consists of a central core of 60 E2 subunits, with 20 E1 and 5 E3 subunits arranged around the core (Wieland, 1983). In addition to the E1, E2 and E3 components described above, the complex also contains two other associated-enzyme activities, a kinase

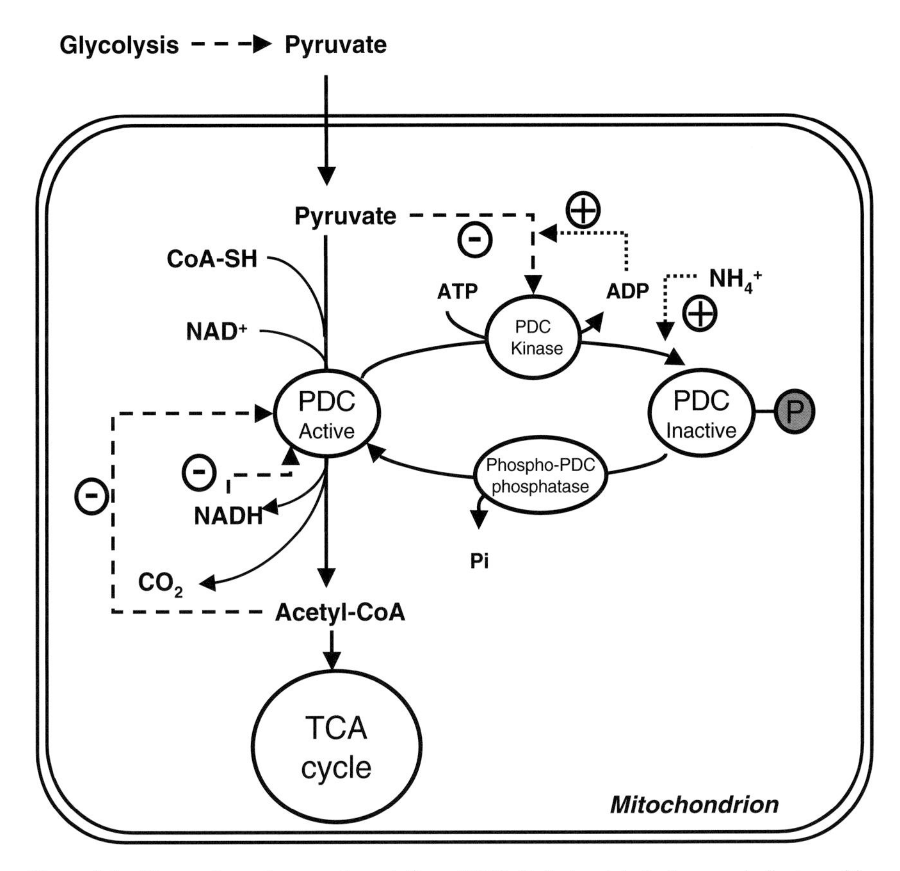

Figure 8.2 The reaction scheme and regulation of PDC. Dotted and dashed arrows indicate positive and negative regulatory interactions, respectively.

that phosphorylates the E1α subunit at specific sites (the PDC-kinase [PDK]) and a phosphatase that dephosphorylates the protein (phosphor-PDC phosphatase). These subunits are involved in the regulation of PDC activity by reversible phosphorylation. The PDC from animal mitochondria also contains a protein, termed component X, which is believed to be involved in attaching the E3 component to the E2 core. Whilst the structure of plant PDC has not been determined *in vivo*, all the components have been detected immunologically (Taylor *et al.*, 1992) and cDNA clones have been isolated that encode E1α (Grof *et al.*, 1995), E1β (Luethy *et al.*, 1994) and E2 (Guan *et al.*, 1995). In addition, a cDNA clone encoding the lipoamide dehydrogenase from pea glycine decarboxylase complex (GDC) has been isolated, and evidence presented that this gene also encodes the E3 subunit of PDC (Bourguignon *et al.*, 1992). In addition, the assembly of the *Arabidopsis* E1 complex following expression in insect cells has recently provided strong support for an $\alpha2\beta2$ heterotetramer (Szurmak *et al.*, 2003).

Regulation of mitochondrial PDC activity, by compartmentalisation, product inhibition and metabolite effectors, has been covered in previous reviews (Randall *et al.*, 1989, 1996; Luethy *et al.*, 1996). Pea mitochondrial PDC activity is also developmentally regulated (Luethy *et al.*, 1996). Mitochondrial PDC activity is highest in etiolated plants, and in the youngest leaves of light-grown plants. The activity then declines as the leaves mature and is almost absent in senescing leaves. Changes in activity correlate with the abundances of E1, and E2 mRNA, and protein (Luethy *et al.*, 1996). The expression pattern of the E3 subunit does not follow that of E1 or E2, although this is perhaps not surprising as a single E3 in pea is shared among all the α-ketoacid dehydrogenase complexes (Douce *et al.*, 2001). A number of plant species show maximum mitochondrial PDC activity early in development (Lernmark and Gardestrom, 1994; Luethy *et al.*, 1996; Thelen *et al.*, 1999), reflecting the biochemical and structural changes during membrane expansion, and remodelling.

The most striking level of regulation of mitochondrial PDC is achieved by reversible phosphorylation (Randall *et al.*, 1977); it is by this mechanism that mitochondrial PDC is inhibited in a light-dependent manner. PDK activity is stimulated by 20–80 μM NH_4^+. Stimulation by NH_4^+ is part of the mechanism of light-dependent inactivation of mitochondrial PDC (Schuller and Randall, 1989). Photorespiratory glycine metabolism produces NH_4^+ in the mitochondria, which in turn stimulates the kinase, causing mitochondrial PDC inactivation in the light; however, as described below this does not fully disrupt operation of the TCA cycle in the light. In support of this conclusion are the findings that pea plants illuminated in a low-O_2/high-CO_2 atmosphere showed reduced light-dependent inactivation of PDC, and inhibitors of photorespiration also prevented the light-dependent inactivation. The NADH from photorespiration can drive oxidative phosphorylation, conserving carbon, and making photorespiration less wasteful (Budde and Randall, 1990; Gemel and Randall, 1992).

Previous study on reversible phosphorylation used a partially purified mitochondrial PDC to study the intrinsic kinase, and phosphatase. Recently, three cDNAs for plant mitochondria PDKs have been cloned. Two mitochondria PDKs, with 77% amino acid identity, were cloned from maize (Thelen *et al.*, 1998a) and one from

Arabidopsis (Thelen *et al.*, 1998b; Zou *et al.*, 1999). Maize PDK1 is constitutively expressed, whereas maize PDK2 is upregulated in leaves, possibly allowing an acute response to high mitochondrial ATP concentrations during photosynthesis. Recombinant maize PDK2 inactivated maize PDC with concomitant phosphorylation of the E1α subunit (Thelen *et al.*, 1998a). Moreover, when a conserved histidine residue in the H-box (H121 of *Arabidopsis* PDK) was mutated, autophosphorylation of the recombinant kinase and transphosphorylation of the E1α subunit of PDC were slower, although not abolished (Thelen *et al.*, 2000), suggesting that whilst H121 might be involved in catalytic activity it was not an essential phosphotransfer histidine residue. Additionally, when all histidine residues in *Arabidopsis* PDK were chemically modified, both auto- and trans-phosphorylation were abolished (Mooney *et al.*, 2000). However, more recent data prove that neither of the two conserved histidine residues in PDK (H121 and H168 in the *Arabidopsis* sequence) are directly involved in the catalytic mechanism (Mooney *et al.*, 2002; Tovar-Mendez *et al.*, 2002, 2005). These results reveal that although PDK has sequence similarity with prokaryotic histidine kinases, it is in fact not a histidine kinase but a unique type of serine kinase (Tovar-Mendez *et al.*, 2005). The controversy surrounding interpretation of these results serves to emphasize potential problems in assigning protein function based exclusively on *in silico* analyses (Mooney *et al.*, 2002).

Both constitutive and seed-specific partial silencing of the *Arabidopsis* PDK resulted in increased seed oil content and seed weight at maturity (Marillia *et al.*, 2003). Feeding 3-C-14 pyruvate to bolts bearing siliques (constitutive transgenics), or to isolated siliques, or immature seeds (seed-specific transgenics), revealed a higher rate of incorporation of radiolabel into all seed-lipid species. These findings suggest that a partial reduction of the repression of mitochondrial PDC by antisense PDK expression can alter carbon flux and, in particular, the contribution from pyruvate to fatty acid biosynthesis and storage lipid accumulation in developing seeds, thereby implicating a role for mitochondrial PDC in fatty-acid biosynthesis in seeds (Marillia *et al.*, 2003). However, the generality of such a role for this enzyme is not yet established.

Whilst much research effort has focussed on PDK, the activity responsible for dephosphorylation/reactivation of the mitochondrial PDC, an E1-specific phosphopyruvate dehydrogenase phosphatase (PDP), is relatively poorly characterised. Mammalian PDPs are members of the PP2C class. PDP1 is an $\alpha\beta$ heterodimer consisting of both catalytic and regulatory subunits, whereas mammalian PDP2 is monomeric (Roche *et al.*, 2001). To date, no such activities have been cloned from plants, and consequently most data on plant PDP is supplied by experiments with a partially purified mitochondrial PDC. PDP requires divalent cations for activity and is potently inhibited by 10 mM Pi, whereas TCA cycle intermediates, nucleotides, NAD^+, NADH, acetyl-CoA, CoASH and polyamines had no effect in mitochondria from light-grown pea seedlings. The ratio of PDK/PDP activity is approximately 5:1 (Miernyk and Randall, 1987). Our current knowledge of the two regulatory enzymes suggests that control of the phosphorylation state of the mitochondrial PDC must be achieved through regulation of PDK; however, it does not extend to understanding how this is achieved, mechanistically, *in vivo*. That said, the multi-level regulation of mitochondrial PDC allows fine (and tight) control of its activity – a fact that highlights the importance of this enzyme to metabolism.

8.2.1.2 NAD-malic enzyme

NAD-malic enzyme (EC1.1.1.39) is located exclusively in mitochondria and catalyses the oxidative decarboxylation of malate:

$$\text{Malate} + \text{NAD}^+ \rightarrow \text{pyruvate} + \text{NADH} + \text{H}^+ + \text{CO}_2$$

The higher plant enzyme consists of two subunits – α and β – and exists in a configuration of $(\alpha\beta)_n$, as a dimmer, tetramer or octamer (Artus and Edwards, 1985). Both the α and β subunits are required for activity (Willeford and Wedding, 1987). Complemenrary DNA clones encoding these subunits have been isolated from potato (Winning *et al.*, 1994), confirming the existence of two separate subunits. The subunits from potato have molecular weights of 59 and 62 kDa and show 65% identity in their amino acid sequence (Winning *et al.*, 1994). The kinetic properties of NAD-malic enzyme are complex. Activity of the enzyme is dependent on the presence of a divalent cation, either Mn^{2+} or Mg^{2+}. Divalent cation identity affects both the V_{max} and K_m of the enzyme for both malate and NAD^+; for example, in the presence of Mn^{2+} both K_m(malate) and V_{max} are lower than in the presence of Mg^{2+} (Grover *et al.*, 1981). Sulphate ions, CoA and fumarate activate NAD-malic enzyme, whilst chloride is an inhibitor (Grissom *et al.*, 1983; Canellas and Wedding, 1984; Day *et al.*, 1984) and *in vitro* studies have provided evidence that both subunits are required for activity (Grover and Wedding, 1984). Although the reaction catalysed by NAD-malic enzyme is reversible, its equilibrium position likely renders the reverse reaction metabolically insignificant (Wedding, 1989). $^{14}CO_2$-labelling experiments have provided evidence that the decarboxylation of malate by NAD-malic enzyme occurs *in vivo* (Bryce and ap Rees, 1985). However, the importance of this enzyme in determining respiratory flux, and even the physiological significance of the above *in vitro* mechanisms, are largely unknown.

Attempts to measure the extent of flux through NAD-malic enzyme *in vivo* suggest that this varies greatly from one plant organ to another. In maize kernels (Day and Hanson, 1977), the flux through NAD-malic enzyme is approximately 30% of that through pyruvate kinase, whereas in root tips and potato tubers it is at maximum 3% of the flux to pyruvate through pyruvate kinase (Dieuaide-Noubhani *et al.*, 1995; Edwards *et al.*, 1998; Jenner *et al.*, 2001). When potato plants were transformed with a cDNA encoding the 59-kDa subunit of the potato tuber NAD-malic enzyme in the antisense orientation, the transformants exhibited a range of reductions in the activity of this enzyme, down to 40% of wild-type activity (Jenner *et al.*, 2001). However, surprisingly there were no detrimental effects on plant growth or tuber yield despite biochemical analyses of developing tubers indicating that a reduction in NAD-malic enzyme activity had an effect on glycolytic metabolism, with significant increases in the concentration of 3-phosphoglycerate and phosphoenolpyruvate. However, there was no detectable effect on flux through the TCA cycle. Despite the universal presence of NAD-malic enzyme in higher plants, and its position in mitochondrial metabolism, the research described above demonstrates that under defined conditions, plants are able to cope with substantial reductions in the activity of NAD-malic enzyme without any detrimental effects on growth and development. A surprising discovery was that tubers with reduced NAD-malic enzyme activity have increased starch content. That said, it should be noted that this increase is minor with respect to others reported recently within the tuber (Regierer *et al.*, 2002; Geigenberger *et al.*, 2005). A recent genome-wide survey of malic enzyme isoforms in the *Arabidopsis* genome (Wheeler *et al.*, 2005) provides a framework in which their function can be systematically analysed.

8.2.1.3 Citrate synthase

Citrate synthase (EC4.1.3.7) catalyses the aldol condensation of acetyl CoA and oxaloacetate to form citroyl CoA, which spontaneously hydrolyses to form citrate and CoA. The protein has been purified from pea leaves and consists of a single subunit of 50 kDa (Unger and Vasconcelos, 1989). This enzyme, by comparison with citrate synthase from animal sources, is thought to exist as a homodimer with a molecular mass of 100 kDa. Citrate synthase is confined to the mitochondrial matrix, except in tissues converting fatty acids into sugars where there is also a glyoxysomal isoform (Pracharoenwattana *et al.*, 2005). Complementary DNA clones encoding mitochondrial citrate synthase have been isolated from a range of species including *Arabidopsis* (Unger *et al.*, 1989), potato (Landschütze *et al.*, 1995a), strawberry (Iannetta *et al.*, 2004) and pumelo (Canel *et al.*, 1996). Analysis of the *Arabidopsis* genome suggests that there are two genes, but only one of these is mitochondrially localised. Structural and expression characteristics of mitochondrial isoforms from diverse species have been subject to detailed study (Landschütze *et al.*, 1995a; LaCognata *et al.*, 1996), and the identification of potential inter-domain disulfides in higher plant mitochondrial citrate synthases suggest paradoxical differences in redox-sensitivity relative to the animal enzyme (Stevens *et al.*, 1997). Morevover, functional studies have suggested that citrate synthase has an important role in floral development (Landschütze *et al.*, 1995b); antisense plants are characterised by a specific disintegration of ovary tissue. Other studies suggest that citrate synthase is of key importance in the process of acid excretion, which facilitates nutrient uptake (de la Fuente *et al.*, 1997; Delhaize *et al.*, 2003). These functions are discussed in more detail in later sections.

8.2.1.4 Aconitase

Aconitase (EC4.2.1.3) catalyses the isomerisation of citrate into isocitrate, via the bound intermediate *cis*-aconitate. Aconitase activity is found in both the mitochondrial and cytosolic compartments in plants, with approximately equal activities in each compartment (Brouquisse *et al.*, 1987). The enzyme appears to be absent from peroxisomes, suggesting that cytosolic aconitase is involved in the glyoxylate cycle (Courtois-Verniquet and Douce, 1993). The cytosolic and mitochondrial forms of the enzyme have been separated from *Acer* suspension cells and have similar kinetic, and physical, properties (Brouquisse *et al.*, 1987). Etiolated pumpkin cotyledons contain three isoforms, again with very similar properties, althought in this case it is not clear with which compartment each form is associated (De Bellis *et al.*, 1993). Two aconitase isoforms with the same N-terminal amino acid sequence have been isolated from melon cotyledons (Peyret *et al.*, 1995). Aconitase cDNA and genomic DNA clones have been isolated from *Arabidopsis* libraries using antibodies raised against the proteins from melon. This analysis has revealed that in *Arabidopsis*, there is only one gene encoding aconitase, raising the possibility that a single gene encodes the mitochondrial and cytosolic isoforms (Peyret *et al.*, 1995). Confirmation that this is the case awaits demonstration that there is more than one isoform of aconitase in *Arabidopsis*. In yeast, there is a single gene that encodes both the cytosolic and mitochondrial isoforms; it has been suggested that import of the gene product into mitochondria is inefficient so that some activity remains in the cytosol (Gangloff *et al.*, 1990). Aconitase consists of a single subunit of 90–100 kDa containing an Fe–S cluster in the active site (Verniquet *et al.*, 1991; De Bellis *et al.*, 1993). This cluster has distinct properties from that found in mammalian aconitase and may be either an Fe_3S_4 cluster, or the more normal Fe_4S_4 cluster with an associated Fe centre (Jordanov *et al.*, 1992). Aconitase shows Michaelis-Menton kinetics with respect to citrate: the K_mS for the *Acer* enzymes are 130 and 120 μM for the mitochondrial and cytosolic forms, respectively. In contrast, two of the three forms found in pumpkin cotyledons have K_mS of 480 and 370 μM. Wild species tomato (*Solanum pennellii*) plants bearing a genetic lesion in the gene encoding aconitase (*Aco-1*) exhibit a restricted flux through the TCA cycle, and reduced levels of TCA cycle intermediates but were characterized by elevated adenylate levels, and an enhanced rate of CO_2 assimilation. Furthermore, the analysis of both steady-state metabolite levels and metabolic fluxes revealed that this tomato accession also exhibited elevated rates of photosynthetic sucrose synthesis, and a corresponding increase in fruit yield. Crystal structures exist for both human (Dupuy *et al.*, 2006) and *Escherichia coli* enzymes (Williams *et al.*, 2002), and these offer some support to the metabolon theory of metabolite channelling first put forward by Srere (1987).

8.2.1.5 Isocitrate dehydrogenase

Isocitrate dehydrogenase catalyses the oxidative decarboxylation of isocitrate:

$$\text{Isocitrate} + \text{NAD(P)}^+ \rightarrow \text{2-oxoglutarate} + \text{NAD(P)H} + \text{H}^+ + \text{CO}_2$$

Plants contain both NAD^+- and $NADP^+$-specific isoforms of isocitrate dehydrogenase (EC1.1.1.41 and EC1.1.1.42, respectively), with the NAD^+ form being specific to mitochondria. $NADP^+$-specific forms are found in the plastid, and the cytosol, and recent evidence suggests that there is also a mitochondrial isoform. A $NADP^+$-specific form co-purifies with mitochondria isolated from potato tubers, and has been shown to be contained within the mitochondrial matrix (Rasmusson and Moller, 1991). This enzyme can be separated from the NAD^+-specific form by ammonium sulphate fractionation, demonstrating that this activity represents a distinct enzyme rather than a lack of specificity of the NAD^+-dependent form. The activity of the $NADP^+$-specific enzyme is around 15% of that of the NAD^+-specific form in potato tuber mitochondria (Rasmusson and Moller, 1990), whilst mitochondria isolated from pea leaves contain equivalent activities of the NAD^+ and $NADP^+$ forms (McIntosh and Oliver, 1992). It has been suggested that the $NADP^+$-dependent form is involved in the provision of NADPH required for the reduction of glutathione, which protects against oxidative damage in mitochondria (Rasmusson and Moller, 1990).

The kinetics of NAD^+-isocitrate dehydrogenase are sigmoidal with respect to isocitrate, with $S_{0.5}$ of 0.2 mM and 0.025 mM for the potato and pea enzymes, respectively. The kinetics with respect to NAD^+ are Michaelis-Menton, and the K_mS are 0.1 mM for the potato enzyme and between 0.2 and 0.8 mM for the pea-leaf enzyme. The enzyme isolated from pea-leaf mitochondria demonstrates product inhibition by NADH ($K_i = 0.2$ mM) and is also inhibited by NADPH ($K_i = 0.3$ mM; McIntosh and Oliver, 1992). The latter inhibition is not competitive with NAD^+ or NADH, suggesting that NADPH may be an allosteric regulator; the K_i is within the physiological range of mitochondrial NADPH concentrations (Wigge *et al.*, 1993). In contrast to the enzyme from non-plant sources, the activity of which is affected by ATP, ADP and AMP, pea NAD^+-ICDH is unaffected by adenylates (McIntosh and Oliver, 1992). The $NADP^+$-dependent isocitrate dehydrogenase from potato tuber mitochondria shows Michaelis-Menton kinetics with respect to both $NADP^+$ ($K_m = 5$ μM) and isocitrate ($K_m = 10$ μM; Rasmusson and Moller, 1990). The K_m for isocitrate is low relative to the $S_{0.5}$ of NAD^+-isocitrate dehydrogenase, but this may not be of physiological significance; in isolated mitochondria the isocitrate concentration is of the order of 2 mM (MacDougall and ap Rees, 1991), sufficient to saturate both forms of isocitrate dehydrogenase.

NAD^+-isocitrate dehydrogenase has been purified to homogeneity from pea leaves (McIntosh and Oliver, 1992). It consists of a single subunit of molecular mass 47 kDa. The freshly isolated enzyme is found in three forms corresponding to an octamer (the major active species), and complexes of two and four octamers. If the preparation is freeze-thawed, NAD^+-isocitrate dehydrogenase disaggregates forming a tetramer, and a dimmer, which both retain activity. This disaggregation correlates with decreased sigmoidicity of the isocitrate kinetics and can be reversed by the addition of citrate.

Mitochondrial isocitrate dehydrogenase has not yet been purified from a plant source and its functional role remains obscure despite studies in maize, potato, sunflower and Norway spruce (Curry and Ting, 1976; Rasmusson and Moller, 1990; Attucci *et al.*, 1994; Cornu *et al.*, 1996). However, a possible role in the production of NADPH for redox-regulation of cellular metabolism has recently been suggested (Moller, 2001; Hodges *et al.*, 2003). Gray *et al.* (2004) employed a range of approaches to elucidate the role of mitochondrial isocitrate dehydrogenase in the mediation of cysteine reduction, and subsequent covalent modification of the mitochondrial alternative oxidase (AOX). First, the contributions of the isocitrate dehydrogenase isoenzymes in stable–transformed suspension cells containing reduced levels of AOX were examined. Second, the enzymatic activity of mitochondrial isocitrate dehydrogenase was determined in a developmental system where the regulation of alternative pathway respiration has been extensively characterised. Finally, tobacco plants were transformed to overexpress a cDNA encoding mitochondrial isocitrate dehydrogenase. These data support the hypothesis that mitochondrial isocitrate dehydrogenase may be a regulatory switch involved in TCA-cycle flux, and the reductive modulation of AOX. However, the exact mechanism by which this is achieved remains unclear. In addition, various isocitrate dehydrogenases have been implicated in nitrate assimilation (Hodges, 2002; Hodges *et al.*, 2003) as detailed in the following sections.

8.2.1.6 2-oxoglutarate dehydrogenase complex

2-oxoglutarate dehydrogenase (2-OGDH; previously known as α-ketoglutarate dehydrogenase) catalyses a reaction that is essentially analogous to that carried out by PDC:

$$\text{2-oxoglutarate} + NAD^+ + CoA \rightarrow \text{succinyl CoA} + NADH + H^+ + CO_2$$

The complex from animal mitochondria has a similar structure and organisation to PDC, differing primarily in the nature of the E1 subunit. In particular, 2-OGDC is not regulated by reversible phosphorylation. As mentioned previously, there is evidence to suggest that the E3 subunit (lipoamide dehydrogenase) is shared between PDC, 2-OGDH and glycine decarboxylase in plants (Bourguignon *et al.*, 1992). The enzyme has been purified from cauliflower florets and has a native molecular mass of around 2 MDa (Karam and Bishop, 1989). In common with other TCA cycle dehydrogenases, 2-OGDH is inhibited by NADH, which acts competitively with NAD^+ (Pascal *et al.*, 1990). The partially purified enzyme from potato tubers has a higher K_m for NAD^+ than the K_i for NADH, suggesting that the activity is particularly sensitive to the NADH/NAD ratio (Hill, 1997). The 2-OGDH isolated from cauliflower mitochondria is activated by 1 mM AMP (Wedding and Black, 1971; Millar *et al.*, 1999a). The binding of AMP to 2-OGDH E1 lowers the K_m for 2-oxoglutarate and increases the V_{max} (Craig and Wedding, 1980a,b). Furthermore, the mammalian enzyme has been proposed to be under exquisite control both at allosteric and redox-mediated levels (Strumilo, 2005). The reaction catalysed by 2-OGDC is likely regulated by the activities of mitochondrial PDC, and citrate synthase through competition for the mitochondrial pool of CoASH (Dry and Witsche, 1987). Like all the α-ketoacid dehydrogenase complexes, the 2-OGDH complex is substrate specific. The rate of NAD^+ reduction by potato 2-OGDH using pyruvate and isovalerate is less than 2% of the rate with 2-oxoglutarate (Millar *et al.*, 1999a). The mammalian 2-OGDH complex also has a molecular mass of approximately 2 MDa, with the E1 and E3 homodimers arranged around the 24mer E2 core (Koike and Koike, 1976). 2-OGDH from potato has an S coefficient of 25 $\pm$ 2, corresponding to a molecular mass of approximately 1.7 MDa. 2-OGDH is significantly smaller than mitochondrial PDC as assessed by gel-filtration chromatography (Millar *et al.*, 1999b). Although the structure of plant 2-OGDH E2 has not been studied to any significant extent, and the fact that 2-OGDH is the least well understood of all the plant complexes, the similarity in molecular weight of the mammalian and plant complexes suggests conservation of the E2 24mer structure (Hill, 1997). However, the recent publication of specific inhibitors of microbial 2-OGDH (Santos *et al.*, 2006), alongside functional gene annotation efforts (A. Sienkiewicz and A.R.

Fernie, unpublished data) should catalyse attempts to gain more information on this enzyme.

8.2.1.7 *Succinyl-CoA ligase*

Succinyl-CoA ligase (EC6.2.1.5) catalyses the hydrolysis of succinyl CoA, and the concomitant synthesis of ATP:

$$\text{Succinyl CoA} + \text{ADP} + \text{P}_\text{i} \rightarrow \text{succinate} + \text{ATP}$$

Little is known about the plant enzyme. All succinyl-CoA ligases examined consist of two types of subunit: α, with a molecular mass of 29–34 kDa, and β, with a molecular mass of 41–45 kDa. The eukaryotic enzyme is an $\alpha\beta$ dimmer. There is now evidence that mammalian cells contain two ligases, one specific for ADP and one for GDP, and that the latter catalyses the synthesis of succinyl CoA during ketone body metabolism. Thus, we can expect plants to contain only the ADP-specific enzyme, and this expectation holds for the few plant enzymes studied (Palmer and Wedding, 1966).

Two tomato cDNAs encoding α-subunits and one encoding a β-subunit of succinyl CoA ligase complemented budding yeast (*Saccharomyces cerevisiae*) mutants deficient in the respective subunits, demonstrating that the tomato cDNAs encode functionally active polypeptides (Studart-Guimarães *et al.*, 2005). In tomato, the three genes were expressed in all tissues, but most strongly in floral and leaf tissues, with the two α-subunit genes being expressed to equivalent levels in all tissues. Subcellular localisation of green fluorescent protein (GFP) fusions confirmed the expected mitochondrial location of all three subunits. Following the development of a novel assay to measure succinyl CoA ligase activity, in the direction of succinate formation, evaluation of the maximal catalytic activities of the enzyme in a range of tissues revealed that these paralleled steady-state mRNA abundance in the same tissues (Studart-Guimarães *et al.*, 2005). Studart-Guimarães *et al.* (2005) also utilised this assay to perform a preliminary characterisation of the regulatory properties of the enzyme, which suggested that allosteric control of this enzyme may regulate flux through the TCA cycle in a manner consistent with its position therein; metabolites upstream of succinyl-CoA ligase appear to activate it, and those downstream inhibit it.

8.2.1.8 *Succinate dehydrogenase*

Succinate dehydrogenase (EC1.3.99.1) is a component of both the TCA cycle, and, as complex II, the mitochondrial electron transport chain (ETC). It catalyses the conversion of succinate into fumarate, with the hydrogen atoms generated being passed to ubiquinone.

$$\text{Succinate} + \text{UQ} \rightarrow \text{fumarate} + \text{UQH}_2$$

Electrons and hydrogen ions are passed to ubiquinone via a bound FAD co-factor. The enzyme isolated from sweet potato root tissue exists in two forms of identical molecular masses but with distinct charges (Hattori and Asahi, 1982). Both forms consist of two subunits of molecular mass 65 and 26 kDa – the former contains covalently bound FAD. By analogy with the enzyme purified from animal systems, it is likely that a number of smaller subunits are weakly associated with the two large subunits and the 26-kDa subunit spans the inner-mitochondrial membrane. The K_m for the sweet potato enzyme for succinate is 290 μM. Mitochondrial respiratory complex II contains four subunits: a flavoprotein (SDH1), an Fe–S subunit (SDH2) and two membrane anchor subunits (SDH3 and SDH4). In *Arabidopsis*, SDH1 and SDH3 are encoded by two and SDH4 by one nuclear gene, respectively (Figueroa *et al.*, 2002). All four subunits are imported into isolated plant mitochondria. Whilst both SDH1 proteins are highly conserved relative to their homolgues in other organisms, SDH3 and SDH4 share little similarity with non-plant homologues. *SDH1-1*, *SDH3* and *SDH4* gene expression was detected in all tissues analysed, with the highest steady-state mRNA abundances found in open flowers and inflorescences of unopened buds. In contrast, the second SDH1 gene (SDH1-2) is expressed at a low level (Figueroa *et al.*, 2002). SDH2 is encoded by a single-copy nuclear gene in mammals and fungi and by a mitochondrial gene in some protists. In *Arabidopsis*, the SDH2 orthologs are encoded by three nuclear genes, each containing the essential conserved sequences required for function (Figueroa *et al.*, 2001).

The regulation of potato tuber SDH has been investigated using tightly coupled mitochondria by simultaneously measuring the oxygen uptake rate, and the ubiquinone (Q) reduction level (Affourtit *et al.*, 2001). These studies revealed that the activation level of the enzyme is unambiguously reflected in the kinetic dependence of the succinate oxidation rate upon the Q-redox poise. Kinetic results indicated that SDH is activated by both ATP ($K_{1/2} \sim 3$ μM) and ADP (Affourtit *et al.*, 2001). The carboxyatractyloside insensitivity of these stimulatory effects indicated that they occur at the cytoplasmic side of the mitochondrial inner membrane. Importantly, this approach also revealed that the enzyme is also activated by oligomycin ($K_{1/2} \sim 16$ nM). Time-resolved kinetic measurements of SDH activation by succinate revealed that the activity of the enzyme is negatively affected by potassium. In addition, the succinate-induced activation (± potassium ions) is prevented by the presence of the uncoupler. Considered together, these results demonstrate that the activity of succinate dehydrogenase *in vitro* is modulated by the protonmotive force, leading the authors to speculate that the widely recognised activation of the enzyme by adenine nucleotides in plants is mediated in this manner (Affourtit *et al.*, 2001).

8.2.1.9 Fumarase

This enzyme (EC4.2.1.2) catalyses the addition of water across the double bond of fumarate, forming malate. The addition is stereospecific so that only the L-form of malate is produced. Little is known about the plant enzyme, except that it is confined to the mitochondrial matrix. The enzyme from pig heart is a homotetramer with subunits of 48.5 kDa.

Carbon flux through the TCA cycle can follow two different routes. Fumarase is, however, required for both the normal state, characterised by the decarboxylative energy-producing reactions of the cycle, and the carbon-conserving state during which carbon is shunted through the glycoxylate cycle (Canvin and Beevers, 1961; Cornah *et al.*, 2004; Pracharoenwattana *et al.*, 2005). In early stages of oil-seed germination, the majority of carbon flux has been reported to be via the glycoxylate cycle, and the respective expression, and activity of fumarase, and NAD-dependent isocitrate dehydrogenase have been postulated to be diagnostic for this switch (Falk *et al.*, 1998). Moreover, fumarase activities have been reported to be very high in guard cells relative to mesophyll cells in both broad bean and pea (Hampp *et al.*, 1982; Vani and Raghavendra, 1994). The allosteric properties of purified pea fumarase, which revealed inhibition by physiological concentrations of pyruvate, 2-oxogluterate and the adenine nucleotides ATP, ADP, and AMP, are consistent with this step being an important control point in the cycle (Behal and Oliver, 1997). Furthermore, the importance of the fumarase reaction has recently been demonstrated in correlative studies showing that this enzyme activity is a reliable diagnostic for the degree of seed dormancy in seeds of several tree species (Shen and Odén, 2002). Moreover, studies with cultured carrot cells suggest that fumarase can be exuded from cells and, at least under these conditions, is important for the utilisation of extracellular malate (Kim and Lee, 2002).

There has been very little analysis of fumarase at the molecular genetic level. Although genes encoding fumarase have been cloned from many plant species including *Arabidopsis*, potato, apple, rice and soybean, only a single, cursory, transgenic analysis, restricted to analysis of the protein content, has been carried out to date (Nast and Müller-Röber, 1996).

8.2.1.10 Malate dehydrogenase

The TCA cycle is completed by malate dehydrogenase (MDH, EC1.1.1.37), which catalyses the oxidation of malate to form oxaloacetate. The equilibrium position favours malate and NAD^+, but *in vivo* the removal of oxaloacetate by citrate synthase, and removal of NADH by the respiratory chain, causes the reaction to function in the direction of malate oxidation in most tissues. In addition to the mitochondrial forms, there are isozymes of NAD^+-MDH in the cytosol and peroxisomes, and an $NADP^+$-utilising form in plastids. The mitochondrial enzyme isolated from watermelon is a homodimer of 38-kDa subunits (Walk *et al.*, 1977). Genetic studies have indicated that there are three genes encoding mitochondrial subunits of malate dehydrogenase in maize (Newton and Schwartz, 1980). This is supported by the purification of six distinct isoforms of MDH from maize mitochondria; three subunits, with molecular masses of 37, 38 and 39 kDa, have been shown to exist as the three possible homodimers, and the three possible heterodimers (Hayes *et al.*, 1991). The majority of the MDH activity is found in enzymes containing the 39-kDa subunit, suggesting that this subunit plays the major catalytic role *in vivo*. Detailed kinetic analyses of these isoforms have yet to be carried out, and it is not yet clear whether the complexity observed in maize is found in other species. However, molecular genetic and simple enzymatic analysis has been carried out. Higher plants contain multiple forms of MDH that differ in co-enzyme specificity and subcellular localisation (Gietl, 1992; Miller *et al.*, 1998). Chloroplasts contain an NADP-dependent MDH that plays an important role in balancing redox equivalents between the cytosol and the stroma. Plants also contain NAD-dependent MDHs in the mitochondria as a component of the Krebs cycle, in the cytosol and peroxisomes where they function in malate–aspartate shuttles, and in glyoxysomes wherein they function in β-oxidation (Gietl, 1992; Miller *et al.*, 1998). The mitochondrial MDH is believed to be important not only in oxidising NADH but also as a component of the malate–aspartate shuttle for the exchange of substrate, and reducing equivalent across the mitochondrial membrane (Gietl, 1992; Scheibe, 2004). This exchange is of particular importance during photorespiration when it is responsible for the export of malate from the mitochondria (see below for details). However, direct studies on the role of the mitochondrial MDH were attempted only recently when an antisense approach in tomato produced plants with higher levels of ascorbate, and a more efficient photosynthesis relative to wild type (Nunes-Nesi *et al.*, 2005). However, given that it is highly unlikely for the *in vivo* function of a protein to be the restriction of primary energy metabolism, these transgenic lines require further testing under suboptimal growth conditions to define the important of mitochondrial MDH within cellular metabolism.

8.3 Mitochondrial electron transport

The above sections detail the TCA cycle in its entirety; however, for a fuller understanding of cycle function, these reactions must be studied in the context of mitochondrial, and even cellular, metabolism as a whole. For this reason, the following sections detail the next stage of respiration, mitochondrial electron transport as well as the anabolic and catabolic process that the TCA cycle fuels, or is fuelled by, respectively. In addition, the communication between mitochondrial and extramitochondrial metabolism as mediated by transport proteins of the inner-mitochondrial membrane is discussed.

The plant mitochondrial ETC has been studied in many species including potato, pea, spinach and soybean with a large portion of this research focused on plant-specific bypasses of the classic respiratory chain (Douce and Neuberger, 1989; Fernie *et al.*, 2004; Millar *et al.*, 2004a). These bypasses include the rotenone-insensitive NADH dehydrogenases (see Chapter 7) and AOX. Biochemical analysis in several plant species has provided considerable insight into the components of the classical ETC (Douce *et al.*, 1977; Hoefnagel *et al.*, 1995; Millar *et al.*, 1998). However, the scarcity of gene sequence information for these species has limited the degree to which these preparations could be compared with the well-characterised complexes in mammals, and fungi (Millar *et al.*, 2004b). However, tools developed for genomic and post-genomic analysis of *Arabidopsis* have recently facilitated detailed insight into both classical and plant-specific components of respiration.

8.3.1 Complex I

Complex I, NADH-ubiquinone oxireductase, is composed of both nuclear- and mitochondrial-encoded subunits, which require strictly coordinated expression to form an active complex with the correct stochiometry of subunits (Rasmusson *et al.*, 1998). *Arabidopsis* contains approximately 30 putative proteins with high similarity to those of mammalian complex I (Rasmusson *et al.*, 1998; Heazlewood *et al.*, 2003a), whilst up to 30 subunits were separated by SDS-PAGE in potato, broad bean and wheat. Blue native-PAGE separation of complex I components identified a set of 20 of these and highlighted a range of novel complex I proteins in plants (Heazlewood *et al.*, 2003a). The newly identified components include a series of ferripyochelin-like proteins, and intriguingly the final enzyme in ascorbate synthesis pathway in plants, galactonolactone dehydrogenase (Bartoli *et al.*, 2000). Subsequent experiments implied that ascorbate synthesis by *Arabidopsis* mitochondria is sensitive to complex I function because both are strongly inhibited by rotenone (Millar *et al.*, 2003). This implication was recently confirmed by experiments revealing that mitochondria isolated from plants displaying a reduced activity of mitochondrial MDH were, by oxidising L-galacto-1,4-lactone, able to maintain respiratory rates comparable to mitochondria isolated from the wild type (Nunes-Nesi *et al.*, 2005). The transgenic lines from which these mitochondria were isolated were additionally characterised by elevated levels of ascorbate, and increased rates of photosynthesis (discussed in detail in Section 8.7). Of all the complexes of the cytochrome pathway, complex I is probably the best characterised (Dutilleul *et al.*, 2003, 2005). The *Nicotiana sylvestris* cytoplasmic male sterile mutant, CMS, which lacks the mitochondrial gene *nad7* and as such a functional complex I, is associated with a profound modification of foliar carbon-nitrogen balance. CMS leaves are characterised by a greater abundance of total free amino acids relative to either wild-type plants or CMS plants in which complex I function has been restored by nuclear transformation with the nad7 cDNA. The metabolite profile of CMS leaves is enriched in amino acids with low carbon/nitrogen and depleted in starch and 2-oxoglutarate. Deficiency in 2-oxoglutarate occurred despite increased citrate, and malate, and a higher capacity of key anaplerotic enzymes, notably the mitochondrial NAD-dependent isocitrate dehydrogenase. Analysis of pyridine nucleotides in this line revealed that both NAD and NADH were increased twofold in CMS leaves. The growth retardation of CMS relative to the wild type was highly dependent on photoperiod, but the link between high amino acid and high NADH content was observed under all photoperiodic regimes. Together, the data provide strong evidence that NADH availability is a critical factor in influencing the rate of nitrate assimilation and NAD status plays a crucial role in co-ordinating ammonia assimilation with the anaplerotic production of carbon skeletons. In addition, these data suggest the general importance of a functional complex I in co-ordinating mitochondrial, and indeed cellular, metabolism (Dutilleul *et al.*, 2005).

8.3.2 Complex II

Complex II, succinate dehydrogenase, is an enzyme that participates in both the citric acid cycle (see Section 9.2.1.8) and the respiratory ETC. Using blue native-PAGE, Eubel *et al.* (2003) successfully separated an intact SDH complex from *Arabidopsis* mitochondrial membranes. This complex contained SDH1–3 as well as four other proteins that comigrated with the SDH complex (Eubel *et al.*, 2003; Millar *et al.*, 2004b). Thus, a novel composition of SDH seems very likely in plants, although no functional information for these additional subunits is available as yet.

8.3.3 Complex III

Complex III, ubiquinone-cytochrome *c* oxidoreductase, consists of 10 subunits, including the bifunctional matrix processing peptidase proteins (Millar *et al.*, 2004b). In blue native-PAGE separations from *Arabidopsis* mitochondria, most of these subunits were identified and linked back to a set of single-copy nuclear genes (Kruft *et al.*, 2001; Werhahn and Braun, 2002), with only one subunit of this complex, cytochrome *b* (COB), encoded in the *Arabidopsis* mitochondrial genome (Unseld *et al.*, 1997).

8.3.4 Complex IV

Complex IV, cytochrome *c* oxidase, contains 13 distinct subunits in mammals. Three of these are mitochondrial encoded whilst 10 are nuclear encoded. Early purifications from plants found only 7 or 8 subunits in the plant complex (Peiffer *et al.*, 1990), but more recently, Eubel *et al.* (2003) separated a COX complex containing up to 12 protein bands from *Arabidopsis*, proposed the identity of COX1 and COX3 based on size, and identified COX2 and COX5b by mass spectrometry. A series of other proteins from the purified *Arabidopsis* COX complex awaits identification (Millar *et al.*, 2004a).

8.3.5 Complex V

Complex V is the membrane-bound F_1F_O-type H^+-ATP synthase of mitochondria that catalyses the terminal step in oxidative respiration, converting the electrochemical gradient into ATP for cellular biosynthesis. The general structure and the core subunits of the enzyme are highly conserved across organisms (see Chapter 6). Its conserved structure is composed of the hydrophilic F_1 domain containing the nucleotide-binding site, and the F_O domain, which channels protons through the membrane. The plant mitochondrial ATP synthase also displays the classical F_1 five-subunit structure for the F_1 domain with identifiable homologues of all five of the F_1 subunits within the *Arabidopsis* mitochondrial and nuclear genomes. The number of subunits associated with the F_O complex, however, varies between plants, mammals and yeast. *Arabidopsis* has orthologs for the F_O-core subunits 6, and 9, together with the peripheral components oligomycin sensitivity conferring protein (OSCP, sometimes referred to as δ' in plants) and subunit d. Using blue native-PAGE and mass spectrometry, Heazlewood *et al.* (2003b) characterized the protein components of the entire ATP synthase complex in *Arabidopsis.* Furthermore, Healzelwood and colleagues identified two additional proteins which had not previously been characterised as associated with the ATP synthase complex and which seem likely to represent the plant equivalents of the F_O components,

AL6 (or subunit 8) and the b subunit (or subunit 4). These two proteins match the predicted products of the mitochondrial *orfB* and *orf25* genes, respectively. Confirmation of these identifications is provided by the discovery of sunflower ORFB as a putative ATP8 subunit (Sabar *et al.*, 2003) and identification of ORF ymf39 (an orf25 orthologue) as a probable ATP4 subunit in protists (Burger *et al.*, 2003).

8.3.6 Alternative oxidase

The alternative terminal oxidase in higher plant mitochondria was initially identified as a thermogenic curiosity during anthesis (reviewed in Meeuse, 1975; Skubatz *et al.*, 1991). It has since been recognized as an important part of the machinery responsible for the regulation of energy/carbon balance in response to a changing environment (McIntosh, 1994; Van der Straeten *et al.*, 1995). The alternative pathway of mitochondrial respiration branches from the cytochrome pathway in the inner-mitochondrial membrane at the ubiquinone pool and passes electrons to a single terminal oxidase, AOX (Figure 8.3). AOX apparently reduces molecular oxygen to water in a single four-electron transfer step (for review, see Day *et al.*, 1991; Moore and Siedow, 1991; Siedow and Moore, 1993; Day and Wiskich, 1995). The alternative pathway is non-phosphorylating. AOX is resistant to cyanide, and inhibited by substituted hydroxamic acids such as salicylhydroxamic acid and *n*-propyl gallate (Schonbaum *et al.*, 1971; Siedow and Bickett, 1981). All angiosperms, many algae and some fungi contain the genetic capacity to express this pathway (Henry and Nyns, 1975; Ordentlich *et al.*, 1991; McIntosh, 1994). In *Arabidopsis*, AOX is encoded by five genes – four *Aox1*-type and one *Aox2*-type (Considine *et al.*, 2002). The expression of the different genes varies with organ type, developmental age and environmental condition (Saisho *et al.*, 1997, 2001; Thirkettle-Watts *et al.*, 2003; Zimmermann *et al.*, 2004). AOX1a is the most prominantly expressed isoform in most tissues, with AOX1c expressed in young cotyledons and, more strongly, in floral tissue. AOX1b, 1d and AOX2 are expressed, but at much lower levels. Comparison of *Arabidopsis* AOX expression data with that obtained using other plants, in which AOX expression and function have been more intensively studied, shows that putatively orthologous genes in different species have very diverse expression patterns. Thus, although the overall pattern of AOX protein expression is similar, the patterns of expression by the various family members differ.

8.3.7 Additional NADH-dehydrogenases

Although rotenone-insensitive NADH dehydrogenases that bypass complex I have only recently been studied in plants (see Moller, 2002; Escobar *et al.*, 2004, 2006), the published results have greatly improved our understanding of the nature of these enzymes. These enzymes are further discussed in Chapter 7.

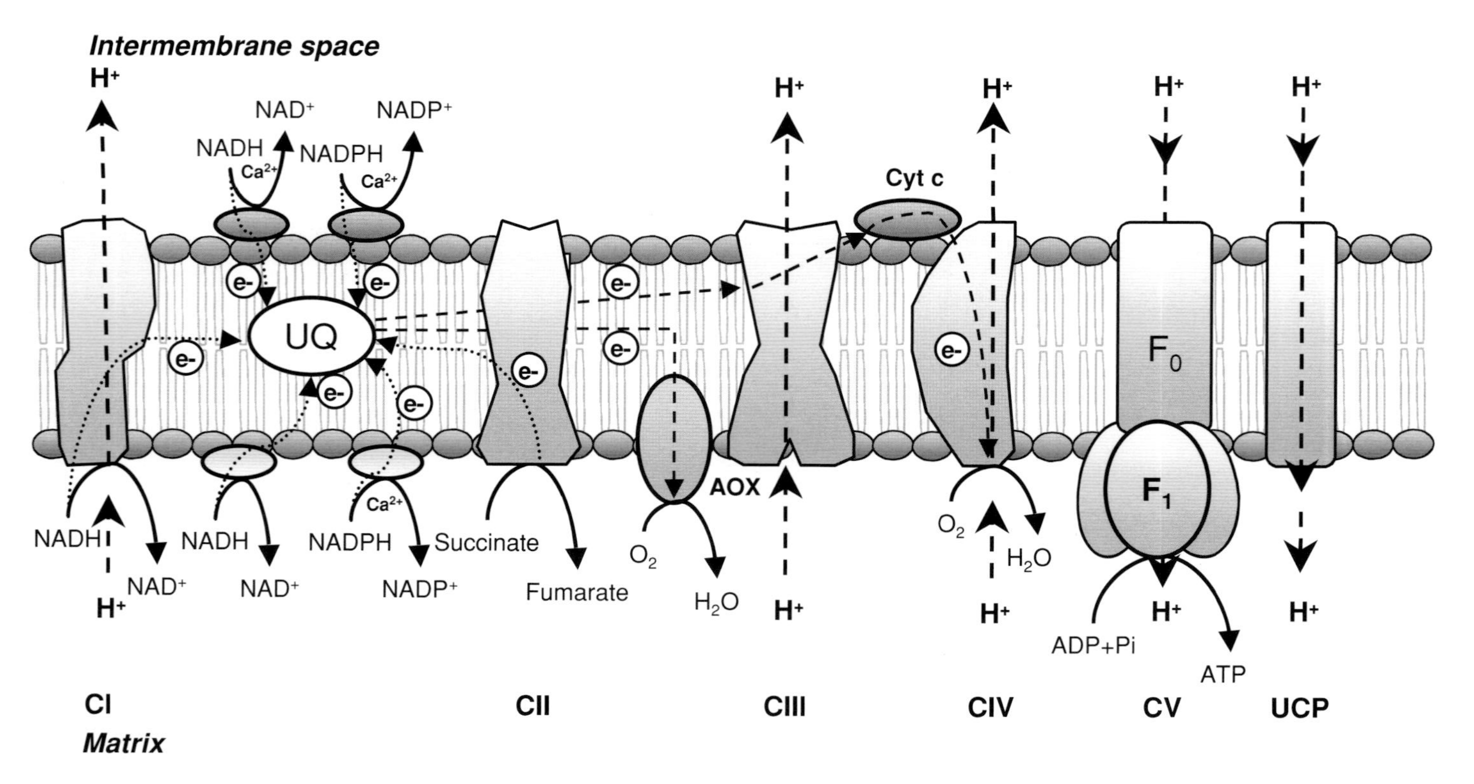

Figure 8.3 Organisation of the ETC and ATP synthesis in the inner membrane of plant mitochondria. Abbreviations: CI, complex I or NADH dehydrogenase; UQ, ubiquinone, a pool of quinine and quinol molecules; CII, complex II or succinate dehydrogenase; AOX, alternative oxidase; CIII, complex III or cytochrome bc1 complex; CIV, complex IV or cytochrome oxydase; CV, complex V or ATP synthetase; UCP, uncoupling protein.

8.4 Carriers

The double membrane system of the mitochondrion allows the relatively non-specific transport of small molecules from the cytosol into the inter-membrane space, whilst the transport of small molecules across the inner membrane into the matrix is more selective (Mannella, 1992; Mannella *et al.*, 2001). This allows a complex set of carrier functions to have great influence on the catabolic, biosynthetic and anapleurotic functions of mitochondria (Laloi, 1999; Millar *et al.*, 2004a). Transport across the outer membrane is predominantly controlled by porins (otherwise known as voltage-dependent anion channels – VDAC), which facilitate the movement of solutes of up to 1 MDa (Mannella and Tedeschi, 1987). Six genes encoding putative isoforms of these porins exist in *Arabidopsis*, and four of these have been identified at the protein level in mitochondrial preparations (Heazlewood *et al.*, 2004).

A clearly defined family of mitochondrial inner membrane carriers operates for the transport of metabolically important compounds into mitochondria, including organic acids, fatty acid-carnitine esters, amino acids, inorganic phosphate and adenine di- and trinucleotides (Picault *et al.*, 2004). The members of this family of ~30-kDa basic proteins have six transmembrane domains (Laloi, 1999). In *Arabidopsis*, a family of 38 genes can be readily identified from genome sequence data. A subset of 10 genes comprises clear orthologs of yeast mitochondrial carriers of known function, whereas 24 have yet to be assigned even a putative function (Millar and Heazlewood, 2003). Eight different members of this family have been directly identified in *Arabidopsis* mitochondrial preparations (Millar and Heazlewood, 2003; Heazlewood *et al.*, 2004). Transporters that have been functionally characterised are presented in Table 8.1. Recently, a carrier previously annotated as a specific oxoglutarate/malate transporter has been experimentally analysed and shown to be a

Table 8.1 Functionally characterised mitochondrial carriers from plants

Carrier	Gene(s)	Function	Reference(s)
DTC	At5g19760	Dicarboxylate-tricarboxylate carrier	Picault *et al.* (2002)
	D45073/4/5[a]	2-oxoglutarate/malate translocator	Taniguchi and Sugiyama (1996)
SFC	At5g01340	Succinate-fumarate carrier	Catoni *et al.* (2003b)
CAC	At5g46800	Carnitine carrier	Lawand *et al.* (2002)
BAC	At2g33820	Basic amino acid carrier	Hoyos *et al.* (2003)
	At1g79900		Catoni *et al.* (2003a)
PiC	At5g14040	Phosphate carrier	Millar and Heazlewood (2003)
AAC	At5g13490	ADP/ATP carrier	Millar and Heazlewood (2003)
UCP	At3g54110	Uncoupling protein	Watanabe *et al.* (1999)
	At5g58970		
PYR	*YIL006w*[b]	Pyruvate carrier	Hildyard and Halestrap (2004)

Table 8.1 shows only those transporters that have been functionally characterised; these are only a subset of the 50 putative members of the mitochondrial carrier family identified in the *Arabidopsis* genome by *in silico*, and proteomic approaches (Millar and Heazlewood, 2003; Picault *et al.*, 2004). Unless otherwise stated, the genes described are from *Arabidopsis*. Although no plant gene encoding a pyruvate has been functionally identified, the pyruvate transporter from *Sac. cerevisiae* is included (in italics) because of its vital importance in linking glycolysis with the TCA cycle.

[a] Gene from *Panicum miliaceum*.

[b] Gene from *S. cerevisiae*.

general carrier able to transport both di- and tricarbocylic acids (Picault *et al.*, 2002). A basic amino acid carrier has also been identified, and functionally characterised independently by two groups (Catoni *et al.*, 2003a; Hoyos *et al.*, 2003). The *Arabidopsis* mitochondrial succinate-fumarate carrier homologue has also been identified by complementation of the yeast *acr*1 mutant (Catoni *et al.*, 2003b). However, despite the fact that these carriers are relatively well characterised at the expression and even kinetic level, the precise importance of these proteins is yet to be revealed. In the case of the succinate-fumarate carrier, expression peaked in 2-day-old dark grown seedlings and declined steadily during further development, consistent with a role in export of fumarate for gluconeogenesis during lipid mobilisation that accompanies early germination of *Arabidopsis* seeds. Similarly, the basic amino acid carrier is also highly expressed in young seedlings consistent with its involvement in arginine breakdown (via mitochondrial arginases) during early seedling development. Key study on the inorganic phosphate carrier (Takabatake *et al.*, 1999), the uncoupling protein (UCP) (Maia *et al.*, 1998; Watanabe *et al.*, 1999) and the adenine di- and trinucleotide carriers (Haferkamp *et al.*, 2002; Millar and Heazlewood, 2003) has also been undertaken in *Arabidopsis*, but these studies were limited with respect to assessing the *in vivo* roles of these proteins.

UCPs dissipate the electrochemical proton gradient generated by respiration, releasing the stored energy as heat (for details, see Nicholls and Rial, 1999). In the presence of fatty acids, they facilitate the re-entry of protons, extruded by the respiratory chain, into the matrix bypassing the ATP synthetase (Nicholls and Locke, 1984). Whilst a large body of biochemical evidence suggested the presence of plant proteins with similar properties to UCPs (Vercesi *et al.*, 1995; Jezek *et al.*, 1996), definitive evidence for their presence in plants was supplied by the cloning of a gene encoding a UCP in potato (Laloi *et al.*, 1997). UCPs are well characterised at the structural, expression and kinetic levels. Three of the six *Arabidopsis* UCP genes are constitutively expressed: one is root specific, one green-silique specific and one is not expressed (Vercesi *et al.*, 2006). UCP expression is strongly enhanced during the senescence stages of mango ripening, and tomato fruit development (Considine *et al.*, 2001; Holtzapffel *et al.*, 2002). In addition, upregulated gene expression has been reported in response to low temperature, drought, wounding and pathogen attack (Vercesi *et al.*, 2006), with downregulation reported under conditions of high-salinity stress (Seki *et al.*, 2002). At the activity level, plant UCPs have been demonstrated to be activated by fatty acids (Winkler and Klingenberg, 1994; Borecky *et al.*, 2001) and reactive oxygen species (Echtay *et al.*, 2002; Considine *et al.*, 2003; Smith *et al.*, 2004), and inhibited by purine nucleotides (Vercesi *et al.*, 1995).

Other classes of carriers are present on the mitochondrial inner membrane, including ABC transporters (Kolukisaoglu *et al.*, 2002). One of these is known to be an Fe–S transporter, and mutation of this gene leads to dwarfism and chlorosis in *Arabidopsis* (Kushnir *et al.*, 2001). These data support the view that plant mitochondria possess an evolutionarily conserved Fe–S cluster biosynthesis pathway, which is linked to intracellular iron homeostasis by ABC transporters (Kushnir *et al.*, 2001; Balk and Lobreaux, 2005).

A further carrier of note, named *bout de souffle* (*BOU*), was recently identified in *Arabidopsis* (Lawand *et al.*, 2002). The *bou* mutant stopped developing after germination and degraded storage lipids, but did not proceed to autotrophic growth. However, externally supplied sugar or germination in the dark could bypass this developmental block and allowed mutant plants to develop roots and leaves. The mutant gene was cloned and found to show sequence similarity to the mitochondrial carnitine acyl carriers (CACs) or CAC-like proteins (Van Roermund *et al.*, 1995). In animals and yeast, these transmembrane proteins are involved in the transport of lipid-derived molecules across mitochondrial membranes for energy and carbon supply (Van Roermund *et al.*, 1995, 1999; DeLucas *et al.*, 1999). In keeping with the *Arabidopsis* work, a carnithine acetyl transferase has been purified from pumpkin mitochondria (Schwadbedissen-Gebling and Gerhardt, 1995). Lawand *et al.* (2002) suggested that the BOU pathway would be an effective alternative to the glyoxylate pathway. However, this proposal is, at least partially, contentious since mutation of peroxisomal citrate synthase led to severe phenotypes (Pracharoenwattana *et al.*, 2005) suggesting that these two pathways are not fully redundant. Further analysis of mitochondrial transporter proteins, despite the high level of redundancy that they often exhibit, is a pre-requisite for a comprehensive understanding of mitochondrial metabolism.

8.5 Amino acid metabolism

Whilst a great deal of amino acid metabolism is localised within the plastid, considerable synthesis and degradation is also associated with mitochondria. The logic in such a spatial organisation is fairly transparent given that the TCA cycle exhibits an anapleurotic function and plants degrade and respire protein (and lipid) in periods of carbohydrate deficiency. Here we will concentrate our discussion on cysteine, proline, photorespiratory amino acids and the still controversial subject of the branched-chain amino acids.

8.5.1 Cysteine

Cysteine metabolism in *Arabidopsis* mitochondria has received significant attention (Jost *et al.*, 2000; Wirtz *et al.*, 2001; Levitan *et al.*, 2004; Riemenschneider *et al.*, 2005). Cysteine is not only an essential amino acid for protein structure and function but also the central connection between sulphur and nitrogen metabolism in plants. The *Arabidopsis* cysteine synthase pathway, involving serine acetyltransferase (SAT) and O-acetylserine (thiol) lyase (OAS-TL), has been characterised by genomic and functional analyses, and the genes responsible for the mitochondrial pathway differentiated from the chloroplastic and cytosolic pathways (Noji *et al.*, 1998; Jost *et al.*, 2000). Structural analysis of *Arabidopsis* mitochondrial SAT has also revealed insights into its C-terminal bifunctional domain which is involved in both catalysis and the binding of OAS-TL (Wirtz *et al.*, 2001). Sulphur-based metabolism in mitochondria links cysteine to Fe–S cluster formation, and cyanide metabolism. A series of researchers has investigated the controversial roles of mercaptopyruvate, thiosulfate sulfurtransferases and β-cyano-alanine synthase in *Arabidopsis*, providing clear evidence for mitochondrial forms of these enzymes

(Hatzfeld *et al.*, 2000; Hatzfeld and Saito, 2000; Papenbrock and Schmidt, 2000). However, their *in vivo* role remains unresolved, despite a probing analysis of knock-out mutants (Nakamura *et al.*, 2000).

8.5.2 Proline

Proline accumulation occurs via rapid synthesis from glutamate and can protect plants from osmotic stress (Verslues *et al.*, 2006). Accumulated proline is rapidly oxidised back to glutamate when osmotic stress is relieved. The first step in this process is catalysed by a mitochondrial proline dehydrogenase. Studies in *Arabidopsis* have shown rapid transcript response of proline oxidase to changing osmotic pressure, suggesting that it is a regulatory link in proline homeostasis in plants (Kiyosue *et al.*, 1996; Verbruggen *et al.*, 1996). A subsequent step in this pathway is catalysed by Δ^1-pyrroline-5-carboxylate dehydrogenase (P5CDH). The expression of this mitochondrial enzyme is regulated by osmotic stress in *Arabidopsis* (Deuschle *et al.*, 2001), and it has recently been shown that it can act in tandem with the stress-related gene transcript to generate two types of siRNAs (small interfering RNAs) that participate in a complex manner to regulate salt tolerance (Borsani *et al.*, 2005). An essential function of P5CDH was recently demonstrated by analysis of a T-DNA insertion mutant of this gene. This study revealed that P5CDH is essential for proline degradation, but not required for vegetative plant growth. External proline application caused programmed cell death, with callose deposition, reactive oxygen species production and DNA laddering, involving a salicylic acid signal transduction pathway. The *p5cdh* mutants were hypersensitive towards proline and other molecules producing P5C, such as arginine and orthornitine (Deuschle *et al.*, 2004).

8.5.3 Glycine

Glycine metabolism, which is central to the photorespiratory cycle (Figure 8.4), occurs within the matrix of mitochondria (Douce *et al.*, 2001). Sommerville and Ogren (1982) highlighted the potential use of *Arabidopsis* mutants as tools to investigate this pathway through the identification and analysis of a glycine decarboxylase mutant over 20 years ago. Detailed assessment of the co-ordinated expression of mitochondrial photorespiratory genes and their co-expression with Calvin cycle enzymes in *Arabidopsis* has helped to elucidate both light-responsive promoter elements (Srinivasan and Oliver, 1995) and circadian expression patterns (McClung *et al.*, 2000). A recent analysis of the genomes of *Arabidopsis* and rice demonstrated that even in these very distantly related plant species, the genes for glycine decarboxylase and serine hydroxymethyltransferase are similarly organised, suggesting a strong functional and evolutionary link (Bauwe and Kolukisaoglu, 2003). Whilst functional analysis of the last enzyme of the phosphorespiratory cycle revealed, as expected, that the majority of the pathway is extra-mitochondrial (Boldt *et al.*, 2005), as is detailed below, the precise pattern of metabolite transport between compartments is still a matter of debate (Taira *et al.*, 2004; Linka and Weber, 2005).

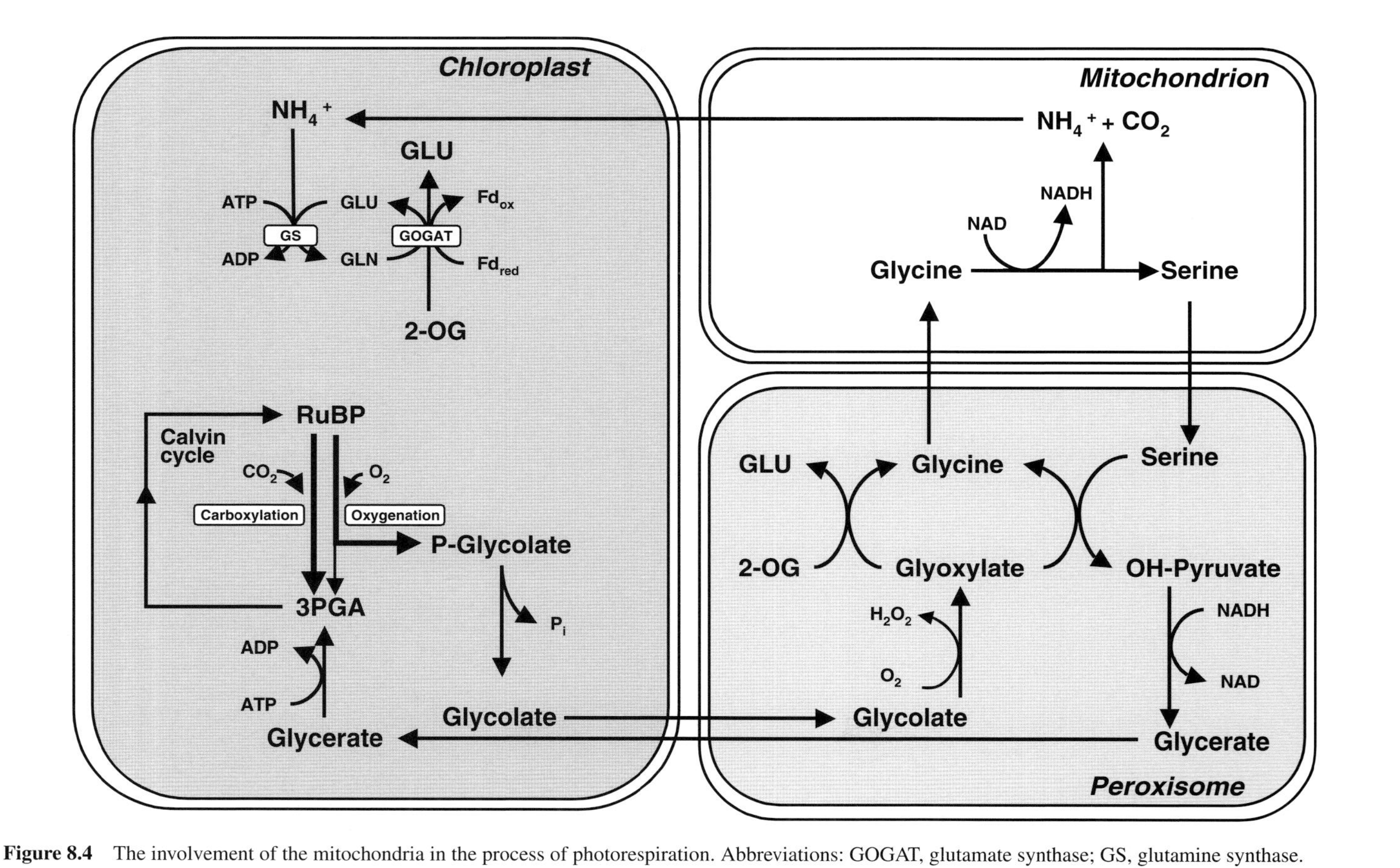

Figure 8.4 The involvement of the mitochondria in the process of photorespiration. Abbreviations: GOGAT, glutamate synthase; GS, glutamine synthase.

8.5.4 Branched-chain amino acids

The intracellular location of branched-chain amino acid catabolism in plants has been controversial, with early evidence suggesting a peroxisomal location (Gerhardt, 1992), as opposed to the well-established mitochondrial location in mammals and yeast (Graham and Eastmond, 2002). The three branched-chain amino acids, valine, leucine and isoleucine, are initially transaminated to their respective branched-chain α-keto acids by the branched-chain aminotransferase (BCAT). This reversible reaction is also the final step of the biosynthesis of these amino acids. Seven putative BCAT genes are present in *Arabidopsis*, and GFP-targeting studies have shown that separate members are targeted to chloroplasts and mitochondria (Diebold *et al.*, 2002). Following this transamination, the α-keto acids are decarboxylated and esterified to CoA by the branched-chain α-keto acid dehydrohenase complex (BCKDC). Early reports suggested that BCKDC activity was located in peroxisomes (Gerhardt, 1992), but when genes for the *Arabidopsis* proteins were sequenced, it was found that each has an N-terminal extension, indicative of mitochondrial targeting (Fujiki *et al.*, 2000). The CoA esters are then oxidised by an acyl-CoA dehydrogenase. An *Arabidopsis* mitochondrial acyl-CoA dehydrogenase (isovaleryl-CoA dehydrogenase, IVD), with activity towards both isovaleryl-CoA (from leucine) and isobutyryl-CoA (from valine), has been identified using gene cloning, *in vivo* targeting studies and biochemical analysis in mitochondria (Daschner *et al.*, 2001). Following this step, the three amino-acid pathways diverge in a series of separate reactions leading to propionyl-CoA in the case of isoleucine and valine metabolism, and to acetyl-CoA and acetoacetate in the case of leucine metabolism. Molecular analysis in *Arabidopsis* has been critical in demonstrating the mitochondrial location of this pathway. Work on the induction of BCKDC genes using luciferase constructs linked to their promoters has shown that sugar starvation drives expression in cell cultures (Fujiki *et al.*, 2002). In this system, induction of the BCKDC gene promoters was influenced by protein kinase and phosphatase inhibitors, indicating the presence of a phosphorylation-based signal transduction pathway in the perception, and transduction of the sugar-starvation response (Fujiki *et al.*, 2002). Similar study on 3-methylcrotonyl-coenzyme A carboxylase expression in *Arabidopsis* has complemented the BCKDC study, indicating the coordinated regulation of the leucine catabolism pathway (Che *et al.*, 2002). Overall, these data suggest that branched-chain amino acids promote their own catabolism but only when cells are sugar starved. Recently, the mitochondrial components of this pathway were directly identified in *Arabidopsis* by mass spectrometry, pinpointing the members of gene families responsible for the mitochondrial pathway, and also revealing a gap in the pathways for leucine and isoleucine degradation that is likely to be localised in peroxisomes rather than mitochondrion (Taylor *et al.*, 2004). Recent analysis of T-DNA insertion mutants of the electron transfer flavoprotein – ubiquinone oxidoreductase (ETFQO) – during dark-induced starvation revealed a marked accumulation in both leucine and isovaleryl CoA confirming the above studies (Ishizaki *et al.*, 2005). In addition, radiolabel tracing experiments in which ^{14}C-leucine was fed to isolated pea mitochondria suggested a mitochondrial location for the IVD reaction. Intriguingly, *etfqo* mutants did not display accumulation of intermediates of either isoleucine or valine catabolism although it is known that ETFQO participates in these processes in mammals (Ikeda *et al.*, 1983). In plants peroxisomal enzymes capable of catalysing these pathway have been characterised (Hayashi *et al.*, 1999; Lange *et al.*, 2004); however, the recent finding that BCAT-1 is capable of initiating degredation of leucine, isoleucine and valine (Schuster and Binder, 2005) suggest that the pathway of branched-chain amino acid degradation, like that of photorespiration, is shared across metabolic compartments.

8.6 Biosynthesis of vitamins and lipids

8.6.1 Lipoic acid

Lipoic acid is an essential co-factor in a number of decarboxylating, dehydrogenase reactions and is synthesised in the mitochondria in plants. The first plant mitochondrial lipoic acid synthase was identified, cloned and analysed in *Arabidopsis* (Yasuno and Wada, 1998). The lipoic acid synthesised by this enzyme is fixed to the acyl-carrier protein that is attached to complex I of the respiratory chain (Shintani and Ohlrogge, 1994), and subsequently transferred to the dehydrogenase complexes by a mitochondrial lipoyltransferase (Wada *et al.*, 2001). Subsequently, an additional *Arabidopsis* lipoic acid synthase isoform was identified in plastids (Yasuno and Wada, 2002), showing that both mitochondria and plastids contain this pathway of co-factor assembly.

8.6.2 Biotin

The co-factor biotin is also synthesised, at least ultimately, in the mitochondria. Biotin is an essential water-soluble vitamin found in all living cells (Dakshinamurti and Cauhan, 1989). Plants, and most micro-organisms, have the ability to synthesise biotin. In contrast, other multi-cellular eukaryotic organisms are biotin auxotrophs. Biotin acts as a co-factor for a set of enzymes that catalyse carboxylation, decarboxylation and trans-carboxylation reactions in a number of crucial metabolic processes (Knowles, 1989). In all known microbes, the co-factor is synthesised from pimeloyl-CoA through four enzymatic steps comprising 7-keto-8-aminopelargonic acid (KAPA) synthase, 7,8-diaminopelargonic acid (DAPA) aminotransferase, dethiobiotin synthase and biotin synthase coded by *bio*F, *bio*A, *bio*D and *bio*B genes, respectively. Enzymes required for biotin synthesis in *E. coli* and *Bacillus sphaericus* have been purified and their activities characterised *in vitro* (for review, see Streit and Entcheva, 2003). KAPA synthase, the first enzyme of this pathway, catalyses the decarboxylative condensation of pimeloyl-CoA and L-alanine to produce KAPA, CoASH and CO_2 (Figure 8.5). The structure and reaction mechanism of KAPA synthase places it in the subfamily of α-oxoamine synthases, a small group of pyridoxal 5′-phosphate (PLP)-dependent enzymes of the α-family (Ploux and Marquet, 1996). In plants, the biosynthetic pathway beginning with pimeloyl-CoA appears to follow the same pattern as identified for bacteria. This was deduced by

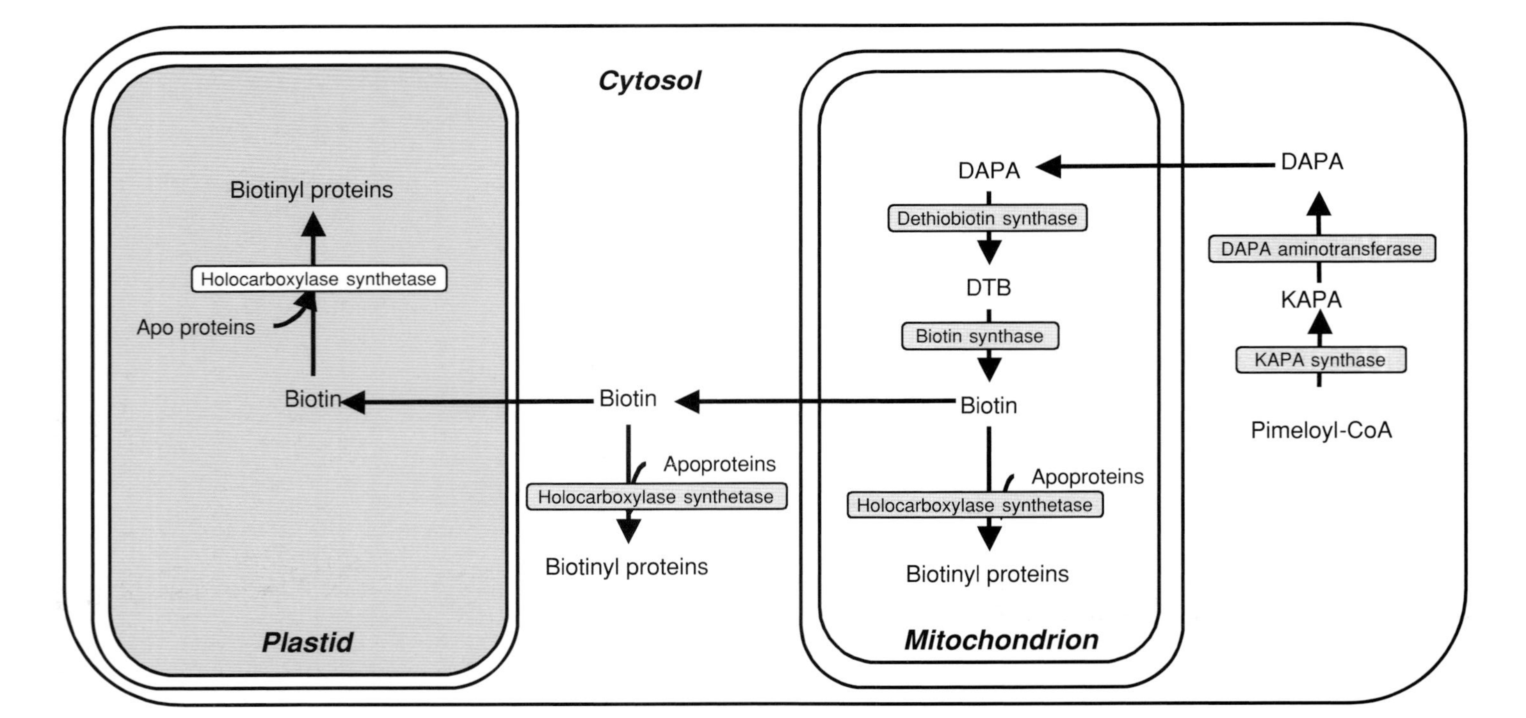

Figure 8.5 Compartmentalisation of biotin metabolism in the plant cell. Abbreviations: KAPA, 7-keto-8-aminopelargonic acid; DAPA, 7,8-diaminopelargonic acid; DTB, dethiobiotin.

measuring pools of the different intermediates of biotin biosynthesis after feeding lavender (*Lavandula vera*) cell cultures with radiolabelled precursors (Baldet *et al.*, 1993). The first genetic information on biotin synthesis in higher plants came from analysis of the *bio1* biotin auxotroph of *Arabidopsis* (Schneider *et al.*, 1989). Plant growth was rescued by biotin, dethiobiotin or DAPA, but not KAPA supply, or by genetic complementation with the *E. coli bioA* gene coding DAPA aminotransferase, demonstrating that mutant plants are defective in this enzyme (Shellhammer and Meinke, 1990; Patton *et al.*, 1996a). The *Arabidopsis* cDNA for biotin synthase, named *BIO2*, was cloned and characterised in several laboratories (Baldet and Ruffet, 1996; Patton *et al.*, 1996b; Weaver *et al.*, 1996; Baldet *et al.*, 1997) before the corresponding mutant was identified. In the *bio2* mutant, a second biotin auxotroph of *Arabidopsis*, arrested embryos are defective in the final step of biotin synthesis, i.e. the conversion of dethiobiotin to biotin (Patton *et al.*, 1998). The enzyme encoded by the *BIO2* gene is a homodimer of 78 kDa with a complex Fe–S centre organisation (Baldet *et al.*, 1997; Picciocchi *et al.*, 2001). Like its bacterial counterparts, BIO2 needs additional protein factors to function. These accessory proteins are as yet unidentified but present in a mitochondrial matrix fraction (Picciocchi *et al.*, 2003), in keeping with the specific location of enzyme activity (Baldet *et al.*, 1997; Picciocchi *et al.*, 2001). Characterisation of plant enzymes situated upstream of biotin synthase, and elucidation of their subcellular distribution have recently revealed that the first reaction of this sequence is localised in the cytosol (Pinon *et al.*, 2005) indicating a complex compartmentation of the pathway of biotin synthesis.

8.6.3 Thiamine

Thiamine synthesis in plants was thought to be plastid specific, whereas in yeast it is a mitochondrial process. Identification of the *Arabidopsis* thiamine synthase gene showed that it encodes a precursor protein with tandem N-terminal signal sequences that direct the protein to chloroplast and mitochondria *in vivo* (Chabregas *et al.*, 2001). Further analysis revealed two translation products from this transcript, leading to a longer form destined for chloroplasts and a shorter form addressed to mitochondria (Chabregas *et al.*, 2003). However, as yet there is no functional evidence for thiamine synthesis in plant mitochondria (Millar *et al.*, 2004a).

8.6.4 Folate

Folate coenzymes mediate C1 transfer reactions involved in several major cellular processes, including the synthesis of purines and thymidylate, amino acid metabolism, pantothenate synthesis, mitochondrial and chloroplastic protein biogenesis and methionine synthesis. Methionine is the direct precursor of *S*-adenosylmethionine, which in turn is the source of methyl units for the synthesis of a myriad of molecules such as choline, chlorophyll and lignin (for reviews, see Cossins, 2000; Scott *et al.*, 2000; Hanson and Roje, 2001). In plants, folate is also involved in the photorespiratory cycle where it is essential for the reactions catalysed by the GDC and serine hydroxymethyltransferase. Folate is composed of three distinct parts, namely a pterin ring, a *p*-aminobenzoic acid and a γ-linked polyglutamyl chain of variable length. Its function is to bind, transport and donate C1 units that differ in their oxidation state (methyl, methylene, methenyl or 10-formyl, from the most reduced to the most oxidised). Thus, the co-factor exists with diverse chemical forms, and these various derivatives are collectively termed folate or vitamin B9 (Cossins, 2000; Scott *et al.*, 2000; Hanson and Roje, 2001). From dihydropterin and *p*-aminobenzoic acid, the biosynthesis of tetrahydrofolate (THF) in plants and micro-organisms requires the sequential operation of five reactions (Figure 8.6), the first three being absent in animals (for reviews, see Scott *et al.*, 2000; Hanson and Gregory, 2002). In plants, mitochondria play a central role in this synthesis (Neuburger *et al.*, 1996; Rébeillé *et al.*, 1997; Ravanel *et al.*, 2001). Leaf mitochondria contain all of the required enzymes, and the first three steps are presumably exclusively localised in this compartment. In contrast, the last step involved in the formation of the polyglutamyl chain is present in the cytosol, and in

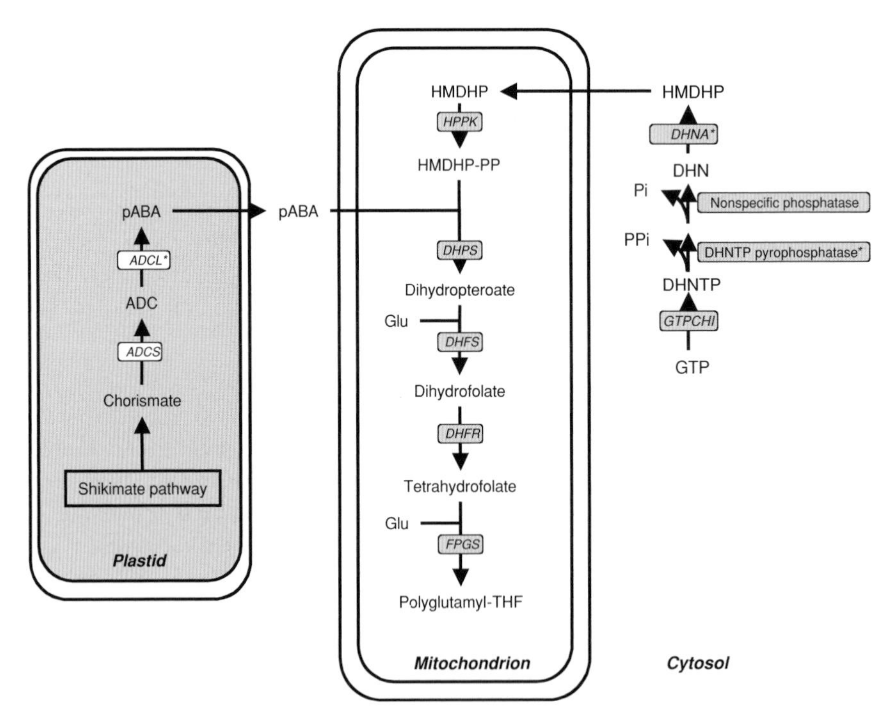

Figure 8.6 The tetrahydrofolate biosynthesis pathway in plants. Abbreviations: ADC, aminodeoxychorismate; DHN, dihydroneopterin; DHNTP, dihydroneopterin triphosphate; Glu, glutamate; GTP, guanosine 5′-triphosphate; HMDHP, hydroxymethyldihydropterin; HMDHP-PPi, hydroxymethyldihydropterin pyrophosphate; *p*ABA, *p*-aminobenzoate; PPi, pyrophosphate; Pi, inorganic phosphate; ADCS, aminodeoxychorismate synthase; ADCL, aminodeoxychorismate lyase; GTPCHI, GTP cyclohydrolase I; DHFR, dihydrofolate reductase; DHFS, dihydrofolate synthase; DHNA, dihydroneopterin aldolase; DHPS, dihydropteroate synthase; FPGS, folylpolyglutamate synthetase; HPPK, hydroxymethyldihydropterin pyrophosphokinase. *Enzymes that have not yet been cloned from plants.

the chloroplasts in addition to the mitochondria (Ravanel *et al.*, 2001). Despite its low concentration in plant tissues (Cossins, 1984), folate is likely to be of major importance during seedling development due to the housekeeping functions mediated by folate co-enzymes (Figure 8.6). In this regard, it is noteworthy that the pool of folate in pea cotyledons increased during germination and the inhibition of *de novo* synthesis of THF using folate analogs blocked seedling development (Roos and Cossins, 1971; Gambonnet *et al.*, 2001). In addition, a continuous synthesis of THF is essential to maintain high rates of serine synthesis through the mitochondrial activities of GDC and SHMT (serine hydroxymethyltransferase) in *Arabidopsis* (Prabhu *et al.*, 1996). Recent identification of an *Arabidopsis* homolog of the mammalian mitochondrial folate transporter, which mediates folate import into chloroplasts (Bedhomme *et al.*, 2005), explains how folate metabolism can be efficiently separated between compartments (Sahr *et al.*, 2005), and may explain how its abundance can be dramatically altered during developmental processes.

8.6.5 Isoprenoids

Several isoforms of enzymes involved in isoprenoid biosynthesis have also been localised to mitochondria in *Arabidopsis*, including a farnesyl-diphosphate synthase (Cunillera *et al.*, 1997) and a geranylgeranyl pyrophosphate synthase (Zhu *et al.*, 1997). However, the isoprenoid end products in mitochondria are poorly defined. Recent metabolic engineering studies were successful in modifying isoprenoid emissions, by altering the location of expression of two genes of terpene synthesis, in such a way that *Arabidopsis* attracted carnivorous predatory mites (*Phytoseiulus persimilis*) that aid the plants defence mechanisms (Kappers *et al.*, 2005). This example highlights the immense importance, and indeed potential, of mitochondrial metabolism.

8.6.6 Lipids

The lipid bilayer composition of mitochondria has also been probed in *Arabidopsis*. The limited ability of mitochondria to synthesise lipids means that they import the majority of their lipid content (Daum and Vance, 1997). A phosphatidylglycerophosphate synthase was identified and functionally characterised in yeast (Muller and Frentzen, 2001) and shown to be dual targeted to plastids and mitochondria in *Arabidopsis* (Babiychuk *et al.*, 2003). Examination of a knockout line revealed a plant with pigment deficiency and lack of thylakoid membrane biosynthesis. This gene is essential for the biosynthesis of phosphatidylglycerol in plastids, but mitochondria in the mutant still accumulated phosphatidylglycerol and its derivative, cardiolipin. This analysis suggested that mitochondria, unlike plastids, may be able to import phosphatidylglycerol from the endoplasmic reticulum, by an as-yet-unidentified route (Babiychuk *et al.*, 2003). In contrast, *Arabidopsis* mutants deficient in omega-6-oleate desaturase have mitochondria with an oleic acid content of more than 70% of the total fatty acids and a greatly increased lipid/protein ratio (Caiveau *et al.*, 2001). Oxygen consumption rate and inner-membrane proton permeability were decreased, and the temperature response of respiration increased in these mutants, suggesting a changed lipid composition that altered the lateral mobility of lipids and the bioenergetic capabilities of these mitochondria (Caiveau *et al.*, 2001).

8.7 Role of mitochondrial metabolism in biological processes

The above sections have focussed exclusively on the properties of independent enzymes, or at most, simple pathways. However, in recent years plant metabolic studies have moved away from such approaches and begun to look at coordinated responses involving many enzymes, and even multiple pathways (Sweetlove and Fernie, 2005). In the following section, the role of enzymes of the TCA cycle in particular, and mitochondrial metabolism in general, will be described with respect to the biological processes of energy metabolism, ripening, thermogenesis and plant fertility.

8.7.1 Respiratory activity in the light

Mitochondrial respiratory activity in the dark includes the oxidation of carbon compounds in the TCA cycle, and of NAD(P)H in the mitochondrial ETC. The TCA cycle releases carbon dioxide and reducing equivalents. During oxidation of these reducing equivalents, oxygen is consumed and a proton electrochemical gradient is created across the inner-mitochondrial membrane providing the energy for ATP synthesis in the mitochondria. In the light, the oxidative decarboxylation of glycine to serine occurs in the mitochondrial matrix as a part of the photorespiratory cycle (Ögren *et al.*, 1984). Plant mitochondria contain a malic enzyme (Section 8.2.1.2) allowing them to operate the TCA cycle independently of glycolysis. The respiratory chain also has several unique features: a cyanide-resistant non-phosphorylating pathway, a rotenone-insensitive oxidation site and the ability to oxidise external NAD(P)H (Moller and Lin, 1986; Douce and Neuberger, 1989; Rasmusson *et al.*, 1999; Michalecka *et al.*, 2003; see Chapter 7). However, despite much biochemical and reverse genetic experimentation (Millar *et al.*, 1998; Fiorani *et al.*, 2005; Umbach *et al.*, 2005), the physiological importance of these features is not entirely clear.

The operation of the Krebs cycle in the light has been shown to differ from that in the dark in at least two ways: the reversible inactivation of the mitochondrial PDC in the light (Budde and Randall, 1990) and the rapid export of TCA cycle intermediates out of the mitochondria (Hanning and Heldt, 1993; Atkin *et al.*, 2000a), for utilisation in glutamate synthesis (Hodges, 2002). However, the mitochondrial ETC continues to be active in the light, despite the limitations that the above modifications must impose on the TCA cycle (Atkin *et al.*, 2000b; Padmasree *et al.*, 2002; Yoshida *et al.*, 2006). Indeed, a wide range of evidence has accumulated and suggests that mitochondrial function is not only merely active in illuminated leaves, but is an integral part of photosynthetic metabolism. Much of the early evidence of a link between mitochondrial and plastid function came from studies using specific inhibitors of the cytochrome and alternative respiratory pathways (Krömer *et al.*,

1988, 1993; Padmaresee and Raghavendra, 1999a). However, several further lines of evidence support a considerable mitochondrial component within photosynthetic leaf metabolism. These include the data from the transgenic studies mentioned above (Fiorani *et al.*, 2005; Umbach *et al.*, 2005), the observation of light-dependent expression of mitochondrial genes (Svensson and Rasmusson, 2001) and the decreased rate of photosynthesis under photorespiratory conditions that was observed in the *cmsII* mutant of *N. sylvestris*, which exhibits a loss of function of mitochondrial complex I (Dutilleul *et al.*, 2003). However, the operation of the TCA cycle in the light remains somewhat contentious with recent studies in broad bean suggesting an almost complete reduction of TCA cycle activity in the light (Tcherkez *et al.*, 2005).

As stated by Krömer (1995), the direct determination of respiratory oxygen uptake and of carbon dioxide release is a complicated task due to the multiplicity of oxygen liberating/consuming and carbon dioxide liberating/consuming reactions in the leaf. For this reason, gas exchange measurements have classically been used in conjuncture with radiolabel studies and mass spectrometry to differentiate rates of evolution and fixation of these gases. Using such approaches, the rate of respiration in the light has been estimated from carbon dioxide release to be around 25% to 100% of the respiratory rate in darkness (Sharp *et al.*, 1984; Kirschbaum and Farquhar, 1987; Cashin *et al.*, 1988; Rebeille *et al.*, 1988; Avelange *et al.*, 1991). Whereas some reports based on oxygen consumption suggest an inhibition of respiration in the light (Canvin *et al.*, 1980; Bate *et al.*, 1988), others suggest that the rate of respiration was invariant irrespective of illumination (Gerbaud and Andre, 1980; Peltier and Thibauld, 1985; Weger and Turpin, 1989). Analysis of microarray datasets shows a clear trend of reduced expression of genes associated with respiratory processes in the light (Bläsing *et al.*, 2005; Urbanczyk-Wochniak *et al.*, 2005, 2006); however, it should be noted that this does not necessarily imply a downregulation of the flux through the pathway. A recent study by Tcherkez *et al.* (2005) investigated the metabolism of ^{13}C-enriched glucose or pyruvate by leaves of French bean by determining the $^{13}CO_2$ production in the light. Using differently positioned ^{13}C-enrichments, these authors estimated that the TCA cycle is reduced by 95% in the light, whereas the pyruvate dehydrogenase reaction is reduced by ~27%. Given that the inhibition of PDC is particularly well characterised (Budde and Randall, 1990; Tovar-Mendez *et al.*, 2003), this result is at first sight rather surprising. However, it should be borne in mind that many of the dehydrogenase enzymes of the TCA cycle would likely be inhibited by the high mitochondrial NADH level that is a consequence of photorespiratory glycine decarboxylation (Atkin *et al.*, 2000c). In a similar vein, experimental evidence has been provided that the mitochondrial isocitrate dehydrogenase is inhibited by the high NADPH/NADP ratios that occur in the light (Igamberdiev and Gardestrom, 2003). Moreover, that there is such a large reduction in the TCA cycle activity in the light as reported by Tcherkez and co-workers seems to somewhat contradict observations of considerable labelling of organic and amino acids following incubation of isolated leaf discs of a range of species in $^{14}CO_2$ (Zeeman and ap Rees, 1999; Lytovchenko *et al.*, 2002a,b; Nunes-Nesi *et al.*, 2005). These data and the clear phenotypes produced by the mutation or antisense inhibition of enzymes of the TCA cycle (Carrari *et al.*, 2003; Nunes-Nesi *et al.*, 2005), as well as studies utilising specific inhibitors of the TCA cycle (Vani *et al.*, 1990), suggest that the cycle is indeed of considerable biological importance during photosynthesis. The exact reason for the lack of consensus between our two approaches is unclear, although as stated above, determining such fluxes is technically very difficult and consensus positions are rarely reached. Whilst it is possible that the study of Tcherkez *et al.* (2005) underestimated the flux through the TCA cycle by using a method that did not facilitate the acknowledgement of dilution factors, it is equally possible that the differences in light inhibition of dark respiration are species dependent. However, given the body of evidence suggesting a biologically significant operation of the TCA cycle in the illuminated leaf, the relative flux through the pathway is relatively academic. For this reason, we will focus the rest of this chapter on addressing the question – what is (are) the major function(s) of this pathway in the illuminated leaf? Four possibilities are discussed: (i) the provision of ATP to aid photosynthetic sucrose synthesis, (ii) the export of redox equivalents required for nitrate reduction in the cytosol, (iii) the export of redox equivalents required for hydroxypyruvate reduction in the peroxisomes and (iv) a role in the energy metabolism *per se* of the illuminated leaf.

8.7.2 Provision of mitochondrial ATP to support cytosolic sucrose synthesis

It has previously been suggested that light may exert an inhibitory effect on the mitochondrial ETC via the cytosolic adenylate pool (Heber and Heldt, 1981). However, the cytosolic ATP/ADP ratio is actually lower in the light during steady-state photosynthesis at saturating carbon dioxide concentration than it is in the dark (Hampp *et al.*, 1982; Stitt *et al.*, 1982; Gardestrom, 1987), and it is probable that even conditions that increase the cytosolic ATP/ADP ratio are unlikely to affect the respiratory oxygen uptake (Dry and Wiskich, 1982). However, given that the cytosolic in illuminated leaves ATP pool can, in principle, be sustained either by photosynthetic ATP synthesis or by oxidative phosphorylation, it is often reported that a primary function of the TCA cycle operation during the day is the production of ATP to support cytosolic sucrose synthesis. Several lines of evidence offer strong support for the theory that mitochondria supply a large portion of the cytosolic ATP. First, plant mitochondria possess a highly active ATP/ADP translocator that efficiently exports matrix-produced ATP to the cytosol (Heldt, 1969). Second, specific inhibition of the mitochondrial ATP synthase by oligomycin decreases the cytosolic ATP/ADP ratio in barley leaf protoplasts by $\sim$60% (Krömer and Heldt, 1991; Krömer *et al.*, 1993), whereas aminoacetonitrile, which inhibits photorespiratory conversion of glycine to serine and thus impairs oxidative phosphorylation, caused a decrease in cytosolic ATP/ADP ratio by 25% and 45% at limiting and strictly limiting carbon dioxide concentrations, respectively (Gardestrom and Wigge, 1988). Whether this supply of mitochondrial-derived ATP is necessary for sucrose biosynthesis is, however, not so clear. Received wisdom holds that cytosolic sucrose synthesis is highly dependent on the supply of UTP, which could be met by the conversion of mitochondrially produced ATP via the action of a cytosolic nucleoside-5′-diphosphate kinase (NDK) (Figure 8.7). Based on kinetic arguments, it can be hypothesised that a drop in mitochondrial ATP synthesis would have knock-on effects mediated via NDK, and consequently UDP glucose pyrophosphorylase (UGPase), and sucrose phosphate synthase (SPS). Experimental evidence for such a scenario is available given that inhibition of oxidative phosphorylation has been demonstrated to result in a decrease in both the activity and activation state of SPS (Krömer *et al.*, 1993). Moreover, the result of at least one genetic experiment is in keeping with this theory: when oxidative phosphorylation was inhibited in both the wild type and a starch-less mutant of *N. sylvestris*, the mutant, but not the wild type, was impaired in photosynthetic activity. This result indicated that in the wild type, the cytosolic ATP level did not limit sucrose synthesis, but it did in the mutant, which required an elevated rate of sucrose synthesis to compensate for the block in starch synthesis in order to maintain comparable rates of photosynthate usage (Hanson, 1992). However, somewhat contradictory to this, albeit more direct, is the fact that impairment of the flux through the TCA cycle in the *Aco-1* accesssion of *S. pennellii* (see Section 8.2.1.4) resulted in a considerable decrease in the cellular ATP/ADP ratio, and an enhanced rate of photosynthetic sucrose synthesis (Carrari *et al.*, 2003). This finding thus raises the possibility that the production of ATP in the mitochondria is not a pre-requisite for the maintenance of high rates of photosynthetic sucrose synthesis, and suggests that further experimentation is required to clarify this point. Recently, experiments to determine the function of respiration in photosynthetic metabolism have suggested that it is not merely to improve sucrose metabolism but also to modulate metabolites related to redox status (Igamberdiev *et al.*, 1998; Padmasree and Raghavendra, 1999b; Padmasree *et al.*, 2002; Dutilleul *et al.*, 2005; Scheibe *et al.*, 2005), and it is this modulation that will form the focus of the majority of the remainder of this chapter.

8.7.3 A role for the reactions converting acetyl CoA to α-ketoglutarate in nitrogen assimilation

Nitrate assimilation includes the reduction of nitrate to nitrite in the cytosol as well as the subsequent reduction of nitrite to ammonium, and the subsequent assimilation of this into amino acids by the activities of glutamine synthase (GS), and glutamate synthase (GOGAT) in the chloroplast (Figure 8.8). Since plant mitochondria mainly export citrate (Hanning and Heldt, 1993), and given that cytosolic isoforms, or at least a cytosolic location, of the enzymes required for the conversion of citrate to 2-oxoglutarate (Chen and Gadal, 1990; Carrari *et al.*, 2003; Hodges *et al.*, 2003; Abiko *et al.*, 2005) as well as an efficient mechanism for 2-oxoglutarate uptake into plastids (Rennè *et al.*, 2003) have been identified, it is frequently postulated that these reactions are involved in nitrate assimilation (Gálvez *et al.*, 1999; Fernie *et al.*, 2004). This theory is supported by an accumulation of experimental evidence ranging from the fact that the regulatory properties of PDC would allow it to operate in support of such a pathway, even under conditions where the rest of the TCA cycle were non-operational (Schuller and Randall, 1989; Gemell and Randall, 1992;

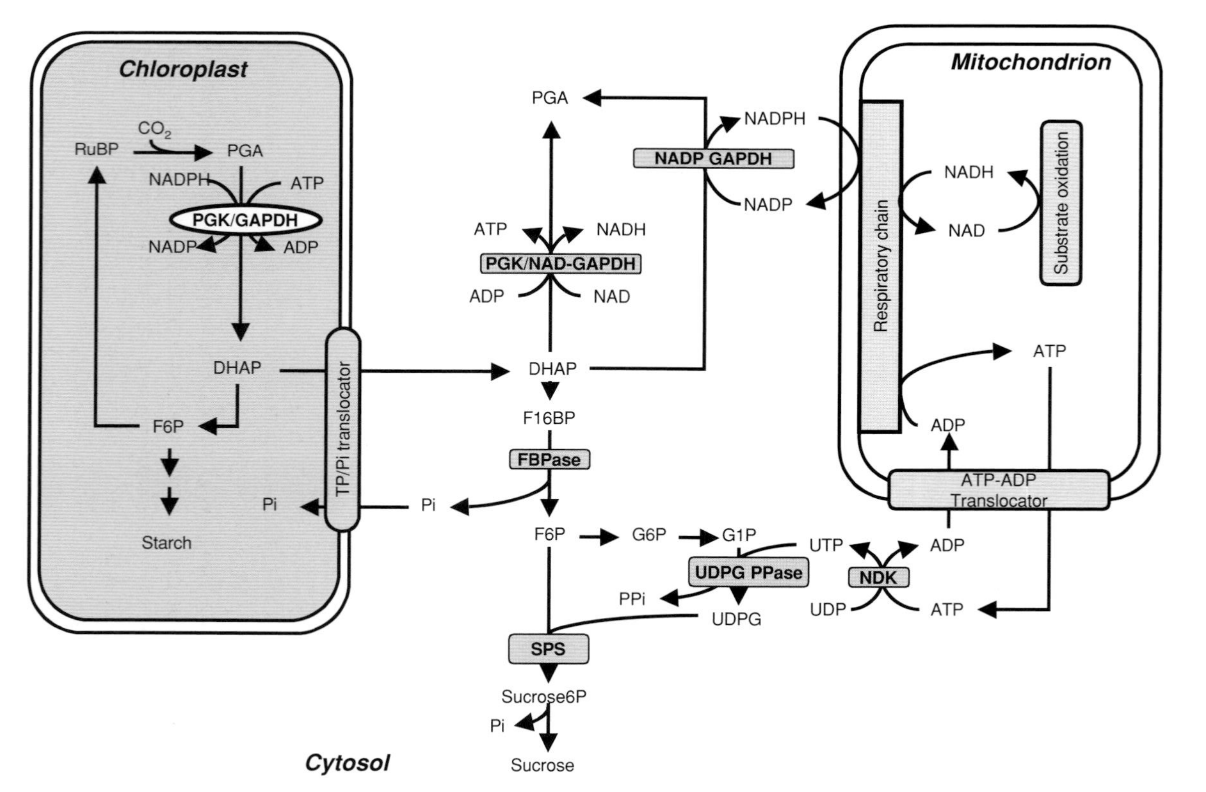

Figure 8.7 Suggested interactions between chloroplasts and mitochondria in the light. Abbreviations: NDK, nucleoside-5′-diphosphate kinase; UDPG PPase, UDP glucose pyrophosphatase; FBPase, fructose-1,6-bisphosphatase; RuBP, ribulose-1,5-bisphosphate; F6P, fructose-6-phosphate; G6P, glucose-6-phosphate; G1P, glucose-1-phosphate; F16BP, fructose-1,6-bisphosphate; F26BP, fructose-2,6-bisphosphate; UDPG, UDP-glucose; PPi, pyrophosphate; TP, triose phosphate; Pi, inorganic phosphate.

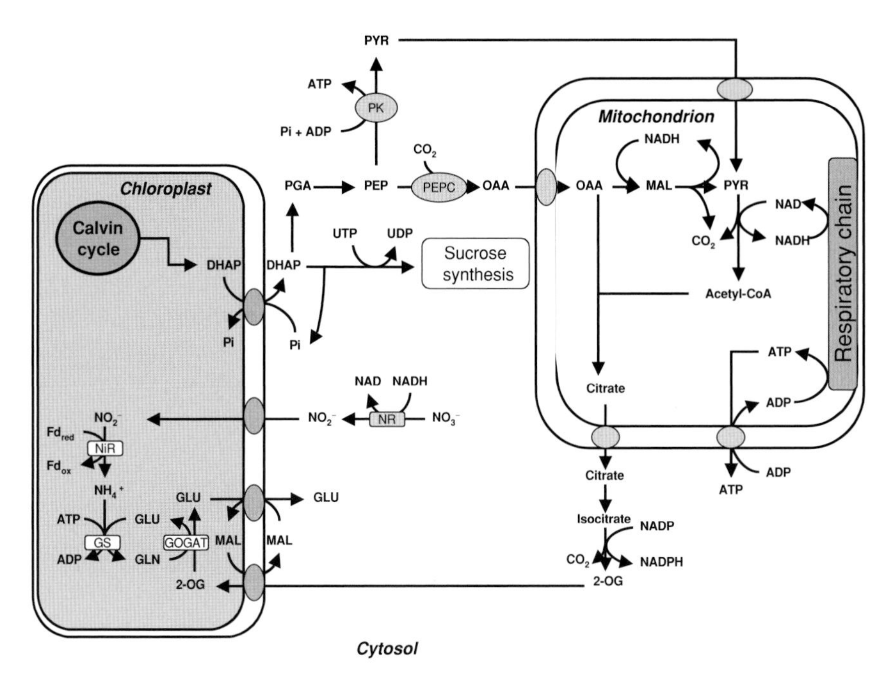

Figure 8.8 Suggested supply of carbon skeletons for photosynthetic nitrate assimilation. Abbreviations: NR, nitrate reductase; NiR, nitrite reductase; Pyr, pyruvate; Mal, malate; glu, glutamate; gln, glutamine; GS, glutamine synthase; GOGAT, glutamate synthase; PK, pyruvate kinase; Pi, inorganic phosphate; PEPC, phosphoenolpyruvate carboxylase; DHAP, dihydroxyacetone phosphate; OAA, oxaloacetate; 2-OG, 2-oxoglutarate.

Krömer, 1995; Hoefnagel *et al.*, 1998). Several other lines of correlative evidence also support a role for the TCA cycle in nitrate assimilation (Fieuw *et al.*, 1995; Scheible *et al.*, 1997; Stitt, 1999; Masclaux *et al.*, 2000); however, there are also many experimental observations that appear to contradict such a role (Gálvez *et al.*, 1996; Lancien *et al.*, 1999; Kruse *et al.*, 1998). The evidence has been critically assessed in an excellent review focussed on the metabolism of 2-oxoglutarate (Hodges, 2002); therefore, we will discuss the major points briefly. Importantly, the demand for 2-oxoglutarate needed for ammonium assimilation could be met either by cytosolic or mitochondrial isocitrate dehydrogenases or by aspartate amino transferases. Whilst direct evidence exists for a role of the aspartate transferase – a mutant of the cytoslic isoform of this enzyme exhibits a retarded growth phenotype and altered aspartate biosynthesis in the light (Schultz *et al.*, 1998) – such direct evidence does not exist for any of the isoforms of isocitrate dehydrogenase (Hodges, 2002). Having said that this does not necessarily imply that there is no role for TCA cycle enzymes in nitrogen assimilation since the *Aco-1* mutant of *Sol. pennellii*, described above, exhibits dramatic changes in amino acid contents relative to wild type, suggesting that the TCA cycle is involved. The identification and preliminary characterisation of *Arabidopsis* isocitrate dehydrogenase isoforms will facilitate reverse genetic analysis of the role(s) of the various isocitrate dehydrogenase isoforms (Hodges *et al.*, 2003). Such experiments would conceivably also clarify the localisation of the reactions catalysing the conversion of citrate to 2-oxoglutarate, since the dual targeting of aconitase to both cytosol and mitochondria (Millar *et al.*, 2001) complicates this issue. In summary, experimental evidence to date supports a role for both aspartate amino transferase and a partial TCA cycle within nitrogen assimilation. Further experiments are required to evaluate the relative importance of these routes of 2-oxogluterate production in this process.

8.7.4 Export of redox equivalents to support photorespiration

The oxygenase activity of Rubisco produces phosphoglycolate under atmospheric conditions. In order to prevent wasteful loss of the carbon, phosphoglycolate is converted to phosphoglycerate in the photorespiratory pathway. The reactions of this pathway are distributed across three distinct subcellular compartments: the chloroplasts, mitochondria and peroxisomes. Within the mitochondrial matrix, two molecules of glycine are converted to one molecule of serine with the simultaneous evolution of carbon dioxide and ammonium, and the production of NADH. Thus 75% of the carbon in a phosphoglycolate molecule is re-introduced into metabolism on the operation of this cycle. Oxidation of NADH produced upon glycine oxidation occurs preferentially over that from other substrates such as malate or succinate (Bergman and Ericson, 1983; Dry *et al.*, 1983). These findings have led to suggestions that protein complexes located in the vicinity of the respiratory chain complexes may give reducing equivalents, formed from the oxidation of given substrates, facilitated access to the respiratory chain (Krömer, 1995). Recent evidence in support of such a theory has been supplied by the identification, using proteomics, of many supercomplexes consisting of a wide range of biosynthetic enzymes and components of the ETC (Eubel *et al.*, 2004; see Chapter 6).

Irrespective of how mitochondria prioritise the oxidation of glycine, the further operation of the photorespiratory cycle requires the delivery of redox equivalents to the peroxisomal matrix for the reduction of hydroxypyruvate. The peroxisomal membrance contains porins, which allow the diffusion of NADH. However, cytosolic concentrations of NADH (Heineke *et al.*, 1991) would support only 1% of the respiratory flux (Reumann *et al.*, 1994). An alternative source of NADH could be the peroxisomal oxidation of malate. Considering that malate is present in concentrations that are tenfold higher than ocaloacetate (Heineke *et al.*, 1991) and the activity of peroxisomal MDH is sufficient to catalyse observed photorespiratory fluxes (Heldt and Flugge, 1987), it is likely that in the illuminated leaf under photorespiratory conditions the demands of peroxisomal hydroxypyruvate reductase for reductant are met by internal oxidation of malate, and not by import of NADH (Reumann *et al.*, 1994). The NADH requirement of peroxisomal hydroxypyruvate reductase is equal to the NADH produced upon glycine oxidation in the mitochondrial matrix, leading to the proposal that the NADH produced in the mitochondria is utilised in the peroxisome (Journet *et al.*, 1981). However, given that the export of only 25–50% of the redox equivalents produced in the mitochondrial matrix is insufficient to sustain hydroxypyruvate reduction, another source of NADH is required.

Experimental evidence suggests that the activity of the malate–oxaloacetate shuttle in the chloroplast envelope is high enough to support this reaction (Hatch *et al.*, 1984). However, it remains likely that both chloroplasts and mitochondria simultaneously allocate NADH to the peroxisomes; however, the relative contributions of the two sources remain to be quantified (Krömer and Scheibe, 1996). Despite the fact that the final gene encoding a component of the photorespiratory pathway has been cloned and functionally analysed (Boldt *et al.*, 2005), there remain many gaps in our understanding of the functioning of this crucial pathway. Even fundamental aspects of the photorespiratory pathway, such as the nature of the metabolite shuttles which facilitate it (both ornithine–citrulline and glutamate–glutamine shuttles are proposed) remain unanswered (Taira *et al.*, 2004; Linka and Weber, 2005). Once such questions have been addressed, it will be far easier to assess the importance of mitochondrial function in general, and the mitochondrial TCA cycle in particular, within the photorespiratory process.

8.7.5 Improved photosynthetic performance by modification of TCA cycle activity

Improving plant productivity is now of unprecedented importance given the twin problems of environmental deterioration and the world population explosion (Miyagawa *et al.*, 2001). Whilst many successful manipulations of agronomic yield have involved the modification of sink metabolism (Stark *et al.*, 1992; Regierer *et al.*, 2002; Fridman *et al.*, 2004; Davuluri *et al.*, 2005; Geigenberger *et al.*, 2005), perhaps the major determinant of crop growth and yield is photosynthetic carbon metabolism (Galtier *et al.*, 1993; Sweetlove *et al.*, 1998; Miyagawa *et al.*, 2001; Schauer *et al.*, 2006). However, very few molecular approaches aimed at improving the efficiency of these pathways have proved successful. Those strategies that did work generally involved expression of cyanobacterial enzymes in higher plants. The expression of fructose-1,6-/sedoheptulose-1,7-bisphosphatase in tomato (*S. lycopersicum*) (Miyagawa *et al.*, 2001), or of *ict*B, a gene involved in HCO_3^- accumulation, in tobacco and *Arabidopsis* (Lieman-Hurwitz *et al.*, 2003), resulted in increased photosynthetic rates and growth under normal and limiting carbon dioxide concentrations, respectively. A recent report showed that heterologous expression in tobacco of the *ots*A gene from *E. coli*, encoding trehalose phosphate synthase, resulted in elevated rates of photosynthesis and growth (Pellny *et al.*, 2004). Interestingly, however, none of the successful strategies described above have been the result of alterations in the levels of the endogenous plant proteins. As mentioned above, as part of an ongoing project to determine the function of the TCA cycle in the illuminated leaf, we previously comprehensively phenotyped the tomato wild species mutant *Aco-1*, which exhibits a deficiency in expression of one of the two isoforms of aconitase present in the tomato (Tanksley *et al.*, 1992; Carrari *et al.*, 2003; see Sections 8.2.1.4 and 8.7.2). Biochemical analyses of the *Aco-1* mutant revealed that it exhibited a decreased flux through the TCA cycle and decreased levels of TCA cycle intermediates, but was characterised by elevated adenylate levels, and an increased rate of carbon dioxide assimilation. In addition, although it must be taken into account that wild species tomato is a green-fruited species bearing very small fruits (Schauer *et al.*, 2005), these plants were characterised by a dramatically increased fruit weight. In order to test the generality of this effect, we generated and characterised plants exhibiting a reduced expression of the mitochondrial isoform of MDH (Nunes-Nesi *et al.*, 2005). These studies revealed that transformants exhibiting significantly reduced activities of MDH also show enhanced photosynthetic activity and aerial growth under atmospheric conditions. Accumulation of carbohydrates and redox-regulated components such as ascorbate were also markedly elevated in the transformants. Interestingly, re-analysis of the GC-MS chromatograms of the *Aco-1* mutant revealed that this was also characterised by elevated ascorbate levels (Urbanczyk-Wochniak *et al.*, 2006), whilst metabolic profiling of lines expressing an acetyl CoA hydrolase, and exhibiting reduced rates of photosynthesis and a stunted phenotype, revealed that these lines had normal ascorbate levels (Bender-Machado *et al.*, 2004). These data are in close agreement with those from other researchers (Raghavendra and Padmasree, 2003) in hinting that the importance of the respiratory pathways in photosynthetic metabolism is greater than once imagined. That ascorbate levels are increased following a restriction of flux through the TCA cycle is fascinating, particularly in light of recent reports suggesting that the terminal enzyme of ascorbate biosynthesis, GLDH is coupled to the cytochrome pathway (Bartoli *et al.*, 2000). Intriguingly, the MDH transformants exhibit an elevated capacity for the synthesis of ascorbate from L-galactono-lactone. Much is known concerning the importance of ascorbate within photosynthesis. Ascorbate acts in the Mehler peroxidase reaction with ascorbate peroxidase to regulate the redox state of photosynthetic electron carriers and acts as a co-factor for violaxanthin deepoxidase, an enzyme involved in xanthophyll cycle-mediated photoprotection (Smirnoff and Wheeler, 2000; Danna *et al.*, 2003). The ascorbate-redox state has also been shown recently to control guard cell signalling and stomatal movement (Chen and Gallie, 2004), as well as the expression levels of both nuclear and chloroplastic components of the photosynthetic apparatus (Smirnoff, 2000; Kiddle *et al.*, 2003; Pastori *et al.*, 2003). An additional function of ascorbate is that of co-factor for prolyl hydroxylase, an enzyme that post-translationally hydroxylates proline residues in the cell wall Hyp-rich glycoproteins required for cell division and expansion (Smirnoff and Wheeler, 2000). The exact mechanism linking ascorbate to the rate of photosynthesis is currently unclear. Feeding experiments revealed that photosynthetic assimilation could be elevated within 2 h (Nunes-Nesi *et al.*, 2005); however, microarray analysis of a transgenic MDH line, and of the *Aco-1* mutant, revealed large-scale modification of the transcription of genes associated with both the Calvin cycle and the light reactions of photosynthesis (Urbanczyk-Wochniak *et al.*, 2006). It is conceivable that the increase in photosynthesis occurs in compensation for the reduced flux through the TCA cycle in order to elevate chloroplastic energy and redox production (Figure 8.9). However, many more experiments will be required to clarify such a hypothesis. Not least of these will be an elucidation of the precise mechanism by which ascorbate or any other mitochondrially derived signal affects the rate of photosynthesis.

Positive effects of respiratory processes on photosynthesis are not confined to manipulation of the TCA cycle. Indeed, there are many examples interlinking the

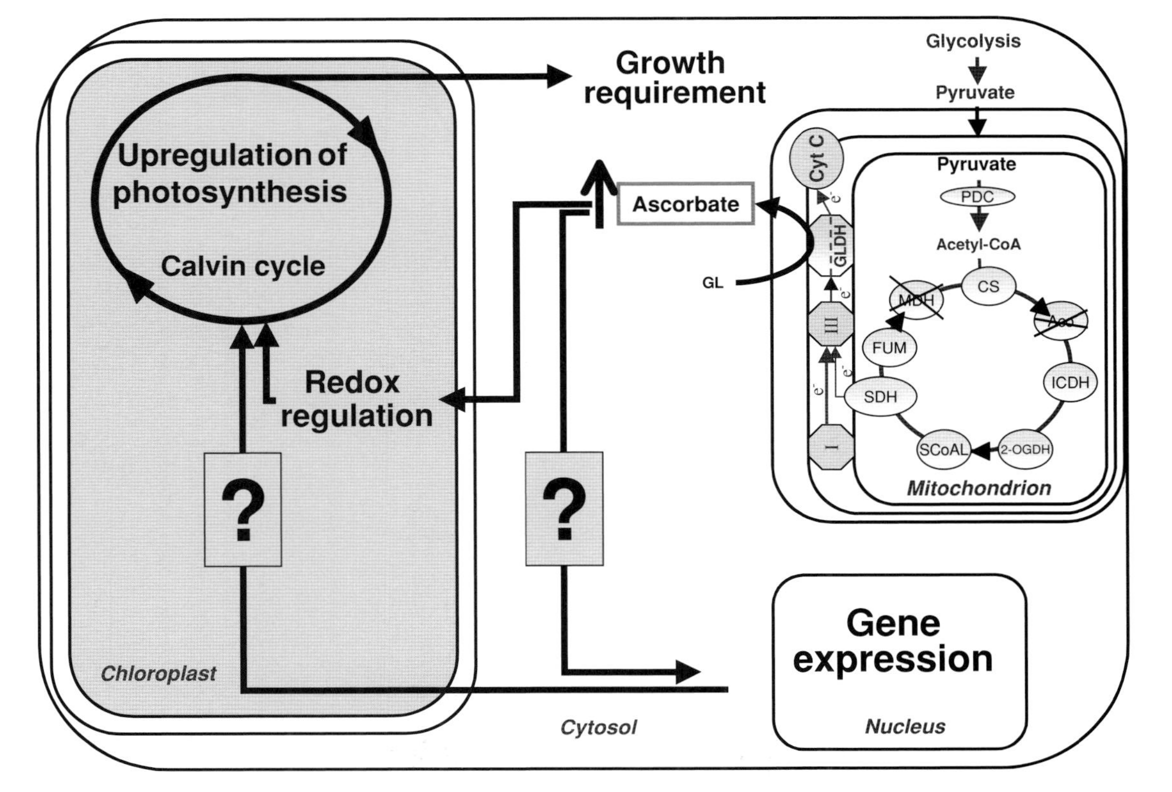

Figure 8.9 Schematic representation of the consequences for photosynthesis of deficiency in either of the TCA-cycle enzymes Aco or MDH. We hypothesise that the observed increased *in vivo* activity of GLDH (L-galactono-1,4-lactone dehydrogenase), which catalyses the conversion of galactono-lactone to ascorbate and is coupled to the mitochondrial ETC, led to an upregulation of photosynthesis by an undefined mechanism that is likely to involve either modulation of gene expression, redox regulation or merely efficient removal of photosynthate to support enhanced growth. (Abbreviations are as defined in Figure 8.1 and as follows: cyt c, cytocrome *c*; GL, L-galactono-1,4-lactone.)

mitochondrial ETC and photosynthetic efficiency (Krömer *et al.*, 1988; Padmasree and Raghavendra, 1999a,b; Bartoli *et al.*, 2005; Ishizaki *et al.*, 2005). Whilst current knowledge suggests several different mechanisms linking mitochondrial and chloroplast metabolism, the most common are redox homeostasis (such as described above for MDH) and the malate valve. Given that the malate value has been reviewed recently (Scheibe, 2004; Scheibe *et al.*, 2005), we will focus this discussion on other mechanisms linking mitochondrial and chloroplast function. That said, a few properties of malate are particularly noteworthy. First, transporters that allow facile transport of malate between compartments have recently been identified (Taniguchi *et al.*, 2002; Emmerlich *et al.*, 2003; Rennè *et al.*, 2003), and further analysis of these should provide a more comprehensive understanding of the regulation of this process. Second, it has been long known that mitochondria contain an effective capacity to use up excess reducing equivalents produced by the chloroplast, and as such are able to prevent thylakoid membrane damage (Gilmore, 1997; Niyogi *et al.*, 1998; Niyogi, 1999). The mechanisms whereby mitochondrial function facilitates optimum rates of photosynthesis under stress conditions have recently received a lot of attention. One example is provided by the recent discovery that AOX expression is highly upregulated under drought conditions (Bartoli *et al.*, 2005). This upregulation provides an effective sink for reducing power, thereby preventing the accumulation of excess reducing equivalents and decreasing the probability of loss of photosynthetic function. Similarly, mitochondria have been reported to oxidise excess photosynthetic reducing equivalents under cold, bright conditions (Raghavendra *et al.*, 1994; Saradadevi and Raghavendra, 1994; Hurry *et al.*, 1995; Atkin *et al.*, 2000a). Moreover, the use of inhibitors of both cytochrome and alternative pathways leads to a large decrease in the rate of photosynthesis (Krömer *et al.*, 1988, 1993; Padmasree and Raghavendra, 1999a,b) and also highlights a major role of these pathways in protecting photosynthesis against photoinhibition (Saradadevi and Raghavendra, 1992). A recent study in broad bean using the same inhibitors revealed that inhibition of AOX causes a reduction in the rate of photosynthetic oxygen evolution even under low irradiances and suggest that this pathway provides chloroplasts with flexible strategies against photoinhibition (Yoshida *et al.*, 2006). Meanwhile, strong evidence for the critical role of the cytochrome pathway in modulating photosynthesis was provided by the characterisation of *N. sylvestris* CMS mutants lacking one of the subunits of complex I, and displaying markedly reduced rates of photosynthesis (Sabar *et al.*, 2000; Dutilleul *et al.*, 2005). A final example provided by the discovery, as detailed in Section 8.5.4, is that ETFQO, the mitochondrial electron-transfer flavoprotein complex, plays an important role in the degradation of chlorophyll during dark-induced senescence (Ishizaki *et al.*, 2005). However, whilst not underestimating the importance of this observation it appears to be of relevance only under extreme environmental circumstances.

When taken together, the multiplicity of mechanisms linking mitochondrial and plastid (and cytosolic and peroxisomal) function serves to reinforce the high degree of metabolic regulation inherent in the plant cell. In addition, these examples highlight the essentiality of the mitochondria to plant photosynthetic metabolism. Intriguingly, the fact that modifying the mitochondrial ETC has different effects on photosynthesis relative to the effects of modifying TCA cycle further suggests that these effects of mitochondrial function are likely independent of one another.

In the final sections of this chapter, we highlight some technical and biological questions that need to be answered to further our understanding of these important and complex interactions. For example, we discuss the demonstrated roles of mitochondrial metabolism in floral development, root exudation and interaction with the rhizosphere and climacteric fruit ripening.

8.7.6 Mitochondrial metabolism and floral development

The plant mitochondrial genome plays a fundamental role during male gametogenesis, and changes in mitochondrial gene expression are known to cause specific types of CMS (Schnable and Wise, 1998; see Chapter 9). CMS has been found to occur in over 150 species and has often led to the assertion that mitochondrial function is of particular importance during flower development (Hanson, 1991). We will not discuss this in detail here since it is the subject of several excellent recent reviews (Schnable and Wise, 1998; Linke and Börner, 2005). Studies on CMS mutants have revealed that, at least in many cases, the sterile phenotype results from lesions in the mitochondrial DNA which causes the production of new chimeric open reading frames composed of fragments derived from other genes, and/or non-coding sequences (Schnable and Wise, 1998; Hanson and Bentolila, 2004; Linke and Börner, 2005). These novel polypeptides may lead to the specific disruption of pollen development (see Linke and Börner, 2005). However, various studies have also suggested that CMS may be a consequence of impaired respiratory energy production, or a defect in some other biosynthetic function of the mitochondria (Landschütze *et al.*, 1995a; Bergman *et al.*, 2000). Confirmation that mitochondrial function plays an important role in flower development has been provided by studies correlating the CMS phenotype with changes in the expression of genes encoding mitochondrial proteins leading to reduced ATP production (Bergman *et al.*, 2000; Ducos *et al.*, 2001; Sabar *et al.*, 2003) or a reduced carbohydrate accumulation (Datta *et al.*, 2002). Increased steady-state abundance of mRNAs encoding NAD-malic enzyme (Winning *et al.*, 1994), the E1α subunit of PDC (Grof *et al.*, 1995), citrate synthase (Landschütze *et al.*, 1995a) and aconitase (Peyret *et al.*, 1995) have been detected in developing flowers, which provides circumstantial evidence in support of increased TCA cycle enzyme activities, and consequently flux, at this developmental stage – a fact that is consistent with a role for these enzymes in normal floral development. In keeping with these results, it has been demonstrated recently that mitochondrial gene expression is differentially regulated during male gametophyte development in *Arabidopsis*, possibly reflecting a changing respiratory status of the mitochondria during development of stamens and pollen (Kalantidis *et al.*, 2002). However, in no case has an increased maximum catalytic activity been demonstrated, and there are no data available that demonstrate particularly high respiratory rates in developing flowers. Our understanding of the importance of TCA cycle function during flower development has been strengthened by the demonstration that expression of a citrate synthase antisense gene in tobacco leads to sterility (Landschütze *et al.*, 1995b). A recent study involving the antisense expression of a PDC E1α subunit under the control of a tapetum-specific promoter in tobacco indicated that inhibition of pyruvate dehydrogenase activity in the tapetum is sufficient to cause male sterility in tobacco (Yui *et al.*, 2003).

A regulatory mitochondrial disulphide thioredoxin has been shown to be essential for normal pollen development in *Brassica* (Cabrillac *et al.*, 2001). In a recent proteomic survey, mitochondrial aldehyde dehydrogenase, one of the genes responsible for the restoration of maize sterility (*Rf2* restorer) (Schnable and Wise, 1998), has been identified as a potential target for thioredoxin (Balmer *et al.*, 2004) giving further credence to the theory that male gametophyte development is particularly sensitive to perturbations in mitochondrial gene expression and metabolism. Furthermore, recent studies in petunia highlighted a crucial role for ethanol fermentative metabolism in pollen style development (Gass *et al.*, 2005). The fact that mutation of alcohol dehyrogenase does not affect male gametophyte function but mutation of pyruvate decarboxylase does led Gass *et al.* (2005) to propose that pyruvate decarboxylase rather than alcohol dehyrogenase is the critical enzyme in a novel pollen-specific pathway and this pathway serves to bypass pyruvate dehydrogenase enzymes thereby maintaining biosynthetic capacity and energy production under the unique conditions prevailing during pollen. The precise role of mitochondrial metabolism in this situation, however, remains to be fully evaluated. The above-mentioned proteomic survey of Balmer *et al.* (2004) also revealed that many of the TCA cycle enzymes are redox regulated, hinting at, at least occasional, co-ordinate control during gametophyte development. However, given that β-oxidation (Baker *et al.*, 2006) and flavonoid formation (van der Meer *et al.*, 1992; Fischer *et al.*, 1997) have also been heavily implicated in flower development, it is clear that much more research is required before the metabolic control underlying this complex process is understood beyond a cursory level.

8.7.7 *Role of mitochondrial metabolism in root nutrient uptake*

Mitochondrial metabolism plays an important role on root growth and development under normal nutrient supply. A constitutive reduction of TCA cycle activity has been shown to diminish root dry weight in tomato plants (Carrari *et al.*, 2003; Nunes-Nesi *et al.*, 2005); however, the importance of mitochondrial metabolism to nutrient uptake is more evident under stress conditions, such as limited phosphate supply, or under conditions where aluminium concentrations become toxic for the plant, inhibiting root growth and development. The rhizosphere is defined as the zone surrounding plant roots that is modified by root activity and is the area within which the plant perceives and responds to their environment, exchanging organic for inorganic substances with the soil system (Ryan *et al.*, 2001; Walker *et al.*, 2003; Bais *et al.*, 2006). Nutrient deficiency, toxic cations and anoxia are all associated with alterations in organic acid exudation (Ryan *et al.*, 2001). Organic anions can benefit the plants under these conditions by increasing the possibility of nutrient uptake by roots, or by reducing the accumulation of potentially toxic metabolites in the cytoplasm (Ryan *et al.*, 2001; Bais *et al.*, 2006). Citrate and malate have been implicated in both aluminium tolerance and enhanced phosphate nutrition. Enzymes involved in citrate metabolism have been manipulated genetically to enhance organic anion exudation in the roots (Neumann and Martinoia, 2002), whilst aluminium tolerance was successfully obtained in transgenic alfalfa plants over-expressing either MDH (Tesfaye *et al.*, 2001) or mitochondrial citrate synthase (Anoop *et al.*, 2003). Moreover, carrot cell lines and *Arabidopsis* plants over-expressing mitochondrial citrate synthase also displayed increased aluminium tolerance (Koyama *et al.*, 1999, 2000). Shane *et al.* (2004) demonstrated that during cluster root maturation in harsh hakea (a member of the Proteaceae) the expression of AOX protein increased prior to the time when citrate and malate exudation peaked and also revealed that citrate and isocitrate synthesis and accumulation contributed markedly to the subsequent burst of citrate and malate exudation. However, despite the apparent consistency and reproducibility of these results, there is currently no clear consensus on the mechanism by which tolerance to aluminium and other toxic metals is achieved. Many authors conclude that it is the transport of organic anions, rather than their synthesis that confers this tolerance, and that citrate synthase activity is not directly related to export (Delhaize *et al.*, 2003; Hayes and Ma, 2003; Yang *et al.*, 2003; Zhu *et al.*, 2003). However, this theory is very much contested (Koyama *et al.*, 2000; Li *et al.*, 2002; Rengel, 2002; Anoop *et al.*, 2003).

8.7.8 Metabolic activity of mitochondria during climacteric fruit ripening

Climacteric fruit differ from non-climacteric by exhibiting a characteristic burst of ethylene biosynthesis, and respiration rate during ripening (Giovannoni, 2001; Alexander and Grierson, 2002; White, 2002). Many studies have been performed to try to understand the mechanisms underpinning the respiratory burst in climacteric fruits (Almeida *et al.*, 1999; Costa *et al.*, 1999; Sluse and Jarmuszkiewicz, 2000; Considine *et al.*, 2001; Holtzapffel *et al.*, 2002; Navet *et al.*, 2003). In mango fruits, changes in several respiratory chain proteins have been observed within this developmental time frame, revealing a post-climacteric increase in the expression profile, and protein accumulation of AOX and UCP (Considine *et al.*, 2001). Similarly, the respiration rates of tomato display a modest peak during the transition period from green/orange to orange/red fruits. However, the respiratory capacity of mitochondria isolated from tomato fruit does not increase during the climacteric burst, despite the change in abundance of many mitochondrial proteins during ripening (Holtzapffel *et al.*, 2002). In the same study, the authors observed a significant increase in the abundance of UCP and AOX but a reduction in the abundance of ATP synthase and Reiske iron sulfur protein during ripening (Holtzapffel *et al.*, 2002). Almeida *et al.* (1999) observed that AOX abundance, ATP-synthesis-sustained respiration and cyanide-resistant respiration of mitochondria isolated from tomato fruit decreased with ripening from green to stage. In contrast, UCP protein abundance and the total UCP-sustained respiration of isolated mitochondria decreased from the yellow stage onwards (Almeida *et al.*, 1999). It has been suggested that AOX and UCP could be responsible for the respiration increase at the end of ripening and the cytochrome pathway could be implicated in the climacteric respiratory burst before the onset of ripening (Navet *et al.*, 2003). Together, these results indicate a shift in mitochondrial activity in fruit during ripening, from the tightly coupled mitochondria during the early stages of ripening to the less constrained (more uncoupled) mitochondria during the climacteric and subsequent senescence stage (Holtzapffel *et al.*, 2002). In comparison with proteins of the respiratory chain, little is known about the activity of TCA cycle enzymes during fruit ripening. It has been reported that the protein abundance of components of the pyruvate dehydrogenase and 2-oxoglutarate dehydrogenase complexes decline at later stages of fruit ripening (Holtzapffel *et al.*, 2002). Similarly, Jeffery *et al.* (1984) and Goodenough *et al.* (1985) reported decreases in the specific activities of malate dehydrogenase, citrate synthase, NAD-dependent isocitrate dehydrogenase and NAD-malic enzyme in breaker stage tomato relative to green tomatoes. In contrast, analyses of the expression of fumarase, the α- and β-subunits of succinyl-CoA ligase and mitochondrial MDH indicate an increase in expression in tomato from green to orange/red stage (60–65 days after flowering) with a subsequent decrease during the post-climacteric senescent stage (70 days after flowering; F. Carrari, C. Guimaraes-Studart, A. Nunes-Nesi and A. R. Fernie, unpublished results). When taken together, these results provide weak correlative evidence that the TCA cycle enzymes play a role in, and contribute substantially to, the climacteric in tomato. To build on this correlative evidence, more direct analyses, including the analysis of transgenic plants with altered expression of these enzymes under the control of fruit-specific, or chemically inducible promoters, are required in order to achieve a complete understanding of the physiological and biochemical roles of TCA cycle enzymes during fruit ripening.

8.8 Concluding remarks – future prospects for improved understanding of the interaction between mitochondrial and extra-mitochondrial metabolism

Improved subcellular resolution of metabolism by improvements to current techniques of non-aqueous fractionation (Farre *et al.*, 2001; Tiessen *et al.*, 2002), alongside more sensitive flux profiling methodologies (Schwender *et al.*, 2003; Roessner-Tunali *et al.*, 2004), and the adaptation of global profiling technologies will likely enable pursuit of new avenues of research in order to increase our understanding of the complex networks governing the role of mitochondrial function within cellular metabolism (Sweetlove and Fernie, 2005). The continued identification and characterisation of carrier proteins (Picault *et al.*, 2002; Millar and Heazlewood, 2003), alongside further analysis of classical (Sabar *et al.*, 2000; Dutilleul *et al.*, 2005; Ishizaki *et al.*, 2005) and plant-specific (Considine *et al.*, 2003; Smith *et al.*, 2004; Escobar *et al.*, 2004, 2006; Umbach *et al.*, 2006) features of the mitochondrial ETC, should allow a more comprehensive analysis of the interactions of the major respiratory pathways. In addition, several recently published reports suggest that it is likely that many aspects of the regulation of mitochondrial metabolism have yet to be uncovered, for example (i) the recent demonstration that several of the polypeptides that constitute the mitochondrial TCA cycle are potential targets for redox regulation (Balmer *et al.*, 2004), (ii) the preliminary characterisation of the allosteric properties of succinyl-CoA ligase (Studart-Guimaraes *et al.*, 2005) and (iii) the observation that enzymes of glycolysis are functionally associated to the outer mitochondrial membrane (Giege *et al.*, 2003). Importantly, contextual assessments of the role of mitochondrial metabolism under various biotic, and abiotic stress conditions, and within various developmental processes are likely research targets for the next few years. Specific examples of processes in which mitochondrial function is believed to be important include photorespiration, cellular redox balancing, flower development, plant–rhizobia interactions and climacteric fruit ripening. However, the evidence for mitochondrial involvement in some of these processes is rather fragmentary. Recently developed chemically inducible promoters have been used successfully to facilitate the dissection of primary, from pleiotropic phenotypic effects (Junker *et al.*, 2004) and are likely to be of considerable use in the future. In addition, retrograde signalling, from the mitochondrial and controlling extra-mitochondrial gene expression, is a poorly understood phenomenon that is likely to attract attention from researchers seeking novel regulatory mechanisms (Dojcinovic *et al.*, 2005; Zarkovic *et al.*, 2005). However, despite the fact that our knowledge of mitochondrial metabolism is far from complete, we hope that the information in this chapter has made it clear that mitochondrial metabolism plays a central role in a wide range of essential metabolic pathways and biological processes, a fact that goes some way to justify the assertion of Lane (2005) that life on earth would not exist as we know it in the absence of these organelles.

References

Abiko T, Obara M, Ushioda A, Hayakawa T, Hodges M and Yamaya T (2005) Localization of NAD-isocitrate dehydrogenase and glutamate dehydrogenase in rice roots: Candidates for providing carbon skeletons to NADH-glutamate synthase. *Plant Cell Physiol* **46**, 1724–1734.

Affourtit C, Krab K, Leach GR, Whitehouse DG and Moore AL (2001) New insights into the regulation of plant succinate dehydrogenase on the role of the protonmotive force. *J Biol Chem* **276**, 32567–32574.

Alexander L and Grierson D (2002) Ethylene biosynthesis and action in tomato: a model for climacteric fruit ripening. *J Exp Bot* **53**, 2039–2055.

Almeida AM, Jarmuszkiewicz W, Khomsi H, Arruda P, Vercesi AE and Sluse FE (1999) Cyanide-resistant, ATP-synthesis-sustained, and uncoupling-protein-sustained respiration during postharvest ripening of tomato fruit. *Plant Physiol* **119**, 1323–1329.

Anoop VM, Basu U, McCammon MT, McAlister-Henn L and Taylor GJ (2003) Modulation of citrate metabolism alters aluminum tolerance in yeast and transgenic canola overexpressing a mitochondrial citrate synthase. *Plant Physiol* **132**, 2205–2217.

ap Rees T (1987) Compartmentation of plant metabolism. In: *The Biochemistry of Plants*, Vol. 12 (ed. Davies DD), pp. 87–115. Academic Press, New York.

Artus N and Edwards G (1985) NAD-malic enzyme from plants. *FEBS Lett* **182**, 225–233.

Atkin OK, Evans JR, Ball MC, Lambers H and Pons TL (2000b) Leaf respiration of snow gum in the light and dark: interactions between temperature and irradiance. *Plant Physiol* **122**, 915–923.

Atkin OK, Holly C and Ball MC (2000a) Acclimation of snow gum (*Eucalyptus pauciflora*) leaf respiration to seasonal and diurnal variations in temperature: the importance of changes in the capacity and temperature sensitivity of respiration. *Plant Cell Environ* **23**, 15–26.

Atkin OK, Millar AH, Gardestrom P and Day DA (2000c) Photosynthesis, carbohydrate metabolism and respiration in leaves of higher plants. In: *Photosynthesis: Physiol and Metab*, Vol. 9 (eds Leegood RC, Sharkey TD and von Caemmerer S), pp. 153–185. Kluwer Academic Publishers, Dordrecht, The Netherlands.

Attucci S, Rivoal J, Brouquisse R, Carde JP, Pradet A and Raymond P (1994) Characterization of a mitochondrial NADP-dependent isocitrate dehydrogenase in axes of germinating sunflower seeds. *Plant Sci* **102**, 49–59.

Avelange MH, Thiery JM, Sarrey F, Gans P and Rébeillé F (1991) Mass-spectrometric determination of O_2 and CO_2 gas-exchange in illuminated higher plant cells. Evidence for light-inhibition of substrate decarboxylations. *Planta* **183**, 150–157.

Babiychuk E, Muller F, Eubel H, Braun HP, Frentzen M and Kushnir S (2003) *Arabidopsis* phosphatidylglycerophosphate synthase 1 is essential for chloroplast differentiation, but is dispensable for mitochondrial function. *Plant J* **33**, 899–909.

Bais HP, Weir TL, Perry LG, Gilroy S and Vivanco JM (2006) The role of root exudates in rhizosphere interactions with plants and other organisms. *Annu Rev Plant Biol* **57**, 233–266.

Baker A, Graham IA, Holdsworth M, Smith SM and Theodoulou FL (2006) Chewing the fat: β-oxidation in signalling and development. *Trends Plant Sci* **11**(3), 124–132.

Baldet P, Alban C and Douce R (1997) Biotin synthesis in higher plants: purification and characterization of bioB gene product equivalent from *Arabidopsis thaliana* overexpressed in *Escherichia coli* and its subcellular localization in pea leaf cells. *FEBS Lett* **419**, 206–210.

Baldet P, Gerbling H, Axiotis S and Douce R (1993) Biotin biosynthesis in higher plant cells. Identification of intermediates. *Eur J Biochem* **217**, 479–485.

Baldet P and Ruffet ML (1996) Biotin synthesis in higher plants: isolation of a cDNA encoding *Arabidopsis thaliana* bioB-gene product equivalent by functional complementation of a biotin auxotroph mutant bioB105 of *Escherichia coli* K12. *C R Acad Sci III* **309**, 99–106.

Balk J and Leaver CJ (2001) The PET1-CMS mitochondrial mutation in sunflower is associated with premature programmed cell death and cytochrome *c* release. *Plant Cell* **13**, 1803–1818.

Balk J and Lobreaux S (2005) Biogenesis of iron–sulfur proteins in plants. *Trends Plant Sci* **10**, 324–331.

Balmer Y, Vensel WH, Tanaka CK, *et al.* (2004) Thioredoxin links redox to the regulation of fundamental processes of plant mitochondria. *Proc Natl Acad Sci USA* **101**, 2642–2647.

Bartoli CG, Gomez F, Gergoff G, Guiamét JJ and Puntarulo S (2005) Up-regulation of the mitochondrial alternative oxidase enhances photosynthetic electron transport under drought conditions. *J Exp Bot* **56**, 1269–1276.

Bartoli CG, Pastori GM and Foyer CH (2000) Ascorbate biosynthesis in mitochondria is linked to the electron transport chain between complexes III and IV. *Plant Physiol* **123**, 335–344.

Bate GC, Siiltemeyer DF and Fock HP (1988) $^{16}O_2/^{18}O_2$ analysis of oxygen exchange in *Dunaliella tertiolecta.* Evidence for the inhibition of mitochondrial respiration in the light. *Photosynth Res* **16**, 219–231.

Bauwe H and Kolukisaoglu U (2003) Genetic manipulation of glycine decarboxylation: *J Exp Bot* **54**, 1523–1535.

Bedhomme M, Hoffmann M, McCarthy EA, *et al.* (2005) Folate metabolism in plants: an *Arabidopsis* homolog of the mammalian mitochondrial folate transporter mediates folate import into chloroplasts. *J Biol Chem* **280**, 34823–34831.

Beevers H (1961) *Respiratory Metabolism in Plants*, pp. 185–197. Harper and Row, New York.

Behal RH and Oliver DJ (1997) Biochemical and molecular characterization of fumarase from plants. Cloning, sequencing, and expression of the *Arabidopsis thaliana* fumarase gene; purification and characterization of *Pisum sativum* fumarase enzyme. *Arch Biochem Biophys* **347**, 65–74.

Bender-Machado L, Bäuerlein M, Carrari F, *et al.* (2004) Expression of a yeast acetyl CoA hydrolase in the mitochondrion of tobacco plants inhibits growth and restricts photosynthesis. *Plant Mol Biol* **55**, 645–662.

Bergman P, Edqvist J, Farbos I and Glimelius K (2000) Male-sterile tobacco displays abnormal mitochondrial *atp1* transcript accumulation and reduced floral ATP/ADP ratio. *Plant Mol Biol* **42**, 531–544.

Bergman A and Ericson I (1983) Effects of NADH, succinate and malate on the oxidation of glycine in spinach leaf mitochondria. *Physiol Plant* **59**, 421–427.

Bläsing OE, Gibon Y, Gunther M, *et al.* (2005) Sugars and circadian regulation make major contributions to the global regulation of diurnal gene expression in *Arabidopsis*. *Plant Cell* **17**, 3257–3281.

Boldt R, Edner C, Kolukisaoglu U, *et al.* (2005) D-Glycerate 3-kinase, the last unknown enzyme in the photorespiratory cycle in *Arabidopsis*, belongs to a novel kinase family. *Plant Cell* **17**, 2413–2420.

Borecky J, Maia IG, Costa ADT, *et al.* (2001) Functional reconstitution of *Arabidopsis thaliana* plant uncoupling mitochondrial protein (AtPUMP1) expressed in *Escherichia coli*. *FEBS Lett* **505**, 240–244.

Borsani O, Zhu JH, Verslues PE and Zhu JK (2005) Endogenous siRNAs derived from a pair of natural cis-antisense transcripts regulate salt tolerance in *Arabidopsis*. *Cell* **123**, 1279–1291.

Bourguignon J, Macherel D, Neuburger M and Douce R (1992) Isolation, characterization, and sequence analysis of a cDNA clone encoding L-protein, the dihydrolipoamide dehydrogenase component of the glycine cleavage system from pea-leaf mitochondria. *Eur J Biochem* **204**, 865–873.

Brouquisse R, Nishimura M, Gaillard J and Douce R (1987) Characterization of a cytosolic aconitase in higher plant cells. *Plant Physiol* **84**, 1402–1407.

Bryce JH and ap Rees T (1985) Rapid decarboxylation of the products of dark fixation of CO_2 in roots of *Pisum* and *Plantago*. *Phytochem* **24**, 1635–1638.

Budde RJA and Randall DD (1990) Pea leaf mitochondrial pyruvate dehydrogenase complex is inactivated *in vivo* in a light-dependent manner. *Proc Natl Acad Sci USA* **87**, 673–676.

Burger G, Lang BF, Braun HP and Marx S (2003) The enigmatic mitochondrial ORF ymf39 codes for ATP synthase chain b. *Nuc Acids Res* **31**, 2353–2360.

Cabrillac D, Cock JM, Dumas C and Gaude T (2001) The S-locus receptor kinase is inhibited by thioredoxins and activated by pollen coat proteins. *Nature* **410**, 220–223.

Caiveau O, Fortune D, Cantrel C, Zachowski A and Moreau F (2001) Consequences of omega-6-oleate desaturase deficiency on lipid dynamics and functional properties of mitochondrial membranes of *Arabidopsis* thaliana. *J Biol Chem* **276**, 5788–5794.

Canel C, BaileySerres JN and Roose ML (1996) Molecular characterization of the mitochondrial citrate synthase gene of an acidless pummelo (*Citrus maxima*). *Plant Mol Biol* **31**, 143–147.

Canellas PF and Wedding RT (1984) Kinetic properties of NAD malic enzyme from caulilower. *Arch Biochem Biophys* **229**, 414–425.

Canvin DT and Beevers H (1961) Sucrose synthesis from acetate in the germinating castor bean: kinetics and pathway. *J Biol Chem* **236**, 988–995.

Canvin DT, Berry JA, Badger MR, Fock H and Osmond CB (1980) O_2 exchange in leaves in the light. *Plant Physiol* **66**, 302–307.

Carrari F, Nunes-Nesi A, Gibon Y, Lytovchenko A, Ehlers Loureiro M and Fernie AR (2003) Reduced expression of aconitase results in an enhanced rate of photosynthesis and marked shifts in carbon partitioning in illuminated leaves of wild species tomato. *Plant Physiol* **133**, 1322–1335.

Cashin McBG, Cossins EA and Canvin DT (1988) Dark respiration during photosynthesis in wheat leaf slices. *Plant Physiol* **87**, 155–161.

Catoni E, Desimone M, Hilpert M, *et al.* (2003a) Expression pattern of a nuclear encoded mitochondrial arginine-ornithine translocator gene from *Arabidopsis*. *BMC Plant Biol* **3**, 1.

Catoni E, Schwab R, Hilpert M, *et al.* (2003b) Identification of an *Arabidopsis* mitochondrial succinate-fumarate translocator. *FEBS Lett* **534**, 87–92.

Chabregas SM, Luche DD, Farias LP, *et al.* (2001) Dual targeting properties of the N-terminal signal sequence of *Arabidopsis thaliana* THI1 protein to mitochondria and chloroplasts. *Plant Mol Biol* **46**, 639–650.

Chabregas SM, Luche DD, Van Sluys MA, Menck CFM and Silva-Filho MC (2003) Differential usage of two in-frame translational start codons regulates subcellular localization of *Arabidopsis thaliana* THI1. *J Cell Sci* **116**, 285–291.

Che P, Wurtele ES and Nikolau BJ (2002) Metabolic and environmental regulation of 3-methylcrotonyl-coenzyme A carboxylase expression in *Arabidopsis*. *Plant Physiol* **129**, 625–637.

Chen RD and Gadal P (1990) Do the mitochondria provide the 2-oxoglutarate needed for glutamate synthesis in higher plant chloroplasts? *Plant Physiol Biochem* **28**, 141–145.

Chen Z and Gallie DR (2004) The ascorbic acid redox state controls guard cell signalling and stomatal movement. *Plant Cell* **16**, 1143–1162.

Considine MJ, Daley DO and Whelan J (2001) The expression of alternative oxidase and uncoupling protein during fruit ripening in mango. *Plant Physiol* **126**, 1619–1629.

Considine MJ, Goodman M, Echtay KS, *et al.* (2003) Superoxide stimulates a proton leak in potato mitochondria that is related to the activity of uncoupling protein. *J Biol Chem* **278**, 22298–22302.

Considine MJ, Holtzapffel RC, Day DA, Whelan J and Millar AH (2002) Molecular distinction between alternative oxidase from monocots and dicots. *Plant Physiol* **129**, 949–953.

Cornah, JE, Germain V, Ward JL, Beale MH and Smith SM (2004) Lipid utilization, gluconeogenesis, and seedling growth in *Arabidopsis* mutants lacking the glyoxylate cycle enzyme malate synthase. *J Biol Chem* **279**, 42916–42923.

Cornu S, Pireaux JC, Gerard J and Dizengremel P (1996) $NAD(P)^+$-dependent isocitrate dehydrogenases in mitochondria purified from *Picea abies* seedlings. *Physiol Plant* **96**, 312–318.

Cossins EA (1984) In: *Folates and Pterins*, Vol. 1 (eds Blakeley RL and Benkovic SJ), pp. 1–59. Wiley-Interscience, New York.

Cossins E (2000) The fascinating world of folate and one-carbon metabolism. *Can J Bot* **78**, 691–708.

Costa ADT, Nantes IL, Jezek P, Leite A, Arruda P and Vercesi AE (1999) Plant uncoupling mitochondrial protein activity in mitochondria isolated from tomatoes at different stages of ripening. *J Bioenerg Biomembr* **31**, 527–533.

Courtois-Verniquet F and Douce R (1993) Lack of aconitase in glyoxysomes and peroxisomes. *Biochem J* **294**, 103–107.

Craig DW and Wedding RT (1980a) Regulation of the 2-oxoglutarate dehydrogenase lipoate succinyltransferase complex from cauliflower by nucleotide. Pre-steady state kinetics and physical studies. *J Biol Chem* **255**, 5769–5775.

Craig DW and Wedding RT (1980b) Regulation of the 2-oxoglutarate dehydrogenase lipoate succinyltransferase complex from cauliflower by nucleotide. Steady state kinetic studies. *J Biol Chem* **255**, 5763–5768.

Cunillera N, Boronat A and Ferrer A (1997) The *Arabidopsis thaliana* FPS1 gene generates a novel mRNA that encodes a mitochondrial farnesyl-diphosphate synthase isoform. *J Biol Chem* **272**, 15381–15388.

Curry RA and Ting IP (1976) Purification, properties, and kinetic observations on the isoenzymes of NADP isocitrate dehydrogenase of maize. *Arch Biochem Biophys* **176**, 501–509.

Danna CH, Bartoli CG, Sacco F, *et al.* (2003) Thylakoid bound ascorbate peroxidase mutant exhibits impaired electron transport and photosynthetic activity. *Plant Physiol* **132**, 2116–2125.

Dakshinamurti K and Cauhan J (1989) Biotin. *Vitam Horm* **45**, 337–384.

Daschner K, Couee I and Binder S (2001) The mitochondrial isovaleryl-coenzyme a dehydrogenase of *Arabidopsis* oxidizes intermediates of leucine and valine catabolism. *Plant Physiol* **126**, 601–612.

Datta R, Chamusco KC and Chourey PS (2002) Starch biosynthesis during pollen maturation is associated with altered patterns of gene expression in maize. *Plant Physiol* **130**, 1645–1656.

Daum G and Vance JE (1997) Import of lipids into mitochondria. *Prog Lipid Res* **36**, 103–130.

Davuluri GR, van Tuinen A, Fraser PD, *et al.* (2005) Fruit-specific RNAi-mediated suppression of DET1 enhances carotenoid and flavonoid content in tomatoes. *Nat Biotech* **23**, 890–895.

Day DA, Dry IB, Soole KL, Wiskich JT and Moore AL (1991) Regulation of alternative pathway activity in plant mitochondria: deviation from Q-pool behavior during oxidation of NADH and quinols. *Plant Physiol* **95**, 948–953.

Day DA and Hanson JB (1977) Pyruvate and malate transport and oxidation in corn mitochondria. *Plant Physiol* **59**, 630–635.

Day DA, Neuberger M and Douce R (1984) Activation of NAD-linked malic enzyme in intact mitochondria by exogenous CoA. *Arch Biochem Biophys* **231**, 233–234.

Day DA and Wiskich JT (1995) Regulation of alternative oxidase activity in higher plants. *J Bioenerg Biomem* **27**, 379–385.

De Bellis L, Tsugeki R, Alpi A and Nishimura M (1993) Purification and characterization of aconitase isoforms from etiolated pumpkin cotyledons. *Physiol Plant* **88**, 485–492.

Delhaize E, Ryan PR, Hocking PJ and Richardson AE (2003) Effects of altered citrate synthase and isocitrate dehydrogenase expression on internal citrate concentrations and citrate efflux from tobacco (*Nicotiana tabacum* L.) roots. *Plant Soil* **248**, 137–144.

de la Fuente JM, Ramirez-Rodrigues V, Cabrera-Ponce JL and Herrera-Estrella L (1997) Aluminum tolerance in transgenic plants by alteration of citrate synthase. *Science* **276**, 1566–1568.

DeLucas JR, Dominguez AI, Valenciano S, Turner G and Laborda F (1999) The acuH gene of Aspergillus nidulans, required for growth on acetate and long chain fatty acids, encodes a putative homologue of the mammalian carnitine/acylcarnitine carrier. *Arch Microbiol* **171**, 386–396.

Deuschle K, Funck D, Forlani G, *et al.* (2004) The role of delta-1-pyrroline-5-carboxylate dehydrogenase in proline degradation. *Plant Cell* **16**, 3413–3425.

Deuschle K, Funck D, Hellmann H, Daeschner K, Binder S and Frommer WB (2001) A nuclear gene encoding mitochondrial Δ^1-pyrroline-5-carboxylate dehydrogenase and its potential role in protection from proline toxicity. *Plant J* **27**, 345–356.

Diebold R, Schuster J, Daschner K and Binder S (2002) The branched-chain amino acid transaminase gene family in *Arabidopsis* encodes plastid and mitochondrial proteins. *Plant Physiol* **129**, 540–550.

Dieuaide-Noubhani M, Raffard G, Canioni P, Pradet A and Raymond P (1995) Quantification of compartmented metabolic fluxes in maize root tips using isotope distribution from ^{13}C- or ^{14}C-labeled glucose. *J Biol Chem* **270**, 13147–13159.

Dojcinovic D, Krosting J, Harris AJ, Wagner DJ and Rhoads DM (2005). Identification of a region of the *Arabidopsis* AtAOX1a promoter necessary for mitochondrial retrograde regulation of expression. *Plant Mol Biol* **58**, 159–175.

Douce R (1985) Mitochondria in higher plants. In: *Functions of Plant Mitochondrial Membranes* (American Society of Plant Physiologists Monograph Series), pp. 77–153. Academic Press, London.

Douce R, Bourguignon J, Neuburger M and Rébeillé F (2001) The glycine decarboxylase system: a fascinating complex. *Trends Plant Sci* **6**, 167–176.

Douce R, Moore AL and Neuburger M (1977) Isolation and oxidative properties of intact mitochondria isolated from spinach leaves. *Plant Physiol* **60**, 625–628.

Douce R and Neuberger M (1989) The uniqueness of plant mitochondria. *Annu Rev Plant Physiol Plant Mol Biol* **40**, 371–414.

Dry IB, Day DA and Wiskich JT (1983) Preferential oxidation of glycine by the respiratory chain in pea leaf mitochondria. *FEBS Lett* **158**, 154–158.

Dry IB and Wiskich JT (1982) Role of the external ATP/ADP ratio in the control of plant mitochondrial respiration. *Arch Biochem Biophys* **217**, 72–79.

Dry IB and Wiskich JT (1987) 2-Oxoglutarate dehydrogenase and pyruvate dehydrogenase activities in plant mitochondria: interaction via a common coenzyme a pool. *Arch Biochem Biophys* **257**, 92–99.

Ducos E, Touzet P and Boutry M (2001) The male sterile *G* cytoplasm of wild beet displays modified mitochondrial respiratory complexes. *Plant J* **26**, 171–180.

Dupuy J, Volbeda A, Carpentier P, Darnault C, Moulis JM and Fontecilla-Camps JC (2006) Crystal structure of human iron regulatory protein 1 as cytosolic aconitase. *Structure* **14**, 129–139.

Dutilleul C, Driscoll S, Cornic G, De Paepe R, Foyer CH and Noctor G (2003) Functional mitochondrial complex I is required for optimal photosynthetic performance in photorespiratory conditions and during transients. *Plant Physiol* **131**, 264–275.

Dutilleul C, Lelarge C, Prioul JL, De Paepe R, Foyer CH and Noctor G (2005) Mitochondria-driven changes in leaf NAD status exert a crucial influence on the control of nitrate assimilation and the integration of carbon and nitrogen metabolism. *Plant Physiol* **139**, 64–78.

Echtay KS, Roussel D, St-Pierre J, *et al.* (2002) Superoxide activates mitochondrial uncoupling proteins. *Nature* **415**, 96–99.

Edwards S, Nguyen BT, Do B and Roberts JKM (1998) Contribution of malic enzyme, pyruvate kinase, phosphoenolpyruvate carboxylase and the Krebs cycle to respiration and biosynthesis and to intracellular pH regulation during hypoxia, in maize root tips observed by nuclear magnetic resonance imaging and gas chromatography-mass spectrometry. *Plant Physiol* **116**, 1073–1081.

Emmerlich V, Linka N, Reinhold T, *et al.* (2003) The plant homolog to the human sodium/dicarboxylic cotransporter is the vacuolar malate carrier. *Proc Natl Acad Sci USA* **100**, 11122–11126.

Escobar MA, Franklin KA, Svensson AS, Salter MG, Whitelam GC and Rasmusson AG (2004) Light regulation of the *Arabidopsis* respiratory chain. Multiple discrete photoreceptor responses contribute to induction of type II NAD(P)H dehydrogenase genes. *Plant Physiol* **136**, 2710–2721.

Escobar MA, Geisler DA and Rasmusson AG (2006) Reorganization of the alternative pathways of the *Arabidopsis* respiratory chain by nitrogen supply: opposing effects of ammonium and nitrate. *Plant J* **45**, 775–788.

Eubel H, Heinemeyer J, Sunderhaus S and Braun HP (2004) Respiratory chain supercomplexes in plant mitochondria. *Plant Physiol Biochem* **42**, 937–942.

Eubel H, Jansch L and Braun HP (2003) New insights into the respiratory chain of plant mitochondria. Supercomplexes and a unique composition of complex II. *Plant Physiol* **133**, 274–286.

Falk KL, Behal RH, Xiang C and Oliver DJ (1998) Metabolic bypass of the tricarboxylic acid cycle during lipid mobilization in germinating oilseeds. Regulation of NAD^+-dependent isocitrate dehydrogenase versus fumarase. *Plant Physiol* **117**, 473–481.

Farre EM, Tiessen A, Roessner U, Geigenberger P, Trethewey RN and Willmitzer L (2001) Analysis of the compartmentation of glycolytic intermediates, nucleotides, sugars, organic acids, amino acids and sugar alcohols in potato tubers using a non-aqueous fractionation method. *Plant Physiol* **127**, 685–700.

Fernie AR, Carrari F and Sweetlove LJ (2004) Respiratory metabolism: glycolysis, the TCA cycle and mitochondrial electron transport chain. *Curr Opin Plant Biol* **7**, 254–261.

Fiehn O (2002) Metabolomics – the link between genotypes and phenotypes. *Plant Mol Biol* **48**, 155–171.

Fieuw S, Müller-Röber B, Gálvez S and Willmitzer L (1995) Cloning and expression analysis of the cytosolic $NADP^+$-dependent isocitrate dehydrogenase from potato. *Plant Physiol* **107**, 905–913.

Figueroa P, León G, Elorza A, Holuigue L and Jordana X (2001) Three different genes encode the iron-sulfur subunit of succinate dehydrogenase in *Arabidopsis thaliana. Plant Mol Biol* **46**, 241–250.

Figueroa P, León G, Elorza A, Holuigue L, Araya A and Jordana X (2002) The four subunits of mitochondrial respiratory complex II are encoded by multiple nuclear genes and targeted to mitochondria in *Arabidopsis* thaliana. *Plant Mol Biol* **50**, 725–734.

Fiorani F, Umbach AL and Siedow JN (2005) The alternative oxidase of plant mitochondria is involved in the acclimation of shoot growth at low temperature: a study of *Arabidopsis AOX1a* transgenic plants. *Plant Physiol* **139**, 1795–1805.

Fischer R, Budde I and Hain R (1997) Stilbene synthase gene expression causes changes in flower colour and male sterility in tobacco. *Plant J* **11**, 489–498.

Fridman E, Carrrari F, Liu YS, Fernie AR and Zamir D (2004) Zooming-in on a quantitative trait nucleotide (QTN) for tomato yield using wild species introgression lines. *Science* **305**, 1786–1789.

Fujiki Y, Ito M, Itoh T, Nishida I and Watanabe A (2002) Activation of the promoters of *Arabidopsis* genes for the branched-chain alpha-keto acid dehy-drogenase complex in transgenic tobacco BY-2 cells under sugar starvation. *Plant Cell Physiol* **43**, 275–280.

Fujiki Y, Sato T, Ito M and Watanabe A (2000) Isolation and characterization of cDNA clones for the E1b and E2 subunits of the branched-chain a-ketoacid dehydrogenase complex in *Arabidopsis. J Biol Chem* **275**, 6007–6013.

Galtier N, Foyer CH, Huber J, Voelker TA and Huber SC (1993) Effects of elevated sucrose phosphate synthase activity on photosynthesis, assimilate partitioning, and growth in tomato (*Lycoperscion esculentum* var UC82B). *Plant Physiol* **101**, 535–543.

Gálvez S, Hodges M, Decottignies P, *et al.* (1996) Identification of a tobacco cDNA encoding a cytosolic NADP-isocitrate dehydrogenase. *Plant Mol Biol* **30**, 307–320.

Gálvez S, Lancien M and Hodges M (1999) Are isocitrate dehydrogenases and 2-oxoglutarate involved in the regulation of glutamate synthesis? *Trends Plant Sci* **4**, 484–490.

Gambonnet B, Jabrin S, Ravanel S, Karan M, Douce R and Rébeillé F (2001) Folate distribution during higher plant development. *J Sci Food Agric* **1**(81), 835–841.

Gangloff SP, Marguet D and Lauquin GJM (1990) Molecular cloning of the yeast aconitase gene (ACO-1) and evidence of the synergistic regulation of expression by glucose plus glutamate. *Mol Cell Biol* **10**, 3551–3561.

Gardestrom P (1987) Adenylate ratios in the cytosol, chloroplasts and mitochondria of barley leaf protoplasts during photosynthesis at different carbon dioxide concentrations. *FEBS Lett* **212**, 114–118.

Gardestrom P and Wigge B (1988) Influence of photorespiration on ATP/ADP ratios in the chloroplasts, mitochondria, and cytosol, studied by rapid fractionation of barley (*Hordeum vulgare*) protoplasts. *Plant Physiol* **88**, 69–76.

Gass N, Glagotskaia T, Mellema S, *et al.* (2005) Pyruvate decarboxylase provides growing pollen tubes with a competitive advantage in petunia. *Plant Cell* **17**, 2355–2368.

Geigenberger P, Regierer B, Nunes-Nesi A, *et al.* (2005) Inhibition of de novo pyrimidine synthesis in growing potato tubers leads to a compensatory stimulation of the pyrimidine salvage pathway and a subsequent increase in biosynthetic performance. *Plant Cell* **17**, 2077–2088.

Gemel J and Randall DD (1992) Light regulation of leaf mitochondrial pyruvate dehydrogenase complex. Role of photorespiratory carbon metabolism. *Plant Physiol* **100**, 908–914.

Gerbaud A and Andre M (1980) Effect of CO_2, O_2 and light on photosynthesis and photorespiration in wheat. *Plant Physiol* **66**, 1032–1036.

Gerhardt B (1992) Fatty acid degradation in plants. *Prog Lipid Res* **31**, 417–446.

Giege P, Heazlewood JL, Roessner-Tunali U, *et al.* (2003) Enzymes of glycolysis are functionally associated with the mitochondrion in *Arabidopsis* cells. *Plant Cell* **15**, 2140–2151.

Gietl C (1992) Malate dehydrogenase isoenzymes: cellular locations and role in the flow of metabolites between the cytoplasm and cell organelles. *Biochim Biophys Acta* **1100**, 217–234.

Gilmore AM (1997) Mechanistic aspects of xanthophyll cycle-dependent photoprotection in higher plant chloroplast and leaves. *Physiol Plant* **99**, 197–209.

Giovannoni J (2001) Molecular biology of fruit maturation and ripening. *Ann Rev Plant Physiol Plant Mol Biol* **52**, 725–749.

Goodenough PW, Prosser IM and Young K (1985) NADP-linked malic enzyme and malate metabolism in ageing tomato fruit. *Phytochem* **24**, 1157–1162.

Graham IA and Eastmond PJ (2002) Pathways of straight and branched chain fatty acid catabolism in higher plants. *Prog Lipid Res* **41**, 156–181.

Gray GR, Villarimo AR, Whitehead CL and McIntosh L (2004) Transgenic tobacco (*Nicotiana tabacum* L.) plants with increased expression levels of mitochondrial $NADP^+$-dependent isocitrate dehydrogenase: evidence implicating this enzyme in the redox activation of the alternative oxidase. *Plant Cell Physiol* **45**, 1413–1425.

Grissom CB, Canellas PF and Wedding RT (1983) Allosteric regulation of the NAD malic enzyme from cauliflower: activation by fumarate and coenzyme A. *Arch Biochem Biophys* **220**, 133–144.

Grof CPL, Winning BM, Scaysbrook TP, Hill SA and Leaver CJ (1995) Mitochondrial pyruvate dehydrogenase: molecular cloning of the E1α subunit and expression analysis. *Plant Physiol* **108**, 1623–1629.

Grover SD, Canellas PF and Wedding RT (1981) Purification of NAD malic enzyme from potato and investigation of some physical and kinetic properties. *Arch Biochem Biophys* **209**, 396–407.

Grover SD and Wedding RT (1984) Modulation of the activity of NAD-malic enzyme from *Solanum tuberosum* by changes in oligomeric state. *Arch Biochem Biophys* **234**, 418–425.

Guan Y, Rawsthorne S, Scofield G, Shaw P and Doonan J (1995) Cloning and characterization of a dihydrolipoamide acetyltransferase (E2) subunit of the pyruvate dehydrogenase complex from *Arabidopsis* thaliana. *J Biol Chem* **270**, 5412–5417.

Gueguen V, Macherel D, Jaquinod M, Douce R and Bourguignon J (2000) Fatty acid and lipoic acid biosynthesis in higher plant mitochondria. *J Biol Chem* **275**, 5016–5025.

Haferkamp I, Hackstein J, Voncken F, Schmit G and Tjaden J (2002) Functional integration of mitochondrial and hydrogenosomal ADP/ATP carriers in the *Escherichia coli* membrane reveals different biochemical characteristics for plants, mammals and anaerobic chytrids. *Eur J Biochem* 269, 3172–3181.

Hampp R, Outlaw WH, Jr and Tarczynski MC (1982) Profile of basic carbon pathways in guard cells and other leaf cells of *Vicia faba* L. *Plant Physiol* **70**, 1582–1585.

Hanning I and Heldt HW (1993) On the function of mitochondrial metabolism during photosynthesis in spinach leaves (*Spinacia oleracea* L.). Partitioning between respiration and export of redox equivalents and precursors for nitrate assimilation products. *Plant Physiol* **103**, 1147–1154.

Hanson MR (1991) Plant mitochondrial mutations and male sterility. *Annu Rev Genet* **25**, 461–486.

Hanson KR (1992) Evidence for mitochondrial regulation of photosynthesis by a starchless mutant of *Nicotiana sylvestris. Plant Physiol* **99**, 276–283.

Hanson M and Bentolila S (2004) Interactions of mitochondrial and nuclear genes that affect male gametophyte development. *Plant Cell* **16**, S154–S169.

Hanson AD and Gregory JF, III (2002) Synthesis and turnover of folates in plants. *Curr Opin Plant Biol* **5**, 244–249.

Hanson AD and Roje S (2001) One-carbon metabolism in higher plants. *Annu Rev Plant Physiol Plant Mol Biol* **52**, 119–137.

Hatch MD, Dröscher L, Flügge UI and Heldt HW (1984) A specific translocator for oxaloacetate transport in chloroplasts. *FEBS Lett* **178**, 15–19.

Hattori T and Asahi T (1982) The presence of 2 forms of succinate-dehydrogenase in sweet-potato root mitochondria. *Plant Cell Physiol* **23**, 515–523.

Hatzfeld Y, Maruyama A, Schmidt A, Noji M, Ishizawa K and Saito K (2000) ß-Cyanoalanine synthase is a mitochondrial cysteine synthase-like protein in spinach and *Arabidopsis thaliana. Plant Physiol* **123**, 1163–1171.

Hatzfeld Y and Saito K (2000) Evidence for the existence of rhodanese (thiosulfate: cyanide sulfurtransferase) in plants: preliminary characterization of two rhodanese cDNAs from *Arabidopsis* thaliana. *FEBS Lett* **470**, 147–150.

Hayashi H, De Bellis L, Ciurli A, Kondo M, Hayashi M and Nishimura M (1999) A novel acyl-CoA oxidase that can oxidize short-chain acyl-CoA in plant peroxisomes. *J Biol Chem* **274**, 12715–12721.

Hayes MK, Luethy MH and Elthon TE (1991) Mitochondrial malate dehydrogenase from corn. *Plant Physiol* **97**, 1381–1387.

Hayes JE and Ma JF (2003) Al-induced efflux of organic acid anions is poorly associated with internal organic acid metabolism in triticale roots. *J Exp Bot* **54**, 1753–1759.

Heazlewood JL, Howell KA and Millar AH (2003a) Mitochondrial complex I from *Arabidopsis* and rice: orthologs of mammalian and fungal components coupled with plant-specific subunits. *Biochim Biophys Acta* **1604**, 159–169.

Heazlewood JL, Tonti-Filippini JS, Gout AM, Day DA, Whelan J and Millar AH (2004) Experimental analysis of the *Arabidopsis* mitochondrial proteome highlights signaling and regulatory components, provides assessment of targeting prediction programs, and indicates plant-specific mitochondrial proteins. *Plant Cell* **16**, 241–256.

Heazlewood JL, Whelan J and Millar AH (2003b) The products of the mitochondrial orf25 and orfB genes are F_O components in the plant F_1F_O ATP synthase. *FEBS Lett* **540**, 201–205.

Heber U and Heldt HW (1981) The chloroplast envelope: structure, function and role in leaf metabolism. *Annu Rev Plant Physiol* **32**, 139–68.

Heineke D, Riens B, Gross H, *et al.* (1991) Redox transfer across the inner chloroplast envelope membrane. *Plant Physiol* **95**, 1131–1137.

Heldt HW (1969) Adenine nucleotide translocation in spinach chloroplasts. *FEBS Lett* **5**, 11–14.

Heldt HW and Flügge UI (1987) Subcellular transport of metabolites in plant cells. In: *The Biochemistry of Plants* (eds Stumpf PK and Conn EE), pp. 49–85. Academic Press, New York.

Henry MF and Nyns EJ (1975) Cyanide-insensitive respiratory: an alternative mitochondrial pathway. *Sub-Cell Biochem* **4**, 1–65.

Hildyard JCW and Halestrap AP (2004) Identification of the mitochondrial pyruvate carrier in *Saccharomyces cerevisiae*. *Biochem J* **374**, 607–611.

Hill SA (1997) Carbon metabolism in mitochondria. In: *Plant Metabolism* (eds Dennis DT, Turpin DH, Lefebvre DD and Layzell DB), pp. 181–199. Addison Welsey Longman, London.

Hodges M (2002) Enzyme redundancy and the importance of 2-oxoglutarate in plant ammonium assimilation. *J Exp Bot* **53**, 905–916.

Hodges M, Flesch V, Gálvez S and Bismuth E (2003) Higher plant $NADP^+$-dependent isocitrate dehydrogenases, ammonium assimilation and NADPH production. *Plant Physiol Biochem* **41**, 577–585.

Hoefnagel MHN, Atkin OK and Wiskich JT (1998) Interdependence between chloroplasts and mitochondria in the light and the dark. *Bioch Biophy Acta* **1366**, 235–255.

Hoefnagel MHN, Millar AH, Wiskich JT and Day DA (1995) Cytochrome and alternative respiratory pathways compete for electrons in the presence of pyruvate in soybean mitochondria. *Arch Biochem Biophys* **318**, 394–400.

Holtzapffel RC, Finnegan PM, Millar AH, Badger MR and Day DA (2002) Mitochondrial protein expression in tomato fruit during on-vine ripening and cold storage. *Funct Plant Biol* **29**, 827–834.

Hoyos ME, Palmieri L, Wertin T, Arrigoni R, Polacco JC and Palmieri F (2003) Identification of a mitochondrial transporter for basic amino acids in *Arabidopsis thaliana* by functional reconstitution into liposomes and complementation in yeast. *Plant J* **33**, 1027–1035.

Hurry VM, Tobiæson M, Krömer S, Gardestrom P and Öquist G (1995) Mitochondria contribute to increased photosynthetic capacity of leaves of winter rye (*Secale cereale* L.) following cold-hardening. *Plant Cell Environ* **18**, 69–76.

Iannetta PPM, Escobar NM, Ross HA, *et al.* (2004) Identification, cloning and expression analysis of strawberry (Fragaria x ananassa) mitochondrial citrate synthase and mitochondrial malate dehydrogenase. *Physiol Plant* **121**, 15–26.

Igamberdiev AU and Gardestrom P (2003) Regulation of NAD- and NADP-dependent isocitrate dehydrogenases by reduction levels of pyridine nucleotides in mitochondria and cytosol of pea leaves. *Biochim Biophys Acta* **1606**, 117–125.

Igamberdiev AU, Hurry V, Krömer S and Gardestrom P (1998) The role of mitochondrial electron transport during photosynthetic induction. A study with barley (*Hordeum vulgare*) protoplasts incubated with rotenone and oligomycin. *Physiol Plant* **104**, 431–439.

Ikeda Y, Dabrowski C and Tanaka K (1983) Separation and properties of five distinct acyl-CoA dehydrogenases from rat liver mitochondria. Identification of a new 2-methyl branched chain acyl-CoA dehydrogenase. *J Biol Chem* **258**, 1066–1076.

Ishizaki KM, Larson TR, Schauer N, Fernie AR, Graham IA and Leaver CJ (2005) The critical role of *Arabidopsis* electron-transfer flavoprotein:ubiquinone oxidoreductase during dark-induced starvation. *Plant Cell* **17**, 2587–2600.

Jeffery D, Smith C, Goodenough P, Prosser I and Grierson D (1984) Ethylene-independent and ethylene-dependent biochemical changes in ripening tomatoes. *Plant Physiol* **74**, 32–38.

Jenner H, Winning B, Millar AH, Tomlinson K, Leaver CJ and Hill SA (2001) NAD malic enzyme and the control of carbohydrate metabolism in potato tubers. *Plant Physiol* **126**, 1139–1149.

Jezek P, Costa ADT and Vercesi AE (1996) Evidence for anion-translocating plant uncoupling mitochondrial protein in potato mitochondria. *J Biol Chem* **271**, 32743–32748.

Jordanov J, Courtois-Verniquet F, Neuburger M and Douce R (1992) Structural investigations by extended X-ray absorption fine structure spectroscopy of the iron center of mitochondrial aconitase in higher plant cells. *J Biol Chem* **267**, 16775–16778.

Jost R, Berkowitz O, Wirtz M, Hopkins L, Hawkesford MJ and Hell R (2000) Genomic and functional characterization of the *oas* gene family encoding *O*-acetylserine(thiol)lyases, enzymes catalysing the final step in cysteine biosynthesis in *Arabidopsis thaliana. Genetics* **253**, 237–247.

Journet EP, Neuburger M and Douce R (1981) The role of glutamate oxaloacetate transaminase and malate dehydrogenase in the regeneration of NAD^+ for glycine oxidation by spinach leaf mitochondria. *Plant Physiol* **67**, 467–469.

Junker BJ, Wuttke R, Tiessen A, *et al.* (2004) Temporally regulated expression of a yeast invertase in potato tubers allows dissection of the complex metabolic phenotype obtained following its constitutive expression. *Plant Mol Biol* **56**, 91–110.

Kalantidis K, Wilson ZA and Mulligan BJ (2002) Mitochondrial gene expression in stamens is differentially regulated during male gametogenesis in *Arabidopsis*. *Sex Plant Reprod* **14**, 299–304.

Kappers IF, Aharoni A, van Herpen TWJM, Luckerhoff LLP, Dicke M and Bouwmeester HJ (2005) Genetic engineering of terpenoid metabolism attracts bodyguards to *Arabidopsis*. *Science* **309**, 2070–2072.

Karam GA and Bishop SH (1989) α-ketoglutarate dehydrogenase from cauliflower mitochondria: preparation and reactivity with substrates. *Phytochem* **28**, 3291–3293.

Kiddle G, Pastori GM, Bernard S, *et al.* (2003) Effects of leaf ascorbate content on defence and photosynthesis gene expression in *Arabidopsis thaliana*. *Antioxid Redox Signal* **5**, 23–32.

Kim SH and Lee WS (2002) Participation of extracellular fumarase in the utilization of malate in cultured carrot cells. *Plant Cell Rep* **20**, 1087–1092.

Kirschbaum MUF and Farquhar GD (1987) Investigation of the CO_2 dependence of quantum yield and respiration in *Eucalyptus pauciflora*. *Plant Physiol* **83**, 1032–1036.

Kiyosue T, Yoshiba Y, Yamaguchi-Shinozaki K and Shinozaki K (1996) A nuclear gene encoding mitochondrial proline dehydrogenase, an enzyme involved in proline metabolism, is upregulated by proline but downregulated by dehydration in *Arabidopsis*. *Plant Cell* **8**, 1323–1335.

Knowles JR (1989) The mechanism of biotin-dependent enzymes. *Annu Rev Biochem* **58**, 195–221.

Koike M and Koike K (1976) Structure, assembly and function of mammalian alpha-keto acid dehydrogenase complexes. *Adv Biophys* **9**, 187–227.

Kolukisaoglu Ñ, Bovet L, Klein M, *et al.* (2002) Family business: The multidrug-resistance related protein (MRP) ABC transporter genes in *Arabidopsis* thaliana. *Planta* **216**, 107–119.

Koyama H, Kawamura A, Kihara T, Hara T, Takita E and Shibata D (2000) Overexpression of mitochondrial citrate synthase in *Arabidopsis thaliana* improved growth on a phosphorus-limited soil. *Plant Cell Physiol* **41**, 1030–1037.

Koyama H, Takita E, Kawamura A, Hara T and Shibata D (1999) Overexpression of mitochondrial citrate synthase gene improves the growth of carrot cells in Al-phosphate medium. *Plant Cell Physiol* **40**, 482–488.

Krebs HA and Johnson WA (1937) The role of the citric acid cycle in intermediary metabolism in animal tissues. *Enzymologia* **4**, 148–156.

Krömer S (1995) Respiration during photosynthesis. *Annu Rev Plant Physiol Plant Mol Biol* **46**, 45–70.

Krömer S and Heldt HW (1991) On the role of mitochondrial oxidative phosphorylation in photosynthesis metabolism as studied by the effect of oligomycin on photosynthesis in protoplasts and leaves of barley. *Plant Physiol* **95**, 1270–1276.

Krömer S, Malmberg G and Gardestrom P (1993) Mitochondrial contribution to photosynthetic metabolism. *Plant Physiol* **102**, 947–955.

Krömer S and Scheibe R (1996) Function of the chloroplastic malate valve for respiration during photosynthesis. *Biochem Soc Trans* **24**, 761–766.

Krömer S, Stitt M and Heldt HW (1988) Mitochondrial oxidative phosphorylation participating in photosynthetic metabolism of a leaf cell. *FEBS Lett* **226**, 352–356.

Kruft V, Eubel H, Jänsch L, Werhahn W and Braun HP (2001) Proteomic approach to identify novel mitochondrial proteins in *Arabidopsis*. *Plant Physiol* **127**, 1694–1710.

Kruse A, Fieuw S, Heineke D and Müller-Röber B (1998) Antisense inhibition of cytosolic NADP-dependent isocitrate dehydrogenase in transgenic potato plants. *Planta* **205**, 82–91.

Kushnir S, Babiychuk E, Storozhenko S, et al. (2001) A mutation of the mitochondrial ABC transporter Sta1 leads to dwar.sm and chlorosis in the *Arabidopsis* mutant starik. *Plant Cell* **13**, 89–100.

LaCognata U, Landschütze V, Willmitzer L and Müller-Röber B (1996) Structure and expression of mitochondrial citrate synthases from higher plants. *Plant Cell Physiol* **37**, 1022–1029.

Laloi M (1999) Plant mitochondrial carriers: an overview. *Cell Mol Life Sci* **56**, 918–944.

Laloi M, Klein M, Riesmeier JW, *et al.* (1997) A plant cold-induced uncoupling protein. *Nature* **389**, 135–136.

Lancien M, Ferrario-Mery S, Roux V, *et al.* (1999) Simultaneous expression of NAD-dependent isocitrate dehydrogenase and other Krebs cycle genes after nitrate resupply to short-term nitrogen-starved tobacco. *Plant Physiol* **120**, 717–725.

Landschütze V, Müller-Röber B and Willmitzer L (1995a) Mitochondrial citrate synthase from potato: predominant expression in mature leaves and young flower buds. *Planta* **196**, 756–764.

Landschütze V, Willmitzer L and Müller-Röber B (1995b) Inhibition of flower formation by anti-sense repression of mitochondrial citrate synthase in transgenic potato plants leads to a specific disintegration of the ovary tissues of flowers. *EMBO J* **14**, 660–666.

Lane N (2005) *Power, Sex, Suicide: Mitochondria and the Meaning of Life*, 368 p. Oxford University Press Inc., New York.

Lange PR, Eastmond PJ, Madagan K and Graham IA (2004) An *Arabidopsis* mutant disrupted in valine catabolism is also compromised in peroxisomal fatty acid beta-oxidation. *FEBS Lett* **571**, 147–153.

Lawand S, Dorne AJ, Long D, Coupland G, Mache R and Carol P (2002) *Arabidopsis A BOUT DE SOUFFLE*, which is homologous with mammalian carnitine acyl carrier, is required for postembryonic growth in the light. *Plant Cell* **14**, 2161–2173.

Lernmark U and Gardestrom P (1994) Distribution of pyruvate dehydrogenase complex activities between chloroplasts and mitochondria from leaves of different species. *Plant Physiol* **106**, 1633–1638.

Levitan A, Danon A and Lisowsky T (2004) Unique features of plant mitochondrial sulfhydryl oxidase. *J Biol Chem* **279**, 20002–20008.

Li XF, Ma JF and Matsumoto H (2002) Aluminium-induced secretion of both citrate and malate in rye. *Plant Soil* **242**, 235–243.

Lieman-Hurwitz J, Rachmilevitch S, Mittler R, Marcus Y and Kaplan A (2003) Enhanced photosynthesis and growth of transgenic plants that express ictB, a gene involved in HCO_3- accumulation in cyanobacteria. *Plant Biotechnol J* **1**, 43–50.

Linka M and Weber AP (2005) Shuffling ammonia between mitochondria and plastids during photorespiration. *Trends Plant Sci* **10**, 461–465.

Linke B and Börner T (2005) Mitochondrial effects on flower and pollen development. *Mitochondrion* **5**, 389–402.

Luethy MH, Miernyk JA, David NR and Randall DD (1996) Plant pyruvate dehydrogenase complexes. In: *Alpha-Keto Acid Dehydrogenase Complexes* (eds Patel MS, Roche TE and Harris RA), pp. 71–92. Birkhauser Verlag, Basel, Switzerland.

Luethy MH, Miernyk JA and Randall DD (1994). The nucleotide and deduced amino acid sequences of a cDNA encoding the E1β-subunit of the *Arabidopsis* thaliana mitochondrial pyruvate dehydrogenase complex. *Biochim Biophys Acta* **1187**, 95–98.

Lytovchenko A, Bieberich K, Willmitzer L and Fernie AR (2002b) Carbon assimilation and metabolism in potato leaves deficient in plastidial phosphoglucomutase. *Planta* **215**, 802–811.

Lytovchenko A, Sweetlove LJ, Pauly M and Fernie AR (2002a) The influence of cytosolic phosphoglucomutase on photosynthetic carbohydrate metabolism. *Planta* **215**, 1013–1021.

MacDougall AJ and ap Rees T (1991) Control of the Krebs cycle in *Arum spadix*. *J Plant Physiol* **137**, 683–690.

Mackenzie S and McIntosh L (1999) Higher plant mitochondria. *Plant Cell* **11**, 571–586.

Maia IG, Benedetti CE, Leite A, Turcinelli SR, Vercesi AE and Arruda P (1998) AtPUMP: an *Arabidopsis* gene encoding a plant uncoupling mitochondrial protein. *FEBS Lett* **429**, 403–406.

Mannella CA (1992) The 'ins' and 'outs' of mitochondrial membrane channels. *Trends Biochem Sci* **17**, 315–320.

Mannella CA, Pfeiffer DR, Bradshaw PC, *et al.* (2001) Topology of the mitochondrial inner membrane: dynamics and bioenergetic implications. *IUBMB Life* **52**, 93–100.

Mannella CA and Tedeschi H (1987) Importance of the mitochondrial outer-membrane channel as a model biological channel. *J Bioenerg Biomembr* **19**, 305–308.

Marillia EF, Micallef BJ, Micallef M, *et al.* (2003) Biochemical and physiological studies of *Arabidopsis thaliana* lines with repressed expression of the mitochondrial pyruvate dehydrogenase kinase. *J Exp Bot* **54**, 259–270.

Masclaux C, Valadier M-H, Brugière N, Morot-Gaudry J-F and Hirel B (2000) Characterization of the sink/source transition in tobacco (*Nicotiana tabacum* L.) shoots in relation to nitrogen management and leaf senescence. *Planta* **211**, 510–518.

McClung CR, Hsu M, Painter JE, Gagne JM, Karlsberg SD and Salome PA (2000) Integrated temporal regulation of the photorespiratory pathway. Circadian regulation of two *Arabidopsis* genes encoding serine hydroxymethyltransferase. *Plant Physiol* **123**, 381–392.

McIntosh CA and Oliver DJ (1992) NAD^+-linked isocitrate dehydrogenase: isolation, purification, and characterization of the protein from pea mitochondria. *Plant Physiol* **100**, 69–75.

McIntosh L (1994) Molecular biology of the alternative oxidase. *Plant Physiol* **105**, 781–786.

Meeuse BJD (1975) Thermogenic respiration in aroids. *Annu Rev Plant Physiol* **26**, 117–126.

Michalecka AM, Svensson AS, Johansson FI, *et al.* (2003) *Arabidopsis* genes encoding mitochondrial type II NAD(P)H dehydrogenases have different evolutionary origin and show distinct responses to light. *Plant Physiol* **133**, 642–652.

Miernyk JA and Randall DD (1987) Some kinetic properties of the pea mitochondrial pyruvate dehydrogenase complex. *Plant Physiol* **83**, 306–310.

Millar AH, Day DA and Whelan J (2004a) Mitochondrial biogenesis and function in *Arabidopsis*. In: *The Arabidopsis Book* (eds Somerville CR and Meyerowitz EM). American Society of Plant Biologists, Rockville, MD. Available at: www.aspb.org/publications/*Arabidopsis*/.

Millar AH, Eubel H, Jänsch L, Kruft V, Heazlewood JL and Braun HP (2004b) Mitochondrial cytochrome *c* oxidase and succinate dehydrogenase complexes contain plant specific subunits. *Plant Mol Biol* **56**, 77–90.

Millar AH and Heazlewood JL (2003) Genomic and proteomic analysis of mitochondrial carrier proteins in *Arabidopsis*. Plant Physiol **131**, 443–453.

Millar AH, Hill SA and Leaver CJ (1999a) Plant mitochondrial 2-oxoglutarate dehydrogenase complex: purification and characterization in potato *Biochem J* **343**, 327–334.

Millar AH, Leaver CJ and Hill SA (1999b) Characterization of the dihydrolipoamide acetyltransferase of the mitochondrial pyruvate dehydrogenase complex from potato and comparisons with similar enzymes in diverse plant species. *Eur J Biochem* **264**, 973–981.

Millar AH, Mittova V, Kiddle G, *et al.* (2003) Control of ascorbate synthesis by respiration and its implications for stress responses. *Plant Physiol* **133**, 443–447.

Millar AH, Owen K, Atkin R, *et al.* (1998) Analysis of respiratory chain regulation in roots of soybean seedlings. *Plant Physiol* **117**, 1083–1093.

Millar AH, Sweetlove LJ and Leaver CJ (2001) Analysis of Arabidospis mitochondrial proteome. *Plant Physiol* **127**, 1711–1727.

Miller SS, Driscoll BT, Gregerson RG, Gantt JS and Vance CP (1998) Alfalfa malate dehydrogenase (MDH): molecular cloning and characterization of five different forms reveals a unique nodule-enhanced MDH. *Plant J* **15**, 173–184.

Miyagawa Y, Tamoi M and Shigeoka S (2001) Overexpression of a cyanobacterial fructose-1,6-/sedoheptulose-1,7-bisphosphatase in tobacco enhances photosynthesis and growth. *Nat Biotechnol* **19**, 965–969.

Moller IM (1986) Membrane-bound NAD(P)H dehydrogenases in plant mitochondria. *Physiol Plant* **67**, 517–520.

Moller IM (2001) Plant mitochondria and oxidative stress. Electron transport, NADPH turnover and metabolism of reactive oxygen species. *Annu Rev Plant Physiol Plant Mol Biol* **52**, 561–591.

Moller IM (2002) A new dawn for plant mitochondrial NAD(P)H dehydrogenases. *Trends Plant Sci* **7**, 235–237.

Moller IM and Lin W (1986) Membrane-bound NAD(P)H dehydrogenases in higher plant cells. *Annu Rev Plant Physiol* **37**, 309–334.

Mooney BP, David NR, Thelen JJ, Miernyk JA and Randall DD (2000) Histidine modifying agents abolish pyruvate dehydrogenase complex kinase activity. *Biochem Biophys Res Commun* **267**, 500–503.

Mooney BP, Miernyk JA and Randall DD (2002) The complex fate of α-ketoacids. *Annu Rev Plant Biol* **53**, 357–375.

Moore AL and Siedow JN (1991) The regulation and nature of the cyanide-resistant alternative oxidase of plant mitochondria. *Biochim Biophys Acta* **1059**, 121–140.

Mouillon JM, Ravanel S, Douce R and Rébeillé F (2002) Folate synthesis in higher-plant mitochondria: coupling between the dihydropterin pyrophosphokinase and the dihydropteroate synthase activities. *Biochem J* **363**, 313–319.

Muller F and Frentzen M (2001) Phosphatidylglycerophosphate synthases from *Arabidopsis* thaliana. *FEBS Lett* **509**, 298–302.

Nakamura T, Yamaguchi Y and Sano H (2000) Plant mercaptopyruvate sulfurtransferases: molecular cloning, subcellular localization and enzymatic activities. *Eur J Biochem* **267**, 5621–5630.

Nast G and Müller-Röber B (1996) Molecular characterization of potato fumarate hydratase and functional expression in *Escherichia coli. Plant Physiol* **112**, 1219–1227.

Navet R, Jarmuszkiewicz W, Almeida AM, Sluse-Goffart C and Sluse FE (2003) Energy conservation and dissipation in mitochondria isolated from developing tomato fruit of ethylene-defective mutants failing normal ripening: the effect of ethephon, a chemical precursor of ethylene. *J Bioenerg Biomembr* **35**, 157–168.

Neuburger M, Rébeillé F, Jourdain A, Nakamura S and Douce R (1996) Mitochondria are a major site for folate and thymidylate synthesis in plants. *J Biol Chem* **271**, 9466–9472.

Neumann G and Martinoia E (2002) Cluster roots – an underground adaptation for survival in extreme environments. *Trends Plant Sci* **7**, 162–167.

Newton KJ and Schwartz D (1980) Genetic basis of the major malate dehydrogenase isozymes in maize. *Genetics* **95**, 425–442.

Nicholls DG and Locke RM (1984) Thermogenic mechanisms in brown fat. *Physiol Rev* **64**, 1–64.

Nicholls DG and Rial E (1999) A history of the first uncoupling protein, UCP1. *J Bioenerg Biomembr* **31**, 399–406.

Niyogi KK (1999) Photoprotection revisited: genetic and molecular approaches. *Annu Rev Plant Physiol Plant Mol Biol* **50**, 333–359.

Niyogi KK, Grossman AR and Björkman O (1998) *Arabidopsis* mutants define a central role for the xanthophyll cycle in the regulation of photosynthetic energy conversion. *Plant Cell* **10**, 1121–1134.

Noji M, Inoue K, Kimura N, Gouda A and Saito K (1998) Isoform-dependent differences in feedback regulation and subcellular localization of serine acetyltransferase involved in cysteine biosynthesis from *Arabidopsis thaliana. J Biol Chem* **273**, 32739–32745.

Nunes-Nesi A, Carrari F, Lytovchenko A, *et al.* (2005) Enhanced photosynthetic performance and growth as a consequence of decreasing mitochondrial malate dehydrogenase activity in transgenic tomato plants. *Plant Physiol* **137**, 611–622.

Ögren E, Öquist G and Hällgren JE (1984) Photoinhibition of photo-synthesis in *Lemna gibba* as induced by the interaction between light and temperature. I. Photosynthesis *in vivo. Physiol Plant* **62**, 181–186.

Ordentlich A, Linzer RA and Raskin I (1991) Alternative respiration and heat evolution in plants. *Plant Physiol* **97**, 1545–1550.

Padmasree K, Padmavathi L and Raghavendra AS (2002) Essentiality of mitochondrial oxidative metabolism for photosynthesis: optimization of carbon assimilation and protection against photoinhibition. *Crit Rev Biochem Mol Biol* **37**, 71–119.

Padmasree K and Raghavendra AS (1999a). Importance of oxidative electron transport over oxidative phosphorylation in optimizing photosynthesis in mesophyll protoplasts of pea (*Pisum sativum* L.). *Physiol Plant* **105**, 546–553.

Padmasree K and Raghavendra AS (1999b) Response of photosynthetic carbon assimilation in mesophyll protoplasts to restriction on mitochondrial oxidative metabolism: metabolites related to the redox status and sucrose biosynthesis. *Photosynth Res* **62**, 231–239.

Palmer JM and Wedding RT (1966) Purification and properties of succinyl-CoA synthetase from Jerusalem artichoke mitochondria. *Biochim Biophys Acta* **113**, 167–174.

Papenbrock J and Schmidt A (2000) Characterization of a sulfurtransferase from *Arabidopsis* thaliana. *Eur J Biochem* **267**, 145–154.

Pascal N, Dumas R and Douce R (1990) Comparison of the kinetic behaviour towards pyridine nucleotides of NAD^+-linked dehydrogenases from plant mitochondria. *Plant Physiol* **94**, 189–193.

Pastori GM, Kiddle G, Antoniw J, *et al.* (2003) Leaf vitamin C contents modulate plant defence transcripts and regulate genes that control development through hormone signaling. *Plant Cell* **15**, 939–951.

Patton DA, Johnson M and Ward ER (1996b) Biotin synthase from *Arabidopsis thaliana.* cDNA isolation and characterization of gene expression. *Plant Physiol* **112**, 371–378.

Patton DA, Schetter AL, Franzmann LH, Nelson K, Ward ER and Meinke DW (1998) An embryo-defective mutant of *Arabidopsis* disrupted in the final step of biotin synthesis. *Plant Physiol* **116**, 935–946.

Patton DA, Volrath S and Ward ER (1996a) Complementation of an *Arabidopsis thaliana* biotin auxotroph with an *Escherichia coli* biotin biosynthetic gene. *Mol Gen Genet* **251**, 261–266.

Peiffer WE, Ingle RT and Ferguson-Miller S (1990) Structurally unique plant cytochrome *c* oxidase isolated from wheat germ, a rich source of plant mitochondrial enzymes. *Biochemistry* **29**, 8696–8701.

Pellny TK, Ghannoum O, Conroy JP, *et al.* (2004) Genetic modification of photosynthesis with *E. coli* genes for trehalose synthesis. *Plant Biotechnol J* **2**, 71–82.

Peltier G and Thibault P (1985) O_2 uptake in the light in Chlamydomonas: evidence for persistent mitochondrial respiration. *Plant Physiol* **79**, 225–230.

Peyret P, Perez P and Alric M (1995) Structure, genomic organization and expression of the *Arabidopsis thaliana* aconitase gene. *J Biol Chem* **270**, 8131–8137.

Picault N, Hodges M, Palmieri L and Palmieri F (2004) The growing family of mitochondrial carriers in *Arabidopsis*. *Trends Plant Sci* **9**, 138–146.

Picault N, Palmieri L, Pisano I, Hodges M and Palmieri F (2002). Identification of a novel transporter for dicarboxylates and tricarboxylates in plant mitochondria. Bacterial expression, reconstitution, functional characterization, and tissue distribution. *J Biol Chem* **277**, 24204–24211.

Picciocchi A, Douce R and Alban C (2001) Biochemical characterization of the *Arabidopsis* biotin synthase reaction. The importance of mitochondria in biotin synthesis. *Plant Physiol* **127**, 1224–1233.

Picciocchi A, Douce R and Alban C (2003) The plant biotin synthase reaction. Identification and characterization of essential mitochondrial accessory protein components. *J Biol Chem* **278**, 24966–24975.

Pinon V, Ravanel S, Douce R and Alban C (2005) Biotin synthesis in plants. The first committed step of the pathway is catalyzed by a cytosolic 7-Keto-8-Aminopelargonic acid synthase. *Plant Physiol* **139**, 1666–1676.

Ploux O and Marquet A (1996) Mechanistic studies on the 8-amino-7-oxopelargonate synthase, a pyridoxal-5′-phosphate-dependent enzyme involved in biotin biosynthesis. *Eur J Biochem* **236**, 301–308.

Prabhu V, Chatson KB, Abrams GD and King J (1996) 13C nuclear magnetic resonance detection of interactions of serine hydroxymethyltransferase with C1-tetrahydrofolate synthase and glycine decarboxylase complex activities in *Arabidopsis*. *Plant Physiol* **112**, 207–216.

Pracharoenwattana I, Cornah JE and Smith SM (2005) *Arabidopsis* peroxisomal citrate synthase is required for fatty acid respiration and seed germination. *Plant Cell* **17**, 2037–2048.

Raghavendra AS and Padmasree K (2003) Beneficial interactions of mitochondrial metabolism with photosynthetic carbon assimilation. *Trends Plant Sci* **8**, 546–553.

Raghavendra AS, Padmasree K and Saradadevi K (1994) Interdependence of photosynthesis and respiration in plant cells: interactions between chloroplasts and mitochondria. *Plant Sci* **97**, 1–14.

Randall DD, Miernyk JA, David NR, Gemel J and Luethy MH (1996) Regulation of leaf mitochondrial pyruvate dehydrogenase complex activity by reversible phosphorylation. In: *Protein Phosphorylation in Plants* (eds Shewry PR and Halford NG), pp. 87–103. Clarendon Press, Oxford, UK.

Randall DD, Miernyk JA, Fang TK, Budde RJ and Schuller KA (1989) Regulation of the pyruvate dehydrogenase complexes in plants. In: *Alpha-Keto Acid Dehydrogenase Complexes: Organization, Regulation and Biomedical Ramifications* (eds Roche TE and Patel MS), pp. 192–205. The New York Academy of Science, New York.

Randall DD, Rubin PM and Fenko M (1977) Plant pyruvate dehydrogenase complex purification, characterization and regulation by metabolites and phosphorylation. *Biochim Biophys Acta* **485**, 336–349.

Rasmusson AG, Heiser V, Zabaleta E, Brennicke A and Grohmann L (1998) Physiological, biochemical and molecular aspects of mitochondrial complex I in plants. *Biochim Biophys Acta* **1364**, 101–111.

Rasmusson AG and Moller IM (1990) NADP-utilizing enzymes in the matrix of plant mitochondria. *Plant Physiol* **94**, 1012–1018.

Rasmusson AG and Moller IM (1991) NAD(P)H dehydrogenases on the inner surface of the inner mitochondrial membrane studied using inside-out submitochondrial particles. *Physiol Plant* **83**, 357–365.

Rasmusson AG, Svensson AS, Knoop V, Grohmann L and Brennicke A (1999) Homologues of yeast and bacterial rotenone-insensitive NADH dehydrogenases in higher eukaryotes: two enzymes are present in potato mitochondria. *Plant J* **20**, 79–87.

Ravanel S, Cherest H, Jabrin S, *et al.* (2001) Tetrahydrofolate biosynthesis in plants: molecular and functional characterization of dihydrofolate synthetase and three isoforms of folylpolyglutamate synthetase in *Arabidopsis thaliana*. *Proc Natl Acad Sci USA* **98**, 15360–15365.

Rébeillé F, Gans P, Chagvardieff P, Pean M, Tapie P and Thibault E (1988) Mass spectrometric determination of the inorganic carbon species assimilated by photoautotrophic cells of *Euphorbia characias* L. *J Biol Chem* **263**, 12373–12377.

Rébeillé F, Macherel D, Mouillon JM, Garin J and Douce R (1997) Folate biosynthesis in higher plants: purification and molecular cloning of a bifunctional 6-hydroxymethyl-7,8-dihydropterin pyrophosphokinase/7,8-dihydropteroate synthase localized in mitochondria. *EMBO J* **16**, 947–957.

Regierer B, Fernie AR, Springer F, *et al.* (2002) Starch content and yield increase as a result of altering adenylate pools in transgenic plants. *Nat Biotechnol* **20**, 1256–1260.

Rengel Z (2002) Genetic control of root exudation. *Plant Soil* **245**, 59–70.

Renné P, Dreßen U, Hebbeker U, *et al.* (2003) The *Arabidopsis* mutant *dct* is deficient in the plastidic glutamate/malate translocator DiT2. *Plant J* **35**, 316–331.

Reumann S, Heupel R and Heldt HW (1994) Compartmentation studies on spinach leaf peroxisomes. II. Evidence for the transfer of reductant from the cytosol to the peroxisomal compartment via a malate shuttle. *Planta* **193**, 167–173.

Riemenschneider A, Wegele R, Schmidt A and Papenbrock J (2005) Isolation and characterization of a D-cysteine desulfhydrase protein from *Arabidopsis thaliana*. *FEBS J* **272**, 1291–1304.

Robson CA and Vanlerberghe GC (2002) Transgenic plant cells lacking mitochondrial alternative oxidase have increased susceptibility to mitochondria-dependent and -independent pathways of programmed cell death. *Plant Physiol* **129**, 1908–1920.

Roche TE, Baker J, Yan X, *et al.* (2001) Distinct regulatory properties of pyruvate dehydrogenase kinase and phosphatase isoforms. *Prog Nucleic Acids Res Mol Biol* **70**, 33–75.

Roessner-Tunali U, Liu J, Leisse A, *et al.* (2004) Kinetics of labelling of organic and amino acids in potato tubers by gas-chromatography mass spectrometry following incubation in C-13 labelled isotopes. *Plant J* **39**, 668–679.

Roos AJ and Cossins EA (1971) Pteroylglutamate derivatives in *Pisum sativum* L.: biosynthesis of cotyledonary tetrahydropteroylglutamates during germination. *Biochem J* **125**, 17–26.

Ryan PR, Delhaize E and Jones DL (2001) Function and mechanism of organic anion exudation from plant roots. *Annu Rev Plant Physiol Plant Mol Biol* **52**, 527–560.

Sabar M, Balk J and Leaver CJ (2005) Histochemical staining and quantification mitochondrial respiratory chain polyacrylamide gel electrophoresis. *Plant J* **44**, 893–901.

Sabar M, De Paepe R and de Kouchkovsky Y (2000) Complex I impairment, respiratory compensations and photosynthetic decrease in nuclear and mitochondrial male sterile mutants of *Nicotiana sylvestris*. *Plant Physiol* **124**, 1239–1249.

Sabar M, Gagliardi D, Balk J and Leaver CJ (2003) ORFB is a subunit of F(1) F(O)-ATP synthase: insight into the basis of cytoplasmic male sterility in sunflower. *EMBO Rep* **4**, 1–6.

Sahr T, Ravanel S and Rebeille F (2005) Tetrahydrofolate biosynthesis and distribution in higher plants. *Biochem Soc Trans* **33**, 758–762.

Saisho D, Nakazono M, Lee K, Tsutsumi N, Akita S and Hirai A (2001) The gene for alternative oxi-dase-2 (AOX2) from *Arabidopsis* thaliana consists of five exons unlike other AOX genes and is transcribed at an early stage during germination. *Genes Genet Syst* **76**, 89–97.

Saisho D, Nambara E, Naito S, Tsutsumi N, Hirai A and Nakazono M (1997) Characterization of the gene family for alternative oxidase from *Arabidopsis* thaliana. *Plant Mol Biol* **35**, 585–596.

Santos SS, Gibson GE, Cooper AJL, *et al.* (2006) Inhibitors of the α-ketoglutarate dehydrogenase complex alter [1–^{13}C]glucose and [U-^{13}C]glutamate metabolism in cerebellar granule neurons. *J Neurosci Res* **83**, 450–458.

Saradadevi K and Raghavendra AS (1992) Dark respiration protects phtosynthesis against photoinhibition in mesophyll protoplasts od pea (*Pisum sativum*). *Plant Physiol* **99**, 1232–1237.

Saradadevi K and Raghavendra AS (1994) Inhibition of photosynthesis by osmotic stress in pea (*Pisum sativum*) mesophyll protoplasts is intensified by chilling or photoinhibitory light: intriguing responses of respiration. *Plant Cell Environ* **17**, 739–746.

Schauer N, Semel Y, Roessner U, Gur A, Balbo I, Carrari F, Pleban T, Perez-Melis A, Bruedigam C, Kopka J, Willmitzer L, Zamir D, Fernie AR (2006) Comprehensive metabolic profiling and phenotyping of interspecific introgression lines for tomato improvement. *Nature Biotech* **24**, 447–454.

Schauer N, Zamir D and Fernie AR (2005) Metabolic profiling of leaves and fruit of wild species tomato: a survey of the *Solanum lycopersicon* complex. *J Exp Bot* **56**, 297–307.

Scheibe R (2004) Malate valves to balance cellular energy supply. *Physiol Plant* **120**, 21–26.

Scheibe R, Backhausen JE, Emmerlich V and Holtgrefe S (2005) Strategies to maintain redox homeostasis during photosynthesis under changing conditions. *J Exp Bot* **56**, 1481–1489.

Scheible WR, Gonzalez-Fontes A, Lauerer M, Müller-Röber B, Caboche M and Stitt M (1997) Nitrate acts as a signal to induce organic acid metabolism and repress starch metabolism in tobacco. *Plant Cell* **9**, 783–798.

Schnable PS and Wise RP (1998) The molecular basis of cytoplasmic male sterility and fertility restoration. *Trends Plant Sci* **3**, 175–180.

Schneider T, Dinkins R, Robinson K, Shellhammer J and Meinke DW (1989) An embryo-lethal mutant of *Arabidopsis thaliana* is a biotin auxotroph. *Dev Biol* **131**, 161–167.

Schonbaum GR, Bonner WD, Storey RT and Bahr JT (1971) Specific inhibition of the cyanide-insensitive respiratory pathway in plant mitochondria by hydroxamic acids. *Plant Physiol* **47**, 124–128.

Schuller KA and Randall DD (1989) Regulation of pea mitochondrial pyruvate dehydrogenase complex: Does photorespiratory ammonium influence mitochondrial carbon metabolism? *Plant Physiol* **89**, 1207–1212.

Schultz CJ, Hsu M, Miesak B and Coruzzi GM (1998) *Arabidopsis* mutants define an *in vivo* role for isoenzymes of aspartate aminotransferase in plant nitrogen assimilation. *Genetics* **149**, 491–499.

Schuster J and Binder S (2005) The mitochondrial branched-chain aminotransferase (AtBCAT-1) is capable to initiate degradation of leucine, isoleucine and valine in almost all tissues in *Arabidopsis* thaliana. *Plant Mol Biol* **57**, 241–254.

Schwabedissen-Gerbling H and Gerhardt B (1995) Purification and characterization of carnitine acyltransferase from higher plant mitochondria. *Phytochem* **39**, 39–43.

Schwender J, Ohlrogge JB and Shachar-Hill Y (2003) A flux model of glycolysis and the oxidative pentose phosphate pathway in developing *Brassica napus* embryos. *J Biol Chem* **278**, 29442–29453.

Scott J, Rébeillé F and Fletcher J (2000) Folic acid and folates: the feasibility for nutritional enhancement in plant foods. *J Sci Food Agric* **80**, 795–824.

Seki M, Narusaka M, Ishida J, *et al.* (2002) Monitoring the expression profiles of 7000 *Arabidopsis* genes under drought, cold and high-salinity stresses using a full-length cDNA microarray. *Plant J* **31**, 279–292.

Shane MW, Cramer MD, Funayama-Noguchi S, *et al.* (2004) Developmental physiology of cluster-root carboxylate synthesis and exudation in Harsh Hakea. Expression of phospho*enol*pyruvate carboxylase and the alternative oxidase. *Plant Physiol* **135**, 549–560.

Sharp RE, Matthews MA and Boyer JS (1984) Kok effect and the quantum yield of photosynthesis: light partially inhibits dark respiration. *Plant Physiol* **75**, 95–101.

Shellhammer J and Meinke D (1990) Arrested embryos from the *bio1* auxotroph of *Arabidopsis thaliana* contain reduced levels of biotin. *Plant Physiol* **93**, 1162–1167.

Shen TY and Odén PC (2002) Relationships between seed vigour and fumarase activity in *Picea abies, Pinus contorta, Betula pendula* and *Fagus sylvatica. Seed Sci Technol* **30**, 177–186.

Shintani DK and Ohlrogge JB (1994) The characterization of a mitochondrial acyl carrier protein iso-form isolated from *Arabidopsis* thaliana. *Plant Physiol* **104**, 1221–1229.

Siedow JN and Bickett DM (1981) Structural features required for inhibition of cyanide-insensitive electron transfer by propyl gallate. *Arch Biochem Biophys* **207**, 32–39.

Siedow JN and Moore AL (1993) A kinetic model for the regulation of electron transfer through the cyanide-resistant pathway in plant mitochondria. *Biochim Biophys Acta* **1142**, 165–174.

Skubatz H, Nelson TA, Meeuse BJD and Bendich AJ (1991) Heat production in the voodoo lily (*Sauromatum guttatum*) as monitored by infrared thermography. *Plant Physiol* **95**, 1084–1088.

Sluse FE and Jarmuszkiewicz W (2000) Activity and functional interaction of alternative oxidase and uncoupling protein in mitochondria from tomato fruit. *Braz J Med Biol Res* **33**, 259–268.

Smirnoff N (2000) Ascorbate biosynthesis and function in photoprotection. *Philos Trans R Soc Lond B Biol Sci* **355**, 1455–1464.

Smirnoff N and Wheeler GL (2000) Ascorbic acid in plants: biosynthesis and function. *Crit Rev Plant Sci* **19**, 267–290.

Smith AMO, Ratcliffe RG and Sweetlove LJ (2004) Activation and function of mitochondrial uncoupling protein in plants. *J Biol Chem* **279**, 51944–51952.

Somerville CR and Ogren WL (1982) Mutants of the cruciferous plant *Arabidopsis* thaliana lacking glycine decarboxylase activity. *Biochem J* **202**, 373–380.

Srere PA (1987) Complexes of sequencial metabolic enzymes. *Annu Rev Biochem* **56**, 89–124.

Srinivasan R and Oliver DJ (1995) Light-dependent and tissue-specific expression of the H-protein of the glycine decarboxylase complex. *Plant Physiol* **109**, 161–168.

Stark DM, Timerman KP, Barry GF, Preiss J and Kishore GM (1992) Regulation of the amount of starch in plant tissues by ADP glucose pyrophosphorylase. *Science* **258**, 287–292.

Stevens FJ, Dong LA, Salman LS and Anderson LE (1997) Identification of potential inter-domain disulfides in three higher plant mitochondrial citrate syntheses: paradoxical differences in redox-sensitivity as compared with the animal enzyme. *Photosynth Res* **54**, 185–197.

Stitt M (1999) Nitrate regulation of metabolism and growth. *Curr Opin Plant Biol* **2**, 178–186.

Stitt M, Lilley RM and Heldt HW (1982) Adenine nucleotide levels in the cytosol, chloroplasts, and mitochondria of wheat leaf protoplasts. *Plant Physiol* **70**, 971–977.

Streit WR and Entcheva P (2003) Biotin in microbes, the genes involved in its biosynthesis, its biochemical role and perspectives for biotechnological production. *Appl Microbiol Biotechnol* **61**, 21–31.

Strumilo S (2005) Often ignored facts about the control of the 2-Oxoglutarate dehydrogenase complex. *Biochem Mol Biol Edu* **33**, 284–287.

Studart-Guimarães C, Gibon Y, Frankel N, *et al.* (2005) Identification and characterisation of the alpha and beta subunits of succinyl CoA ligase of tomato. *Plant Mol Biol* **59**, 781–791.

Svensson AS and Rasmusson AG (2001) Light-dependent gene expression for proteins in the respiratory chain of potato leaves. *Plant J* **28**, 73–82.

Sweetlove LJ and Fernie AR (2005) Regulation of metabolic networks: understanding metabolic complexity in the systems biology era. *New Phytol* **168**, 9–23.

Sweetlove LJ, Heazlewood JL, Herald V, *et al.* (2002) The impact of oxidative stress on *Arabidopsis* mitochondria. *Plant J* **32**, 891–904.

Sweetlove LJ, Kossmann J, Riesmeier JW, Trethewey RN and Hill SA (1998) The control of source to sink carbon flux during tuber development in potato. *Plant J* **15**, 697–706.

Szurmak B, Strokovskaya L, Mooney BP, Randall DD and Miernyk JA (2003) Expression and assembly of *Arabidopsis* thaliana pyruvate dehydrogenase in insect cell cytoplasm. *Protein Expr Purif* **28**, 357–361.

Taira M, Valtersson U, Burkhardt B and Ludwig RA (2004) *Arabidopsis thaliana GLN2*-encoded glutamine synthetase is dual targeted to leaf mitochondria and chloroplasts. *Plant Cell* **16**, 2048–2058.

Takabatake R, Hata S, Taniguchi M, Kouchi H, Sugiyama T and Izui K (1999) Isolation and characterization of cDNAs encoding mitochondrial phosphate transporters in soybean, maize, rice, and *Arabidopsis*. *Plant Mol Biol* **40**, 479–486.

Taniguchi M and Sugiyama T (1996) Isolation, characterization and expression of cDNA clones encoding a mitochondrial malate translocator from *Panicum miliaceum* L. *Plant Mol Biol* **30**, 51–64.

Taniguchi M, Taniguchi Y, Kawasaki M, *et al.* (2002) Identifying and characterizing plastidic 2-oxoglutarate/malate and dicarboxylate transporters in *Arabidopsis thaliana*. *Plant Cell Environ* **43**, 706–717.

Tanksley SD, Ganal MW, Prince JP, *et al.* (1992) High density molecular linkage maps of the tomato and potato genomes. *Genetics* **132**, 1141–1160.

Taylor AE, Cogdell RJ and Lindsay JG (1992) Immunological comparison of the pyruvate dehydrogenase complexes from pea mitochondria and chloroplasts. *Planta* **188**, 225–231.

Taylor NL, Heazlewood JL, Day DA and Millar AH (2004) Lipoic acid-dependent oxidative catabolism of α-keto acids in mitochondria provides evidence of branched chain amino acid catabolism in *Arabidopsis*. *Plant Physiol* **134**, 838–848.

Tcherkez G, Cornic G, Bligny R, Gout E and Ghashghaie J (2005) *In vivo* respiratory metabolism of illuminated leaves. *Plant Physiol* **138**, 1596–1606.

Tesfaye M, Temple SJ, Allan DL, Vance CP and Samac DA (2001) Overexpression of malate dehydrogenase in transgenic alfalfa enhances organic acid synthesis and confers tolerance to aluminum. *Plant Physiol* **127**, 1836–1844.

Thelen JJ, Miernyk JA and Randall DD (1998b) Nucleotide and deduced amino acid sequences of the pyruvate dehydrogenase kinase from *Arabidopsis thaliana*. *Plant Physiol* **118**, 1533.

Thelen JJ, Miernyk JA and Randall DD (1999) Molecular cloning and expression analysis of the mitochondrial pyruvate dehydrogenase from maize. *Plant Physiol* **119**, 635–644.

Thelen JJ, Miernyk JA and Randall DD (2000) Pyruvate dehydrogenase kinase from *Arabidopsis thaliana*: a protein histidine kinase that phosphorylates serine residues. *Biochem J* **349**, 195–201.

Thelen JJ, Muszynski MG, Miernyk JA and Randall DD (1998a) Molecular analysis of two pyruvate dehydrogenase kinases from maize. *J Biol Chem* **273**, 26618–26623.

Thirkettle-Watts D, McCabe TC, Clifton R, *et al.* (2003) Analysis of the alternative oxidase promoters from soybean. *Plant Physiol* **133**, 1158–1169.

Tiessen A, Hendriks JHM, Stitt M, *et al.* (2002) Starch synthesis in potato tubers is regulated by post-translational redox modification of ADP-glucose pyrophosphorylase: a novel regulatory mechanism linking starch synthesis to the sucrose supply. *Plant Cell* **14**, 2191–2213.

Tovar-Mendez A, Hirani TA, Miernyk JA and Randall DD (2005) Analysis of the catalytic mechanism of pyruvate dehydrogenase Kinase. *Arch Biochem Biophys* **434**, 159–168.

Tovar-Mendez A, Miernyk JA and Randall DD (2002) Histidine mutagenesis of *Arabidopsis* thaliana pyruvate dehydrogenase kinase. *Eur J Biochem* **269**, 2601–2606.

Tovar-Mendez A, Miernyk JA and Randall DD (2003) Regulation of pyruvate dehydrogenase complex activity in plant cells. *Eur J Biochem* **270**, 1043–1049.

Umbach AL, Fiorani F and Siedow JN (2005) Characterization of transformed *Arabidopsis* with altered alternative oxidase levels and analysis of effects on reactive oxygen species in tissue. *Plant Physiol* **139**, 1806–1820.

Umbach AL, Ng VS and Siedow JN (2006) Regulation of plant alternative oxidase activity: a tale of two cysteines. *Biochim Biophys Acta* **1757**, 135–142.

Unger EA, Hand JM, Cashmore AR and Vasconcelos AC (1989) Isolation of a cDNA encoding mitochondrial citrate synthase from *Arabidopsis thaliana*. *Plant Mol Biol* **13**, 411–418.

Unger EA and Vasconcelos AC (1989) Purification and characterization of mitochondrial citrate synthase. *Plant Physiol* **89**, 719–723.

Unseld M, Marienfeld JR, Brandt P and Brennicke A (1997) The mitochondrial genome of *Arabidopsis thaliana* contains 57 genes in 366, 924 nucleotides. *Nat Genet* **15**, 57–61.

Urbanczyk-Wochniak E, Baxter C, Kolbe A, Kopka J, Sweetlove LJ and Fernie AR (2005) Profiling of diurnal patterns of metabolite and transcript abundance in potato (Solanum tuberosum) leaves. *Planta* **221**, 891–903.

Urbanczyk-Wochniak E, Usadel B, Thimm O, *et al.* (2006) Conversion of MapMan to allow the analysis of transcript data from Solanaceous species: effects of genetic and environmental alterations in energy metabolism in the leaf. *Plant Mol Biol* **60**, 773–792.

Van Der Meer IM, Stam ME, Vantunen AJ, Mol JNM and Stuitje AR (1992) Antisense inhibition of flavonoid biosynthesis in petunia anthers results in male sterility. *Plant Cell* **4**, 253–262.

Van Der Straeten D, Chaerle L, Sharkov G, Lambers H and Van Montagu M (1995) Salicylic acid enhances the activity of the alternative pathway of respiration in tobacco leaves and induces thermogenicity. *Planta* **196**, 412–419.

Vani T and Raghavendra AS (1994) High mitochondrial activity but incomplete engagement of the cyanide-resistant alternative pathway in guard cell protoplasts of pea. *Plant Physiol* **105**, 1263–1268.

Vani T, Reddy MM and Raghavendra AS (1990) Beneficial interaction between photosynthesis and respiration in mesophyll protoplasts of pea during short-dark cycles. *Physiol Plant* **80**, 467–471.

Vanlerberghe GC and McIntosh L (1997) Alternative oxidase: from gene to function. *Annu Rev Plant Physiol Plant Mol Biol* **48**, 703–734.

van Roermund CWT, Elgersma Y, Singh N, Wanders RJA and Tabak HF (1995) The membrane of peroxisomes in *Saccharomyces cerevisiae* is impermeable to NAD(H) and acetyl-CoA under *in vivo* conditions. *EMBO J* **14**, 3480–3486.

van Roermund CWT, Hettema EW, van den Berg M, Tabak HF and Wanders RJA (1999) Molecular characterization of carnitine-dependent transport of acetyl-CoA from peroxisome to mitochondria in *Saccharomyces cerevisiae* and identification of a plasma membrane carnitine transporter, Agp2p. *EMBO J* **18**, 5843–5852.

Verbruggen N, Hua XJ, May M and Van Montagu M (1996) Environmental and development signals modulate proline homeostasis: evidence for a negative transcriptional regulator. *Proc Natl Acad Sci USA* **93**, 8787–8791.

Vercesi AE, Borecký J, Maia IG, Arruda P, Cuccovia IM and Chaimovich H (2006) Plant uncoupling mitochondrial proteins. *Annu Rev Plant Biol* **57**, 383–404.

Vercesi AE, Martins IS, Silva MAP, Leite HMF, Cuccovia IM and Chalmovich H (1995) PUMPing plants. *Nature* **375**, 24–24.

Verniquet F, Gaillard J, Neuburger M and Douce R (1991) Rapid inactivation of plant aconitase by hydrogen peroxide. *Biochem J* **276**, 643–648.

Verslues PE, Agarwal M, Katiyar-Agarwal S, Zhu JH and Zhu JK (2006) Methods and concepts in quantifying resistance to drought, salt and freezing, abiotic stresses that affect plant water status *Plant J* **45**, 523–539.

Wada M, Yasuno R, Jordan SW, Cronan JE, Wada H (2001) Lipoic acid metabolism in *Arabidopsis thaliana*: cloning and characterization of a cDNA encoding lipoyltransferase. *Plant Cell Physiol* **42**, 650–656.

Walk RA, Michaeli S and Hock B (1977) Glyoxysomal and mitochondrial malate dehydrogenase of watermelon (*Citrullus vulgaris*) cotyledons. I Molecular properties of the purified isoenzymes. *Planta* **136**, 211–220.

Walker TS, Bais HP, Grotewold E and Vivanco JM (2003) Root exudation and rhizosphere biology. *Plant Physiol* **132**, 44–51.

Watanabe A, Nakazono M, Tsutsumi N and Hirai A (1999) AtUCP2: a novel isoform of the mitochondrial uncoupling protein of *Arabidopsis thaliana. Plant Cell Physiol* **40**, 1160–1166.
Weaver LM, Yu F, Wurtele ES and Nikolau BJ (1996) Characterization of the cDNA and gene coding for the biotin synthase of *Arabidopsis thaliana. Plant Physiol* **110**, 1021–1028.
Wedding RT (1989) Malic enzymes of higher plants. *Plant Physiol* **90**, 367–371.
Wedding RT and Black MK (1971) Nucleotide activation of cauliflower alpha-ketoglutarate dehydrogenase. *J Biol Chem* **246**, 1638–1643.
Weger HG and Turpin DH (1989) Mitochondrial respiration can support NO_3^- and NO_2^- reduction during photosynthesis. *Plant Physiol* **89**, 409–415.
Werhahn W and Braun HP (2002) Biochemical dissection of the mitochondrial proteome from *Arabidopsis thaliana* by three-dimensional gel electrophoresis. *Electrophoresis* **23**, 640–646.
Wheeler MCG, Tronconi MA, Drincovich MF, Andreo CS, Flugge UI and Maurino VG (2005) A comprehensive analysis of the NADP-malic enzyme gene family of *Arabidopsis. Plant Physiol* **139**, 39–51.
White PJ (2002) Recent advances in fruit development and ripening: an overview. *J Exp Bot* **53**, 1995–2000.
Wieland OH (1983) The mammalian pyruvate dehydrogenase complex: structure and regulation. *Rev Physiol Biochem Pharmacol* **96**, 123–170.
Wigge B, Kromer S, Gardestrom P (1993) The redox levels and subcellular-distribution of pyridine-nucleotides in illuminated barley leaf protoplasts studied by rapid fractionation physiologia. *Plantarum* **88**, 10–18.
Willeford KO and Wedding RT (1987) Evidence for a multiple subunit composition of plant NAD-malic enzyme. *J Biol Chem* **262**, 8423–8429.
Williams CH, Stillman TJ, Barynin VV, *et al.* (2002) *E. coli* aconitase B structure reveals a HEAT-like domain with implications for protein-protein recognition. *Nat Struct Biol* **9**, 447–452.
Winkler E and Klingenberg M (1994) Effect of fatty acids on H^+ transport activity of the reconstituted uncoupling protein. *J Biol Chem* **269**, 2508–2515.
Winning BM, Bourguignon J and Leaver CJ (1994) Plant mitochondrial NAD^+-dependent malic enzyme: cDNA cloning, deduced primary structure of the 59- and 62-kDa subunits, import, gene complexity and expression analysis. *J Biol Chem* **269**, 4780–4786.
Wirtz M, Berkowitz O, Droux M and Hell R (2001) The cysteine synthase complex from plants. Mitochondrial serine acetyltransferase from *Arabidopsis* thaliana carries a bifunctional domain for catalysis and protein–protein interaction. *Eur J Biochem* **268**, 686–693.
Yang ZM, Wang J, Wang SH and Xu LL (2003) Salicylic acid-induced aluminium tolerance by modulation of citrate efflux from roots of *Cassia tora* L. *Planta* **217**, 168–174.
Yao N, Eisfelder BJ, Marvin J and Greenberg JT (2004) The mitochondrion – an organelle commonly involved in programmed cell death in *Arabidopsis thaliana. Plant J* **40**, 596–610.
Yasuno R and Wada H (1998) Biosynthesis of lipoic acid in *Arabidopsis*: cloning and characterization of the cDNA for lipoic acid synthase. *Plant Physiol* **118**, 935–943.
Yasuno R and Wada H (2002) The biosynthetic pathway for lipoic acid is present in plastids and mitochondria in *Arabidopsis* thaliana. *FEBS Lett* **517**, 110–114.
Yoshida K, Terashima I and Noguchi K (2006) Distinct roles of the cytochrome pathway and alternative oxidase in leaf photosynthesis. *Plant Cell Physiol* **47**, 22–31.
Yui R, Iketani S, Mikami T and Kubo T (2003) Antisense inhibition of mitochondrial pyruvate dehydrogenase E1α subunit in anther tapetum causes male sterility. *Plant J* **34**, 57–66.
Zarkovic J, Anderson SL and Rhoads DM (2005) A reporter gene system used to study developmental expression of alternative oxidase and isolate mitochondrial retrograde regulation mutants in *Arabidopsis. Plant Mol Biol* **57**, 871–888.
Zeeman SC and Rees T (1999) Changes in carbohydrates metabolism and assimilate export in starch-excess mutants of *Arabidopsis. Plant Cell Environ* **22**, 1445–1453.
Zhu MYY, Pan JW, Wang LL, Gu Q and Huang CY (2003) Mutation induced enhancement of Al tolerance in barley cell lines. *Plant Sci* **164**, 17–23.

Zhu XF, Suzuki K, Saito T, *et al.* (1997) Geranylgeranyl pyrophosphate synthase encoded by the newly isolated gene GGPS6 from *Arabidopsis thaliana* is localized in mitochondria. *Plant Mol Biol* **35**, 331–341.

Zimmermann P, Hirsch-Hoffmann M, Hennig L and Gruissem W (2004) GENEVESTIGATOR. *Arabidopsis* microarray database and analysis toolbox. *Plant Physiol* **136**, 2621–2632.

Zou J, Qi Q, Katavic V, Marillia EF and Taylor DC (1999) Effects of antisense repression of an *Arabidopsis thaliana* pyruvate dehydrogenase kinase cDNA on plant development. *Plant Mol Biol* **41**, 837–849.

9 Cytoplasmic male sterilities and mitochondrial gene mutations in plants

Françoise Budar and Richard Berthomé

9.1 Introduction

Unlike yeast, which has been an extremely useful model for the study of mitochondrial mutations, plants display very few mitochondrial mutations, at least in the homoplasmic state (Chetrit *et al.*, 1992; Yamato and Newton, 1999; Lilly *et al.*, 2001; Ohtani *et al.*, 2002). Most mitochondrial mutations seem to be incompatible with normal plant development (Gu *et al.*, 1994), demonstrating that mitochondrially encoded functions are essential for the viability of developed plants (Goodman *et al.*, 1981). Nevertheless, genetic variants of plant mitochondria are available from various sources.

The mutation of nuclear genes affecting mitochondrial genomic stability may induce mitochondrial mutations via the rearrangement of mtDNA sequences (Abdelnoor *et al.*, 2003; Kuzmin *et al.*, 2005; see Chapter 2). This is the case for the non-chromosomal stripe (NCS) mutations of maize, most of which arose in particular genetic backgrounds (Shumway and Bauman, 1967; Newton and Coe, 1986). However, most of the mitochondrial mutations induced by such genotypes can be maintained only in the heteroplasmic state. *In vitro* culture can be used to recover mitochondrial genomic variations, some of which are associated with remarkable phenotypes (Chetrit *et al.*, 1992; Lilly *et al.*, 2001; Bartoszewski *et al.*, 2004). Again, the scarcity of phenotypes and of homoplasmy among plants recovered from *in vitro* culture suggests that there is a strong need for an unaltered mitochondrial genome in plants. However, *in vitro* culture can also be used to create new mitochondrial genomes via the recombination of two different parental mitochondrial genomes after protoplast fusion (Belliard *et al.*, 1979). When viable plants carrying recombined mitochondrial genomes can be regenerated, the most dramatic phenotypic effects concern flower morphology and pollen production (Belliard *et al.*, 1979; Leino *et al.*, 2003; Zubko *et al.*, 2003). It should be stressed that the recombined mitochondrial genomes generated by protoplast fusion experiments should not be regarded as new mutations resulting from genuine mutagenesis. They correspond instead to the sampling and redistribution of mitochondrial genes from the parent plants, and can therefore be considered as intermediate stages between the normal homoplasmic situation of the parents and the alloplasmic situation obtained following cytoplasm exchange in sexual crosses. Indeed, the phenotypes of plants carrying recombined mitochondrial genomes resulting from protoplast fusion experiments are generally intermediate between normality and severely affected alloplasmic lines (Belliard *et al.*, 1979; Hakansson and Glimelius, 1991; Zubko *et al.*, 2003). However, these variants of the mitochondrial genome demonstrate the sensitivity of the reproductive phenotype and of pollen production in particular, to variations in the mitochondrial genome. They also highlight the strong dependence of this sensitivity on the background nuclear genotype.

The foremost and most fascinating example of the importance of interactions between mitochondria and the nucleus in pollen production is undoubtedly cytoplasmic male sterility (CMS). CMS is a maternally inherited incapacity to produce pollen. This trait results in the generation of purely female individuals in a hermaphrodite species. Darwin described the occurrence of females in natural populations of otherwise hermaphrodite individuals and suggested that the coexistence of females and hermaphrodites is an evolutionary step from hermaphroditism towards dioecy; he suggested that such a coexistence 'of hermaphrodites and females without males' should be described as 'gynodioecy' (Darwin, 1877). The unusual pattern of inheritance of some male sterilities was associated with cytoplasmic organelles some time after the discovery of cytoplasmic (actually plastid) heredity by Baur (Baur, 1909; Hagemann, 2000), but well before the formal demonstration of the presence of genetic material (DNA) in organelles (Gibor and Granick, 1964). In a remarkable but much neglected paper, Chittenden and Pellew (1927) suggested the involvement of a combination of Mendelian segregation of nuclear genes and maternal inheritance of cytoplasmic factors to account for the male sterility phenotype occurring in the progeny of crosses between two flax genotypes. In the same year, Chittenden (1927) provided an experimental demonstration of this hypothesis. Chittenden proposed that a component of the cytoplasm, other than the chloroplasts, affected the reproductive phenotype by interacting with the nucleus, that this constituent could be variable and different between species or races and that it was inherited through the female line. In the early 1930s, Rhoades described the first CMS in maize (Rhoades, 1931, 1933). In his papers, Rhoades provided a large body of data demonstrating the maternal inheritance of male sterility. Although it is unclear whether he was aware of Chittenden's work, there is no doubt that he was convinced that nuclear genes were not involved in maize male sterility, and his work contributed to establishing the term *cytoplasmic male sterility* to describe this phenomenon. Rhoades also later suggested that mitochondria were the cytoplasmic factors responsible for male sterility (Rhoades, 1950).

Over almost half a century, two groups of plant biologists studied cytoplasmic male sterilities from two different points of view, with very little interaction (Budar, 1998). The first group, consisting mostly of population and evolutionary geneticists, followed the route traced by Darwin and focused on gynodioecy in natural populations, the selection and maintenance of this reproductive system and its effect on the evolution of plant species. The second group, comprising mostly breeders and molecular geneticists, focused on the molecular aspects of CMS. This second group focused on CMS in cultivated species, after the widespread adoption of this system for the production of hybrid seeds, first suggested for onions by Jones (Jones and Emsweller, 1937; Jones and Clarke, 1943). Studies of the molecular biology of CMS began after a severe epidemic of southern corn leaf blight devastating maize production in North America in 1970 (Wise *et al.*, 1999a). Almost all of what we know about the genes and gene products involved in CMS and fertility restoration originates from studies performed on CMS systems used for hybrid seed production in cultivated species.

9.2 The identification and characteristics of CMS genes

9.2.1 The problem of identifying CMS genes

All but one of the genetic determinants of CMS identified to date reside in the mitochondrial genome. The only exception is a cytoplasmic viroid-like particle carrying a double-stranded RNA molecule, which has been correlated with the 447 CMS in *Vicia faba* (Lefebvre *et al.*, 1990). The identification of CMS genes is always difficult (Hanson and Bentolila, 2004). The uniparental heredity of mitochondrial genomes prevents the use of a classical genetic approach. Nevertheless, in some cases, fertile revertants, generated by modifications to the mitochondrial genome, have led to the unambiguous identification of CMS genes. This was the case for T-*urf13* in the Texas CMS of maize (Gengenbach *et al.*, 1981; Kemble *et al.*, 1982; Umbeck and Gengenbach, 1983; Fauron *et al.*, 1987), for *pvs-orf239* in the Sprite CMS of bean (Mackenzie *et al.*, 1988; Mackenzie and Chase, 1990) and for *orf138* in the Ogura CMS of *Brassica* (Bonhomme *et al.*, 1991, 1992). In the case of CMS-S in maize, spontaneous reversion has been associated with the integrative recombination of linear episomes into the mitochondrial genome (Pring *et al.*, 1977; Levings *et al.*, 1980; Laughnan and Gabay-Laughnan, 1983). In oilseed rape, *orf224* was confirmed to be responsible for *pol* CMS by genetic association after recombination following protoplast fusion (Wang *et al.*, 1995). However, in the vast majority of cases, strong genetic association is not possible with the available plant material. This makes it difficult to establish a causal link between a mitochondrial gene or gene product and male sterility. Comparisons of the mitochondrial genomes of male sterile and male fertile plants of the same species are inefficient for the identification of CMS genes, because the two types of mitochondrial genomes generally display too many differences, even within a single species (Borck and Walbot, 1982; Makaroff and Palmer, 1988; Spassova *et al.*, 1994; Engelke and Tatlioglu, 2000; Satoh *et al.*, 2004). Thus, in many cases, evidence implicating a mitochondrial gene in male sterility is derived primarily from the differential expression of the gene concerned in maintainer and restorer genetic backgrounds (see Table 9.1). However, such differential expression provides convincing evidence only if the maintainer and restorer nuclear genetic backgrounds are almost isogenic lines or differ at only a small number of loci. This is clearly not the case when the maintainer and restorer genotypes differ by an entire chromosome or more, as is often the case for alloplasmic male sterilities (Banga *et al.*, 2003; Leino *et al.*, 2004). The nuclear genetic background has a very large influence on mitochondrial gene expression, so comparisons of the transcript profiles of mitochondrial genes in nuclear backgrounds differing at a large number of loci cannot reliably be used to identify CMS genes (Hakansson and Glimelius, 1991; Suzuki *et al.*, 1995; Yamasaki *et al.*, 2004; Leino *et al.*, 2005).

Table 9.1 Mitochondrial genes associated with CMS

Mitochondrial gene	CMS (species)	Evidence for involvement in CMS	Reference
T-*urf13*	Texas (maize)	Identification of revertants	Gengenbach *et al.* (1981), Kemble *et al.* (1982), Umbeck and Gengenbach (1983), Fauron *et al.* (1987)
orf355-orf77	S (maize)	Analysis of revertants	Zabala *et al.* (1997)
orf107	A3 (sorghum)	Transcript processing upon restoration	Tang *et al.* (1996)
atp6-orf79	Boro (rice)	Comparison of fertile and CMS cybrids	Akagi *et al.* (1994)
orf48-orf25(atpb)	Photoperiod-sensitive CMS (wheat)	Transcript processing upon restoration	Ogihara *et al.* (1999)
orf256	wheat	Difference in protein accumulation between sterile and restored genotypes	Song and Hedgcoth (1994), Hedgcoth *et al.* (2002)
pvs-orf239	'Sprite' (bean)	Identification of revertants	Mackenzie *et al.* (1988), Mackenzie and Chase (1990), Janska and Mackenzie (1993)
orf224	*pol* (oilseed rape)	Analysis of genetic recombinants after somatic hybridisation	Wang *et al.* (1995)
orf222	*nap* (oilseed rape)	Similarity to *orf224*; differential transcription in maintainer and restorer backgrounds	L'Homme *et al.* (1997)
orf263	*tour* (oilseed rape)	Different transcription patterns in maintainer and restorer backgrounds	Landgren *et al.* (1996)
orf193	Stiewe (oilseed rape)	Depletion of transcript accumulation in near-isogenic restored lines	Dieterich *et al.* (2003)
orf138	Ogura (radish, *Brassica* sp.)	Analysis of revertants (in *Brassica*)	Bonhomme *et al.* (1991, 1992), Krishnasamy and Makaroff (1993)
orf125	Kosena (radish, *B. napus* cybrids)	Similarity to *orf138*, decrease of protein accumulation in restored plants	Iwabuchi *et al.* (1999)

(*Continued*)

Table 9.1 Mitochondrial genes associated with CMS (*Continued*)

Mitochondrial gene	CMS (species)	Evidence for involvement in CMS	Reference
AtpA	*Diplotaxis catholica* (*B. juncea*)	Different transcript patterns in sterile and restored genotypes	Pathania *et al.* (2003)
Orf72	*'mur' (B. oleracea)*	Lower transcript accumulation in restored line and no transcript in fertile cytoplasm donor	Shinada *et al.* (2006)
orf274-atp1	*rep* (tobacco)	Co-transcript accumulated only in sterile plants, not restored ones	Bergman *et al.* (2000)
Pcf	Petunia	Differential expression in restored and maintainer genotypes of transcript and protein from CMS-specific locus identified in recombined cybrid mt genomes	Boeshore *et al.* (1985), Young and Hanson (1987) Nivison and Hanson (1989)
orf522	*pet1* (sunflower)	Tissue-specific decrease in expression in restored genotype	Horn *et al.* (1991), Kohler *et al.* (1991), Laver *et al.* (1991), Moneger *et al.* (1994), Smart *et al.* (1994)
CMS1-specific chimeric gene	CMS1 (chives)	Transcript profile (and possibly protein synthesis) influenced by restoration status	Engelke and Tatlioglu (2004)
preSatp6	Owen (beet)	Polypeptide associated with mt membranes	Yamamoto *et al.* (2005)
Truncated *cox2*	G (beet)	Change in cytochrome oxidase activity	Ducos *et al.* (2001)
orfB-CMS	Carrot	Specific production of the protein in floral organs	Nakajima *et al.* (2001)
ψatp6-2	Pepper	Different transcript profiles according to restoration status	Kim and Kim (2006)

One way of unambiguously attributing male sterility to a given mitochondrial gene would be to produce male sterile transgenic plants in which the gene to be tested was introduced into the mitochondrial genome of a fertile receiver plant. Unfortunately, the genetic transformation of mitochondria is not yet technically possible in plants. As an alternative, some researchers have produced transgenic plants in which the CMS gene was modified by adding a mitochondrial targeting sequence 5′ to the CMS gene in-frame with the CMS gene coding sequence, before introduction into the nucleus. Very few such attempts have been successful (Hernould *et al.*, 1993; He *et al.*, 1996; Wang *et al.*, 2006). Many of the unsuccessful attempts published concerned CMS genes previously demonstrated to be involved in sterility in independent studies (Chaumont *et al.*, 1995; Wintz *et al.*, 1995; Duroc *et al.*, 2006). Many other scientists may have unsuccessfully attempted such experiments with more dubious CMS genes, but never published their results. To date, three mitochondrial genes with mitochondrial targeting sequences introduced into the nucleus have been shown to induce male sterility in transgenic plants. The first reported case was an unedited copy of the *atp9* gene. This finding was highly unexpected, given the presence of the normal *atp9* gene in the mitochondrial genome of receiver plants. The 'unedited *atp9*' seems to act as a dominant mutation and continues to challenge our understanding (Hernould *et al.*, 1993). The other two genes were chimeric genes identified in bean CMS (He *et al.*, 1996) and in rice Boro CMS (Wang *et al.*, 2006) systems.

The information in Table 9.1 summarises studies carried out with the aim of identifying mitochondrial genes responsible for cytoplasmic male sterilities. This summary prompts two main comments. The first is that the strength of the evidence correlating a particular mitochondrial gene to the CMS phenotype varies considerably. The second is that two types of genes emerge: chimeric genes on the one hand, and modified or mutated forms of mitochondrial genes encoding subunits of respiratory complexes on the other. The difference between these two types of genetic factors highlights the contrasting concepts of 'gain-of-function' resulting from the active involvement of a new gene product in a particular phenotype and 'loss-of-function' resulting from a deleterious mutation preventing normal development.

9.2.2 *The origin of CMS genes and its consequences*

Chimeric genes were the first CMS factors to be unambiguously identified (see Table 9.1). They originate from the prodigious recombination activity of plant mitochondrial genomes and the lack of size constraints on these genomes (see Chapter 2). It should be noted that the plant mitochondrial genomes sequenced to date carry a significant number of chimeric genes, not all of which are necessarily associated with CMS (Marienfeld *et al.*, 1997, 1999; Kubo *et al.*, 2000; Notsu *et al.*, 2002; Handa, 2003; Clifton *et al.*, 2004; Ogihara *et al.*, 2005; Sugiyama *et al.*, 2005). Given their origin, it is not surprising that the expression of such CMS genes can be mediated by flanking sequences recruited from essential mitochondrial genes. Moreover, most of these genes are co-transcribed with a flanking essential mitochondrial gene (Schnable and Wise, 1998; Budar and Pelletier, 2001; Hanson and Bentolila, 2004).

Indeed, searches for extended transcripts of essential mitochondrial genes carrying original open reading frames (ORFs) in male sterile plants have often been used to track candidate CMS genes (Singh and Brown, 1991; Landgren *et al.*, 1996; de la Canal *et al.*, 2001; Engelke and Tatlioglu, 2002; Dieterich *et al.*, 2003).

We can make several observations about chimeric CMS genes, some of which remain puzzling after years of research. Firstly, most are constitutively expressed, like the genes from which they obtained their expression sequences or with which they are expressed. In contrast, the associated phenotype is restricted to male reproductive organs. This dichotomy has been called the paradox of CMS. The only exception is the *pvs-orf239* gene product in Sprite CMS in bean, which is specifically degraded by a protease in all parts of the plant except the anthers (Abad *et al.*, 1995; Sarria *et al.*, 1998). It has also been suggested that proteolytic cleavage is necessary for production of the mature *Petunia Pcf* gene product (Hanson *et al.*, 1999) and for the product of the *preS-atp6* gene in the Owen CMS of beet (Yamamoto *et al.*, 2005). Secondly, with a few exceptions, discussed below, the chimeric CMS genes identified to date are not related. This is not really surprising if we consider the origin of CMS genes: in different species, different recombinations of the mitochondrial genome give rise to different new chimeric ORFs. The most surprising aspect is the similarity of the phenotypic effects of these genes. The physiological mechanisms leading to male sterility are very unlikely to be as numerous as the chimeric mitochondrial genes involved in different systems. There are three groups of CMS genes displaying significant homology. Interestingly, they probably illustrate three different situations. In radish the *orf138* and *orf125* genes, associated with Ogura and *kosena* CMS, respectively, are certainly spontaneous variants of the same original CMS gene and belong to the same system. The sterilities induced by these genes are restored by the same allele of the restorer locus, in both radish and *Brassica* cybrids (Brown *et al.*, 2003; Desloire *et al.*, 2003; Koizuka *et al.*, 2003). Furthermore, other variants of *orf138* have been described in wild and cultivated Asian *Raphanus*, making it possible to infer phylogenetic relationships (Yamagishi and Terachi, 2001). The case of *orf224* and *orf222*, associated with *pol* and *nap* CMS, respectively, in *Brassica napus*, is more interesting because these genes are closely related, with one probably derived from the other (L'Homme *et al.*, 1997). However, *pol* and *nap* CMS have different, albeit allelic, restorers (Li *et al.*, 1998). These two CMS systems therefore seem to be different, but related. In addition, a third gene, *orf220*, genetically related to *orf222* and *orf224*, has been identified in a tuber mustard CMS (Zhang *et al.*, 2003). Sequence similarities have also been found between the sorghum A3 CMS-associated *orf107* and the rice Boro CMS-associated *orf79* (Tang *et al.*, 1996; Wang *et al.*, 2006). However, it is unclear whether these two CMS genes are phylogenetically related, in which case, the last common ancestor of these two species would probably have contained a CMS gene, or whether their similarity reflects convergent evolution, suggesting selection for a particular function. If these two CMS genes are the result of convergent evolution, then combined studies of both the corresponding CMS proteins should enable us to increase our understanding of mechanism underlying sterility. Alternatively, if these two CMS genes were derived from an ancestral gene, this would suggest that CMS genes can remain in mitochondrial genomes for long periods of time, compatible with speciation. Several CMS genes seem to be conserved among related species. This is the case for the bean *pvs-orf239* gene, which has been detected in several *Phaseolus* species (Hervieu *et al.*, 1993, 1994), and for the *orf256* gene thought to be responsible for the CMS of wheat and found in various cereal species (Hedgcoth *et al.*, 2002). In addition, sequences related to the *Petunia* CMS gene *Pcf* have been identified in several Solanaceae species (Soferman-Avshalom *et al.*, 1993).

In contrast to the situation discussed above, some of the mitochondrial genes identified as potentially responsible for CMS are variant (mutant) forms of normal genes encoding respiratory complex subunits (see Table 9.1). The evidence implicating the variant gene in CMS is somewhat weak in cases, but the possibility must still be considered. Furthermore, as chimeric CMS genes are co-expressed with essential respiratory complex subunit genes and as their products sometimes contain parts of respiratory complex subunits, it has been suggested that chimeric CMS proteins may function as mutant subunits of these complexes, disrupting complex function and thereby causing CMS by a loss-of-function effect (Heazlewood *et al.*, 2003; Sabar *et al.*, 2003; Hanson and Bentolila, 2004).

9.2.3 The population geneticists' view of CMS

The gain-of-function point of view is entirely consistent with the results of theoretical and population genetics studies on natural CMS. The first such studies were initiated after Lewis (1941) demonstrated mathematically that there would probably be counter-selection against female plants in wild populations if the sterility determinant displayed Mendelian inheritance, because its maintenance would require females to produce twice as many seeds as hermaphrodites. In contrast, if the sterility determinant was maternally inherited and females produced slightly more seeds than hermaphrodites, the maintenance of female plants in natural populations would be theoretically possible (Lewis, 1941). Recent models are based on a genomic conflict between cytoplasmic and nuclear genes (Gouyon and Couvet, 1987; Frank, 1989; Gouyon *et al.*, 1991). Such models rely on several assumptions, some of which have been experimentally verified, at least in some systems. Firstly, the cytoplasm must be variable both within and between natural populations, allowing the emergence of new CMS-inducing cytoplasms. The mode of evolution of mitochondrial genomes appears to be compatible with this assumption. Mitochondrial genome variability has been described in populations with CMS in thyme (Belhassen *et al.*, 1993), beet (Cuguen *et al.*, 1994) and *Silene* (Olson and McCauley, 2002). Secondly, the reproductive success of female plants must be greater than the female reproductive success of hermaphrodites. If this is the case, the male sterility-inducing cytoplasm can spread in the population by selection. This female advantage has been demonstrated for thyme (Thompson and Tarayre, 2000) and *Plantago lanceolata* (Poot, 1997). However, no clear female advantage has been detected in other cases of natural CMS (Boutin *et al.*, 1988).

A number of mitochondrial genes potentially responsible for CMS have been identified in the last 15 years. Much of what is known about chimeric genes inducing

CMS is compatible with the gain-of-function view, although a number of cases do not seem to be consistent with this hypothesis. A completely different phenotype has been associated with the gain-of-function provided by a new mitochondrial ORF in *Citrus*: sensitivity to the fungal pathogen *Alternaria alternata* is due to a mitochondrial gene, ACRS (ACR-toxin sensitivity), located in a t-RNA intron, and expressed only in sensitive genotypes as a result of differential post-transcriptional processing (Ohtani *et al.*, 2002). This situation is reminiscent of the susceptibility to fungal toxin induced by T-*urf13*, the Texas CMS gene in maize (Wise *et al.*, 1999a).

However, only by increasing our understanding of the mechanisms involved in male sterilities induced by mitochondrial genes will we be able to discriminate between situations in which CMS results from gain-of-function and situations in which it results from loss-of-function.

9.3 Current hypotheses concerning sterility mechanisms and the role of mitochondria in plant reproduction

9.3.1 Mitochondria and plant sexual reproduction

Plant mitochondria have many functions, from respiration and energy production to creation of carbon skeletons and the biosynthesis of metabolites (Dennis, 1987; Neuburger *et al.*, 1996; Baldet *et al.*, 1997; Focke *et al.*, 2003; Picciocchi *et al.*, 2003). Plant morphogenesis involves the precise coordination of cell division and cell expansion. Both processes induce an energetic and metabolic burst in tissues undergoing growth for the synthesis of new proteins and other biomolecules. The activities of chloroplasts and mitochondria must be coordinated and tightly regulated to meet these energetic and metabolic demands. The number of mitochondria and the steady-state abundances of mRNAs for several nucleus-encoded mitochondrial proteins have been shown to increase severalfold in developing flowers (Conley and Hanson, 1994; Huang *et al.*, 1994; Mackenzie and McIntosh, 1999) and in certain other cell types, such as the companion cells of phloem (Esau and Cronshaw, 1968). Mutations affecting mitochondrial function have long been known to have particularly dramatic effects in reproductive tissues.

A number of studies have demonstrated that unimpaired carbon and energy metabolism is required for pollen production. For example, plants expressing an antisense mitochondrial pyruvate dehydrogenase subunit gene (Yui *et al.*, 2003) specifically in the anther tapetum are male sterile (Yui *et al.*, 2003). In addition, the *Nicotiana sylvestris* mitochondrial CMS I and II mutants, in which respiratory complex I activity is impaired due to homoplasmic deletions of the mitochondrial gene *nad7*, are male sterile and display impaired vegetative development (Pla *et al.*, 1995; Gutierres *et al.*, 1997; Sabar *et al.*, 2000). Plants of this species with nuclear mutations impairing processing of the *nad4* mitochondrial transcript also fail to produce pollen (De Paepe *et al.*, 1990; Brangeon *et al.*, 2000). However, *N. sylvestris* plants with mutations affecting complex I can produce limited amounts of pollen in long day length, high light-intensity conditions (Budar *et al.*, 2003). Mutations in the nuclear gene encoding mitochondrial aldehyde dehydrogenase in maize result in partial male sterility, suggesting that higher levels of aldehyde dehydrogenase activity are required for normal pollen production than in non-pollen-producing organs (Liu *et al.*, 2001). Interestingly, mitochondrial aldehyde dehydrogenase was identified as the *Rf2* gene required to restore male fertility in plants with Texas CMS (Cui *et al.*, 1996). It therefore seems that T-URF13 increases the need for this mitochondrial activity. It has been inferred that higher levels of mitochondrial activity in male reproductive organs may lead to the production of harmful by-products, such as reactive oxygen species, which must be detoxified (Moller, 2001). Indeed, the abundance of alternative oxidase, which is thought to be a marker of cells in a state of redox stress (Vanlerberghe and McIntosh, 1996), has been found to increase in some mitochondrial mutants (Karpova *et al.*, 2002; Dutilleul *et al.*, 2003) and CMS plants (Ducos *et al.*, 2001; Pring *et al.*, 2006). Furthermore, transgenic tobacco plants producing antisense transcripts for mitochondrial alternative oxidase specifically in the anthers display partial male sterility (Kitashiba *et al.*, 1999).

NCS mutants in maize are particularly interesting. In these mutants, mitochondrial deletions are present heteroplasmically, leading to deleterious phenotypes at both the vegetative and reproductive stages. The *nad4* gene is affected in the NCS2 mutant, which cannot assemble complex I (Karpova and Newton, 1999), whereas the NCS5 and NCS6 mutants have mutations in the *cox2* gene encoding a subunit of complex IV (Lauer *et al.*, 1990; Newton *et al.*, 1990). The NCS3 and NCS4 mutants are even more severely affected, carrying mutations affecting mitochondrial ribosomal protein genes, resulting in impaired mitochondrial protein synthesis (Hunt and Newton, 1991; Newton *et al.*, 1996). The parts of the heteroplasmic plant's leaves that carry the mutated mitochondrial genome have impaired chloroplast biogenesis. In the reproductive organs of the plants, the mitochondrial mutations cause male sterility and kernel abortion.

Another mutation affecting a mitochondrial ribosomal protein gene with a reproductive phenotype has been reported in *Arabidopsis*. The nuclear *huellenlos* (*HLL*) gene encodes a putative mitochondrial L14 ribosomal subunit, as does its paralogue (*HLP*). The *huellenlos* mutant is female sterile because this mutation prevents normal ovule development (Skinner *et al.*, 2001). The ovule-specificity of the phenotype seems to result from a lack of expression of the paralogue *HLP* in the developing ovules, suppressing the redundancy of the two paralogous genes in these organs (Skinner *et al.*, 2001).

There are very few nuclear genes encoding mitochondrial proteins in which mutations specifically affect plant reproduction. These genes include the mitochondrial aldehyde dehydrogenase gene (*Rf2*) discussed above and *GFA2*. *GFA2* encodes a J-domain-containing protein that probably acts as a chaperone in mitochondria. Mutations in *GFA2* delay the death of the synergid cell and cause defective fusion of the polar nuclei at fertilisation (Christensen *et al.*, 2002). The mutant plants are therefore female sterile.

In a post-genomic project to identify nuclear genes encoding proteins targeted to the mitochondria that affect sexual reproduction in *Arabidopsis* (R. Berthomé, unpublished data), we identified a gene that seems to encode the phosphoribosylformylglycinamidine synthase (FGAM) that catalyses the fourth step in the *de novo* purine nucleotide biosynthesis pathway. Purine nucleotides are essential determinants of primary and secondary metabolism and the impairment of their biosynthesis may therefore affect plant growth and development (Boldt and Zrenner, 2003; Zrenner *et al.*, 2006). We have shown that the putative AtFGAM is targeted to mitochondria and required for development of the male, but not the female, gametophyte in heterozygous plants (R. Berthomé, M. Thomasset, N. Bourgeois, M. Maene, N. Froger, A. Martin-Canadell and F. Budar, unpublished results).

In alloplasmic lines, male sterility is often associated with homeotic alterations in floral tissue identity (Gerstel *et al.*, 1978; Nikova and Vladova, 2002; Leino *et al.*, 2003; Linke *et al.*, 2003; Zubko *et al.*, 2003; Linke and Borner, 2005). These changes include morphological transformations, such as the conversion of stamens into petaloid or carpeloid structures. Moreover, the extent to which the morphology of the flower is distorted depends on the proportion of mitochondrial genetic information originating from the cytoplasm donor parent, as shown by protoplast fusion experiments creating recombined mitochondrial genomes in *Nicotiana* and *Brassica* (Vedel *et al.*, 1986; Kofer *et al.*, 1992; Leino *et al.*, 2003). Floral homeotic gene expression has been monitored in deformed CMS flowers, restored lines and their wild-type counterparts in wheat (Murai *et al.*, 2002), carrot (Linke *et al.*, 2003) and male sterile *B. napus* derived from somatic hybrids of *B. napus* and *Arabidopsis* (Leino *et al.*, 2003; Teixeira *et al.*, 2005a). The hypothesis that disruption of the expression of known homeotic genes is linked to CMS-associated abnormal floral morphology was first tested in a tobacco line with *N. repanda* cytoplasm (Bereterbide *et al.*, 2002). Expressing the tobacco homologue of the *Arabidopsis SUPERMAN* gene improved floral morphology in this alloplasmic *N. tabacum* line, which had severely distorted stamens fused to the carpel. In addition, the expression of nuclear genes involved in floral organ identity was shown to be disturbed in *B. napus* cybrids with distorted flowers (Teixeira *et al.*, 2005a). Some authors have suggested that mitochondria are directly involved in controlling the expression of homeotic nuclear genes (Zubko, 2004; Linke and Borner, 2005), but this hypothesis remains unproven in our view. In any case, the link between aberrant flower morphology and possible defects in mitochondrial metabolism due to alloplasmy remains unclear. It is generally assumed that the defects observed in alloplasmic lines are due to the defective assembly of respiratory complexes resulting from imperfect interactions between nuclear- and mitochondrial-encoded subunits. In the alloplasmic tobacco line with *N. repanda* cytoplasm, the floral meristem cells were found to have an abnormal mitochondrial morphology (Farbos *et al.*, 2001). These plants also displayed changes in ATP/ADP ratio (Bergman *et al.*, 2000). In *B. napus* cybrids with mitochondrial genomes recombined from *B. napus* and *Arabidopsis* parents, two populations of mitochondria were observed in disorganised floral meristems (Teixeira *et al.*, 2005a). However, despite having a lower ATP content, except in green leaves, the cybrids showed no significant difference in ATP/ADP ratio from that in the normal *B. napus* line (Teixeira *et al.*, 2005b).

The plethora of evidence that mitochondrial activity plays an important role in plant reproduction provides strong support for the loss-of-function hypothesis for the mechanism of CMS. The facility with which pollen production is impaired by disturbances to mitochondrial metabolism seems to render the search for other types of mechanism superfluous. Indeed, the impairment of mitochondrial activity has been demonstrated in some cases of CMS. In the *G* CMS of beet, a decrease in cytochrome oxidase activity has been measured (Ducos *et al.*, 2001). However, no causal link was established between this observation and male sterility. In the *PET* CMS of sunflower, the level of ATPase activity is low (Sabar *et al.*, 2003). Similarly, the ORF522 protein has not been demonstrated to be involved in decreasing ATPase activity in sterile sunflower. In a sorghum CMS, no effect on any of the respiratory complexes or ATPase was observed, but the *in vitro* kinetic properties of the ATPases of sterile or fertile maintainer and restored plants were found to differ (Sane *et al.*, 1997).

If sterility results from the chimeric CMS gene products behaving as 'aberrant respiratory complex subunits', as suggested for ORF522 in sunflower CMS, then the presence of the normal subunit genes in the mitochondrial genome of the plant would imply that these abnormal peptides correspond to dominant mutations of the subunits. Furthermore, as most CMS genes are constitutively expressed, such dominant mutations should have little or no effect on vegetative growth and female fertility, unlike documented mitochondrial mutants.

9.3.2 T-URF13 and possible mechanisms for maize Texas sterility plus other CMS gene products and their characteristics

The maize CMS-T URF13 protein has been the focus of many studies because it renders the plant sensitive to a toxin produced by *Cochliobolus heterostrophus* race T and to the carbamate insecticide methomyl (for a complete review on Texas CMS, see Wise *et al.*, 1999a). It remains the most widely studied CMS-associated protein, at least in part because the sensitivity phenotype, a feature unique to this system, makes it possible to carry out functional analysis in microbial cells. When produced in *Escherichia coli*, URF13 confers sensitivity to both T-toxin and methomyl and functions as a ligand-gated receptor, specifically binding these molecules and producing hydrophilic pores in the plasma membrane (Dewey *et al.*, 1988; Rhoads *et al.*, 1995). URF13 contains three membrane-spanning domains that are probably involved in pore formation and exists in an oligomeric form in CMS-T maize mitochondria and when expressed in *E. coli* (Rhoads *et al.*, 1995, 1998). In maize CMS-T plants, URF13 was found to accumulate in all tissues tested. However, in the absence of T-toxin, the only tissue affected is the tapetum of the anthers (Warmke and Lee, 1978). Hypothetical mechanisms accounting for the male sterility caused by URF13 in CMS-T maize must resolve this apparent paradox. According to one attractive hypothesis, of the gain-of-function type, a biosynthetic molecule present only in tapetal cells may interact with URF13 in a manner analogous to that used by methomyl or T-toxin, leading to pore opening and cell death (Flavell, 1974). However, no such molecule has yet been identified. Alternatively, one possible loss-of-function hypothesis is that the efficiency of mitochondrial activity is decreased by T-URF13, resulting in an incapacity of the plants to reach a threshold of ATP production required for pollen development (Levings, 1993).

To what extent could Texas CMS be considered an archetype for other types of CMS? The products of many of the chimeric genes associated with CMS are predicted to have membrane-spanning domains, but very few have been subjected to biochemical study. The petunia and sunflower CMS proteins are associated with mitochondrial membranes (Nivison and Hanson, 1989; Horn *et al.*, 1996). The Ogura CMS protein has been shown to be integrated into the inner mitochondrial membrane and to form oligomers (Duroc *et al.*, 2005). Multimeric forms have also been reported for the *Phaseolus* CMS protein (Sarria *et al.*, 1998). These observations are consistent with these proteins being pore-forming proteins, but it remains unclear whether they act by similar mechanisms. No CMS protein has yet been found in association with any mitochondrial respiratory complex.

9.3.3 CMS and PCD

The demonstrated involvement of mitochondria in programmed cell death (PCD; see Chapter 10) suggests a possible gain-of-function mechanism for CMS: triggering cell death in the tapetal cells of microspores. Indeed, in normal pollen development, the tapetal cells degenerate at a precise stage, probably via PCD (Papini *et al.*, 1999; Wang *et al.*, 1999; Varnier *et al.*, 2005). PCD is also thought to be involved in thermosensitive genic male sterility in rice (Ku *et al.*, 2003), in which male sterility is associated with premature death of the tapetum just before pollen maturation. In many types of CMS, the anthers develop normally but the tapetum undergoes premature cellular degradation, preventing the production of viable pollen grains. Only one report, for the sunflower *PET1* CMS system, has provided evidence for PCD in tapetal cells in CMS lines (Balk and Leaver, 2001). In this system, tapetal degeneration is associated with condensation of the cytoplasm, DNA cleavage into nucleosomal units and the release of cytochrome *c* from the mitochondria. The lack of direct evidence for the involvement of a PCD mechanism in various CMS systems may be linked to the technical problems encountered in obtaining sufficient quantities of reproductive tissue for physiological and biochemical studies. It is also unclear how PCD would be triggered in the context of CMS.

9.4 The restoration of fertility by nuclear genes

9.4.1 Rf genes: genetic characteristics and molecular effects

Despite the general use of the term cytoplasmic male sterility to designate the phenomenon, neither population geneticists nor breeders have ever underestimated the role of the nuclear genotype in the phenotypic expression of CMS. Population geneticists have been confronted with the presence of restorers in the populations studied, often complicating the analysis of CMS systems in natural populations (Charlesworth and Laporte, 1998; van Damme *et al.*, 2004). Conversely, breeders have been faced with the problem of selecting good restorer genotypes for the production of self-fertile hybrids for seed crop species.

The restoration of fertility is generally controlled by one or two genetic loci. When several nuclear loci are involved, they may all be required for restoration or they may act independently. Furthermore, fertility may be restored at the sporophytic or gametophytic stage. In gametophytic restoration, only pollen carrying the restorer gene is viable. Consequently, a heterozygous plant produces half the normal amount of pollen and all its pollen carries the restorer allele. In sporophytic restoration, all the pollen of a restored plant is viable, whatever its genotype. As most sporophytic restorers are dominant, a heterozygous plant produces normal amounts of pollen, half of which carries the non-restorer allele.

The sporophytic or gametophytic nature of restoration may provide clues as to which tissue or cell type is targeted by the male sterility mechanism. If male sterility results from pollen abortion due to the premature degeneration of the tapetum, then restoration must be sporophytic, as it must target the tapetum. If male sterility results from defects in pollen development after meiosis, then restoration must be gametophytic. Thus, determining whether restoration is sporophytic or gametophytic can provide information about the target tissue of the male sterility mechanism.

In alloplasmic situations, it is difficult to determine how many genes are required for the recovery of normal floral morphology and male fertility. Restoration is frequently achieved by introducing genes from the fertile cytoplasm donor species into the alloplasmic male sterile line. These genes may be introgressed as parts of chromosomes from the donor genome, or even as an entire chromosome from the donor species (Leino *et al.*, 2004). Even if fertility restoration behaves as a single locus in the restored introgressed line, the lack of genetic recombination between the introgressed chromosome segment and the recipient genome usually precludes the formal genetic analysis of restoration (Delourme *et al.*, 1998; Banga *et al.*, 2003; Giancola *et al.*, 2003).

A summary of current knowledge about fertility restorers and their mode of action is provided in Table 9.2. In a male sterility-inducing cytoplasm, fertility may be restored by a nuclear gene in two ways. Firstly, the nuclear gene may inhibit production of the sterility protein throughout the plant, or specifically in floral organs or tissues. This situation is the most frequently reported, generally with a post-transcriptional effect on the male sterility gene mRNA (see Table 9.2). In recent studies, attempts to identify the male sterility determinant have often been based on the assumption that the restorer genotype affects the amount, or size, of the mRNA produced from the CMS gene (see Table 9.1). This assumption may account for the rarity of cases in which restoration has been shown to affect the CMS protein, but not its RNA (Grelon *et al.*, 1994; Krishnasamy and Makaroff, 1994). Alternatively, the restorer gene may restore fertility by counteracting the physiological effect of the sterility gene product. This is assumed to be the case for the *Rf2* of maize Texas CMS. This restorer gene was the first to be identified and encodes a mitochondrial aldehyde dehydrogenase (Cui *et al.*, 1996; Liu *et al.*, 2001; Moller, 2001). 'Physiological restorers' also include genes introduced into

Table 9.2 Nuclear restorer genes and their effect

Restorer	CMS (species)	Mode of action – molecular effect	Reference
Rf1[a]	Texas (maize)	Sporophytic – processing of T-*urf13* transcript	Wise *et al.* (1999b)
Rf2[b]	Texas (maize)	Sporophytic – no effect on T-*urf13* expression, probable effect by physiological mechanism	Cui *et al.* (1996), Liu *et al.* (2001)
Rf8[a]	Texas (maize)	Sporophytic – processing of T-*urf13* transcript	Dill *et al.* (1997), Wise *et al.* (1999b)
*Rf**[a]	Texas (maize)	Sporophytic – processing of T-*urf13* transcript	Dill *et al.* (1997), Wise *et al.* (1999b)
Rf3[c]	CMS-S (maize)	Gametophytic – processing of *orf355-orf77* transcript	Wen and Chase (1999)
rfl1[d]	CMS-S (maize)	Gametophytic – decrease in abundance of *orf355-orf77* and *atpA* transcripts	Wen *et al.* (2003)
Rf3[e]	A3 (sorghum)	Gametophytic – internal processing of *orf107,* internal processing of *urf 209,* editing of *orf107* transcripts	Tang *et al.* (1996, 1998, 1999)
Rf1–Rf1a[f]	Boro (rice)	Gametophytic – endonucleolytic cleavage of *atp6-orf79* transcript	Kazama and Toriyama (2003), Akagi *et al.* (2004), Komori *et al.* (2004), Wang *et al.* (2006)
Rf1b[f]	Boro (rice)	Gametophytic – degradation of *atp6-orf79* transcript	Wang *et al.* (2006)
Rf[g]	Petunia	Sporophytic – specific decrease in PCF mRNA and protein levels in flowers	Young and Hanson (1987), Hanson *et al.* (1999)
Rfk1/Rfo[h]	Ogura/*kosena* (radish, *Brassica*)	Sporophytic – tissue specific inhibition of ORF138/ORF125 protein accumulation. No effect on transcript	Grelon *et al.* (1994), Koizuka *et al.* (1998, 2000), Bellaoui *et al.* (1999)

Rfp[i]	*pol* (oilseed rape)	Sporophytic – processing of *orf224/atp6* co-transcript in floral organs	Li *et al.* (1998), Menassa *et al.* (1999)
Rfn[i]	*nap* (oilseed rape)	Sporophytic – processing of *orf222* transcript in anthers	Li *et al.* (1998), Brown (1999), Geddy *et al.* (2005)
Restorer gene of line RPA843R	*PET1* (sunflower)	Sporophytic – tissue-specific reduction of *orf522* transcript level	Moneger *et al.* (1994), Smart *et al.* (1994)
Fr[j]	Sprite (bean)	Sporophytic – stoichiometric shift of genomic molecule carrying *pvs-orf239*, decreasing gene copy number	Mackenzie and Chase (1990)
Fr2	Sprite (bean)	Post-transcriptional impairment of *orf239* expression	Chase (1994), Abad *et al.* (1995)
RfX and RfT[k]	CMS1 (chives)	Sporophytic – processing of CMS1 specific transcript in flowers	Engelke and Tatlioglu (2004)

[a] *Rf1*, *Rf8* and *Rf** are not allelic and act independently. *Rf8* and *Rf** give partial fertility with *Rf2* (Dill *et al.*, 1997; Wise *et al.*, 1999b).
[b] *Rf2* is necessary with *Rf1*, *Rf8* and *Rf**. It has been identified and encodes a mitochondrial aldehyde dehydrogenase (Cui *et al.*, 1996). Its nature as a genuine restorer has been discussed (Touzet, 2002).
[c] *Rf3* restorer allele has been identified as dominant in diploid pollen from tetraploid plants (Kamps *et al.*, 1996).
[d] *rfl1* is a recessive, homozygous lethal restorer allele (Wen *et al.*, 2003).
[e] *Rf3* acts together with *Rf4* for restoration of fertility (Tang *et al.*, 1998).
[f] *Rf1a* and *Rf1b* have been identified and encode similar PPR proteins. They are located at the same locus (*Rf-1*), which carries several related genes encoding PPR proteins. *Rf1a* also affects *atp6* transcript editing. It has an epistatic effect on *Rf1b,* modifying its effect on the *atp6-orf79* transcript. The dominance status of Rf1a and Rf1b is not known (Wang *et al.*, 2006).
[g] The *Rf* gene has been identified and encodes a PPR protein (Bentolila *et al.*, 2002).
[h] *Rfo* and *Rfk1* have been identified and encode almost identical PPR proteins (Brown *et al.*, 2003; Desloire *et al.*, 2003; Koizuka *et al.*, 2003).
[i] *Rfp* and *Rfn* are allelic to a *Mmt* gene (modifier of mitochondrial transcript) that modifies *nad4* transcripts (Singh *et al.*, 1996; Li *et al.*, 1998).
[j] Unlike other restorers, the *Fr* gene has a permanent effect, lasting after its genetic segregation.
[k] *RfX* and *RfT* act independently and are not allelic (Engelke and Tatlioglu, 2004).

the nuclear genomes of male sterile plants to compensate for 'loss-of-function' mutations. For example, the *nad7* subunit gene fused to a sequence encoding a mitochondrial targeting peptide has been shown to complement deletion of the NAD7 subunit gene in the CMS II *N. sylvestris* mutant, following its introduction into the nucleus (Pineau *et al.*, 2005).

It is tempting to speculate that the two types of restoration system are related to the two types of male sterility, with loss-of-function sterilities requiring restoration genes complementing the deficient mitochondrial function and gain-of-function sterilities requiring nuclear genes to switch off their expression.

In population genetics models, the occurrence of a male sterility-inducing cytoplasm immediately generates very strong selection pressure for nuclear genes restoring male fertility, because the nuclear genome loses half of its selective value in the presence of such a cytoplasm. Thus, any modification in any nuclear gene that leads to the production of even only small numbers of viable pollen grains in the presence of the sterility-inducing cytoplasm is likely to be selected for in natural populations. Consistent with this line of argument, restorer genes are sometimes redundant in the species in which the sterility they counteract originated, as in the case of maize CMS-S (Gabay-Laughnan *et al.*, 2004) and CMS-T (Edwardson, 1955; Wise *et al.*, 1999b), WA and HL CMS in rice (Li *et al.*, 2005) and Ogura CMS in wild radish (Bett and Lydiate, 2004). The fixation of such restorer alleles effectively leads to the disappearance of the male sterile phenotype in the population. Theoretical studies have suggested possible explanations for the maintenance of sexual polymorphism, and efforts are still being made to integrate all the selection forces involved into comprehensive models (see, for example, Saur Jacobs and Wade, 2003). We have suggested that chimeric CMS genes not expressed in the cytoplasm donor species due to restorer gene fixation, but revealed in a foreign nuclear background containing no restorers, may be responsible for some of the male sterilities that emerge in alloplasmic situations (Budar *et al.*, 2003).

Since the identification of the first restorer gene affecting expression of the corresponding mitochondrial sterility gene in *Petunia* (Bentolila *et al.*, 2002), a handful of restorers of the same type have been identified in radish, rice and *Sorghum* (Brown *et al.*, 2003; Desloire *et al.*, 2003; Kazama and Toriyama, 2003; Koizuka *et al.*, 2003; Akagi *et al.*, 2004; Komori *et al.*, 2004; Klein *et al.*, 2005; Wang *et al.*, 2006). All encode proteins with pentatricopeptide repeat motifs (PPR).

9.5 Restorers as PPR proteins, biochemical and evolutionary implications

Since the discovery of the PPR protein family in plants (Aubourg *et al.*, 2000; Small and Peeters, 2000), evidence has been accumulating that the members of this family are involved in the expression of organellar genomes (Lurin *et al.*, 2004; Nakamura *et al.*, 2004, see Chapter 3). On the basis of their structure, it has been suggested that PPR proteins behave as molecular adaptors, facilitating the editing or processing of a specific RNA molecule targeted (Small and Peeters, 2000). Although their precise molecular mode of action remains unknown, PPR proteins have been shown to bind specific RNAs and to be involved in post-transcriptional events, such as RNA editing and translation in chloroplasts (Kotera *et al.*, 2005; Schmitz-Linneweber *et al.*, 2005). So far, the only mitochondrial PPR proteins with identified functions are fertility restorers. This is consistent both with the hypothesis for the molecular function of PPR proteins, which predicts that a restorer PPR targets the mRNA of the sterility gene, and with the gain-of-function view of CMS, which predicts that the restorer inhibits the expression of the sterility gene. Therefore, we expect that restorer PPR proteins target corresponding sterility gene mRNAs and trigger events that lead to the absence of the sterility proteins. However, the first results reported regarding restorer PPRs suggest a more complex situation. A recent report on the restorers of 'Boro' CMS in rice revealed that two related PPR genes, named *Rf1-A* and *Rf1-B*, act as restorers of the same CMS, but with different effects on the sterility gene mRNA (Wang *et al.*, 2006). In addition, the same PPR protein, encoded by the *Rf1-A* restorer allele, can influence post-transcriptional processing of the *orf79* transcript and editing of the *atp6* mRNA (Wang *et al.*, 2006). This suggests that a given protein may have different effects on different RNA molecules and calls into question the specificity of the interaction between a PPR protein and its RNA target. As *orf79* is co-transcribed with a second copy of the *atp6* gene, it is conceivable that the *Rf1-A*-encoded PPR protein recognises the *atp6* part of the co-transcript. However, this nonetheless suggests that its interaction with different RNA molecules leads to the recruitment of different RNA metabolism activities. A possible ambiguous specificity for restorer target is suggested in the case of a 'lethal' restorer of CMS-S in maize, although this restorer locus has not been identified. In this case, the same allele affects mRNA accumulation for both the CMS gene and the necessary *ATPA* gene, resulting in the death of homozygous plants (Wen *et al.*, 2003). The only viable restored plants are heterozygous for the restorer; only 50% of their pollen is viable (gametophytic restoration) and these pollen grains do not accumulate the ATPA subunit normally (Wen *et al.*, 2003). In oilseed rape, a single genetic locus seems to control the maturation of different mitochondrial transcripts, depending on the allele (L'Homme *et al.*, 1997; Li *et al.*, 1998). Two of the known alleles affect two, probably related, CMS-gene transcripts called *nap* and *pol* (see Section 9.2.2), whereas another, *Mmt1*, controls maturation of the *nad4* mRNA (Li *et al.*, 1998).

A striking feature of the proteins of the PPR family is their impressive proliferation in the plant kingdom, drawing parallels with the importance of organellar RNA editing in plants (Lurin *et al.*, 2004). This proliferation of PPR family proteins may result in part from duplication events and the selection of genes encoding proteins controlling expression of the chimeric mitochondrial ORFs that induce male sterility. This hypothesis is supported by the complex structure of the identified *Rf* loci, resembling that of resistance gene loci (Touzet and Budar, 2004). Among the four PPR restorer loci described so far, three (those from *Petunia*, radish and rice) revealed complex situations, where restorer genes are flanked by very similar, non-restorer paralogues (Bentolila *et al.*, 2002; Brown *et al.*, 2003; Desloire *et al.*, 2003; Koizuka *et al.*, 2003; Akagi *et al.*, 2004; Komori *et al.*, 2004; Wang *et al.*, 2006). These *Rf* genes encode PPR proteins of the P subfamily, which are thought to be the basic archetype for PPR proteins (Lurin *et al.*, 2004). The recent identification of a gene restoring fertility in the A1 CMS of *Sorghum* and encoding a protein of the E subgroup of PPR proteins indicates that the control of mitochondrial CMS gene expression is not restricted to a particular subgroup of PPR proteins (Klein *et al.*, 2005). The *Sorghum Rf* is also the only one described so far not lying in a complex locus. Whether or not the level of complexity of *Rf* loci is correlated with the particular PPR subfamily to which they belong will become apparent after the analysis of additional *Rf* loci. In any case, we face challenging questions about the evolution of restorer genes encoding PPR proteins at both the functional and genomic level.

9.6 Concluding remarks

Mitochondrial activity is undoubtedly of fundamental importance in plant cells and the mitochondrial genome encodes proteins of prime importance for this activity. This fundamental role results in a scarcity of homoplasmic mitochondrial mutants. However, the study of heteroplasmic mitochondrial mutants has provided insight into the integration of plastid and mitochondrial biogenesis in plants (Newton, 1993; Jiao *et al.*, 2005). It has also been shown that sexual reproduction, and pollen production in particular, is particularly sensitive to mitochondrial dysfunction. Alloplasmic situations, in which floral morphology, anther organ identity and pollen production are affected in a remarkably similar way in different plant taxa, illustrate this sensitivity. The specific defects in plant development induced by mitochondrial mutations are comparable with human mitochondrial diseases (Mazat *et al.*, 2001; Houstek *et al.*, 2004). Understanding these defects will provide insight into the metabolic, energetic and redox homeostasis of cells.

After decades of studies on CMS carried out independently by molecular geneticists and population geneticists, we can now develop a synthetic view, in which mitochondria play an unusual role in plants, and genomic conflict between the cytoplasm and nucleus results from the different modes of inheritance of these genomes. Parallels can be drawn between this phenomenon and intracellular bacteria of the genus *Wolbachia*, which distort the sex ratio or manipulate the sex of their hosts (Knight, 2001; Zimmer, 2001). In plants, the recombination of mitochondrial DNA creates new CMS genes, in the probable absence of genome size constraints (Bullerwell and Gray, 2004). Some of these genes may induce male sterility and may confer a selective advantage on the corresponding cytoplasm, giving better seed set, for example. The constraints on the selection of new CMS genes therefore seem to concern the possibilities for impairing pollen production without damaging the vegetative or female fitness of the plant. It is not yet known how this is achieved.

The occurrence of sterility-inducing cytoplasms led to the selection of specific alleles of nuclear genes restoring pollen production, mostly by controlling the expression of CMS genes at the post-transcriptional level. The nuclear response to CMS genes seems to involve diversification of the genes encoding PPR proteins, probably contributing to the proliferation of this family. The last 5 years have seen the identification of some restorer genes and have raised some exciting questions for future research: What role do the PPR proteins encoded by non-restorer alleles of *Rf* genes play? How do *Rf* loci evolve?

CMS remains a fascinating model for the study of the coordination between the nucleus and mitochondria, in terms of gene expression and molecular interactions and in terms of what is the most ancient relationship in eukaryotic evolution.

Acknowledgements

We thank Hakim Mireau for exciting discussions and critical reading of the manuscript. We also thank Georges Pelletier and Philippe Guerche for constant support.

References

Abad AR, Mehrtens BJ and Mackenzie SA (1995) Specific expression in reproductive tissues and fate of a mitochondrial sterility-associated protein in cytoplasmic male-sterile bean. *Plant Cell* **7**, 271–285.

Abdelnoor RV, Yule R, Elo A, Christensen AC, Meyer-Gauen G and Mackenzie SA (2003) Substoichiometric shifting in the plant mitochondrial genome is influenced by a gene homologous to MutS. *Proc Natl Acad Sci USA* **100**, 5968–5973.

Akagi H, Nakamura A, Yokozeki-Misono Y, *et al.* (2004) Positional cloning of the rice *Rf-1* gene, a restorer of BT-type cytoplasmic male sterility that encodes a mitochondria-targeting PPR protein. *Theor Appl Genet* **108**, 1449–1457.

Akagi H, Sakamoto M, Shinjyo C, Shimada H and Fujimura T (1994) A unique sequence located downstream from the rice mitochondrial atp6 may cause male sterility. *Curr Genet* **25**(1), 52–58.

Aubourg S, Boudet N, Kreis M and Lecharny A (2000) In *Arabidopsis*, 1% of the genome codes for a novel protein family unique to plants. *Plant Mol Biol* **42**, 603–613.

Baldet P, Alban C and Douce R (1997) Biotin synthesis in higher plants: purification and characterization of *bioB* gene product equivalent from *Arabidopsis* overexpressed in *Escherichia coli* and its subcellular localization in pea leaf cells. *FEBS Lett* **419**, 206–210.

Balk J and Leaver CJ (2001) The PET1-CMS mitochondrial mutation in sunflower is associated with premature programmed cell death and cytochrome *c* release. *Plant Cell* **13**, 1803–1818.

Banga SS, Deol JS and Banga SK (2003) Alloplasmic male-sterile *Brassica juncea* with *Enarthrocarpus lyratus* cytoplasm and the introgression of gene(s) for fertility restoration from cytoplasm donor species. *Theor Appl Genet* **106**, 1390–1395.

Bartoszewski G, Malepszy S and Havey MJ (2004) Mosaic (MSC) cucumbers regenerated from independent cell cultures possess different mitochondrial rearrangements. *Curr Genet* **45**, 45–53.

Baur E (1909) Das Wesen und die Erblichkeitsverhältnisse der "Varietates albomarginatae hort" von Pelargonium zonale. *Z Vererbungslehre* **1**, 330–351.

Belhassen E, Atlan A, Couvet D, Gouyon P-H and Quétier F (1993) Mitochondrial genome of *Thymus vulgaris* L. (Labiate) is highly polymorphic between and among natural populations. *Heredity* **71**, 462–472.

Bellaoui M, Grelon M, Pelletier G and Budar F (1999) The restorer Rfo gene acts post-translationally on the stability of the ORF138 Ogura CMS-associated protein in reproductive tissues of rapeseed cybrids. *Plant Mol Biol* **40**(5), 893–902.

Belliard G, Vedel F and Pelletier G (1979) Mitochondrial recombination in cytoplasmic hybrids of *Nicotiana tabacum* by protoplast fusion. *Nature* **281**, 401–403.

Bentolila S, Alfonso AA and Hanson MR (2002) A pentatricopeptide repeat-containing gene restores fertility to cytoplasmic male-sterile plants. *Proc Natl Acad Sci USA* **22**, 22.

Bereterbide A, Hernould M, Farbos I, Glimelius K and Mouras A (2002) Restoration of stamen development and production of functional pollen in an alloplasmic CMS tobacco line by ectopic expression of the *Arabidopsis SUPERMAN* gene. *Plant J* **29**, 607–615.

Bergman P, Edqvist J, Farbos I and Glimelius K (2000) Male-sterile tobacco displays abnormal mitochondrial atp1 transcript accumulation and reduced floral ATP/ADP ratio. *Plant Mol Biol* **42**, 531–544.

Bett KE and Lydiate DJ (2004) Mapping and genetic characterization of loci controlling the restoration of male fertility in Ogura CMS radish. *Mol Breed* **13**, 125–133.

Boeshore ML, Lifshitz I, Hanson MR and Izhar S (1985) Novel composition of mitochondrial genomes in *Petunia* somatic hybrids derived from cytoplasmic male sterile and fertile plants. *Mol Gen Genet* **190**(3), 459–467.

Boldt R and Zrenner R (2003) Purine and pyrimidine biosynthesis in higher plants. *Physiol Plant* **117**, 297–304.

Bonhomme S, Budar F, Férault M and Pelletier G (1991) A 2.5 kb *Nco*I fragment of Ogura radish mitochondrial DNA is correlated with cytoplasmic male-sterility in *Brassica* cybrids. *Curr Genet* **19**, 121–127.

Bonhomme S, Budar F, Lancelin D, Small I, Defrance M-C and Pelletier G (1992) Sequence and transcript analysis of the *Nco2.5* Ogura-specific fragment correlated with cytoplasmic male sterility in *Brassica* cybrids. *Mol Gen Genet* **235**, 340–348.

Borck KS and Walbot V (1982) Comparison of the restriction endonuclease digestion patterns of mitochondrial DNA from normal and male sterile cytoplasms of *Zea mays* L. *Genetics* **102**, 109–128.

Boutin V, Jean R, Valero M and Vernet P (1988) Gynodioecy in *Beta maritima*. *Oecol Plant* **9**, 61–66.

Brangeon J, Sabar M, Gutierres S, *et al.* (2000) Defective splicing of the first nad4 intron is associated with lack of several complex I subunits in the *Nicotiana sylvestris* NMS1 nuclear mutant. *Plant J* **21**, 269–280.

Brown GG (1999) Unique aspects of cytoplasmic male sterility and fertility restoration in *Brassica napus*. *J Hered* **90**(3), 351–356.

Brown GG, Formanova N, Jin H, *et al.* (2003) The radish *Rfo* restorer gene of Ogura cytoplasmic male sterility encodes a protein with multiple pentatricopeptide repeats. *Plant J* **35**, 262–272.

Budar F (1998) What can we learn about the interactions between the nuclear and mitochondrial genomes by studying cytoplasmic male sterilities ? In: *International Congress on Plant Mitochondria: From Gene to Function* (eds Moller IM, Gardestrom P, Glimelius K and Glaser E), pp. 49–55. Backhuys, Aronsborg, Sweden.

Budar F and Pelletier G (2001) Male sterility in plants: occurrence, determinism, significance and use. *C R Acad Sci III* **324**, 543–550.

Budar F, Touzet P and De Paepe R (2003) The nucleo-mitochondrial conflict in cytoplasmic male sterilities revisited. *Genetica* **117**, 3–16.

Bullerwell CE and Gray MW (2004) Evolution of the mitochondrial genome: protist connections to animals, fungi and plants. *Curr Opin Microbiol* **7**, 528–534.

Charlesworth D and Laporte V (1998) The male-sterility polymorphism of *Silene vulgaris*: analysis of genetic data from two populations and comparison with *Thymus vulgaris*. *Genetics* **150**, 1267–1282.

Chase CD (1994) Expression of CMS-unique and flanking mitochondrial DNA sequences in *Phaseolus vulgaris* L. *Curr Genet* **25**(3), 245–251.

Chaumont F, Bernier B, Buxant R, Williams ME, Levings CS, III and Boutry M (1995) Targeting the maize T-urf13 product into tobacco mitochondria confers methomyl sensitivity to mitochondrial respiration. *Proc Natl Acad Sci USA* **92**, 1167–1171.

Chetrit P, Rios R, De Paepe R, Vitart V, Gutierres S and Vedel F (1992) Cytoplasmic male sterility is associated with large deletions in the mitochondrial DNA of two *Nicotiana sylvestris* protoclones. *Curr Genet* **21**, 131–137.

Chittenden RJ (1927) Cytoplasmic inheritance in flax. *J Hered* **18**, 337–343.

Chittenden RJ and Pellew C (1927) A suggested interpretation of certain cases of anisogeny. *Nature* **119**, 10–12.

Christensen CA, Gorsich SW, Brown RH, *et al.* (2002) Mitochondrial GFA2 is required for synergid cell death in *Arabidopsis*. *Plant Cell* **14**, 2215–2232.

Clifton SW, Minx P, Fauron CM, *et al.* (2004) Sequence and comparative analysis of the maize NB mitochondrial genome. *Plant Physiol* **136**, 3486–3503.

Conley CA and Hanson MR (1994) Tissue-specific protein expression in plant mitochondria. *Plant Cell* **6**, 85–91.

Cuguen J, Wattier R, Saumitou-laprade P, *et al.* (1994) Gynodioecy and mitochondrial DNA polymorphism in natural populations of *Beta vulgaris* ssp *maritima*. *Genet Sel Evol* **26**, 87–101.

Cui X, Wise RP and Schnable PS (1996) The *rf2* nuclear restorer gene of male-sterile T-cytoplasm maize. *Science* **272**, 1334–1336.

Darwin C (1877) *The Different Forms of Flowers on Plants of the Same Species*. John Murray, London.

de la Canal L, Crouzillat D, Quetier F and Ledoigt G (2001) A transcriptional alteration in the *atp9* gene is associated with a sunflower male-sterile cytoplasm. *Theor Appl Genet* **102**, 1185–1189.

De Paepe R, Chetrit P, Vitart V, Ambard-Bretteville F, Prat D and Vedel F (1990) Several nuclear genes control both male sterility and mitochondrial protein synthesis in *Nicotiana sylvestris* protoclones. *Mol Gen Genet* **222**, 206–210.

Delourme R, Foisset N, Horvais R, *et al.* (1998) Characterisation of the radish introgression carrying the *Rfo* restorer gene for the *Ogu*-INRA cytoplasmic male sterility in rapeseed (*Brassica napus* L.). *Theor Appl Genet* **97**, 129–134.

Dennis DT (1987) *The Biochemistry of Energy Utilization in Plants*. Chapman and Hall, New York, 145 pp.

Desloire S, Gherbi H, Laloui W, *et al.* (2003) Identification of the fertility restoration locus, *Rfo*, in radish, as a member of the pentatricopeptide-repeat protein family. *EMBO Rep* **4**, 588–594.

Dewey RE, Siedow JN, Timothy DH and Lewings CS, III (1988) A 13-kilodalton maize mitochondrial protein in *E. coli* confers sensitivity to *Bipolaris maydis* toxin. *Science* **239**, 293–294.

Dieterich JH, Braun HP and Schmitz UK (2003) Alloplasmic male sterility in *Brassica napus* (CMS 'Tournefortii-Stiewe') is associated with a special gene arrangement around a novel *atp9* gene. *Mol Genet Genomics* **269**, 723–731.

Dill CL, Wise RP and Schnable PS (1997) Rf8 and Rf* mediate unique T-urf13-transcript accumulation, revealing a conserved motif associated with RNA processing and restoration of pollen fertility in T-cytoplasm maize. *Genetics* **147**(3), 1367–1379.

Ducos E, Touzet P and Boutry M (2001) The male sterile G cytoplasm of wild beet displays modified mitochondrial respiratory complexes. *Plant J* **26**, 171–180.

Duroc Y, Gaillard C, Hiard S, Defrance MC, Pelletier G and Budar F (2005) Biochemical and functional characterization of ORF138, a mitochondrial protein responsible for Ogura cytoplasmic male sterility in Brassiceae. *Biochimie* **87**, 1089–1100.

Duroc Y, Gaillard C, Hiard S, Tinchant C, Berthomé R, Pelletier G and Budar F (2006) Nuclear expression of a cytoplasmic male sterility gene modifies mitochondrial morphology in yeast and plant cells. *Plant Science* **170**, 755–767.

Dutilleul C, Garmier M, Noctor G, *et al.* (2003) Leaf mitochondria modulate whole cell redox homeostasis, set antioxidant capacity, and determine stress resistance through altered signaling and diurnal regulation. *Plant Cell* **15**, 1212–1226.

Edwardson JR (1955) The restoration of fertility to cytoplasmic male-sterile corn. *Agron J* **47**, 457–461.

Engelke T and Tatlioglu T (2000) Mitochondrial genome diversity in connection with male sterility in *Allium schoenoprasum* L. *Theor Appl Genet* **100**, 942–948.

Engelke T and Tatlioglu T (2002) A PCR-marker for the CMS1 inducing cytoplasm in chives derived from recombination events affecting the mitochondrial gene *atp9*. *Theor Appl Genet* **104**, 698–702.

Engelke T and Tatlioglu T (2004) The fertility restorer genes *X* and *T* alter the transcripts of a novel mitochondrial gene implicated in CMS1 in chives (*Allium schoenoprasum* L.). *Mol Genet Genomics* **271**(2), 150–160.

Esau K and Cronshaw J (1968) Plastids and mitochondria in the phloem of *Cucurbita*. *Can J Bot* **46**, 877–880.

Farbos I, Mouras A, Bereterbide A and Glimelius K (2001) Defective cell proliferation in the floral meristem of alloplasmic plants of *Nicotiana tabacum* leads to abnormal floral organ development and male sterility. *Plant J* **26**, 131–142.

Fauron CM-R, Abbott AG, Brettell RIS and Gesteland RF (1987) Maize mitochondrial DNA rearrangements between the normal type, the Texas male sterile cytoplasm, and a fertile revertant CMS-T regenerated plant. *Curr Genet* **11**, 339–342.

Flavell R (1974) A model for the mechanism of cytoplasmic male sterility in plants, with special reference to maize. *Plant Sci Lett* **3**, 259–263.

Focke M, Gieringer E, Schwan S, Jansch L, Binder S and Braun HP (2003) Fatty acid biosynthesis in mitochondria of grasses: malonyl-coenzyme A is generated by a mitochondrial-localized acetyl-coenzyme A carboxylase. *Plant Physiol* **133**, 875–884.

Frank SA (1989) The evolutionary dynamics of cytoplasmic male sterility. *Am Nat* **133**, 345–376.

Gabay-Laughnan S, Chase CD, Ortega VM and Zhao L (2004) Molecular-genetic characterization of CMS-S restorer-of-fertility alleles identified in Mexican maize and teosinte. *Genetics* **166**, 959–970.

Geddy R, Mahe L and Brown GG (2005) Cell-specific regulation of a *Brassica napus* CMS-associated gene by a nuclear restorer with related effects on a floral homeotic gene promoter. *Plant J* **41**(3), 333–345.

Gengenbach BG, Connelly JA, Pring DR and Conde MF (1981) Mitochondrial DNA variation in maize plants regenerated during tissue culture selection. *Theor Appl Genet* **59**, 161–167.

Gerstel DU, Burns JA and Burk LG (1978) Cytoplasmic male sterility in *Nicotiana*, restoration of fertility, and the nucleolus. *Genetics* **89**, 157–169.

Giancola S, Marhadour S, Desloire S, *et al.* (2003) Characterization of a radish introgression carrying the Ogura fertility restorer gene *Rfo* in rapeseed, using the *Arabidopsis* genome sequence and radish genetic mapping. *Theor Appl Genet* **107**, 1442–1451.

Gibor A and Granick S (1964) Plastids and mitochondria: inheritable systems. *Science* **145**, 890–897.

Goodman MM, Newton KJ and Stuber CW (1981) Malate dehydrogenase: viability of cytosolic nulls and lethality of mitochondrial nulls in maize. *Proc Natl Acad Sci USA* **78**, 1783–1785.

Gouyon PH and Couvet D (1987) A conflict between two sexes, females and hermaphrodites. *Experientia Suppl* **55**, 245–261.

Gouyon PH, Vichot F and Van Damme JMM (1991) Nuclear-cytoplasmic male sterility: single point equilibria versus limit cycles. *Am Nat* **137**, 498–514.

Grelon M, Budar F, Bonhomme S and Pelletier G (1994) Ogura cytoplasmic male-sterility (CMS)-associated *orf138* is translated into a mitochondrial membrane polypeptide in male-sterile *Brassica* cybrids. *Mol Gen Genet* **243**, 540–547.

Gu J, Dempsey S and Newton KJ (1994) Rescue of a maize mitochondrial cytochrome oxidase mutant by tissue culture. *Plant J* **6**, 787–794.

Gutierres S, Sabar M, Lelandais C, *et al.* (1997) Lack of mitochondrial and nuclear-encoded subunits of complex I and alteration of the respiratory chain in *Nicotiana sylvestris* mitochondrial deletion mutants. *Proc Natl Acad Sci USA* **94**, 3436–3441.

Hagemann R (2000) Erwin Baur or Carl Correns: who really created the theory of plastid inheritance? *J Hered* **91**, 435–440.

Hakansson G and Glimelius K (1991) Extensive nuclear influence on mitochondrial transcription and genome structure in male-fertile and male-sterile alloplasmic *Nicotiana* materials. *Mol Gen Genet* **229**, 380–388.

Handa H (2003) The complete nucleotide sequence and RNA editing content of the mitochondrial genome of rapeseed (*Brassica napus* L.): comparative analysis of the mitochondrial genomes of rapeseed and *Arabidopsis*. *Nucleic Acids Res* **31**, 5907–5916.

Hanson MR and Bentolila S (2004) Interactions of mitochondrial and nuclear genes that affect male gametophyte development. *Plant Cell* **16** (Suppl.), S154–S169.

Hanson MR, Wilson RK, Bentolila S, Kohler RH and Chen HC (1999) Mitochondrial gene organization and expression in petunia male fertile and sterile plants. *J Hered* **90**, 362–368.

He S, Abad AR, Gelvin SB and Mackenzie SA (1996) A cytoplasmic male sterility-associated mitochondrial protein causes pollen disruption in transgenic tobacco. *Proc Natl Acad Sci USA* **93**, 11763–11768.

Heazlewood JL, Whelan J and Millar AH (2003) The products of the mitochondrial *orf25* and *orfB* genes are F(O) components in the plant F(1)F(O) ATP synthase. *FEBS Lett* **540**, 201–205.

Hedgcoth C, el-Shehawi AM, Wei P, Clarkson M and Tamalis D (2002) A chimeric open reading frame associated with cytoplasmic male sterility in alloplasmic wheat with *Triticum timopheevi* mitochondria is present in several *Triticum* and *Aegilops* species, barley, and rye. *Curr Genet* **41**, 357–365.

Hernould M, Suharsono S, Litvak S, Araya A and Mouras A (1993) Male-sterility induction in transgenic tobacco plants with an unedited *atp9* mitochondrial gene from wheat. *Proc Natl Acad Sci USA* **90**, 2370–2374.

Hervieu F, Bannerot H and Pelletier G (1994) A unique cytoplasmic male sterility (CMS) determinant is present in three *Phaseolus* species characterized by different mitochondrial genomes. *Theor Appl Genet* **88**, 314–320.

Hervieu F, Charbonnier L, Bannerot H and Pelletier G (1993) The cytoplasmic male-sterility (CMS) determinant of common bean is widespread in *Phaseolus coccineus* L. and *Phaseolus vulgaris* L. *Curr Genet* **24**, 149–155.

Horn R, Hustedt J, Horstmeyer A, Hahnen J, Zetsche K and Friedt W (1996) The CMS-associated 16 kDa protein encoded by *orfH522* in the PET1 cytoplasm is also present in other male-sterile cytoplasms of sunflower. *Plant Mol Biol* **30**, 523–538.

Horn R, Kohler RH and Zetsche K (1991) A mitochondrial 16 kDa protein is associated with cytoplasmic male sterility in sunflower. *Plant Mol Biol* **17**(1), 29–36.

Houstek J, Mracek T, Vojtiskova A and Zeman J (2004) Mitochondrial diseases and ATPase defects of nuclear origin. *Biochim Biophys Acta* **1658**, 115–121.

Huang J, Struck F, Matzinger DF and Levings CS, III (1994) Flower-enhanced expression of a nuclear-encoded mitochondrial respiratory protein is associated with changes in mitochondrion number. *Plant Cell* **6**, 439–448.

Hunt MD and Newton KJ (1991) The NCS3 mutation: genetic evidence for the expression of ribosomal protein genes in *Zea mays* mitochondria. *EMBO J* **10**, 1045–1052.

Janska H and Mackenzie SA (1993) Unusual mitochondrial genome organization in cytoplasmic male sterile common bean and the nature of cytoplasmic reversion to fertility. *Genetics* **135**, 869–879.

Jiao S, Thornsberry JM, Elthon TE and Newton KJ (2005) Biochemical and molecular characterization of photosystem I deficiency in the NCS6 mitochondrial mutant of maize. *Plant Mol Biol* **57**, 303–313.

Jones HA and Clarke AE (1943) Inheritance of male sterility in the onion and the production of hybrid seed. *Proc Am Soc Hort Sci* **43**, 189–194.

Jones HA and Emsweller SL (1937) A male-sterile onion. *Proc Am Soc Hort Sci* **34**, 582–585.

Kamps TL, McCarty DR and Chase CD (1996) Gametophyte genetics in *Zea mays* L.: dominance of a restoration-of-fertility allele (*Rf3*) in diploid pollen. *Genetics* **142**(3), 1001–1007.

Karpova OV, Kuzmin EV, Elthon TE and Newton KJ (2002) Differential expression of alternative oxidase genes in maize mitochondrial mutants. *Plant Cell* **14**, 3271–3284.

Karpova OV and Newton KJ (1999) A partially assembled complex I in NAD4-deficient mitochondria of maize. *Plant J* **17**, 511–521.

Kazama T and Toriyama K (2003) A pentatricopeptide repeat-containing gene that promotes the processing of aberrant atp6 RNA of cytoplasmic male-sterile rice. *FEBS Lett* **544**, 99–102.

Kemble RJ, Flavell RB and Brettell RIS (1982) Mitochondrial DNA analysis of fertile and sterile maize plants derived from tissue culture with the Texas male sterile cytoplasm. *Theor Appl Genet* **62**, 213–217.

Kim DH and Kim BD (2006) The organization of mitochondrial *atp6* gene region in male fertile and CMS lines of pepper (*Capsicum annuum* L.). *Curr Genet* **49**(1), 59–67.

Kitashiba H, Kitazawa E, Kishitani S and Toriyama K (1999) Partial male sterility in transgenic tobacco carrying an antisense gene for alternative oxidase under the control of a tapetum-specific promoter. *Mol Breed* **5**, 209–218.

Klein RR, Klein PE, Mullet JE, Minx P, Rooney WL and Schertz KF (2005) Fertility restorer locus *Rf1* of sorghum (*Sorghum bicolor* L.) encodes a pentatricopeptide repeat protein not present in the colinear region of rice chromosome 12. *Theor Appl Genet* **111**(6), 994–1012.

Knight J (2001) Meet the Herod bug. *Nature* **412**, 12–14.

Kofer W, Glimelius K and Bonnett HT (1992) Fusion of male sterile tobacco causes modifications of mtDNA leading to changes in floral morphology and restoration of fertility in cybrid plants. *Physiol Plant* **85**, 334–338.

Kohler RH, Horn R, Lossl A and Zetsche K (1991) Cytoplasmic male sterility in sunflower is correlated with the co-transcription of a new open reading frame with the *atpA* gene. *Mol Gen Genet* **227**(3), 369–376.

Koizuka N, Fujimoto H, Sakai T and Imamura J (1998) Translational control of *ORF125* expression by a radish fertility-restoration gene in *Brassica napus* kosena CMS cybrids. In: *Plant Mitochondria: From Gene to Function* (eds Moller IM, Gardestrom P, Glimelius K and Glaser E), pp. 83–86. Blackhuys Publishers, Leiden.

Koizuka N, Imai R, Fujimoto H, *et al.* (2003) Genetic characterization of a pentatricopeptide repeat protein gene, *orf687*, that restores fertility in the cytoplasmic male-sterile Kosena radish. *Plant J* **34**, 407–415.

Koizuka N, Imai R, Iwabuchi M, Sakai T and Imamura J (2000) Genetic analysis of fertility restoration and accumulation of ORF125 mitochondrial protein in the kosena radish (*Raphanus sativus* cv. Kosena) and a *Brassica napus* restorer line. *Theor Appl Genet* **100**, 949–955.

Komori T, Ohta S, Murai N, *et al.* (2004) Map-based cloning of a fertility restorer gene, *Rf-1*, in rice (*Oryza sativa* L.). *Plant J* **37**, 315–325.

Kotera E, Tasaka M and Shikanai T (2005) A pentatricopeptide repeat protein is essential for RNA editing in chloroplasts. *Nature* **433**, 326–330.

Krishnasamy S and Makaroff CA (1993) Characterization of the radish mitochondrial orfB locus: possible relationship with male sterility in Ogura radish. *Curr Genet* **24**(1–2), 156–163.

Krishnasamy S and Makaroff CA (1994) Organ-specific reduction in the abundance of a mitochondrial protein accompanies fertility restoration in cytoplasmic male-sterile radish. *Plant Mol Biol* **26**, 935–946.

Ku S, Yoon H, Suh HS and Chung YY (2003) Male-sterility of thermosensitive genic male-sterile rice is associated with premature programmed cell death of the tapetum. *Planta* **217**(4), 559–565.

Kubo T, Nishizawa S, Sugawara A, Itchoda N, Estiati A and Mikami T (2000) The complete nucleotide sequence of the mitochondrial genome of sugar beet (*Beta vulgaris* L.) reveals a novel gene for tRNA(Cys)(GCA). *Nucleic Acids Res* **28**, 2571–2576.

Kuzmin EV, Duvick DN and Newton KJ (2005) A mitochondrial mutator system in maize. *Plant Physiol* **137**, 779–789.

Landgren M, Zetterstrand M, Sundberg E and Glimelius K (1996) Alloplasmic male-sterile *Brassica* lines containing *B. tournefortii* mitochondria express an ORF 3′ of the *atp6* gene and a 32 kDa protein. off. *Plant Mol Biol* **32**, 879–890.

Laver HK, Reynolds SJ, Moneger F and Leaver CJ (1991) Mitochondrial genome organization and expression associated with cytoplasmic male sterility in sunflower (*Helianthus annuus*). *Plant J* **1**(2), 185–193.

Lauer M, Knudsen C, Newton KJ, Gabay-Laughnan S and Laughnan JR (1990) A partially deleted mitochondrial cytochrome oxidase gene in the NCS6 abnormal growth mutant of maize. *New Biol* **2**, 179–186.

Laughnan JR and Gabay-Laughnan S (1983) Cytoplasmic male sterility in maize. *Ann Rev Genet* **17**, 27–48.

Lefebvre A, Scalla R and Pfeiffer P (1990) The double-stranded RNA associated with the '447' cytoplasmic male sterility in *Vicia faba* is packaged together with its replicase in cytoplasmic membranous vesicles. *Plant Mol Biol* **14**, 477–490.

Leino M, Landgren M and Glimelius K (2005) Alloplasmic effects on mitochondrial transcriptional activity and RNA turnover result in accumulated transcripts of *Arabidopsis* orfs in cytoplasmic male-sterile *Brassica napus*. *Plant J* **42**, 469–480.

Leino M, Teixeira R, Landgren M and Glimelius K (2003) *Brassica napus* lines with rearranged *Arabidopsis* mitochondria display CMS and a range of developmental aberrations. *Theor Appl Genet* **106**, 1156–1163.

Leino M, Thyselius S, Landgren M and Glimelius K (2004) *Arabidopsis* chromosome III restores fertility in a cytoplasmic male-sterile *Brassica napus* line with *Arabidopsis* mitochondrial DNA. *Theor Appl Genet* **109**, 272–279.

Levings CS, III (1993) Thoughts on cytoplasmic male sterility in maize. *Plant Cell* **5**, 1285–1290.

Levings CS, III, Kim BD, Pring DR, *et al.* (1980) Cytoplasmic reversion of *CMS-S* in maize: association with a transpositional event. *Science* **209**, 1021–1023.

Lewis D (1941) Male sterility in natural populations of hermaphrodite plants. The equilibrium between females and hermaphrodites to be expected with different types of inheritance. *New Phytol* **40**, 56–63.

L'Homme Y, Stahl RJ, Li XQ, Hameed A and Brown GG (1997) *Brassica* nap cytoplasmic male sterility is associated with expression of a mtDNA region containing a chimeric gene similar to the *pol* CMS-associated *orf224* gene. *Curr Genet* **31**, 325–335.

Li S, Yang G, Li Y, Chen Z and Zhu Y (2005) Distribution of fertility-restorer genes for wild-abortive and Honglian CMS lines of rice in the AA genome species of genus *Oryza*. *Ann Bot (Lond)* **96**, 461–466.

Li XQ, Jean M, Landry BS and Brown GG (1998) Restorer genes for different forms of *Brassica* cytoplasmic male sterility map to a single nuclear locus that modifies transcripts of several mitochondrial genes. *Proc Natl Acad Sci USA* **95**, 10032–10037.

Lilly JW, Bartoszewski G, Malepszy S and Havey MJ (2001) A major deletion in the cucumber mitochondrial genome sorts with the MSC phenotype. *Curr Genet* **40**, 144–151.

Linke B and Borner T (2005) Mitochondrial effects on flower and pollen development. *Mitochondrion* **5**, 389–402.

Linke B, Nothnagel T and Borner T (2003) Flower development in carrot CMS plants: mitochondria affect the expression of MADS box genes homologous to GLOBOSA and DEFICIENS. *Plant J* **34**, 27–37.

Liu F, Cui X, Horner HT, Weiner H and Schnable PS (2001) Mitochondrial aldehyde dehydrogenase activity is required for male fertility in maize. *Plant Cell* **13**, 1063–1078.

Lurin C, Andres C, Aubourg S, *et al.* (2004) Genome-wide analysis of *Arabidopsis* pentatricopeptide repeat (PPR) proteins reveals their essential role in organelle biogenesis. *Plant Cell* **16**, 2089–2103.

Mackenzie S and McIntosh L (1999) Higher plant mitochondria. *Plant Cell* **11**, 571–586.

Mackenzie SA and Chase S (1990) Fertility restoration is associated with a loss of a portion of the mitochondrial genome in cytoplasmic male sterile common bean. *Plant Cell* **2**, 905–912.

Mackenzie SA, Pring DR, Bassett M and Chase CD (1988) Mitochondrial DNA rearrangement associated with fertility restoration and reversion to fertility in cytoplasmic male sterile *Phaseolus vulgaris*. *Proc Natl Acad Sci USA* **85**, 2714–2717.

Makaroff CA and Palmer JD (1988) Mitochondrial DNA rearrangements and transcriptional alterations in the male sterile cytoplasm of Ogura radish. *Mol Cell Biol* **8**, 1474–1480.

Marienfeld J, Unseld M and Brennicke A (1999) The mitochondrial genome of *Arabidopsis* is composed of both native and immigrant information. *Trends Plant Sci* **4**, 495–502.

Marienfeld JR, Unseld M, Brandt P and Brennicke A (1997) Mosaic open reading frames in the *Arabidopsis* mitochondrial genome. *Biol Chem* **378**, 859–862.

Mazat JP, Rossignol R, Malgat M, Rocher C, Faustin B and Letellier T (2001) What do mitochondrial diseases teach us about normal mitochondrial functions . . . that we already knew: threshold expression of mitochondrial defects. *Biochim Biophys Acta* **1504**, 20–30.

Menassa R, L'Homme Y and Brown GG (1999) Post-transcriptional and developmental regulation of a CMS-associated mitochondrial gene region by a nuclear restorer gene. *Plant J* **17**(5), 491–499.

Moller IM (2001) A general mechanism of cytoplasmic male fertility? *Trends Plant Sci* **6**, 560.
Moneger F, Smart CJ and Leaver CJ (1994) Nuclear restoration of cytoplasmic male sterility in sunflower is associated with the tissue-specific regulation of a novel mitochondrial gene. *EMBO J* **13**(1), 8–17.
Murai K, Takumi S, Koga H and Ogihara Y (2002) Pistillody, homeotic transformation of stamens into pistil-like structures, caused by nuclear-cytoplasm interaction in wheat. *Plant J* **29**, 169–181.
Nakamura T, Schuster G, Sugiura M and Sugita M (2004) Chloroplast RNA-binding and pentatricopeptide repeat proteins. *Biochem Soc Trans* **32**, 571–574.
Nakajima Y, Yamamoto T, Muranaka T and Oeda K (2001) A novel orfB-related gene of carrot mitochondrial genomes that is associated with homeotic cytoplasmic male sterility (CMS). *Plant Mol Biol* **46**(1), 99–107.
Neuburger M, Rebeille F, Jourdain A, Nakamura S and Douce R (1996) Mitochondria are a major site for folate and thymidylate synthesis in plants. *J Biol Chem* **271**, 9466–9472.
Newton KJ (1993) Nonchromosomal stripe mutants of maize. In: *Plant Mitochondria* (eds Brennicke A and Kück U). VCH, Weinheim, Germany, 341–345.
Newton KJ and Coe EHJ (1986) Mitochondrial DNA changes in abnormal growth (non chromosomal stripe) mutants of maize. *Proc Natl Acad Sci USA* **83**, 7363–7366.
Newton KJ, Knudsen C, Gabay-Laughnan S and Laughnan JR (1990) An abnormal growth mutant in maize has a defective mitochondrial cytochrome oxidase gene. *Plant Cell* **2**, 107–113.
Newton KJ, Mariano JM, Gibson CM, Kuzmin E and Gabay-Laughnan S (1996) Involvement of S2 episomal sequences in the generation of NCS4 deletion mutation in maize mitochondria. *Dev Genet* **19**, 277–286.
Nikova V and Vladova R (2002) Wild *Nicotiana* species as a source of cytoplasmic male sterility in *Nicotiana tabacum*. *Contrib Tob Res* **20**, 301–311.
Nivison HT and Hanson MR (1989) Identification of a mitochondrial protein associated with cytoplasmic male sterility in petunia. *Plant Cell* **1**, 1121–1130.
Notsu Y, Masood S, Nishikawa T, *et al.* (2002) The complete sequence of the rice (*Oryza sativa* L.) mitochondrial genome: frequent DNA sequence acquisition and loss during the evolution of flowering plants. *Mol Genet Genomics* **268**, 434–445.
Ogihara Y, Yamazaki Y, Murai K, *et al.* (2005) Structural dynamics of cereal mitochondrial genomes as revealed by complete nucleotide sequencing of the wheat mitochondrial genome. *Nucleic Acids Res* **33**, 6235–6250.
Ohtani K, Yamamoto H and Akimitsu K (2002) Sensitivity to *Alternaria alternata* toxin in citrus because of altered mitochondrial RNA processing. *Proc Natl Acad Sci USA* **99**, 2439–2444.
Olson MS and McCauley DE (2002) Mitochondrial DNA diversity, population structure, and gender association in the gynodioecious plant *Silene vulgaris*. *Evol Int J Org Evol* **56**, 253–262.
Papini A, Mosti S and Brighgna L (1999) Programmed-cell-death events during tapetim development of angiosperms. *Protoplasma* **207**, 213–221.
Picciocchi A, Douce R and Alban C (2003) The plant biotin synthase reaction. Identification and characterization of essential mitochondrial accessory protein components. *J Biol Chem* **278**, 24966–24975.
Pineau B, Mathieu C, Gerard-Hirne C, De Paepe R and Chetrit P (2005) Targeting the NAD7 subunit to mitochondria restores a functional complex I and a wild type phenotype in the *Nicotiana sylvestris* CMS II mutant lacking *nad7*. *J Biol Chem* **280**, 25994–26001.
Pla M, Mathieu C, De Paepe R, Chetrit P and Vedel F (1995) Deletion of the last two exons of the mitochondrial *nad7* gene results in lack of the NAD7 polypeptide in a *Nicotiana sylvestris* CMS mutant. *Mol Gen Genet* **248**, 79–88.
Poot P (1997) Reproductive allocation and resource compensation in male-sterile and hermaphroditic plants of *Plantago lanceolata* (Plantaginaceae). *Am J Bot* **84**, 1256–1265.
Pring DR, Levings CS, Hu WW and Timothy DH (1977) Unique DNA associated with mitochondria in the 'S'-type cytoplasm. *Proc Natl Acad Sci USA* **74**, 2904–2908.
Pring DR, Tang HV, Chase CD and Siripant MN (2006) Microspore gene expression associated with cytoplasmic male sterility and fertility restoration in sorghum. *Sex Plant Reprod* **19**, 25–35.
Rhoades MM (1931) Cytoplasmic inheritance of male sterility in *Zea mays*. *Science* **73**, 340–341.

Rhoades MM (1933) The cytoplasmic inheritance of male sterility in *Zea mays*. *J Genet* **27**, 71–93.

Rhoades MM (1950) Gene induced mutation of a heritable cytoplasmic factor producing male sterility in maize. *Proc Natl Acad Sci USA* **36**, 634–635.

Rhoads DM, Brunner-Neuenschwander B, Levings CS III and Siedow JN (1998) Cross-linking and disulfide bond formation of introduced cysteine residues suggest a modified model for the tertiary structure of URF13 in the pore-forming oligomers. *Arch Biochem Biophys* **354**, 158–164.

Rhoads DM, Levings CS, III and Siedow JN (1995) URF13, a ligand-gated, pore-forming receptor for T-toxin in the inner membrane of CMS-T mitochondria. *J Bioenerg Biomembr* **27**, 437–445.

Sabar M, De Paepe R and de Kouchkovsky Y (2000) Complex I impairment, respiratory compensations, and photosynthetic decrease in nuclear and mitochondrial male sterile mutants of *Nicotiana sylvestris*. *Plant Physiol* **124**, 1239–1250.

Sabar M, Gagliardi D, Balk J and Leaver CJ (2003) ORFB is a subunit of F(1)F(O)-ATP synthase: insight into the basis of cytoplasmic male sterility in sunflower. *EMBO Rep* **4**, 1–6.

Sane AP, Nath P and Sane PV (1997) Differences in kinetics of F1-ATPases of cytoplasmic male sterile, maintainer and fertility restored lines of sorghum. *Plant Sci* **130**, 19–25.

Sarria R, Lyznik A, Vallejos CE and Mackenzie SA (1998) A cytoplasmic male sterility-associated mitochondrial peptide in common bean is post-translationally regulated. *Plant Cell* **10**, 1217–1228.

Satoh M, Kubo T, Nishizawa S, Estiati A, Itchoda N and Mikami T (2004) The cytoplasmic male-sterile type and normal type mitochondrial genomes of sugar beet share the same complement of genes of known function but differ in the content of expressed ORFs. *Mol Genet Genomics* **272**, 247–256.

Saur Jacobs M and Wade MJ (2003) A synthetic review of the theory of gynodioecy. *Am Nat* **161**, 837–851.

Schmitz-Linneweber C, Williams-Carrier R and Barkan A (2005) RNA immunoprecipitation and microarray analysis show a chloroplast pentatricopeptide repeat protein to be associated with the 5′ region of mRNAs whose translation it activates. *Plant Cell* **17**, 2791–2804.

Schnable PS and Wise RP (1998) The molecular basis of cytoplasmic male sterility and fertility restoration. *Trends Plant Sci* **3**, 175–180.

Shinada T, Kikuchi Y, Fujimoto R and Kishitani S (2006) An alloplasmic male-sterile line of *Brassica oleracea* harboring the mitochondria from *Diplotaxis muralis* expresses a novel chimeric open reading frame, orf72. *Plant Cell Physiol* **47**(4), 549–553.

Shumway LK and Bauman LF (1967) Nonchromosomal stripe of maize. *Genetics* **55**, 33–38.

Singh M and Brown GG (1991) Suppression of cytoplasmic male sterility by nuclear genes alters expression of a novel mitochondrial gene region. *Plant Cell* **3**, 1349–1362.

Singh M, Hamel N, Menassa R, Li XQ, Young B, Jean M, Landry BS and Brown GG (1996) Nuclear genes associated with a single *Brassica* CMS restorer locus influence transcripts of three different mitochondrial gene regions. *Genetics* **143**(1), 505–516.

Skinner DJ, Baker SC, Meister RJ, Broadhvest J, Schneitz K and Gasser CS (2001) The *Arabidopsis HUELLENLOS* gene, which is essential for normal ovule development, encodes a mitochondrial ribosomal protein. *Plant Cell* **13**, 2719–2730.

Small I and Peeters N (2000) The PPR motif: a TPR-related motif prevalent in plant organellar proteins. *Trends Biochem Sci* **25**, 46–47.

Smart CJ, Moneger F and Leaver CJ (1994) Cell-specific regulation of gene expression in mitochondria during anther development in sunflower. *Plant Cell* **6**(6), 811–825.

Soferman-Avshalom O, Yesodi V, Tabib Y, Gidoni D, Izhar S and Firon N (1993) Detection of an open reading frame related to the CMS-associated *urf-s* in fertile *Petunia* lines and species and in other fertile Solanaceae species. *Theor Appl Genet* **86**, 308–311.

Song J and Hedgcoth C (1994) Influence of nuclear background on transcription of a chimeric gene (*orf256*) and coxI in fertile and cytoplasmic male sterile wheats. *Genome* **37**(2): 203–9.

Spassova M, Moneger F, Leaver CJ, *et al.* (1994) Characterisation and expression of the mitochondrial genome of a new type of cytoplasmic male-sterile sunflower. *Plant Mol Biol* **26**, 1819–1831.

Sugiyama Y, Watase Y, Nagase M, *et al.* (2005) The complete nucleotide sequence and multipartite organization of the tobacco mitochondrial genome: comparative analysis of mitochondrial genomes in higher plants. *Mol Genet Genomics* **272**, 603–615.

Suzuki T, Nakamura C, Mori N and Kaneda C (1995) Overexpression of mitochondrial genes in alloplasmic common wheat with a cytoplasm of wheatgrass (*Agropyron trichophorum*) showing depressed vigor and male sterility. *Plant Mol Biol* **27**, 553–565.

Tang HV, Chang R and Pring DR (1998) Cosegregation of single genes associated with fertility restoration and transcript processing of sorghum mitochondrial orf107 and urf209. *Genetics* **150**(1), 383–391.

Tang HV, Chen W and Pring DR (1999) Mitochondrial *orf107* transcription, editing, and nucleolytic cleavage conferred by the *Rf3* are expressed in sorghum pollen. *Sex Plant Reprod* **12**, 53–59.

Tang HV, Pring DR, Shaw LC, *et al.* (1996) Transcript processing internal to a mitochondrial open reading frame is correlated with fertility restoration in male-sterile sorghum. *Plant J* **10**, 123–133.

Teixeira RT, Farbos I and Glimelius K (2005a) Expression levels of meristem identity and homeotic genes are modified by nuclear-mitochondrial interactions in alloplasmic male-sterile lines of *Brassica napus*. *Plant J* **42**, 731–742.

Teixeira RT, Knorpp C and Glimelius K (2005b) Modified sucrose, starch, and ATP levels in two alloplasmic male-sterile lines of *B. napus*. *J Exp Bot* **56**, 1245–1253.

Thompson JD and Tarayre M (2000) Exploring the genetic basis and proximate causes of female fertility advantage in gynodioecious *Thymus vulgaris*. *Evol Int J Org Evol* **54**, 1510–1520.

Touzet P (2002) Is the *rf2* a restorer gene of CMS-T in maize? *Trends Plant Sci* **7**, 434.

Touzet P and Budar F (2004) Unveiling the molecular arms race between two conflicting genomes in cytoplasmic male sterility? *Trends Plant Sci* **9**, 568–570.

Umbeck PF and Gengenbach BG (1983) Reversion of male-sterile T-cytoplasm maize to male fertility in tissue culture. *Crop Sci* **23**, 584–588.

van Damme JM, Hundscheid MP, Ivanovic S and Koelewijn HP (2004) Multiple CMS-restorer gene polymorphism in gynodioecious *Plantago coronopus*. *Heredity* **93**, 175–181.

Vanlerberghe GC and McIntosh L (1996) Signals regulating the expression of the nuclear gene encoding alternative oxidase of plant mitochondria. *Plant Physiol* **111**, 589–595.

Varnier AL, Mazeyrat-Gourbeyre F, Sangwan RS and Clement C (2005) Programmed cell death progressively models the development of anther sporophytic tissues from the tapetum and is triggered in pollen grains during maturation. *J Struct Biol* **152**, 118–128.

Vedel F, Chétrit P, Mathieu C, Pelletier G and Primard C (1986) Several different mitochondrial DNA regions are involved in intergenomic recombination in *Brassica napus* cybrid plants. *Curr Genet* **11**, 17–24.

Wang HM, Ketela T, Keller WA, Gleddie SC and Brown GG (1995) Genetic correlation of the *orf224/atp6* gene region with Polima CMS in *Brassica* somatic hybrids. *Plant Mol Biol* **27**, 801–807.

Wang M, Hoekstra S, van Bergen S, *et al.* (1999) Apoptosis in developing anthers and the role of ABA in this process during androgenesis in *Hordeum vulgare* L. *Plant Mol Biol* **39**, 489–501.

Wang Z, Zou Y, Li X, *et al.* (2006) Cytoplasmic male sterility of rice with Boro II cytoplasm is caused by a cytotoxic peptide and is restored by two related PPR motif genes via distinct modes of mRNA silencing. *Plant Cell* **18**, 676–687.

Warmke HE and Lee S-HJ (1978) Pollen abortion in T cytoplasmic male-sterile corn (*Zea mays*): a suggested mechanism. *Science* **200**, 561–563.

Wen L, Ruesch KL, Ortega VM, Kamps TL, Gabay-Laughnan S and Chase CD (2003) A nuclear restorer-of-fertility mutation disrupts accumulation of mitochondrial ATP synthase subunit alpha in developing pollen of S male-sterile maize. *Genetics* **165**, 771–779.

Wintz H, Chen HC, Sutton CA, *et al.* (1995) Expression of the CMS-associated urfS sequence in transgenic petunia and tobacco. *Plant Mol Biol* **28**, 83–92.

Wise RP, Bronson CR, Schnable PS and Horner HT (1999a) The genetics, pathology and molecular biology of T-cytoplasm male sterility in maize. *Adv Agronomy* **65**, 79–131.

Wise RP, Gobelman-Werner K, Pei D, Dill CL and Schnable PS (1999b) Mitochondrial transcript processing and restoration of male fertility in T-cytoplasm maize. *J Hered* **90**, 380–385.

Yamagishi H and Terachi T (2001) Intra- and inter-specific variations in the mitochondrial gene *orf138* of Ogura-type male sterile cytoplasm from *Raphanus sativus* and *Raphanus raphanistrum*. *Theor Appli Genet* **103**, 725–732.

Yamamoto MP, Kubo T and Mikami T (2005) The 5′-leader sequence of sugar beet mitochondrial *atp6* encodes a novel polypeptide that is characteristic of Owen cytoplasmic male sterility. *Mol Genet Genomics* **273**, 342–349.

Yamasaki S, Konno N and Kishitani S (2004) Overexpression of mitochondrial genes is caused by interactions between the nucleus of *Brassica rapa* and the cytoplasm of *Diplotaxis muralis* in the leaves of alloplasmic lines of *B. rapa*. *J Plant Res* **117**, 339–344.

Yamato KT and Newton KJ (1999) Heteroplasmy and homoplasmy for maize mitochondrial mutants: a rare homoplasmic *nad4* deletion mutant plant. *J Hered* **90**, 369–373.

Young EG and Hanson MR (1987) A fused mitochondrial gene associated with cytoplasmic male sterility is developmentally regulated. *Cell* **50**(1), 41–49.

Yui R, Iketani S, Mikami T and Kubo T (2003) Antisense inhibition of mitochondrial pyruvate dehydrogenase E1a subunit in anther tapetum causes male sterility. *Plant J* **34**, 57–66.

Zabala G, Gabay-Laughnan S and Laughnan JR (1997) The nuclear gene *Rf3* affects the expression of the mitochondrial chimeric sequence R implicated in S-type male sterility in maize. *Genetics* **147**(2), 847–860.

Zhang Q, Liu Y and Sodmergen (2003) Examination of the cytoplasmic DNA in male reproductive cells to determine the potential for cytoplasmic inheritance in 295 angiosperm species. *Plant Cell Physiol* **44**, 941–951.

Zimmer C (2001) Wolbachia: a tale of sex and survival. *Science* **292**, 1093–1095.

Zrenner R, Stitt M, Sonnewald U and Boldt R (2006) Pyrimidine and purine biosynthesis and degradation in plants. *Annu Rev Plant Biol*, **57**, 805–836.

Zubko MK (2004) Mitochondrial tuning fork in nuclear homeotic functions. *Trends Plant Sci* **9**, 61–64.

Zubko MK, Zubko EI, Adler K, Grimm B and Gleba YY (2003) New CMS-associated phenotypes in cybrids *Nicotiana tabacum* L. (+ *Hyoscyamus niger* L.). *Ann Bot* **92**, 281–288.

10 The mitochondrion and plant programmed cell death

Mark Diamond and Paul F. McCabe

10.1 Introduction

As explained in any standard undergraduate biology textbook, mitochondria are the powerhouse of the cell whose principal function is energy generation. Recently however, a central role for mitochondria as mediators of programmed cell death (PCD) has been established in both plants and animals (Green and Read, 1998; Jones, 2000; Blackstone and Kirkwood, 2003). PCD is an essential process in multicellular organisms controlling the elimination of cells during development, defence and stress responses. Mitochondria have been shown to be involved in PCD in most, if not all, multicellular organisms (Blackstone and Green, 1999), and the role of the mitochondria as the cellular executioner may be a shared derived feature for eukaryotes (Blackstone and Kirkwood, 2003). The conserved role of mitochondria in PCD may stem from the single endosymbiotic event involving an α-proteobacterium endosymbiont and archaebacterial host, creating the first eukaryotic cell. The endosymbiont, or proto-mitochondrion, possessed an electron transport chain (ETC) capable of producing reactive oxygen species (ROS) and it has been hypothesised that production of ROS meant the proto-mitochondria would have the capability to signal to, or kill, the host cell. This facility to produce large amounts of ROS is thought to have evolved into PCD. There are two theories as to why mitochondria became PCD executioners. The first theory proposes that PCD is a vestige of an early host–parasite relationship where a free-living α-proteobacterium invaded a host cell, and replicated. Eventually the host was no longer able to sustain the parasites and the parasites triggered PCD and were thus released to become free-living again (Kroemer, 1997). In this hypothetical scenario, death would have been induced by the generation of ROS by the invader. Large amounts of ROS can be generated by mitochondria following the release of cytochrome *c* from the organelle; loss of cytochrome *c* disrupts the ETC leading to an upregulation of ROS production. An alternative theory proposed by Blackstone and Kirkwood (2003) posits that PCD arose from symbiont-host signalling mechanisms that relied on ROS. Again ROS are generated from mitochondria, and ROS production can be upregulated by cytochrome *c* release. This ROS signalling mechanism could have evolved into PCD. However it evolved, PCD is essential to the development and survival of multicellular organisms, and mitochondria not only are the providers of the energy to sustain life but also provide the means to destroy it.

10.2 Programmed cell death

In 1972, Kerr and co-workers published their seminal work on PCD. They identified a type of cell death that they named apoptosis, which was morphologically distinct from the necrotic death that, at that time, was typically associated with dying cells. This research led to a reevaluation of how we view cell death and to the categorisation of cell death into two forms, one regulated and one unregulated. Under this classification system, necrosis can be considered a violent, or unregulated, form of death that is initiated by environmental stimuli, and results in the rapid disruption of cellular homeostasis (Bras *et al.*, 2005), whereas apoptosis (from the Greek for the 'dropping off' of leaves or petals from a healthy plant; Bras *et al.*, 2005) can be considered as an alternative, regulated form of death (Kerr *et al.*, 1972). This classification system has since been reassessed, and apoptosis is now considered as just one form of PCD that is unique to animals and characterised by a specific set of morphological and biochemical features.

PCD can be defined in general terms as a highly controlled active cellular suicide that is an indispensable part of normal growth and development in all eukaryotes (McCabe and Leaver, 2000). During development, many cells are produced in excess or tissues are produced with transitory functions, and these eventually undergo PCD and thereby contribute to the natural development and sculpting of the organism (Meier *et al.*, 2000; Rathmell and Thompson, 2002; Gunawardena *et al.*, 2004). A good example of the role of PCD in animal development occurs during the formation of independent digits by massive cell death in the interdigital mesenchymal tissue during foetal development (Zuzarte-Luis and Hurle, 2002). Another example is the widespread cell death that occurs during development of the adult brain, where half of the neurons initially created will die (Hutchins and Barger, 1998).

PCD is also activated during normal plant growth and development, and there are many examples of PCD during the plant life cycle. For instance, in reproductive tissue, PCD is vital for the production of viable pollen, for sex selection in unisexual flowers and for ensuring successful pollination and gamete fusion (Wu and Cheung, 2000). In poppy, self-incompatibility is promoted through the inhibition of pollen tube growth, which is followed by the death of the pollen via PCD (Thomas and Franklin-Tong, 2004). Tracheary element (TE) cells die at maturity so they can realise their function as the water-conducting vessels of the xylem. Differentiation of TE cells requires secondary wall synthesis, PCD to kill the cell and cellular autolysis to remove the cell contents (Fukuda, 2000; Yu *et al.*, 2002). In certain plants, PCD is important for the formation of complex leaf shapes. Gunawardena *et al.* (2004) found that the complex lace-like leaf pattern in *Aponogeton madagascariensis* is caused by PCD occurring in discrete linear patches positioned between the transverse and longitudinal veins. The nuclei, cytoplasm and cell walls are then degraded, leaving a gap that widens as the leaf expands.

Senescence, a process that is associated with the remobilisation of nutrients from dying tissue, has long been linked with PCD, and it is known that PCD occurs at the very end of senescence. However, it is unclear as to whether PCD is the final component of the senescent programme or whether it is initiated following the termination of the process (Delorme *et al.*, 2000). PCD is also important for embryogenesis. In angiosperm embryogenesis the zygote divides to form two sister cells. One of these cells becomes the embryo while the other develops to form the suspensor. The suspensor functions as a growth promoter during early embryogenesis; however, it becomes functionally redundant after this and dies. Indeed, McCabe *et al.* (1997b) demonstrated that following the initiation of somatic embryogenesis in carrot suspension cell cultures the suspensor cell died via PCD, and that this PCD was likely important for normal embryogenesis. In addition, PCD was found to be vital during the formation of somatic embryos in Norway spruce. Filonova *et al.* (2000) established that two successive waves of PCD were involved in the elimination of proembryogenic masses and subsequently in the removal of embryo-suspensors, and concluded that these cell death events were needed to ensure the normal progression of somatic embryogenesis.

In addition to its role in development, PCD is also an important part of a plants defensive arsenal against fungal, bacterial and viral pathogens. The hypersensitive response to avirulent pathogens often terminates in the rapid death of infected or challenged cells, which can result in the arrest of pathogen growth (Heath, 2000). A recent study of oat infected with a broad variety of pathogens (viruses, bacteria and fungi) demonstrated that many virulent pathogens induce PCD (Yao *et al.*, 2002). Cell death with apoptotic-like features occurred in infected cells and sometimes neighbouring cells at various time points depending on the type of infectious agent (Yao *et al.*, 2002). It is also thought that PCD has a role in promoting the growth of some pathogens, especially those that secrete toxins in order to kill host cells rapidly (Greenberg and Yao, 2004). AAL, the toxin produced by the tomato pathogen *Alternaria alternata* f. sp. *Lycopersici*, induces PCD, and pathogens lacking the ability to produce this toxin have severely reduced growth on susceptible plants (Akamatsu *et al.*, 1997).

10.3 Death programmes

Cell death programmes can be separated into three distinct phases: (1) an induction phase, involving the perception of the death-inducing stimulus by the cell and initiation of the death programme; (2) an effector phase, where the cell commits irrevocably to death by activating the cell death machinery; and (3) a degradation phase, where the cell is dismantled via the action of the cell death machinery (McCabe and Leaver, 2000). In animals, the mitochondrion is involved in the coordination of death signals in the induction phase and in the initiation of the death programme through the release of pro-apoptotic molecules in the effector phase.

The most extensively characterised form of PCD is mammalian apoptosis. This is usually mediated by the caspase family of cysteine proteases as well as certain proteins from the mitochondria, and leads to a stereotypical set of morphological and biochemical changes including plasma membrane blebbing, cell shrinkage, oligonucleosomal DNA fragmentation, chromatin condensation leading to the appearance of pyknotic nuclei and regulated disintegration of the cell via the formation of apoptotic bodies (for review, see Bras *et al.*, 2005). Many of these morphological changes have also been recorded during PCD in plant systems (McCabe *et al.*, 1997a). Much of the research into the involvement of mitochondria in plant PCD has focused on possible similarities with mammalian apoptosis.

10.4 The mitochondrion and mammalian apoptosis

Much is known in animal systems about the death machinery responsible for a cell's fate (Figure 10.1). Intrinsic or extrinsic signalling pathways can switch on the death

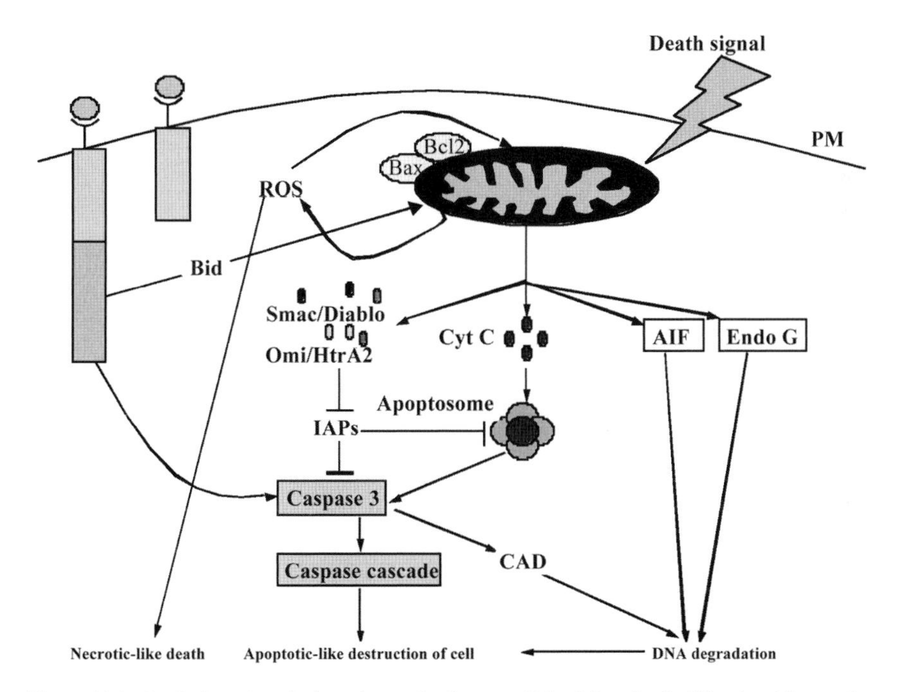

Figure 10.1 Intrinsic and extrinsic pathways leading to cell death in animals. Mitochondria can play a central role in the cell death response. In the intrinsic pathway the mitochondrion perceives the death stimulus and releases a number of factors via the action of Bcl family proteins or via permeabilisation/rupture of the mitochondrial membrane. In particular, cytochrome *c* is released and associates with Apaf-1, procaspase-9 and dATP to form the apoptosome. The apoptosome then sets in motion the activation of downstream caspases, which leads to the apoptotic destruction of the cell. IAP proteins can block the action of the apoptosome, but these can be regulated by other factors (Smac/Diablo and Omi/HtrA2) released from the mitochondria. Other molecules released from the mitochondria are involved in the degradation of DNA. AIF and Endo G may work in cooperation or independently towards this aim, but can bring about a caspase-independent PCD. The release of ROS from the mitochondria can lead to further permeabilisation and release of more cytochrome *c* to amplify the apoptotic signal. However, ROS can also lead to necrotic-like cell death. In the extrinsic pathway, death receptors can either interact with the caspases via caspase-8 and caspase-3 or interact with the mitochondria via Bid, which can lead to the release of pro-apoptogenic molecules. (Figure compiled from data summarised in Leist and Jaattela, 2001, and Bras *et al.*, 2005.)

machinery. The mitochondrion is a central component of the intrinsic pathway. Mitochondria release pro-apoptotic proteins into the cytosol, and it is believed that the mitochondria are central in integrating the plethora of cell death signals (Bras *et al.*, 2005). The caspase pathway leading to classical apoptosis is triggered by the release of cytochrome *c* from the mitochondrial intermembrane space (Leist and Jaattela, 2001). Cytochrome *c* normally takes part in energy production in the mitochondria, but upon its release it converges and complexes in the cytosol with the apoptosis protease-activating factor1 (Apaf-1), procaspase-9 and dATP to form the apoptosome. The apoptosome then initiates a caspase-activation cascade that subsequently brings about the apoptotic destruction of the cell. The activation of caspases appears to be negatively regulated by caspase-inhibitory factors (also known as inhibitor of apoptosis proteins – IAPs) as a safeguard mechanism (Leist and Jaattela, 2001). However, these IAPs can be inhibited by the antagonising function of two other mitochondrial proteins SMAC/DIABLO and Omi/HtrA2, which are released upon apoptotic stimulation. Omi/HtrA2 can also induce caspase-independent cell death via its serine protease activity and via collaboration with death receptors at the cell surface (Bras *et al.*, 2005).

A second mitochondrial signalling pathway involves apoptosis-inducing factor (AIF). This is a PCD mechanism that is conserved in nematode (*Caenorhabditis elegans*; Wang *et al.*, 2002), slime mould (*Dictyostelium discoideum*; Arnoult *et al.*, 2001), and mammals (Susin *et al.*, 1999). AIF is a 57-kDa flavoprotein that resides in the mitochondrial intermembrane space and resembles bacterial oxidoreductase. Upon release from the mitochondria, AIF migrates to the nucleus where it instigates chromatin condensation and cleavage of DNA into large ($\sim$50 kbp) fragments (Susin *et al.*, 1999). AIF is believed to control caspase-independent PCD because inhibition of caspase activation or caspase activity does not abolish the cell death-promoting activity of the protein (Cande *et al.*, 2004).

A third pathway, which is also caspase-independent, is mediated by the mitochondrial endonuclease G (Endo G) (Li *et al.*, 2001). Endo G is a 30-kDa nuclease that has been proposed to participate in mitochondrial DNA synthesis, which takes place in the mitochondrial matrix. However, it is likely that most Endo G colocalises with cytochrome *c* in the intermembrane space, and is released from the mitochondrion after death signal perception (Li *et al.*, 2001). It then brings about internucleosomal DNA fragmentation, and appears to work in conjunction with AIF and the caspase-activated DNase in chromatin condensation and nuclear degradation (Bras *et al.*, 2005). The role of Endo G in apoptosis may also be an evolutionarily conserved pathway as it is apparently conserved in nematodes (*C. elegans*; Parrish *et al.*, 2001) and mammals (Wang, 2001).

Different models for the release of pro-apoptotic proteins from mitochondria have been proposed (for review, see Bras *et al.*, 2005). In one model the permeability transition pore (PTP) plays an important role. In this system the PTP opens causing the mitochondria to swell, and this leads to the rupturing of the outer mitochondrial membrane (OMM). While the full structure of the PTP has not yet been revealed, this pore is a polyprotein complex that includes OMM-localized proteins such as the voltage-dependent anion channel (VDAC), inner membrane proteins such as the adenine nucleotide translocator (ANT), and cyclophilin D from the matrix (Green and Reed, 1998; Saviani *et al.*, 2002). The opening of the PTP is promoted by cell stress (e.g. oxidative stress, a reduction in adenine nucleotides or overloading of matrix Ca^{2+}; Lemasters *et al.*, 1998; Crompton, 1999) and via the action of 'death' signals (e.g. pro-apoptotic Bcl-2 family proteins and caspases; Marzo *et al.*, 1998; Pastorino *et al.*, 1999; Guo *et al.*, 2002; Curtis and Wolpert, 2004). PTP opening likely causes a mitochondrial permeability transition (MPT), which leads to the collapse of mitochondrial transmembrane potential ($\Delta\Psi_m$), the generation of ROS, the cessation of ATP synthesis, and the release of matrix solutes (e.g. Ca^{2+}) and apoptogenic proteins (Curtis and Wolpert, 2004).

The second model for the release of pro-apoptotic proteins from mitochondria is a variation on the first. Bax, a pro-apoptotic Bcl-2 family protein, interacts with VDAC, and ANT to induce the permeabilisation of the mitochondria. Other Bcl-2 family members are also involved in the release of mitochondrial factors. The pro-apoptotic channel forming proteins Bax and Bak are regulated by Bid and are negatively regulated by Bcl-2 and Bcl-X_L (for review, see Bras *et al.*, 2005). Furthermore, the Bcl-2 family members appear to be able to alter not only mitochondrial permeability but also mitochondrial morphology. In addition to regulating the activities of Bax and Bak, Bid can cause a transformation of mitochondrial configuration, characterised by an induction of membrane curvature and mobilisation of $\sim$85% of the cytochrome *c* stores in the cristae (Scorrano *et al.*, 2002).

One other factor that is important when considering mitochondrial involvement in mammalian PCD is the production of ROS. The formation of ROS is an unavoidable consequence of aerobic metabolism (see Chapter 7). Elevated intracellular ROS levels can cause significant damage, resulting in the physiological turnover of organelles through autophagy, or can result in death of the cell via activation of one or more of the PCD pathways (Bras *et al.*, 2005). ROS can also act as a signalling molecule and cause the opening of the PTP. This would cause the release of more cytochrome *c*, and the production of more ROS, as the release of cytochrome *c* causes further disruption to the ETC. This would result in a feedback loop, which amplifies the original PCD-inducing death signal (Jabs, 1999). It has been suggested that AIF could also add a further level of complexity to ROS signalling as it can have a dual function. AIF, while contained in the mitochondrion, can act as an ROS scavenger to promote cellular viability via its oxidoreductase activity. However, the release of AIF into the cytosol can enhance ROS accumulation in the mitochondrion (for review, see Bras *et al.*, 2005).

Extrinsic apoptosis signalling is mediated by so-called death receptors. These are cell surface receptors that disseminate death signals after association with a particular ligand. These death receptors belong to the tumour necrosis factor receptor (TNFR) gene family. This family includes TNFR-1, Fas/CD95 and the TRAIL receptors DR-4 and DR-5 (Ashkenazi, 2002). After binding with the ligand, subsequent signalling is mediated by the cytoplasmic part of the death receptor, which has a conserved area called the death domain. When the death receptors are activated, adapter molecules (e.g. FADD or TRADD) bind to them on the cytosolic side to form a death-inducing signalling complex (DISC). The DISC brings about the autocatalytic activation of procaspase-8, which is locally concentrated around the adaptor/receptor complex. Active caspase-8 then activates downstream effector caspases, which subsequently bring about apoptosis (Scaffidi *et al.*, 1998). In some cases, the signal coming from the death receptors does not generate a strong enough caspase activation cascade to bring about cell death on its own. If this eventuality arises, the signal is amplified via mitochondrial-dependent apoptotic pathways. Caspase-8 cleaves Bid (a Bcl-2 interacting protein), which can initiate the complete release of cytochrome *c* stores from the mitochondria, without gross mitochondrial swelling or permeability transition (Luo *et al.*, 1998).

10.5 Mitochondrial structural and physiological changes during PCD

Changes to the gross morphology, ultrastructure and physiology of mitochondria often occur during PCD and necrosis. Physiological changes include a reduced membrane potential ($\Delta\Psi_m$), increased ROS accumulation and a reduction in ATP production (usually during necrosis). However, there does not appear to be a characteristic type of structural change that occurs during all forms of PCD. Instead, it appears that different degrees of mitochondrial structural change occur in relation to the cell type and the nature and strength of the death signal. In animal systems it is possible that mitochondrial structural changes mediate the decision to enter PCD or necrosis and that PCD and necrosis occur at two ends of a death response continuum (Lemasters, 1999). In other words, numerous death pathways may be initiated simultaneously and the fate of the cell could depend on the nature and strength of the death stimulus, the state of the cell and whether or not inhibitors of the death machinery are present. The rate dependency of PCD pathways may also be a factor. This means that the death pathway that is most strongly activated may govern the type of death the cell experiences, and when the PCD process is blocked, or only weakly stimulated, necrosis may be the final outcome.

In plants, various studies have catalogued changes in mitochondrial structure during both developmental and stress-induced PCD. Yu *et al.* (2002) examined the mitochondrial structural aberrations that occurred prior to TE PCD in *Zinnia elegans*. Mesophyll cells were isolated from *Zinnia* leaves and cultured in a liquid suspension. The isolated cells expand approximately 24 h after isolation and then begin to synthesise a secondary wall. At around 72 h the cells finish synthesising their secondary walls and undergo a type of PCD at terminal differentiation (distinct from apoptotic-like PCD) termed autolysis, which is mediated by the rupture of the vacuole (Ca^{2+}-dependent), and the release of hydrolases, which degrade the protoplast, leaving a hollow corpse. However, as a result of the cell isolation procedure around 20% of the cells died, and these were designated ‘necrotic’ (even though they displayed a retraction of the plasma membrane away from the cell wall, which is characteristic of many forms of plant PCD). Differences between mitochondrial morphology were observed between ‘necrotic’ (24 h), non-necrotic (24 h) and terminally differentiated cells (72 h). The authors found that non-necrotic (24 h) cells contained mitochondria with well-organised ultrastructure and a normal membrane potential. By assaying $\Delta\Psi_m$ with the fluorescent dye JC-1 (5,5′,6,6′-tetrachloro-1,1′,3,3′-tetraethylbenzimidazo-carbocyanine iodide), the authors determined that in these cells mitochondrial membranes retained high voltage. Through examining these cells via electron microscopy they also determined that the OMM remained intact, cristae were regularly organised, and the electron density of the matrix was uniform and similar to the cytoplasm. Conversely, in 'necrotic' cells the mitochondria exhibited a more electron dense matrix and the internal membranes of these mitochondria exhibited a more angular contour (in comparison to non-necrotic cells). Terminally differentiated cells displayed subtle differences in mitochondrial ultrastructure and changes to their physiological state. In these cells, many of the mitochondria had lost their $\Delta\Psi_m$ (as assayed with JC-1), mitochondrial cristae were swollen and disorganised, and the OMM was discontinuous in some instances. Although matrix swelling was not observed, the authors concluded that the PTP and mitochondria were involved in TE PCD, as cyclosporine A (CsA), an inhibitor of the PTP, blocked TE differentiation.

Self-incompatibility (SI) is known to trigger PCD in pollen (Thomas and Franklin-Tong, 2004), and Geitmann *et al.* (2004) observed mitochondrial ultrastructural changes during the SI response in pollen of *Papaver rhoeas*. The authors found that within 1 h of SI induction most pollen grains demonstrated dramatic alterations in mitochondrial structure. These included swelling, the loss of cristae, loss of electron density and blebbing. At this time point generally all mitochondria within a single pollen grain showed similar symptoms. This suggested that once a certain threshold was passed a simultaneous swelling of the complete mitochondrial population in an individual cell was triggered. At 2 h after SI challenge, mitochondria showed two types of symptoms: either extreme swelling and reduction in cristae number (and quite often mitochondria were fused together), or blebbing and local ballooning of the cristae and/or OMM. By 4 h after SI induction most mitochondria had become unrecognisable. Interestingly, in many cases the endoplasmic reticulum (ER) became wrapped around other organelles, including the mitochondria as early as 1 h after SI induction. In animals it is known that mitochondria and ER are often closely associated, providing the conditions for communication between the two organelles (Rutter and Rizzuto, 2000; Walter and Hajnóczky, 2005). Indeed, it is known that mitochondria and the ER maintain a local Ca^{2+} communication conduit, which allows highly efficient and effective transfer of Ca^{2+} between the two organelles. It is also known that ER-targeted stress can activate Ca^{2+} efflux from the ER; this Ca^{2+} is then redistributed to the mitochondria and leads to PCD, which can be suppressed by inhibitors of mitochondrial Ca^{2+} uptake such as ruthenium red (for review, see Walter and Hajnóczky, 2005). It is possible that the close proximity of the ER to the mitochondria in pollen grains undergoing SI-induced PCD (Geitmann *et al.*, 2004) could result in the ER inducing the opening of the PTP in those mitochondria via Ca^{2+} efflux.

Casolo *et al.* (2005) examined changes in mitochondrial structure during H_2O_2-induced cell death in soybean cells. Firstly, the authors examined the effects of H_2O_2

(at 5 and 20 mM concentrations) on cell morphology and viability and determined that 5 mM H_2O_2 induces an apoptotic-like PCD, whereas 20 mM H_2O_2 induces a mostly necrotic death (with some cells showing intermediate hallmark features of PCD and necrosis, termed necrapoptotic). They then measured changes in mitochondrial integrity resulting from H_2O_2 treatment and found that 5 mM H_2O_2 caused a loss of mitochondrial functionality and a slight rupture of the OMM. It was also determined that mitochondrial oxygen consumption was only slightly disrupted in these cells and that ATP levels remained constant. In contrast, cells treated with 20 mM H_2O_2 displayed a pronounced rupture of the OMM and a severe reduction in mitochondrial oxygen consumption as well as a reduction in ATP levels.

Hauser *et al.* (2006) studied changes in mitochondrial physiology and anatomy during ovule abortion in *Arabidopsis*. Ovule abortion is known to display characteristics of PCD, including DNA cleavage, and can be induced by environmental stresses such as salt stress. The authors found other hallmark features of PCD during ovule abortion. At 24 h after subjecting plants to a salt stress in a hydroponic medium, nearly all of the ovules accumulated ROS. By using the fluorescent dye JC-1 the authors observed that membrane depolarisation of gametophyte mitochondria occurred 14 h after salt stress. However, when the authors examined mitochondrial structural aberrations following salt stress they found that there were no reproducible changes in mitochondrial structure (although the time point at which mitochondrial structure was examined was unclear). Interestingly, concentric rings of ER were again found encircling many organelles including mitochondria. These results indicated that mitochondrial membrane depolarisation precedes ROS accumulation during ovule abortion, but that changes to mitochondrial structure are perhaps not important. ROS build-up, cytochrome *c* translocation from mitochondria and depolarisation of mitochondria are events that frequently precede the morphological changes associated with PCD (Yao *et al.*, 2004). Changes to mitochondrial membrane potential have been hypothesised to be a key signalling event that promotes PCD. This hypothesis was made after observing that agents that stabilise mitochondrial membrane potential led to an inhibition of PCD (Yao *et al.*, 2004).

Mitochondria are dynamic pleomorphic organelles that in animals and yeast typically exist as a reticulum that often changes shape and subcellular distribution (see Chapter 1; Logan, 2006). This reticulum, which is maintained by a balance of fission and fusion (for review, see Karbowski and Youle, 2003; Okamoto and Shaw, 2005; Logan, 2006), provides an efficient system for delivering energy to different areas of the cell with its optimal functioning determined by the balance between fusion and fission events (Bras *et al.*, 2005). It is now known that many stimuli that induce apoptosis also induce fragmentation of the mitochondrial reticulum. In addition, work with mitochondrial fission mutants has shown that the inhibition of mitochondrial fragmentation (fission) can block apoptosis and lead to increased survival (Karbowski and Youle, 2003). In plants, mitochondria typically exist as a population of physically discrete organelles (Logan, 2006); however, it would be interesting to investigate whether or not mitochondrial restructuring (including fission/fusion) events occur during plant PCD.

10.6 Roles of mitochondrial apoptogenic proteins in plant PCD

10.6.1 Cytochrome c and plant PCD

Cytochrome *c* is an ancient and highly conserved protein found loosely associated with the inner mitochondrial membrane (IMM). It functions within mitochondria as an electron carrier between complexes III and IV and is thus key to aerobic respiration in plants, animals and fungi (Goodsell, 2004). In addition, cytochrome *c* is a key component of mammalian apoptosis. As discussed previously, cytochrome *c* is released from the mitochondrion after the perception of a death signal, and complexes with Apaf-1, procaspase-9 and dATP to form the apoptosome. The apoptosome then initiates a proteolytic cascade of caspase activation, which in turn leads to the apoptotic destruction of the cell through the cleavage of key cellular components. The central role played by cytochrome *c* in mammalian apoptosis together with its high degree of sequence conservation led researchers to try to discover what role, if any, cytochrome *c* has in plant PCD.

In plants, translocation of cytochrome *c* from mitochondria has been shown to occur as a result of several different PCD-inducing stimuli (Table 10.1). For example, cytochrome *c* was shown to migrate from mitochondria to the cytosol upon induction of PCD in cell suspension cultures of maize by treatment with D-mannose (Stein and Hansen, 1999). Balk *et al.* (1999) initiated PCD in cucumber cotyledons using a short 55°C heat treatment and demonstrated that cytochrome *c* began migrating from the mitochondria almost immediately and was undetectable in mitochondria after 3 h. It was further demonstrated that this translocation was selective and was not a result of mitochondrial membrane rupture, since another mitochondrial protein, fumarase, did not relocate to the cytosol (Balk *et al.*, 1999). In addition, treatment of *Arabidopsis* suspension cells with the proteinaceous bacterial elicitor harpin, which

Table 10.1 Studies showing cytochrome *c* release from mitochondria following plant PCD induction

PCD induced by	Species	Reference
D-Mannose	Maize	Stein and Hansen (1999)
Heat	Cucumber	Balk *et al.* (1999)
Menadione	Tobacco	Sun *et al.* (1999)
Anther development	Sunflower	Balk and Leaver (2001)
Betulinic acid	*Zinnia*	Yu *et al.* (2002)
H_2O_2 and salicylic acid in alternative oxidase-lacking cells	Tobacco	Robson and Vanlerberghe (2002)
Oxidative stress	*Arabidopsis*	Tiwari *et al.* (2002)
Victorin	Oat	Curtis and Wolpert (2002)
Heat treatment	*Arabidopsis*	Balk *et al.* (2003)
Cytokinins	Carrot	Carimi *et al.* (2003)
Ozone	Tobacco	Pasqualini *et al.* (2003)
Harpin	Petunia	Krause and Durner (2004)
Perturbation of a telomeric repeat binding factor	Tobacco	Yang *et al.* (2004)
Pathogenic agent	Soybean	Zuppini *et al.* (2005)

is a known inducer of PCD, triggered the release of cytochrome *c* from mitochondria (Krause and Durner, 2004).

Sun *et al.* (1999) demonstrated that cytochrome *c* was translocated from the mitochondria to the cytosol between 2 and 4 h after treatment with the ROS promoter menadione. The authors then showed, using a cell free system incorporating mitochondria and cytosol from tobacco and nuclei from mouse liver, that menadione could induce apoptosis only in the mouse liver nuclei in the presence of mitochondria. Furthermore, the effects of menadione were replicated by adding animal cytochrome *c* to the cell free system (Sun *et al.*, 1999). Indeed, Zhao *et al.* (1999) showed that animal cytochrome *c* could activate caspase-like molecules in carrot cytoplasm, which could then degrade rat liver nuclei in an apoptotic fashion. These results suggested that cytochrome *c* might be an initiator of plant PCD, similar to its role in mammalian apoptosis.

However, Xu and Hanson (2000) demonstrated that cytochrome *c* was not released during the PCD that occurs following pollination-induced petal senescence (PIPS) in petunia. The authors did not detect the translocation of cytochrome *c* during the first 30 h following PIPS initiation, and all petals had either wilted or collapsed at the 30-h time point. Therefore, the authors suggested that cytochrome *c* was not involved in the early signalling cascade following PIPS, which is in contrast to the critical role of cytochrome *c* release in apoptosis. However, PCD occurs only at the very end stage of senescence (Delorme *et al.*, 2000), and it should be pointed out that DNA laddering was observed only at 36 h following PIPS and that there was no data presented as to the location of cytochrome *c* between the 30- and 36-h period, or indeed as to the viability of cells during this time. Cytochrome *c* release usually precedes DNA laddering, so its cellular location during this period is of critical relevance to any assumption made about the role of cytochrome *c* in PIPS-induced PCD, and plant PCD in general.

Yu *et al.* (2002) investigated the role of mitochondria during betulinic acid-induced TE cell death in *Zinnia*. The authors showed that inhibiting the mitochondrial PTP with CsA could block TE death; however, this inhibition did not result in the blocking of cytochrome *c* release. This result suggested that cytochrome *c* relocation was insufficient to trigger PCD in those cells. Further evidence for a different role for cytochrome *c* in plant PCD was obtained by Balk *et al.* (2003). The authors investigated the role of cytochrome *c* in PCD using an *Arabidopsis* cell free system containing purified nuclei incubated with mitochondria, and/or cytosolic extract. Through the addition of broken mitochondria (but not intact mitochondria) to the cell free system they isolated two nuclear degradation activities. One required the presence of cytosol and resulted in DNA laddering after 12 h. The other did not require the presence of a cytosolic extract and resulted in the cleavage of DNA into large ($\sim$30 kbp) fragments within 3 h and chromatin condensation within 6 h. Both activities were insensitive to broad-range caspase inhibitors. However, the addition to the cell free system of purified *Arabidopsis* cytochrome *c*, instead of broken mitochondria, caused no nuclear degradation activity. These results suggest that the release of cytochrome *c* does not result in the activation of cytosolic executioner proteases in plant cells. However, as discussed previously, it may be that cytochrome *c* release could serve to amplify cell death signals by causing an increase in ROS generation (Jabs, 1999).

10.6.2 AIF and Endo G

Once it was determined that cytochrome *c* did not initiate PCD in an *Arabidopsis* cell free system, Balk *et al.* (2003) examined the possibility that functional homologues of the AIF or Endo G could have a role in plant PCD. AIF is a flavoprotein with NADH oxidase activity that is normally resident in the intermembrane space (IMS) of the mitochondria. However, it has been shown to translocate from the mitochondria to the cytosol upon induction of apoptosis (Susin *et al.*, 1999). AIF can bind to DNA via electrostatic interactions and recombinant mouse AIF can cause caspase-independent chromatin condensation and the cleavage of DNA into high-molecular weight fragments ($\sim$50 kbp). Yet AIF has little effect on heat-inactivated nuclei or on naked DNA, indicating that it must interact with at least one other protein to bring about DNA cleavage (Susin *et al.*, 1999; Cande *et al.*, 2004).

There are five close homologues of AIF in the *Arabidopsis* genome, annotated as monodehydroascorbate reductases (MDARs). Balk *et al.* (2003) showed by mitochondrial proteome analysis that one of these homologues was localised in the mitochondrial matrix. The authors cloned and overexpressed this protein in *Escherichia coli* (minus the mitochondrial targeting sequence) and added the purified, recombinant protein to *Arabidopsis* nuclei. The MDAR induced neither high-molecular weight ($\sim$30 kbp) DNA fragmentation nor chromatin condensation. Although this MDAR does not appear to have a role in plant PCD, at least one other MDAR is known to be associated with the mitochondrial IMS. It remains to be seen whether or not this or any other AIF homologues play any part in executing plant PCD.

Balk *et al.* (2003) analysed submitochondrial fractions for the presence of a DNase activity capable of hydrolysing artificial DNA substrate. The results suggested that the mitochondrial IMS contained a strong Mg^{2+}-dependent nuclease activity, which had a direct involvement in high-molecular weight ($\sim$30 kbp) DNA cleavage and chromatin condensation. Both the IMS localisation and the Mg^{2+} dependency are shared with Endo G, which contributes to apoptotic DNA cleavage in mammals and nematodes (Li *et al.*, 2001; Balk *et al.*, 2003). The authors suggested that during the evolution of eukaryotic cell types an endonuclease derived from the α-proteobacterial ancestor of mitochondria was used and elaborated upon in the regulation of PCD in plants.

10.7 Release mechanisms for mitochondrial factors

From the results obtained by Balk *et al.* (2003) it would seem that the release of mitochondrial factors plays an important role in the triggering of plant PCD. As discussed previously there are two main ways in which apoptogenic proteins are released from mitochondria in animal cells. One way is via the action of the Bcl-2 family of proteins. The second way is via PTP opening and membrane rupture. The possible involvement of either the Bcl-2 family of proteins or the PTP in plant PCD is discussed further here.

10.7.1 Bcl-like proteins and plant PCD

In mammals, the Bcl-2 family consists of over 25 proteins that show similar folding structure and contain one or more Bcl-2 homology domains. In living cells, Bcl family members localise on the membranes of organelles and in the cytosol. Functionally, the Bcl-2 family is divided into anti-apoptotic proteins including Bcl-2 and Bcl-xL and pro-apoptotic proteins including Bid, Bad, Bak and Bax. Under normal conditions the OMM does not allow the release of IMS proteins and also limits the translocation of small ions such as Ca^{2+}. However, if the amount of anti-apoptotic Bcl-2 decreases or if the pro-apoptotic Bcl-2s redistribute from other organelles or the cytosol to the mitochondria, the OMM becomes permeable to larger proteins. The mechanism of permeabilisation seems to revolve around the accumulation of Bax oligomers in the OMM. A delicate balance between anti- and pro-apoptotic Bcl-2s is therefore required to maintain the integrity of the OMM (for review, see Walter and Hajnóczky, 2005). Although no homologues of Bax or Bcl-2 have been identified in plant genomes to date, recent evidence suggests that both Bcl-2 and Bax can fulfil similar roles when introduced into plant cells (Jones, 2000).

Lacomme and Cruz (1999) expressed mammalian Bax in tobacco, utilising a tobacco mosaic virus vector. The authors found that Bax expression triggered a PCD with hallmark features similar to the hypersensitive response. The PCD-promoting function of Bax correlated with the induction of the defence protein PR1, and PCD was inhibited by an inhibitor of protein phosphatase activity. These results suggested that Bax activated an endogenous plant cell death programme. Kawai-Yamada *et al.* (2001) investigated the role of BCL-like proteins (BLPs) in the release of apoptogenic factors from the mitochondria of *Arabidopsis*. They overexpressed mammalian Bax in transgenic *Arabidopsis* plants and found that this caused significant levels of PCD in the plants. They also found when they retransformed the transgenics with a vector containing AtBI-1 (*Arabidopsis* Bax inhibitor-1 homologue) that Bax-induced cell death was blocked (Kawai-Yamada *et al.*, 2001).

Kawai-Yamada *et al.* (2004) have demonstrated that AtBI-1 can suppress cell death induced by H_2O_2 in *Arabidopsis*. The authors also showed that AtBI-1 did not prevent the Bax-induced accumulation of ROS and that it was not responsible for scavenging oxidants. This means that AtBI-1 suppresses Bax-induced cell death downstream of ROS generation. More recently, Watanabe and Lam (2006) have shown that AtBI-1 is involved in the attenuation of PCD induced by either the mycotoxin fumonisin B_1 or heat treatment. The authors suggested that this may constitute the first example of a core regulator of PCD. However, because *Arabidopsis* mutants with a functionally silenced version of this gene did not have an abnormal phenotype, AtBI-1 by itself is unlikely to be an absolute requirement for developmental cell death.

AtBI-1 is localised in the ER (Kawai-Yamada *et al.*, 2004). It is related to animal BI-1 proteins and contains a highly conserved C-terminal region and a not so well conserved N-terminal region. BI-1 contains six potential transmembrane helices, so it has been proposed that it (like Bax and Bcl-2) may form ion-conducting channels. It is known that BI-1 can associate with Bcl-2 *in vivo*, but not with Bax. It remains to be seen whether or not AtBI-1 interacts with mitochondria or with a Bax- or Bcl-2-like protein directly. As AtBI-1 is localised in the ER (Kawai-Yamada *et al.*, 2004) and one of the functions of Bcl-2 in mammals is to regulate the ER Ca^{2+} store (Walter and Hajnóczky, 2005), it is possible that AtBI-1 may inhibit cell death by controlling the efflux of Ca^{2+} from the ER, which has been linked to the permeabilisation of the mitochondria through the activation of the PTP (Walter and Hajnóczky, 2005).

It is possible that although BLPs can operate in plants, they do not do so normally, as BLPs have not been found in plant protein database searches. It may be that only the PTP complex is conserved between animals and plants (Jones, 2000).

10.7.2 The PTP and plant PCD

The mitochondrial PTP is a multi-protein complex comprising the OMM, VDAC, IMM-localised ANT, and the matrix-localised protein cyclophilin D, with a proposed involvement for OMM porin and the benzodiazepine receptor (for review, see Forte and Bernardi, 2005). When the pore opens, solutes of up to 1.5 kDa in size can enter the mitochondria and pass through the inner membrane in a process known as the mitochondrial permeability transition (MPT; see Section 10.4). When MPT occurs, $\Delta\Psi_m$ drops, oxidative phosphorylation is dissociated from electron transport, ions and solutes are released and a high amplitude swelling can occur, which causes a rupturing of the OMM, and leads to the release of IMS-localised components (Arpagaus *et al.*, 2002).

Pore opening can be mediated by a number of factors including the build-up of Ca^{2+} in the mitochondrial matrix, phosphate (P_i) levels, $\Delta\Psi_m$, ATP levels, fatty acids, ROS and pH (Arpagaus *et al.*, 2002). PTP opening can be inhibited by CsA and this has led to key advances in the understanding of the physiological role of MPT in animals. Purification of mitochondrial CsA-binding proteins from total mitochondrial extracts using CsA affinity chromatography identified cyclophilin D as the target for CsA. It has therefore been suggested that CsA inhibits PTP opening by modulation of cyclophilin D binding to core components of the PTP complex (for review, see Forte and Bernardi, 2005). It is now known that CsA acts, in disrupting the PTP, by displacing the binding of cyclophilin D to ANT (Crompton, 1999). This highly specific effect has led to the widespread acceptance of the pore theory, and CsA inhibition of PTP opening is used today as the primary diagnostic trait of the MPT (Arpagaus *et al.*, 2002). Much of the research undertaken thus far into the role of the PTP in plant PCD has therefore also utilised CsA as a diagnostic tool.

Arpagaus *et al.* (2002) were one of the first groups to present evidence for the existence of the MPT in plants. They investigated the effects of Ca^{2+} and other factors on potato tuber mitochondrial morphology and the role that the PTP played in these effects. The authors found that the addition of $CaCl_2$ (5 mM) initially caused a fast shrinkage, followed by a lag phase and then a pronounced swelling of mitochondria, but only in the presence of P_i. When $CaCl_2$ was replaced by an equal concentration of $MgCl_2$ shrinkage still occurred but no swelling was evident. This was not surprising because in animal models divalent cations other than Ca^{2+} counteract pore opening. Indeed, if $MgCl_2$ (20 mM) was added at the same time as $CaCl_2$ (5 mM) the time required for the completion of swelling was significantly increased. Ca^{2+}-induced swelling was counteracted effectively when CsA was added (in the presence of dithioerythritol). Because mitochondrial swelling usually occurs as a result of matrix expansion and usually culminates in OMM rupture and the release of IMS proteins, the amount of protein released after Ca^{2+}-induced swelling was measured. Significant amounts of protein were released (including proteins of a similar size to cytochrome *c*), but only in samples in which swelling was promoted. The highly inhibitive effect of CsA on Ca^{2+}-induced swelling was observed not only under standard conditions but also under anoxic conditions, and after pretreatment with the $\Delta\Psi_m$-collapsing linolenic acid. This suggested that CsA acted on the same pore element in each case. Together these results suggested that MPT can occur in plants and that the PTP is probably involved. However, it has not yet been established if either the PTP or indeed an MPT plays any role in plant PCD.

Maxwell *et al.* (2002) presented evidence for the involvement of the mitochondria and the PTP in the transduction of signals required for the induction of genes associated with senescence and pathogen attack. The authors treated tobacco cell cultures with the mitochondrial electron transport inhibitor antimycin A, H_2O_2 or salicylic acid, and identified seven rapidly induced cDNAs (termed NtAIs, for *Nicotiana tabacum* antimycin induced) using a differential display technique. Six of these cDNAs showed similarity to genes upregulated by processes (such as senescence and pathogen attack) that are associated with PCD. It was then demonstrated that another inhibitor of animal PTP formation, bongkrekic acid, could block the induction (by all of the above stimuli) of these NtAIs even in the presence of H_2O_2 (Maxwell *et al.*, 2002). While this is not clear-cut evidence of PTP involvement in PCD, it does show that the PTP was probably involved in a signalling event leading to the transcription of genes with a possible role in plant PCD.

One of the first pieces of evidence suggesting a role for an MPT and the PTP in plant PCD was provided by Yu *et al.* (2002). The authors investigated the role of the mitochondrion in PCD induced by betulinic acid in TE cells of *Zinnia elegans* and, as discussed previously, showed that blocking PTP opening with CsA inhibited PCD even though cytochrome *c* release was not inhibited. At about the same time, Saviani *et al.* (2002) described the participation of the PTP in nitric oxide (NO)-induced cell death. NO is a potent radical oxidant that, in a variety of animal cells, causes apoptosis by triggering PTP opening. The authors treated cell cultures of *Citrus sinensis* with the NO donor sodium nitroprusside (SNP). SNP (10 μM) induced PCD, which was identified by chromatin condensation and nuclear DNA fragmentation. However, when CsA was added to cell cultures 1 h prior to treatment with SNP, chromatin condensation and DNA fragmentation was inhibited. Tiwari *et al.* (2002) investigated the role of the mitochondrion in ROS-induced PCD in *Arabidopsis* cells.

Arabidopsis cells were treated with CsA for 15 min prior to the introduction of a cell death inducing level of Glc/Glc oxidase (which causes an increase in H_2O_2, to levels analogous to those reported during the hypersensitive response-induced oxidative burst). CsA pretreatment effectively halved cell death (from 64% down to 33%) as assayed by Evans blue staining. CsA reduced the amount of H_2O_2 produced by around 50%, and analysis of the ATP contents indicated that exposure to CsA also improved mitochondrial function and restored mitochondrial ATP production to near control levels. In summary, CsA-induced PTP inhibition reduced cell death and stabilised mitochondrial structure after ROS-inducing treatment, suggesting that the MPT is indeed involved in ROS-induced plant PCD.

Yao *et al.* (2004) used protoporphyrin IX (PPIX), a molecule with similar properties to the proposed substrate (red chlorophyll catabolite reductase) of ACD2 (ACCELERATED CELL DEATH2), to induce PCD in *Arabidopsis* leaves and protoplasts. Overexpression of ACD2 had previously been shown to reduce cell death during pathogen infection (Mach *et al.*, 2001), suggesting that the ACD2 substrate accumulated during pathogen attack. Yao *et al.* (2004) showed that PPIX induced macroscopic cell death, in a light-dependent fashion, when infiltrated into wild-type or *acd2* leaves. The PPIX-treated cells at the leaf tips showed chromatin condensation and DNA cleavage. The authors then showed that PPIX induced a loss of $\Delta\Psi_m$ (which preceded any noticeable change in chloroplast/mitochondria ultrastructure) in both wild-type and *acd2* protoplasts, which could be inhibited by the addition of CsA, and that CsA could also partially protect against PPIX-induced cell death.

Three important components of the PTP in animals (VDAC, ANT and cyclophilin D) are present in plants (Yao *et al.*, 2004). Indeed, Lacomme and Roby (1999) reported that genes encoding proteins with similarities to VDACs were upregulated during the hypersensitive response induced by *Xanthomonas campestris* in *Arabidopsis*. Swidzinski *et al.* (2002) reported no such change in the expression of VDAC in response to either heat treatment or senescence-induced PCD; however, a gene encoding ANT was decreased in expression. They postulated that this would lead to an inhibition of ATP synthesis and a reduction in the export of mitochondrial ATP, which may play a role in the initiation of PCD as a disruption of ATP/ADP exchange between the mitochondria and cytosol occurs in the early stages of apoptosis (Vander Heiden *et al.*, 1999; Swidzinski *et al.*, 2002).

The case for the involvement of the PTP in plant PCD seems quite strong. The next step in the investigation of the role of the PTP in plant PCD should involve knocking out individual components (such as VDAC, ANT and cyclophilin D) and assaying the effects on PCD induced by different stimuli. Recently, such investigations into the roles of cyclophilin D and ANT in the PTP have been undertaken in animal systems (for review, see Forte and Bernardi, 2005). Kokoszka *et al.* (2004) used genetic strategies to create mice in which the two isoforms of ANT were selectively eliminated in the liver. The loss of ANT resulted in the selective loss of regulation of the PTP by ANT ligands such as ADP (which can cause pore closure) and atractyloside (which can cause pore opening). Additionally, the PTP in ANT-deficient mitochondria required threefold higher Ca^{2+} concentrations for opening, relative to control mitochondria. These results suggested that ANT has an essential role in animal mitochondria in regulating PTP opening by controlling the sensitivity of the PTP to ANT ligands and Ca^{2+}. Basso *et al.* (2005) then demonstrated that mice mitochondria lacking cyclophilin D were also desensitised to Ca^{2+}. The PTP in these mitochondria required twice the concentration of Ca^{2+} for opening to occur. Conversely, the PTP response to depolarisation, pH, adenine nucleotides and oxidative stress was indistinguishable in cyclophilin D-lacking mitochondria relative to control mitochondria. This demonstrates that cyclophilin D participates in the modulation of PTP opening in response to Ca^{2+}, but not in the response to other important regulators of PTP opening.

10.8 Reactive oxygen species

ROS, as the name implies, are extremely reactive molecules (see Chapter 7). Apart from their stereotypical role as toxic by-products of cellular metabolism that can cause extensive oxidative damage, there is an increasing body of evidence suggesting that they may regulate signal transduction in both animals and plants and that they can promote acclimatisation to environmental conditions (Jabs, 1999; Dat *et al.*, 2000; Hauser *et al.*, 2006). In plants, numerous environmental and developmental conditions lead to ROS formation and subsequent PCD. These include senescence (Pastori and del Rio, 1997), heat (Vacca *et al.*, 2004), cold (Koukalová *et al.*, 1997), UV irradiation (Green and Fluhr, 1995), salt stress-induced ovule abortion (Hauser *et al.*, 2006) and avirulent pathogen-induced hypersensitive response (Levine *et al.*, 1994).

The most direct evidence for the involvement of ROS in plant PCD has come from studies demonstrating that PCD can be induced by the direct application of ROS themselves, or that inhibition of PCD can occur by the direct addition of ROS scavengers. PCD can be induced by low doses of H_2O_2 in soybean (Solomon *et al.*, 1999) or superoxide in *Arabidopsis* (Jabs *et al.*, 1996), and PCD has been attenuated by scavengers of ROS such as superoxide dismutase, ascorbate or catalase (Levine *et al.*, 1994; Vacca *et al.*, 2004). In addition, plants in which the capacity for the endogenous production of cytosolic ascorbate peroxidase or catalase has been compromised have an increased susceptibility to biotic and abiotic stresses. This susceptibility is characterised by the formation of necrotic lesions (for review, see Jabs, 1999).

It is evident from the above that ROS can play an important role in plant PCD but how and why do ROS accumulate? And in what way do they participate in the initiation and execution of plant PCD?

The ultimate electron acceptor in the mitochondrial ETC is oxygen, which is reduced to water by cytochrome *c* oxidase. However, oxygen also accepts electrons at other points in the ETC, notably at complexes I (NADH dehydrogenase) and III (cytochrome *c* reductase), and when this occurs oxygen radicals are formed. These are normally converted by the cell into H_2O_2, or other ROS, including superoxide and hydroxyl radicals (Bras *et al.*, 2005). Under normal circumstances the cell can eliminate these damaging radicals through the production of catalase and other ROS scavengers (see Chapter 7). However, under certain circumstances, such as when the cell is under stress, the levels of ROS produced can exceed the capacity of ROS scavengers, and this can result in damage to cellular proteins, DNA and organelles, which can lead to the destruction of the cell via PCD or necrosis.

The mitochondrial ETC is finely tuned, and any disruption to mitochondrial structure or physiology can lead to a severe disruption in its activity. PCD-inducing agents can cause such a disruption. One way in which this can be achieved is PTP opening, which allows the diffusion of low-molecular-weight compounds from the cytosol into the mitochondrion. This can cause swelling of the matrix and rupture of the OMM, leading to severe disruption of the ETC and a high accumulation of ROS. ROS release can then lead to the opening of PTP in other mitochondria and the build-up of more ROS in a positive feedback loop. Thus, ROS can facilitate the initiation of an apoptotic response in other mitochondria. We have already seen that PCD-inducing agents also induce the release of cytochrome *c* (Balk *et al.*, 2003). The normal role of cytochrome *c* is as a component of the ETC, so when it is released from the chain ROS is produced. Neither superoxide nor H_2O_2 is particularly dangerous to the cell; however, in the presence of metal-containing molecules (such as cytochrome *c*, an iron-containing haem protein) they can interact to form hydroxyl radicals, which are amongst the most reactive and mutagenic molecules known (Blackstone and Kirkwood, 2003; see Chapter 7). Once released, therefore, cytochrome *c* not only disrupts electron transport resulting in the generation of ROS, but can also increase the damage caused by these ROS by converting them to hydroxyl radicals. Thus, minor disruptions to mitochondrial respiration can become amplified, resulting in an expedited induction of cell death (Bras *et al.*, 2005).

Besides facilitating the opening of the PTP, ROS, in animal systems, are also known to participate in signalling events during the effector phase of PCD (e.g. tumour necrosis factor-induced cytotoxicity) downstream of PTP opening, leading to structural changes characteristic of the degradation phase (Jabs, 1999). In plants, it has been demonstrated that ROS can activate signalling pathways that are important for environmental stress tolerance and resistance to parasitic infection.

Kovtun *et al.* (2000) demonstrated that H_2O_2 could activate a specific *Arabidopsis* mitogen-activated protein kinase (MAPK) kinase kinase, ANP1 (*Arabidopsis* NPK1-like protein kinase1), which initiates a phosphorylation cascade involving two stress MAPKs, AtMPK3 and AtMPK6. The authors also demonstrated that transgenic tobacco plants constitutively expressing NPK1 (*Nicotiana* protein kinase 1, a tobacco ANP1 orthologue) displayed an enhanced tolerance to a variety of environmental stresses including freezing, heat shock and salt stress.

Rentel *et al.* (2004) demonstrated that an *Arabidopsis* gene (*OXI1*, oxidative signal-inducible1), which encodes a serine/threonine kinase, is induced in response to a variety of H_2O_2-generating stimuli. The authors found that *OXI1* was expressed in response to wounding, H_2O_2 and cellulase treatment, as well as during normal root growth, and in response to *Peronospora parasitica* infection. The requirement of

OXI1 for resistance to the fungus was confirmed when mutants lacking a functional OXI1 protein (*oxi1*) were shown to have a significantly enhanced susceptibility to virulent *P. parasitica*. The *oxi1* mutant plants also showed severely impaired root growth. The OXI1 serine/threonine kinase was also shown to be required for the full activation of the previously identified MAPKs: AtMPK3 and AtMPK6.

The data of Kovtun *et al.* (2000) and Rentel *et al.* (2004) indicate that ROS can function as signalling molecules in plant responses to a variety of stimuli and during development, including freezing, heat treatment, salt stress, wounding, pathogen attack and root formation. There is therefore a good possibility that ROS, which are strongly induced as a result of PCD-inducing stimuli, could function as signalling molecules during the effector phase of plant PCD. However, a role as effectors of plant PCD has not yet been established, and there is some evidence to suggest that they do not function as 'core' effectors in some forms of PCD. Townley *et al.* (2005) demonstrated that C2-ceramide causes a PCD in *Arabidopsis* cell cultures that occurs after the generation of a Ca^{2+} transient and an accumulation of ROS. The authors showed that inhibiting the Ca^{2+} transient prevented PCD. However, they also showed that inhibiting ROS accumulation (with catalase) had no effect on cell death levels. This indicates that ROS do not function as vital components in the execution of ceramide-induced cell death.

It may be, therefore, that ROS can function as effector molecules in PCD but that they do not always do so. Whatever the case may be, it is unlikely that ROS can induce PCD directly, as evidence obtained by Solomon *et al.* (1999) and Vacca *et al.* (2004) suggests that *de novo* transcription and translation are required for PCD in which ROS are induced. Solomon *et al.* (1999) demonstrated that inhibitors of protein synthesis could block cell death induced by H_2O_2 in soybean cells. Vacca *et al.* (2004) showed that tobacco cells given a heat shock produce ROS prior to the onset of PCD; however, this PCD could be partially blocked by inhibitors of transcription and translation.

10.9 Alternative oxidase

Allied to the ubiquitous cytochrome pathway found in all eukaryotic mitochondria, plants have an additional pathway of electron flow, the alternative pathway (Vanlerberghe and MacIntosh, 1997). Alternative oxidase (AOX) is an IMM enzyme that is present in plant, but not animal, mitochondria. AOX catalyses the flow of electrons from ubiquinol directly to oxygen, thus generating an electron shunt that circumvents complexes III and IV of the ETC and results in a cyanide-insensitive electron-transfer pathway (Lam *et al.*, 2001). Alternative pathway activity and AOX abundance are low in unstressed plants, but both increase when plants are subjected to pathogen attack (Simons *et al.*, 1999) or environmental stresses such as freezing (Vanlerberghe and McIntosh, 1997).

Numerous studies indicate a role for AOX in plant PCD. Vanlerberghe *et al.* (2002) demonstrated that treatment of tobacco cells with cysteine triggered a signalling pathway, which culminated in a significant loss in the capacity of the mitochondrial cytochrome pathway. Concomitant with the loss in capacity of the cytochrome pathway, there was a significant upregulation of AOX protein and AOX activity. The authors showed that transgenic cells lacking the ability to induce AOX (AS8 cells) lost all respiratory capacity upon cysteine treatment, and as a consequence underwent PCD. Induction of AOX allowed wild-type cells to retain a high rate of respiration, and thus cell viability remained high. Robson and Vanlerberghe (2002) further demonstrated that AOX is important as an endogenous inhibitor of PCD. Transgenic AS8 tobacco cells showed increased susceptibility (relative to wild-type cells) to three different death-inducing compounds (salicylic acid, H_2O_2 and the protein phosphatase inhibitor cantharidin) and underwent a PCD characterised by DNA laddering. The authors proposed that the ability of AOX to attenuate PCD induced by those stimuli might relate to its ability to sustain mitochondrial function after the application of the death-inducing insult or might relate to its ability to prevent chronic oxidative stress within the mitochondrion (Robson and Vanlerberghe, 2002). Subsequently, Mizuno *et al.* (2005) showed that *AOX* expression and H_2O_2 accumulation could be strongly induced by a β-glucan elicitor in potato cells. The authors then showed that simultaneous inhibition of catalase and AOX activities in β-glucan-elicited cells led to a significant increase in H_2O_2 accumulation and reduced $\Delta\Psi_m$, and led to PCD. This demonstrated that catalase, as well as AOX, plays a central role in the suppression of mitochondrial $\Delta\Psi_m$ loss and the PCD induced by the β-glucan elicitor.

All of the above data indicate (as previously stated by Lam *et al.*, 2001) that AOX may act as a safety valve, controlling ROS generation from within the mitochondrion during plant PCD.

10.10 The involvement of *de novo* protein synthesis in plant PCD

Another consideration to take into account when comparing plant and animal PCD activation is the role of *de novo* protein synthesis. Martin *et al.* (1990) disproved the widely held view (at the time) that transcription and translation are universal requirements for animal cells dying via PCD. They pointed out that the human promyelocytic leukaemia cell line HL-60, along with numerous other tumour cell types, is susceptible to protein or RNA synthesis inhibitors; these cells undergo apoptosis within several hours of treatment with actinomycin D (Act D – transcriptional inhibitor) or cycloheximide (CHX – translational inhibitor) (Martin *et al.*, 1990; Martin, 1993). Martin (1993) stated that many other studies either had failed to show that the presence of inhibitors of transcription or translation could block apoptosis or, in fact, had demonstrated that inhibition of protein or RNA synthesis can induce apoptosis.

In plant cells, based on current knowledge, the story would seem to be different. Desikan *et al.* (1998) demonstrated that H_2O_2- or harpin-induced cell death in suspension cell cultures of *Arabidopsis* could be blocked either by an inhibitor of transcription (cordycepin) or by the translational inhibitor CHX. This showed that both H_2O_2- and harpin-induced cell death are active processes that require *de novo* protein synthesis. Solomon *et al.* (1999) obtained similar results with soybean, reporting that cycloheximide could reduce the cell death caused by H_2O_2 treatment. In addition, protein synthesis inhibitors blocked heat- or ethanol-induced PCD in *Arabidopsis* cell suspension cultures, confirming that *de novo* protein synthesis is required for the death response in these cases (Diamond, Reape, Rocha, Doohan, McCabe, unpublished results). Interestingly, the inhibitors, while preventing PCD did not prevent translocation of cytochrome *c* from the mitochondria, which occurred within 1 h of heat or ethanol treatment.

One of the reasons *de novo* protein synthesis is not a prerequisite for mammalian apoptosis may be because of the cytochrome *c*/caspase system. Caspases are present in an inactive form in the cytosol, and cytochrome *c* need only be released from the mitochondria to activate them. Given that no such system has been found to operate in plant PCD, there is a possibility that *de novo* protein synthesis could be a requirement for the activation of death effectors (possibly including caspase like-proteases) and plant PCD. It is possible that the mitochondrion signals the nucleus upon death signal perception and that there is an upregulation in the transcription and translation of genes encoding death effectors. However, it may also be the case that this is true only in certain cell types, or perhaps only in response to certain death stimuli.

10.11 Concluding remarks

It is clear that the mitochondrion plays a pivotal role in all eukaryotic PCD systems. In hindsight, the mitochondrion is an obvious organelle as an effector of death: firstly, because it controls the energy supply to the cell, but also because it is a good place to sequester death molecules, away from their site of action in the cytoplasm. However, it is likely that there are multiple pathways through which the mitochondrion executes death (Figure 10.2). For example there may be primordial pathways of mitochondrial execution shared between plants and animals, involving ROS accumulation and export, and the translocation of cytochrome *c*. Other pathways may have evolved after the divergence of plants and animals, for instance the participation of the Bax/Bcl-2 proteins and the cytochrome *c*-initiated activation of caspases in animals. The identification of primordial death mechanisms may be better facilitated through the study of plant PCD, as this can be difficult to study in animal systems due to the highly efficient and rapid cytochrome *c*/caspase execution pathway. On the other hand, it is possible that a more sophisticated mechanism of PCD execution has evolved in plants. It has been demonstrated that the mitochondrion interacts with the ER during plant PCD. In the future it will be interesting to investigate the interaction between the mitochondrion and other organelles, including the nucleus (and the possibility that the mitochondrial pathways trigger nuclear PCD-gene transcription) and the chloroplast that shares an endosymbiotic, prokaryotic, free-living origin.

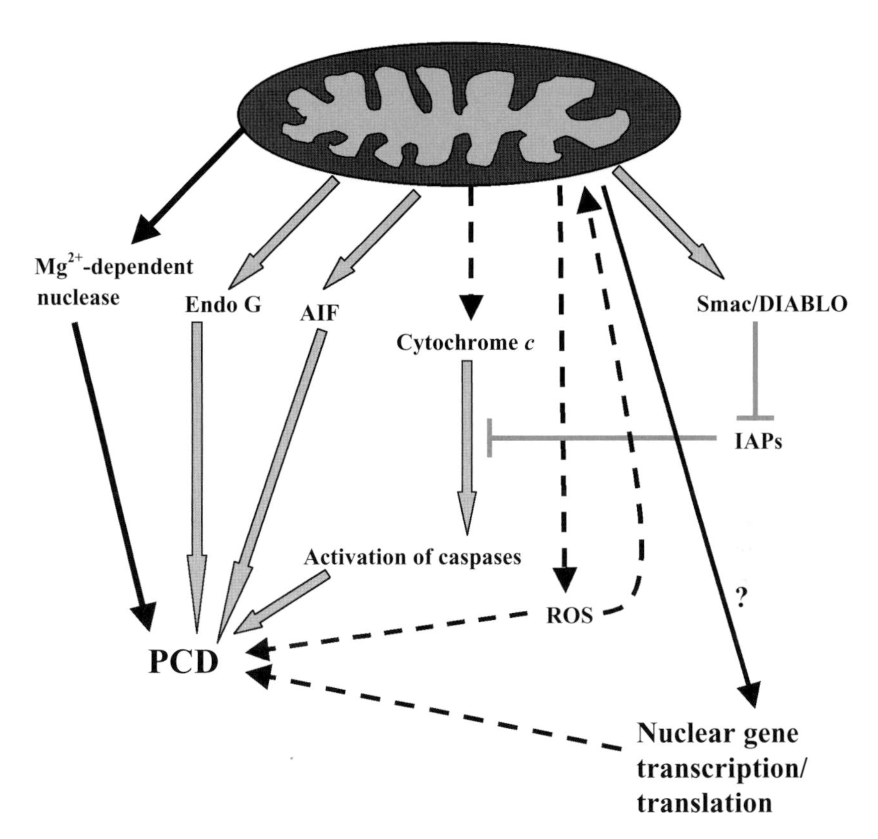

Figure 10.2 Mitochondrial PCD execution pathways in plants and animals. Various input signals can trigger the mitochondrion to release factors that execute PCD. In animals, the proteins released can aid in the activation of caspases (in the case of cytochrome *c* and Smac/DIABLO) or can bring about apoptosis in a caspase-independent fashion (AIF and Endo G). In both animals and plants, ROS accumulation and release from the mitochondrion can amplify the effects of death signals by damaging mitochondria and other organelles and may have a signalling role that possibly leads to the transcription and translation of factors that participate in PCD execution. An Mg^{2+}-dependent nuclease localised in plant mitochondria could also initiate plant PCD through the cleavage of nuclear DNA. However, the execution of plant PCD may be dependent on *de novo* transcription and translation of effector proteins. There is a possibility that mitochondrial signals initiate the transcription of such effectors. Grey lines represent animal-specific pathways, black lines represent plant-specific pathways and dashed lines represent pathways common to plants and animals.

References

Akamatsu H, Itoh Y, Kodama M, Otani H and Kohmoto K (1997) AAL-toxin-deficient mutants of *Alternaria alternata* tomato pathotype by restriction enzyme-mediated integration. *Phytopathology* **87**, 967–972.

Arnoult D, Tatischeff I, Estaquier J, *et al.* (2001) On the evolutionary conservation of the cell death pathway: mitochondrial release of an apoptosis-inducing factor during *Dictyostelium discoideum* cell death. *Mol Biol Cell* **12**, 3016–3030.

Arpagaus S, Rawyler A, Braendle R (2002) Occurrence and characteristics of the mitochondrial permeability transition in plants. *J Biol Chem* **277**, 1780–1787.

Ashkenazi A (2002) Targeting death and decoy receptors of the tumour-necrosis factor superfamily. *Nat Rev Cancer* **2**, 420–430.

Balk J, Chew SK, Leaver CJ and McCabe PF (2003) The intermembrane space of plant mitochondria contains a DNase activity that may be involved in programmed cell death. *Plant J* **34**, 573–583.

Balk J and Leaver CJ (2001) The PET1-CMS mitochondrial mutation in sunflower is associated with premature programmed cell death and cytochrome *c* release. *Plant Cell* **13**, 1803–1818.

Balk J, Leaver CJ and McCabe PF (1999) Translocation of cytochrome *c* from the mitochondria to the cytosol occurs during heat-induced programmed cell death in cucumber plants. *FEBS Lett* **463**, 151–154.

Basso E, Fante L, Fowlkes J, Petronilli V, Forte MA and Bernardi P (2005) Properties of the permeability transition pore in mitochondria devoid of cyclophilin D. *J Biol Chem* **280**, 18558–18561.

Blackstone NW and Green DR (1999) The evolution of a mechanism of cell suicide. *Bioessays* **21**, 84–88.

Blackstone NW and Kirkwood TBL (2003) Mitochondria and programmed cell death: "slave revolt" or community homeostasis? In: *Genetic and Cultural Evolution of Cooperation* (ed. Hammerstein P), pp. 309–325. The MIT Press, Cambridge, MA.

Bras M, Queenan B and Susin SA (2005) Programmed cell death via mitochondria: different modes of dying. *Biochemistry (Mosc)* **70**, 231–239.

Cande C, Vahsen N, Garrido C and Kroemer G (2004) Apoptosis-inducing factor (AIF): caspase-independent after all. *Cell Death Differ* **11**, 591–595.

Carimi F, Zottini M, Formentin E, Terzi M and Lo Schiavo F (2003) Cytokinins: new apoptotic inducers in plants. *Planta* **216**, 413–421.

Casolo V, Petrussa E, Krajnakova J, Macri F and Vianello A (2005) Involvement of the mitochondrial K(+)ATP channel in H_2O_2- or NO-induced programmed death of soybean suspension cell cultures. *J Exp Bot* **56**, 997–1006.

Crompton M (1999) The mitochondrial permeability transition pore and its role in cell death. *Biochem J* **341**, 233–249.

Curtis MJ and Wolpert TJ (2004) The victorin-induced mitochondrial permeability transition precedes cell shrinkage and biochemical markers of cell death, and shrinkage occurs without loss of membrane integrity. *Plant J* **38**, 244–259.

Dat J, Vandenabeele S, Vranova E, Van Montagu M, Inze D and Van Breusegem F (2000) Dual action of the active oxygen species during plant stress responses. *Cell Mol Life Sci* **57**, 779–795.

Delorme VGR, McCabe PF, Kim DJ and Leaver CJ (2000) A matrix metalloproteinase gene is expressed at the boundary of senescence and programmed cell death in cucumber. *Plant Physiol* **123**, 917–927.

Desikan R, Reynolds A, Hancock JT and Neill SJ (1998) Harpin and hydrogen peroxide both initiate programmed cell death but have differential effects on defence gene expression in *Arabidopsis* suspension cultures. *Biochemistry* **330**, 115–120.

Filonova LH, Bozhkov PV, Brukhin VB, Daniel G, Zhivotovsky B and von Arnold, S (2000) Two waves of programmed cell death occur during formation and development of somatic embryos in the gymnosperm, Norway spruce. *J Cell Sci* **113**, 4399–4411.

Forte M and Bernardi P (2005) Genetic dissection of the permeability transition pore. *J Bioenerg Biomembr* **37**, 121–128.

Fukuda H (2000) Programmed cell death of tracheary elements as a paradigm in plants. *Plant Mol Biol* **44**, 245–253.

Geitmann A, Franklin-Tong VE and Emons AC (2004) The self-incompatibility response in *Papaver rhoeas* pollen causes early and striking alterations to organelles. *Cell Death Differ* **11**, 812–822.

Goodsell DS (2004) The molecular perspective: cytochrome *c* and apoptosis. *Oncologist* **9**, 226–227.

Green DR and Reed JC (1998) Mitochondria and apoptosis. *Science* **281**, 1309–1312.

Green R and Fluhr R (1995) UV-B-induced PR-1 accumulation is mediated by active oxygen species. *Plant Cell* **7**, 203–212.

Greenberg JT and Yao N (2004) The role and regulation of programmed cell death in plant–pathogen interactions. *Cell Microbiol* **6**, 201–211.

Gunawardena A, Greenwood JS and Dengler NG (2004) Programmed cell death remodels lace plant leaf shape during development. *Plant Cell* **16**, 60–73.

Guo Y, Srinivasula SM, Druilhe A, Fernandes-Alnemri T and Alnemri ES (2002) Caspase-2 induces apoptosis by releasing proapoptotic proteins from mitochondria. *J Biol Chem* **277**, 13430–13437.

Hauser BA, Sun K, Oppenheimer DG and Sage TL (2006) Changes in mitochondrial membrane potential and accumulation of reactive oxygen species precede ultrastructural changes during ovule abortion. *Planta* **223**, 492–499.

Heath MC (2000) Hypersensitive response-related death. *Plant Mol Biol* **44**, 321–334.

Hutchins JB and Barger SW (1998) Why neurons die: cell death in the nervous system. *Anat Rec* **253**, 79–90.

Jabs T (1999) Reactive oxygen intermediates as mediators of programmed cell death in plants and animals. *Biochem Pharmacol* **57**, 231–245.

Jabs T, Dietrich RA and Dangl JL (1996) Initiation of runaway cell death in an *Arabidopsis* mutant by extracellular superoxide. *Science* **273**, 1853–1856.

Jones A (2000) Does the plant mitochondrion integrate cellular stress and regulate programmed cell death? *Trends Plant Sci* **5**, 225–230.

Karbowski M and Youle RJ (2003) Dynamics of mitochondrial morphology in healthy cells and during apoptosis. *Cell Death Differ* **10**, 870–880.

Kawai-Yamada M, Jin LH, Yoshinaga K, Hirata A and Uchimiya H (2001) Mammalian Bax-induced plant cell death can be down-regulated by overexpression of *Arabidopsis* Bax inhibitor-1 (AtBI-1). *Proc Natl Acad Sci USA* **98**, 12295–12300.

Kawai-Yamada M, Ohori Y and Uchimiya H (2004). Dissection of *Arabidopsis* Bax inhibitor-1 suppressing Bax-, hydrogen peroxide-, and salicylic acid-induced cell death. *Plant Cell* **16**, 21–32.

Kerr JFR, Wyllie AH and Currie AR (1972) Apoptosis – basic biological phenomenon with wide-ranging implications in tissue kinetics. *Br J Cancer* **26**, 239–257.

Kokoszka JE, Waymire KG, Levy SE, *et al.* (2004) The ADP/ATP translocator is not essential for the mitochondrial permeability transition pore. *Nature* **427**, 461–465.

Koukalová B, Kovarik A, Fajkus J and Siroky J (1997) Chromatin fragmentation associated with apoptotic changes in tobacco cells exposed to cold stress. *FEBS Lett* **414**, 289–292.

Kovtun Y, Chiu WL, Tena G and Sheen J (2000) Functional analysis of oxidative stress-activated mitogen-activated protein kinase cascade in plants. *Proc Natl Acad Sci USA* **97**, 2940–2945.

Krause M and Durner J (2004) Harpin inactivates mitochondria in *Arabidopsis* suspension cells. *Mol Plant Microbe Interact* **17**, 131–139.

Kroemer G (1997) Mitochondrial implication in apoptosis. Towards an endosymbiont hypothesis of apoptosis evolution. *Cell Death Differ* **4**, 443–456.

Lacomme C and Cruz SS (1999) Bax-induced cell death in tobacco is similar to the hypersensitive response. *Proc Natl Acad Sci USA* **96**, 7956–7961.

Lacomme C and Roby D (1999) Identification of new early markers of the hypersensitive response in *Arabidopsis thaliana*. *FEBS Lett* **459**, 149–153.

Lam E, Kato N and Lawton M (2001) Programmed cell death, mitochondria and the plant hypersensitive response. *Nature* **411**, 848–853.

Leist M and Jaattela M (2001) Four deaths and a funeral: from caspases to alternative mechanisms. *Nat Rev Mol Cell Biol* **2**, 589–598.

Lemasters JJ (1999) V. Necrapoptosis and the mitochondrial permeability transition: shared pathways to necrosis and apoptosis. *Am J Physiol* **276**, G1–G6.

Lemasters JJ, Nieminen AL, Qian T, *et al.* (1998) The mitochondrial permeability transition in cell death: a common mechanism in necrosis, apoptosis and autophagy. *Biochim Biophys Acta* **1366**, 177–196.

Levine A, Tenhaken R, Dixon R and Lamb C (1994) H_2O_2 from the oxidative burst orchestrates the plant hypersensitive disease resistance response. *Cell* **79**, 583–593.

Li LY, Luo L and Wang XD (2001) Endonuclease G is an apoptotic DNase when released from mitochondria. *Nature* **412**, 95–99.

Logan DC (2006) The mitochondrial compartment. *J Exp Biol* **57**, 1225–1243.

Luo X, Budihardjo I, Zou H, Slaughter C and Wang XD (1998) Bid, a Bcl2 interacting protein, mediates cytochrome *c* release from mitochondria in response to activation of cell surface death receptors. *Cell* **94**, 481–490.

Mach JM, Castillo AR, Hoogstraten R and Greenberg JT (2001) The *Arabidopsis*-accelerated cell death gene *ACD2* encodes red chlorophyll catabolite reductase and suppresses the spread of disease symptoms. *Proc Natl Acad Sci USA* **98**, 771–776.

Martin SJ (1993) Apoptosis: suicide, execution or murder? *Trends Cell Biol* **3**, 141–144.

Martin SJ, Lennon SV, Bonham AM and Cotter TG (1990) Induction of apoptosis (programmed cell death) in human leukemic HL-60 cells by inhibition of RNA or protein synthesis. *J Immunol* **145**, 1859–1867.

Marzo I, Brenner C, Zamzami N, *et al.* (1998) The permeability transition pore complex: a target for apoptosis regulation by caspases and Bcl-2-related proteins. *J Exp Med* **187**, 1261–1271.

Maxwell DP, Nickels R and McIntosh L (2002) Evidence of mitochondrial involvement in the transduction of signals required for the induction of genes associated with pathogen attack and senescence. *Plant J* **29**, 269–279.

McCabe PF and Leaver CJ (2000) Programmed cell death in cell cultures. *Plant Mol Biol* **44**, 359–368.

McCabe PF, Levine A, Meijer PJ, Tapon NA and Pennell RI (1997a) A programmed cell death pathway activated in carrot cells cultured at low cell density. *Plant J* **12**, 267–280.

McCabe PF, Valentine TA, Forsberg LS and Pennell RI (1997b) Soluble signals from cells identified at the cell wall establish a developmental pathway in carrot. *Plant Cell* **9**, 2225–2241.

Meier P, Finch A and Evan G (2000) Apoptosis in development. *Nature* **407**, 796–801.

Mizuno M, Tada Y, Uchii K, Kawakami S and Mayama S (2005) Catalase and alternative oxidase cooperatively regulate programmed cell death induced β-glucan elicitor in potato suspension cultures. *Planta* **220**, 849–853.

Okamoto K and Shaw JM (2005) Mitochondrial morphology and dynamics in yeast and multicellular eukaryotes. *Ann Rev Genet* **39**, 503–536.

Parrish J, Li L, Klotz K, Ledwich D, Wang X and Xue D (2001) Mitochondrial endonuclease G is important for apoptosis in *C. elegans*. *Nature* **412**, 90–94.

Pasqualini S, Piccioni C, Reale L, Ederli L, Della Torre G and Ferranti F (2003) Ozone-induced cell death in tobacco cultivar Bel W3 plants. The role of programmed cell death in lesion formation. *Plant Physiol* **133**, 1122–1134.

Pastori GM and del Rio LA (1997) Natural senescence of pea leaves (an activated oxygen-mediated function for peroxisomes). *Plant Physiol* **113**, 411–418.

Pastorino JG, Tafani M, Rothman RJ, Marcineviciute A, Hoek JB and Farber JL (1999) Functional consequences of the sustained or transient activation by Bax of the mitochondrial permeability transition pore. *J Biol Chem* **274**, 31734–31739.

Rathmell JC and Thompson CB (2002) Pathways of apoptosis in lymphocyte development, homeostasis, and disease. *Cell* **109**, S97–S107.

Rentel MC, Lecourieux D, Ouaked F, *et al.* (2004) OXI1 kinase is necessary for oxidative burst-mediated signalling in *Arabidopsis*. *Nature* **427**, 858–861.

Robson CA and Vanlerberghe GC (2002) Transgenic plant cells lacking mitochondrial alternative oxidase have increased susceptibility to mitochondria-dependent and -independent pathways of programmed cell death. *Plant Physiol* **129**, 1908–1920.

Rutter GA and Rizzuto R (2000) Regulation of mitochondrial metabolism by ER Ca^{2+} release: an intimate connection. *Trends Biochem Sci* **25**, 215–221.

Saviani EE, Orsi CH, Oliveira JFP, Pinto-Maglio CAF and Salgado I (2002) Participation of the mitochondrial permeability transition pore in nitric oxide-induced plant cell death. *FEBS Lett* **510**, 136–140.

Scaffidi C, Fulda S, Srinivasan A, *et al.* (1998) Two CD95 (APO-1/Fas) signaling pathways. *EMBO J* **17**, 1675–1687.

Scorrano L, Ashiya M, Buttle K, *et al.* (2002) A distinct pathway remodels mitochondrial cristae and mobilizes cytochrome *c* during apoptosis. *Dev Cell* **2**, 55–67.

Simons BH, Millenaar FF, Mulder L, Van Loon LC and Lambers H (1999) Enhanced expression and activation of the alternative oxidase during infection of *Arabidopsis* with *Pseudomonas syringae* pv tomato. *Plant Physiol* **120**, 529–538.

Solomon M, Belenghi B, Delledonne M, Menachem E and Levine A (1999) The involvement of cysteine proteases and protease inhibitor genes in the regulation of programmed cell death in plants. *Plant Cell* **11**, 431–443.

Stein JC and Hansen G (1999) Mannose induces an endonuclease responsible for DNA laddering in plant cells. *Plant Physiol* **121**, 71–79.

Sun YL, Zhao Y, Hong X and Zhai ZH (1999) Cytochrome *c* release and caspase activation during menadione-induced apoptosis in plants. *FEBS Lett* **462**, 317–321.

Susin SA, Lorenzo HK, Zamzami N, *et al.* (1999) Molecular characterization of mitochondrial apoptosis-inducing factor. *Nature* **397**, 441–446.

Swidzinski JA, Sweetlove LJ and Leaver CJ (2002) A custom microarray analysis of gene expression during programmed cell death in *Arabidopsis thaliana. Plant J* **30**, 431–446.

Thomas SG and Franklin-Tong VE (2004) Self-incompatibility triggers programmed cell death in *Papaver* pollen. *Nature* **429**, 305–309.

Tiwari BS, Belenghi B and Levine A (2002) Oxidative stress increased respiration and generation of reactive oxygen species, resulting in ATP depletion, opening of mitochondrial permeability transition, and programmed cell death. *Plant Physiol* **128**, 1271–1281.

Townley HE, McDonald K, Jenkins GI, Knight MR and Leaver CJ (2005) Ceramides induce programmed cell death in *Arabidopsis* cells in a calcium-dependent manner. *Biol Chem* **386**, 161–166.

Vacca RA, de Pinto MC, Valenti D, Passarella S, Marra E and De Gara L (2004) Production of reactive oxygen species, alteration of cytosolic ascorbate peroxidase, and impairment of mitochondrial metabolism are early events in heat shock-induced programmed cell death in tobacco Bright-Yellow 2 cells. *Plant Physiol* **134**, 1100–1112.

Vander Heiden MG, Chandel NS, Schumacker PT and Thompson CB (1999) Bcl-xL prevents death following growth factor withdrawal by facilitating mitochondrial ATP/ADP exchange. *Mol Cell* **3**, 159–167.

Vanlerberghe GC and McIntosh L (1997) Alternative oxidase: from gene to function. *Annu Rev Plant Physiol Plant Mol Biol* **48**, 703–734.

Vanlerberghe GC, Robson CA and Yip JYH (2002) Induction of mitochondrial alternative oxidase in response to a cell signal pathway down-regulating the cytochrome pathway prevents programmed cell death. *Plant Physiol* **129**, 1829–1842.

Walter L and Hajnóczky G (2005) Mitochondria and endoplasmic reticulum: the lethal interorganelle cross-talk. *J Bioenerg Biomembr* **37**, 191–206.

Wang X, Yang C, Chai J, Shi Y and Xue D (2002) Mechanisms of AIF-mediated apoptotic DNA degradation in *Caenorhabditis elegans*. *Science* **298**, 1587–1592.

Wang XD (2001) The expanding role of mitochondria in apoptosis. *Genes Dev* **15**, 2922–2933.

Watanabe N and Lam E (2006) *Arabidopsis* Bax inhibitor-1 functions as an attenuator of biotic and abiotic types of cell death. *Plant J* **45**, 884–894.

Wu HM and Cheung AY (2000) Programmed cell death in plant reproduction. *Plant Mol Biol* **44**, 267–281.

Xu Y and Hanson MR (2000) Programmed cell death during pollination-induced petal senescence in petunia. *Plant Physiol* **122**, 1323–1333.

Yang SW, Kim SK and Kim WT (2004) Perturbation of NgTRF1 expression induces apoptosis-like cell death in tobacco BY-2 cells and implicates NgTRF1 in the control of telomere length and stability. *Plant Cell* **16**, 3370–3385.

Yao N, Eisfelder BJ, Marvin J and Greenberg JT (2004) The mitochondrion – an organelle commonly involved in programmed cell death in *Arabidopsis thaliana. Plant J* **40**, 596–610.

Yao N, Imai S, Tada Y, *et al.* (2002) Apoptotic cell death is a common response to pathogen attack in oats. *Mol Plant Microbe Interact* **15**, 1000–1007.

Yu XH, Perdue TD, Heimer YM and Jones AM (2002) Mitochondrial involvement in tracheary element programmed cell death. *Cell Death Differ* **9**, 189–198.

Zhao Y, Jiang ZF, Sun YL and Zhai ZH (1999) Apoptosis of mouse liver nuclei induced in the cytosol of carrot cells. *FEBS Lett* **448**, 197–200.

Zuppini A, Navazio L, Sella L, Castiglioni C, Favaron F and Mariani P (2005) An endopolygalacturonase from *Sclerotinia sclerotiorum* induces calcium-mediated signaling and programmed cell death in soybean cells. *Mol Plant Microbe Interact* **18**, 849–855.

Zuzarte-Luis V and Hurle JM (2002) Programmed cell death in the developing limb. *Int J Dev Biol* **46**, 871–876.

Index